Osswald / Menges

Materials Science of Polymers for Engineers

Tim A. Osswald / Georg Menges

Materials Science of Polymers for Engineers

2nd Edition

HANSER
Hanser Publishers, Munich
Hanser Gardner Publications, Inc., Cincinnati

The Authors:
Prof. Dr. Tim A. Osswald, Polymer Engineering Center, University of Wisconsin – Madison,
Dept. of Mechanical Engineering, 1513 University Avenue, Madison WI 53706, USA
Prof. Dr.–Ing. Georg Menges, Am Beulardstein 19, 52072 Aachen, Germany

Distributed in the USA and in Canada by
Hanser Gardner Publications, Inc.
6915 Valley Avenue, Cincinnati, Ohio 45244-3029, USA
Fax: (513) 527-8801
Phone: (513) 527-8977 or 1-800-950-8977
Internet: http://www.hansergardner.com

Distributed in all other countries by
Carl Hanser Verlag
Postfach 86 04 20, 81631 München, Germany
Fax: +49 (89) 98 48 09
Internet: http://www.hanser.de

Library of Congress Cataloging-in-Publication Data

Osswald, Tim A.
Materials science of polymers for engineers / Tim A. Osswald, Georg Menges.— 2nd ed.
p. cm.
ISBN 1-56990-348-4 (hardcover)
1. Polymers. 2. Plastics. I. Menges, Georg, 1923- II. Title.
TA455.P58.O68 2003
620.1'92—dc21
2003007497

Bibliografische Information Der Deutschen Bibliothek
Die Deutsche Bibliothek verzeichnet diese Publikation in der Deutschen Nationalbibliografie; detaillierte bibliografische Daten sind im Internet über <http://dnb.ddb.de> abrufbar.

ISBN 3-446-22464-5

Production Management: Oswald Immel
Typeset by Angela Ospina-Garcia, USA
Coverdesign: MCP • Susanne Kraus GbR, Holzkirchen, Germany
Printed and bound by Kösel, Kempten, Germany

For Diane, Palitos and Rudi
Tim A. Osswald

Dedicated to my wife in gratitude for her patience
Georg Menges

"Why does this applied science, which saves work and makes life easier, bring us so little happiness? The simple answer runs: Because we have not yet learned to make sensible use of it."

Albert Einstein

Preface to the Second Edition

The first edition of this book was adopted by several universities in North and South America, Europe, and Asia as a textbook to introduce engineering students to the materials science of polymers. The book was also translated into Japanese in 1998. The professors who taught with the first edition as well as their students liked the unified approach we took. Furthermore, it has become a reference for many practicing engineers, most of whom were introduced to the book as students. The changes and additions that were introduced in this edition are based on suggestions from these professors and their students, as well as from our own experience using it as a class textbook.

One major change that we made to this edition is the introduction of examples and problems at the end of each chapter. A new chapter containing the history of polymers replaced a few pages in the first edition. From a polymers course taught at the University of Wisconsin-Madison, we found that introducing polymers from a historical perspective not only made the topic less dry, but it also gave us a chance to show the role polymers have played, for better and for worse, while shaping today's industrial world. The mixing chapter in the first edition was replaced with a comprehensive polymer processing chapter. The first edition was praised because of the vast number of graphs and data that can be used as a reference. We have further strengthened this attribute by adding a table in the appendix that contains material property graphs for several polymers.

With this edition we owe our gratitude to Dr. Christine Strohm, Dr. Wolfgang Glenz and Oswald Immel of Hanser Publishers for their cooperation during the production of this book. We are grateful to Amelia Cosgrove for combing through the manuscript in such detail, finding typos and making suggestions. We thank Lina Lopez and Paul Wilichowski for preparing DSC data for Chapter 3. We also thank Aaron Hade, Soenarto Hadiprajitno, Juan Pablo Hernandez, Juan Rodrigo Sanz-Uribe and Jarrod Schemenauer for their suggestions. We are grateful to Angela Maria Ospina for the superb job preparing the camera-ready manuscript. These projects consume a large number of hours, which results in several months of late nights; and so, we must thank our families for their patience.

Spring 2003

Tim A. Osswald
Madison, Wisconsin, USA

Georg Menges
Aachen, Germany

Preface to the First Edition

This book is designed to provide a polymer materials science background to engineering students and practicing engineers. It is written on an intermediate level for students, and as an introduction to polymer materials science for engineers. The book presents enough information that, in conjunction with a good design background, it will enable the engineer to design polymer components.

Materials Science of Polymers for Engineers is based on the German textbook, *Werkstoffkunde Kunststoffe* (G. Menges, Hanser Publishers, 1989), and on lecture notes from polymer materials science courses taught at the Technical University of Aachen, Germany, and at the University of Wisconsin-Madison. The chapters on thermal and electrical properties are loose translations from *Werkstoffkunde Kunststoffe,* and many figures throughout the manuscript were taken from this book.

We have chosen a unified approach and have divided the book into three major sections: Basic Principles, Influence of Processing on Properties, and Engineering Design Properties. This approach is often referred to as the four P's: polymer, processing, product and performance. The first section covers general topics such as historical background, basic material properties, molecular structure of polymers and thermal properties of polymers. The second section ties processing and design by discussing the effects of processing on properties of the final polymer component. Here, we introduce the reader to the rheology of polymer melts, mixing of polymer blends, development of anisotropy during processing and solidification processes. In essence, in this section we go from the melt (rheology) to the finished product (solidification). The third section covers the different properties that need to be considered when designing a polymer component, and analyzing its performance. These properties include mechanical properties, failure of polymers, electrical properties, optical properties, acoustic properties, and permeability of polymers.

The authors cannot acknowledge everyone who helped in one way or another in the preparation of this manuscript. We would like to thank the students of our polymer materials science courses who in the past few years endured our experimenting and trying out of new ideas. The authors are grateful to the staff and faculty of the Mechanical Engineering Department at the University of Wisconsin-Madison, and the Institut für Kunststoffverarbeitung (IKV) at the Technical University of Aachen for their support while developing the courses which gave the base for this book. We are grateful to Richard Theriault for proofreading the entire manuscript. We also thank the following people who helped proofread or gave suggestions during the preparation of the book: Susanne Belovari, Bruce A. Davis, Jeffrey Giacomin, Paul J. Gramann, Matthew Kaegebein, Gwan-Wan Lai, Maria del Pilar Noriega E., Antoine C. Rios B., Linards U. Stradins and Ester M. Sun. Susanne Belovari and Andrea Jung-Mack are acknowledged for translating portions of *Werkstoffkunde Kunststoffe* from German to English. We also thank Tara Ruggiero for preparing the camera-ready manuscript. Many of the figures were taken from class notes of the mechanical engineering senior elective course *Engineering Design with Polymers.* Special thanks are offered to Lynda Litzkow, Philipp Ehrenstein and Bryan Hutchinson for the superb job of drawing those figures. Matthias Mahlke of Bayer AG in Leverkusen, Germany, Laura Dietsche, Joseph Dooley and Kevin Hughes of Dow Chemical in Midland, Michigan, and Mauricio DeGreif and Juan Diego Sierra of the ICIPC in Medellín, Colombia, are acknowledged for some of the figures. Thanks are due to Marcia Sanders for copy editing the final manuscript. We are grateful to Wolfgang Glenz, Martha Kürzl, Ed Immergut and Carol Radtke of Hanser Publishers for their support throughout the development of this book. Above all, the authors thank their wives for their patience.

Summer 1995

Tim A. Osswald
Madison, Wisconsin, USA

Georg Menges
Aachen, Germany

Table of Contents

Part I

Basic Principles

1

Introduction to Polymers

As the word itself suggests, polymers[1] are materials composed of molecules of very high molecular weight. These large molecules are generally referred to as *macromolecules*. The unique material properties of polymers and the versatility of their processing methods are attributed to their molecular structure. The ease with which polymers and *plastics* [2] are processed makes them, for many applications, the most sought-after material today. Because of their low density and their ability to be shaped and molded at relatively low temperatures compared to traditional materials such as metals, plastics and polymers are the material of choice when integrating several parts into a single component—a design aspect usually called *part consolidation*. In fact, parts and components, which have traditionally been made of wood, metal, ceramic or glass are redesigned with plastics on a daily basis.

1.1 General Properties

Polymers can be placed into either a thermoplastic, thermoset, or elastomer category. Thermoplastics in turn include a special family, which is relatively new called thermoplastic elastomers. However, all these material have in common that they are made of huge molecules. Some of these molecules are uncrosslinked, which means that each molecule can move freely relative to its neighbors, and others are crosslinked, which means that "bridges", or physical links interconnect the polymer molecules. Thermoplastics and un-vulcanized elastomers are uncrosslinked. Vulcanized rubber, or elastomers, and thermosets are cross-linked.

Thermoplastics are those polymers that solidify as they are cooled, no longer allowing the long molecules to move freely. When heated, these materials regain the ability to "flow", as the molecules are able to slide past

1 From the Greek, *poli* which means many, and *meros* which means parts.

2 The term plastics describes a compound of polymers and various additives.

each other with ease. Furthermore, thermoplastic polymers are divided into two classes: amorphous and semi-crystalline polymers. Amorphous thermoplastics are those with molecules that remain in disorder as they cool, leading to a material with a fairly random molecular structure. An amorphous polymer solidifies, or vitrifies, as it is cooled below its glass transition temperature, T_g. Semi-crystalline thermoplastics, on the other hand, solidify with a certain order in their molecular structure. Hence, as they are cooled, they harden when the molecules begin to arrange in a regular order below what is usually referred to as the melting temperature, T_m. The molecules in semi-crystalline polymers that are not transformed into ordered regions remain as small amorphous regions. These amorphous regions within the semi-crystalline domains lose their "flowability" below their glass transition temperature. Most semi-crystalline polymers have a glass transition temperature at subzero temperatures, hence, behaving at room temperature as rubbery or leathery materials. On the other hand, thermosetting polymers solidify by being chemically cured. Here, the long macromolecules cross-link with each other during cure, resulting in a network of molecules that cannot slide past each other. The formation of these networks causes the material to lose the ability to "flow" even after reheating. The high density of cross-linking between the molecules makes thermosetting material stiff and brittle. Thermosets also exhibit a glass transition temperature which is sometimes near or above thermal degradation temperatures. Compared to thermosets, elastomers are only lightly cross-linked which permits almost full extension of the molecules. However, the links across the molecules hinder them from sliding past each other, making even large deformations reversible. One common characteristic of elastomeric materials is that the glass transition temperature is much lower than room temperature. Their ability to "flow" is lost after they are vulcanized or cross-linked. Since cross-linked elastomers at room temperature are significantly above their glass transition temperature, they are very soft and very compliant elastic solids.

Table 1.1 presents the most common amorphous and semi-crystalline thermoplastics, as well as thermosets and elastomers, with some of their applications. The abbreviations used in Table 1.1 are described in Table 1.2. Figure 1.1 presents just one example of where we can find polymers in everyday life. The young skier in the photograph is protected with a helmet that is composed of various polymer components. His skiing equipment is made up of various polymers and polymer composites, and he is keeping warm with several polymer fabrics and insulating materials. It should be pointed out that not all the polymer components the skier is wearing are listed in the figure, and that in addition to the specific polymers listed in Fig. 1.1, there are many more possibilities. For example, some skiing boots are reinforced with Aramid fibers, and there are skiing poles made of carbon fiber reinforced epoxy composites.

Table 1.1 Common Polymers and Some of Their Applications

Polymer	Applications
	Thermoplastics
Amorphous	
Polystyrene	Mass-produced transparent articles, thermoformed packaging, etc. Thermal insulation (foamed), etc.
Polymethyl methacrylate	Skylights, airplane windows, lenses, bulletproof windows, stop lights, etc.
Polycarbonate	Helmets, hockey masks, bulletproof windows, blinker lights, head lights, etc.
Unplasticized polyvinyl chloride	Tubes, window frames, siding, rain gutters, bottles, thermoformed packaging, etc.
Plasticized polyvinyl chloride	Shoes, hoses, roto-molded hollow articles such as balls and other toys, calendered films for raincoats and tablecloths, etc
Semi-crystalline	
High density polyethylene	Milk and soap bottles, mass production of household goods of higher quality, tubes, paper coating, etc.
Low density polyethylene	Mass production of household goods, grocery bags, etc.
Polypropylene	Goods such as suitcases, tubes, engineering application (fiber-glass-reinforced), housings for electric appliances, etc.
Polytetrafluoroethylene	Coating of cooking pans, lubricant-free bearings, etc.
Polyamide	Bearings, gears, bolts, skate wheels, pipes, fishing line, textiles, ropes, etc.
	Thermosets
Epoxy	Adhesive, automotive leaf springs (with glass fiber), bicycle frames (with carbon fiber), etc.
Melamine	Decorative heat-resistant surfaces for kitchens and furniture, dishes, etc.
Phenolics	Heat-resistant handles for pans, irons and toasters, electric outlets, etc.
Unsaturated polyester	Toaster sides, iron handles, satellite dishes, breaker switch housing (with glass fiber), automotive body panels (with glass fiber), etc.
	Elastomers
Polybutadiene	Automotive tires (blended with natural rubber and styrene butadiene rubber), golf ball skin, etc.
Ethylene propylene rubber	Automotive radiator hoses and window seals, roof covering, etc.
Natural rubber (polyisoprene)	Automotive tires, engine mounts, etc.
Polyurethane elastomer	Roller skate wheels, sport arena floors, ski boots, automotive seats (foamed), shoe soles (foamed), etc.
Silicone rubber	Seals, flexible hoses for medical applications, etc.
Styrene butadiene rubber	Automotive tire treads, etc.

Table 1.2 Abbreviations for Common Polymers

Polymer	**Abbrev.**	**Polymer**	**Abbrev.**
Acrylonitrile-butadiene-styrene	ABS	Polyether ether ketone	PEEK
Butadiene rubber	BR	Polyether sulfone	PES
Cellulose acetate	CA	Polyethylene teraphthalate	PET
Cellulose acetate butyrate	CAB	Polyimide	PI
Cellulose nitrate	CN	Polyisobutylene	PIB
Epoxy	EP	Polymethyl methacrylate	PMMA
Ethylene-propylene-diene rubber	EPDM	Polyphenylene sulfide	PPS
Expanded polystyrene	EPS	Polyphenylene sulfone	PPSU
High density polyethylene	PE-HD	Polypropylene	PP
Impact resistant polystyrene	PS-HI (HIPS)	Polypropylene copolymer	PP-CO
Linear low density polyethylene	PE-LLD	Polystyrene	PS
Linear medium density polyethylene	PE-LMD	Polystyrene high impact	PS-HI
Liquid crystalline polymer	LCP	Polysulfone	PSU
Low density polyethylene	PE-LD	Polytetrafluoroethylene	PTFE
Melamine-formaldehyde	MF	Polyurethane	PUR
Metallocene catalyzed polyethylene	mPE	Polyvinylidene acetate	PVAC
Natural rubber	NR	Polyvinyl alcohol	PVAL
Olefinic thermoplastic elastomer	TPO	Polyvinyl carbazole	PVK
Phenol-formaldehyde (phenolic)	PF	Polyvinyl chloride	PVC
PF mineral filled moldings	PF-mf	Polyvinylidene fluoride	PVDF
PF organic filled moldings	PF-of	Rigid PVC	PVC-U
Plasticized PVC	PVC-P	Silicone	SI
Polyacetal (polyoxymethylene)	POM	Sillicone rubber	SI
Polyacrylate	PAR	Styrene-acrylonitrile copolymer	SAN
Polyacrylonitrile	PAN	Thermoplastic elastomer	TPE
Polyamide 6	PA6	Thermoplastic polyurethane elastomer	TPU
Polyamide 66	PA66	Unsaturated polyester	UP
Polybutylene teraphthalate	PBT	Urea-formaldehyde	UF
Polycarbonate	PC	Vinyl ester resin	VE

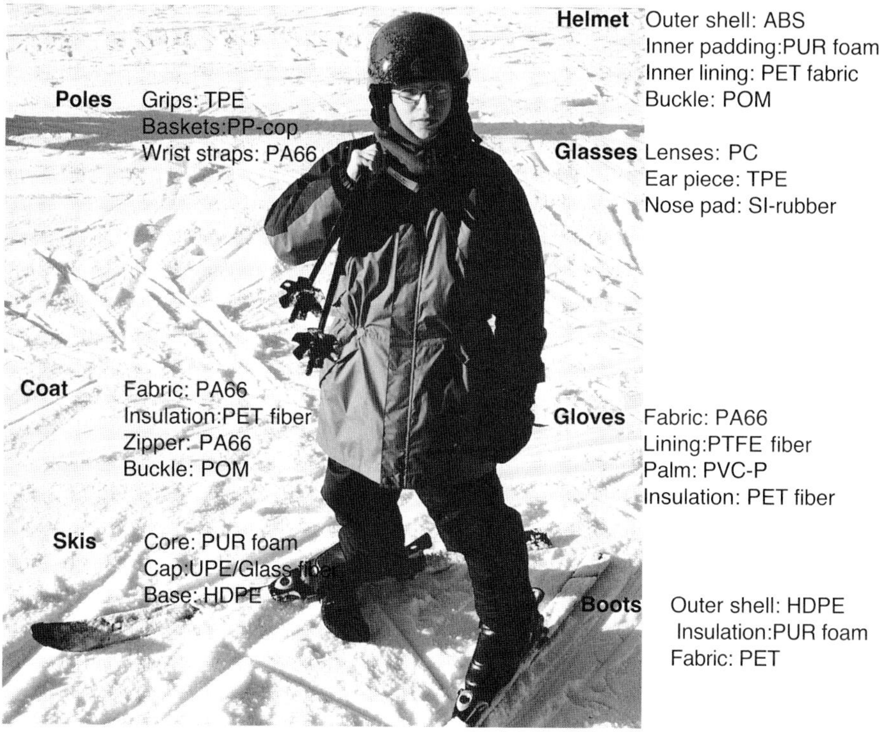

Figure 1.1 Polymers pervade everyday everywhere.

There are thousands of different grades of polymers available to the design engineer. These materials cover a wide range of properties, from soft to hard, ductile to brittle, and weak to tough. Figure 1.2 shows this range by plotting important average properties for selected polymers. The abbreviations used in Fig. 1.2 are described in Table 1.2. The values which correspond to each material in Fig. 1.2 only represent an average. There is a wide range in properties even for materials of the same class. The spread in properties is often increased with the different additives and fillers that a resin may contain. At this point, it is important to point out that the properties listed in Fig. 1.2 are not properties that should be used in design. The properties represent a general trend in stiffness and toughness of the various materials, and should be used only when comparing one material against another. As will be discussed in later chapters, properties that are used in design need to be time-dependent, such that loading time or speed can be included as design parameters. Today, it is very common for engineers to erroneously design a product using strength or stiffness data measured with tests such as ASTM D 638 or ASTM D790. These tests neglect the time dependence reflected in the mechanical behavior of polymers.

The relatively low stiffness of polymeric materials is attributed to their molecular structure, which allows movement of the molecules with relative ease while under stress. However, the strength and stiffness of individual polymer chains are much higher than the measured properties of the bulk polymeric material. For example, polyethylene whose molecules have a theoretical stiffness of 300,000 MPa, has a bulk stiffness of only 1,000 MPa [1, 2]. By introducing high molecular orientation, the stiffness and strength of a polymer can be substantially increased. In the case of *ultra-drawn, ultra high molecular weight high density polyethylene*, UHMHDPE, fibers can reach a stiffness of over 200,000 MPa [2]. Figure 1.2 [3] presents a plot of specific strength versus specific modulus of various polymer fibers and other materials.

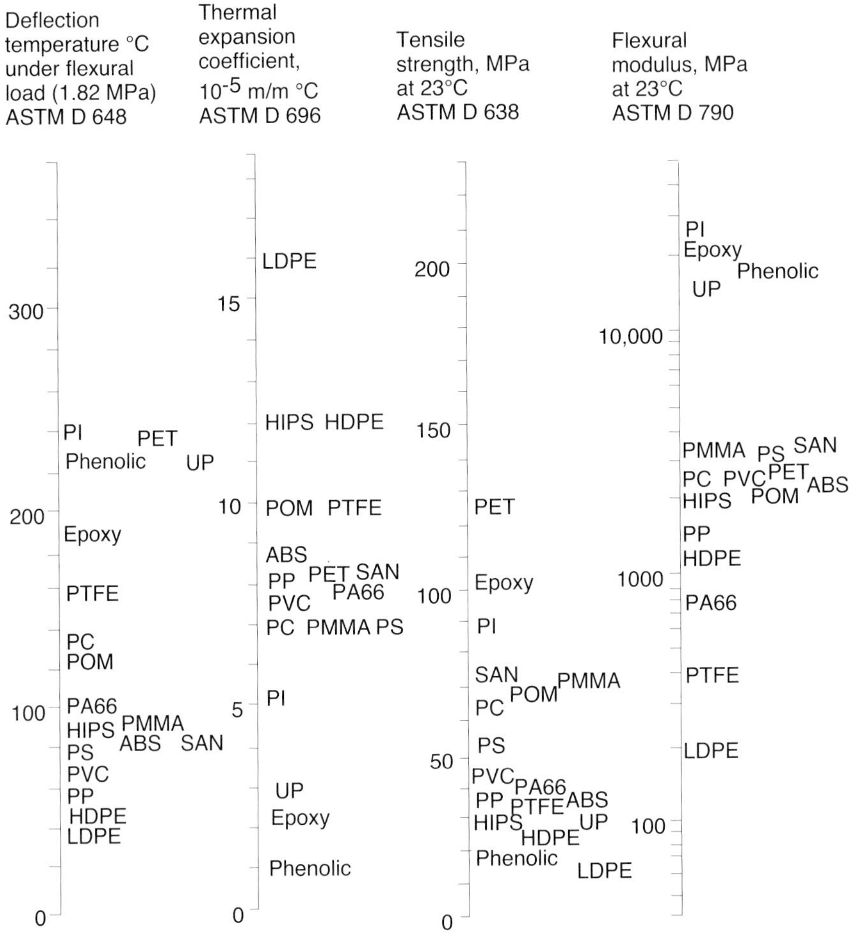

Figure 1.2 Average properties for common polymers. *(Continued)*

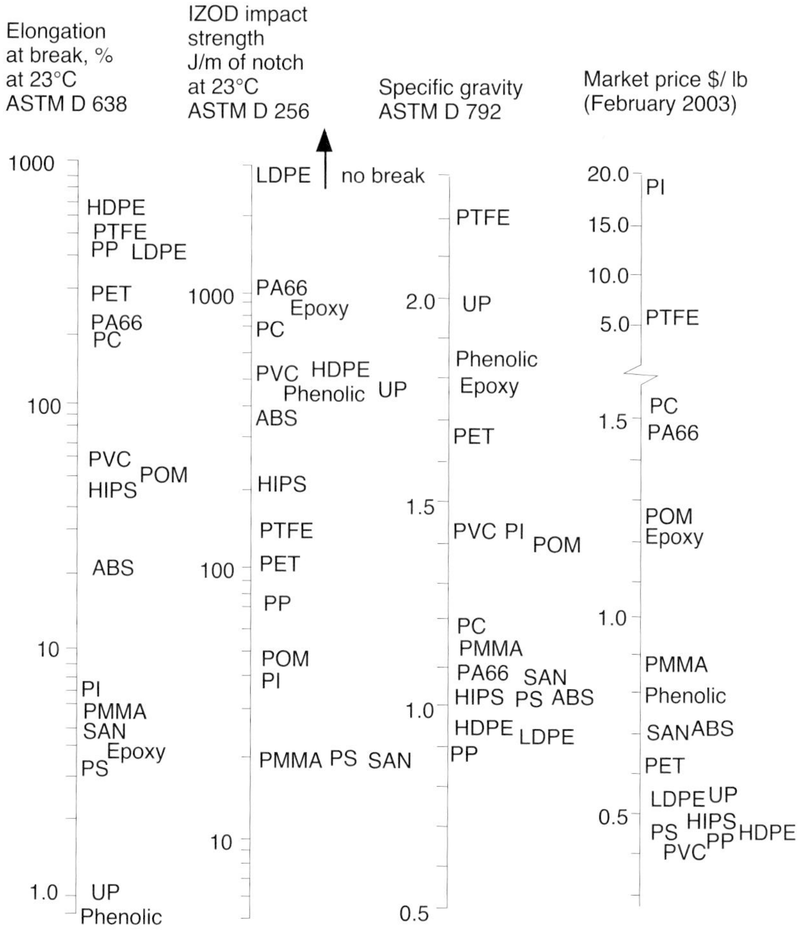

Figure 1.2*(Continued)* Average properties for common polymers.

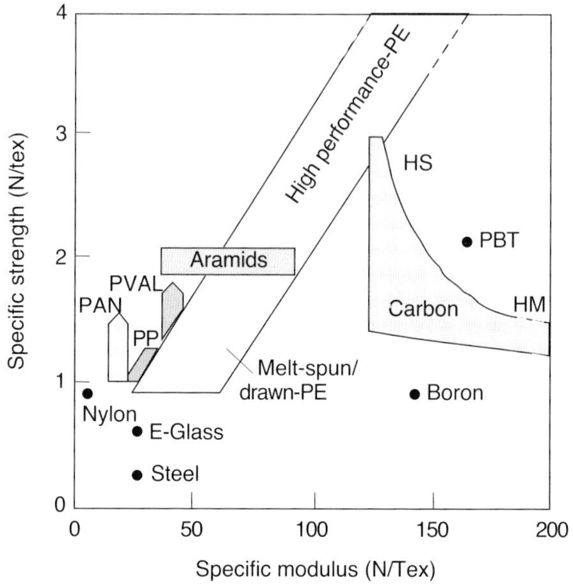

Figure 1.3 Specific strength versus specific modulus of fibers and other materials.

As will be discussed in more detail in Chapter 3, polymers also have a large range of molecular weights and molecular weight distributions, which affect the mechanical properties as well as the viscosity. Chapter 4 will discuss the thermal behavior of polymeric materials, and Chapters 5, 6, 7 and 8 will concentrate on rheology, processing of polymers, and how processing affects the properties of the final component through mixing, orientation, and solidification of the material during manufacturing. The important properties of the finished polymer component, such as mechanical properties, failure mechanisms, electrical properties, optical properties, permeability of polymers, etc. will be discussed in Chapters 9 through 14.

1.2 Identification of Polymers

Perhaps one of the most important skills a plastics design engineer needs to have is the ability to identify a specific polymer or polymer component. An experienced plastics engineers can often identify a polymer by touching, smelling, and tapping it. However, the complete identification of the chemical composition, additives, and fillers of a plastic material is an extremely complicated task. To achieve this, equipment that performs differential scanning calorimetry, infrared spectroscopy, and dynamic mechanical analysis, to name a few, is available[3]. Most process and design engineers do not have these measuring devices at hand, nor do they have the analytical experience to run them and interpret the resulting data. However, to identify a polymer often only simple means are needed in addition to basic knowledge of polymer chemistry. Through simple observation, a burn test and experience, an engineer is able to identify most plastics. Table 1.3 presents a summarized guide to aid in the identification of polymers.

Information presented in tables such as Table 1.3 must be used in conjunction with common sense and engineering insight. For example, the attribute labeled "Appearance" can sometimes be misleading. Depending on the additives or colorants added, even an amorphous thermoplastic such as polystyrene, can be opaque. The stiffness attribute is broken down into three categories: Flexible, semi-rigid and rigid. Flexible are those materials that feel rubbery or leathery, while semi-rigid are these that are strong and stiff but can deflect a substantial amount. The materials that fall under the rigid category are those that are stiff but brittle. During the burn test, as tempting as it may seem to some of us, it is not recommended to inhale the fumes released by a burning plastic, or those released soon after extinguishing the burning polymer. However, it can be said that without directly inhaling the fumes one can still discern the particular smells of the material during the burning test. Due to the benzene ring in their molecular structure, styrenic materials, such as PS, ABS and SAN have a sweet smell, while polyolefin plastics such as PE and PP smell waxy. Because of the nitrogen atoms in their structure, polyamides can smell like burning hair. When burning POM, a strong smell of formaldehyde is released. Burning PVC releases HCl. To no surprise, cellulose plastics smell like burning paper. All burn tests must be performed under a ventilation hood or in a well ventilated room.

3 These tests are discussed in detail in Chapters 3 and 8 of this book.

Table 1.3 Plastics Identification Attributes

	Appearance				Stiffness			Surface			Burn test							
Polymer	Transparent	Transparent (thin film)	Transluscent	Opaque	Flexible (resilient)	Semi-rigid	Rigid (brittle)	Glassy	Waxy	Dull	Black soot	Burns clean	Self extinguishing	Drips	Does not drip	Yellow flame	Blue flame	Green flame
PE-LD		●	●		●				●			●			●		●	
PE-HD		●	●		●				●			●			●		●	
PP		●	●			●			●			●			●		●	
PP-CO		●	●			●			●			●			●		●	
PS	●						●	●			●			●		●		
PS-HI				●		●				●	●			●		●		
SAN	●						●	●			●			●		●		
ABS				●		●				●	●			●		●		
PVC-U	●					●		●			●		●					●
PVC-P	●				●			●			●		●					●
PTFE				●	●			●				●	●				●	●
PVDF	●						●						●					
PVAC					●						●					●		
PVAL	●				●							●				●		
PMMA	●						●	●				●		●			●	
POM				●		●		●				●			●		●	
PA6			●			●		●				●			●		●	
PA66			●			●		●				●			●		●	
PSU	●					●		●			●			●		●		
PI			●				●						●			●		
CA	●					●		●			●			●		●		
CAB	●					●		●				●			●		●	
CN	●						●	●			●					●		
PC	●					●		●			●			●		●		
PET	●					●		●			●				●	●		
PBT	●					●		●			●				●	●		
PF				●			●	●			●		●					
PF-mf				●			●	●			●		●					
PF-of				●			●	●			●		●					
UP				●			●	●			●					●		
EP	●						●	●			●		●			●		
PUR				●	●	●	●							●		●		
SI	●				●					●			●					

Problems

1.1 Determine what material was used to fabricate the samples supplied to you by the instructor.

1.2 An opaque, semi-rigid plastic toy with a smooth surface is subjected to a burn test. It drips as it burns clean with a blue flame. There is a penetrating smell when the flame is extinguished. What material was used to manufacture this toy? Which of the above attributes should usually not be taken into account in the decision process?

1.3 A transparent and stiff cup which is involved in a Hamburger-Burger "cold juice spill" lawsuit is given to you by a group of lawyers. Of the various materials it could be made of, what is your first guess? Explain.

1.4 A competitor of yours makes a toy that is opaque, stiff, and has a smooth surface. You want to know what material it is made of. You burn a corner of the toy. The flame is yellow. What do you conclude from your tables? Which one of the attributes can you eliminate to make your decision easier.

References

1. Termonia, Y., and Smith, P., *High Modulus Polymers*, A.E. Zachariades, and R.S. Porter, Eds., Marcel Dekker Inc., New York, (1988).
2. Ehrenstein, G.W., Faserverbundkunststoffe, Carl Hanser Verlag, München, (1992).
3. Ihm, D.W., Hiltner, A., and Baer, E., *High Performance Polymers*, E. Baer and A. Moet, Eds., Hanser Publishers, Munich, (1991).

2

Historical Background

Most topics are best introduced from a historical perspective. Although synthetic polymers and the plastics industry is a product of the 20th Century, the history of polymers goes back several centuries. This section presents some of the key materials and people involved in making the polymer industry into what it is known today.

2.1 From Natural to Synthetic Rubber

Natural polymeric materials such as rubber have been in use for several millennia. Natural rubber also known as *caoutchouc*[1] or *gummi elasticum*[2] has been used by South American Indians in the manufacture of waterproof containers, shoes, and torches (Fig. 2.1)[1]. When the natives made an incision in the bark of a tree it produced a white colored sap, composed of water and a monomer[3] called isoprene (Fig. 2.2). As the sap dried in air, and in conjunction with the sun's ultraviolet rays, the isopreme polymerizes into long chains or polymers, called polyisoprene (Fig. 2.2). This process turns the liquid into a sticky bouncy mass. Sometimes the natives performed the drying process while dipping a clay container in and out of a bat of rubber tree sap. This way, the container was slowly coated with a layer of polyisoprene, and when the layer was thick enough the clay was broken and washed out. This first dip-coating process produced a light waterproof container to carry liquids. Similarly, the indians produced rubber boots, by dip-coating their feet. A container produced by this method is shown in Fig. 2.3[1].

---

1 From a native South American language *cahuchu: caa* (wood) and *o-chu* (tears). In French (caoutschouc) and German (Kautschuk) this word is still used to describe some elastomers.

2 From the Latin elastic gum.

3 While a polymer is made from many parts, a monomer is a single entity.

Figure 2.1 Tapping the rubber trees (1830).

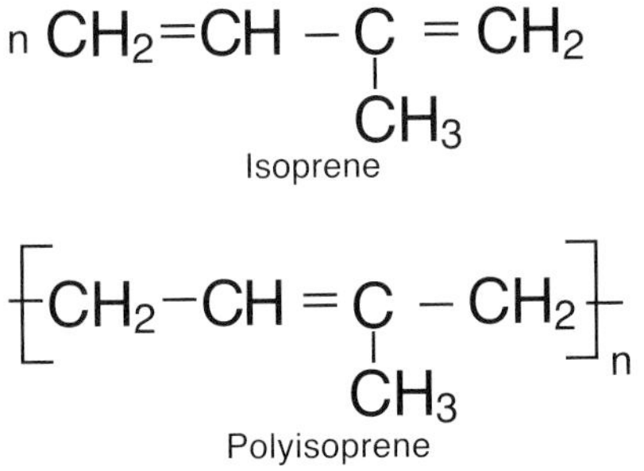

Figure 2.2 Isoprene and polyisoprene before and after polymerization

Figure 2.3 Container from the Amazon produced by an early dip-coating technique.

The first Spanish explorers of Haiti and Mexico reported that natives played games on clay courts with rubber balls [2]. Rubber trees were first mentioned in *De Orbe Novo*, originally published in Latin, by Pietro Martire d'Anghiera in 1516. The French explorer and mathematician Charles Maria de la Condamine, who was sent to Peru by the French *Académie des Sciences*, brought caoutchouc from South America to Europe in the 1740s. In his report [3] he mentions several rubber items made by native South Americans including a pistonless pump composed of a rubber pear with a hole in the bottom. He points out that the most remarkable property of natural rubber is its great elasticity. The first chemical investigations on *gummi elasticum* were published by the Frenchman Macquer in 1761. However, it was not until the 20th Century that the molecular architecture of polymers was well understood. Soon after its introduction to Europe, various uses were found for natural rubber. Gossart manufactured the first polymer tubes in 1768 by wrapping rubber sheets around glass pipes. During the same time period small rubber blocks where introduced to erase lead pencil marks from paper. In fact, the word *rubber* originates from this specific application—*rubbing*.

These new materials slowly evolved from being just a novelty thanks to new applications and processing equipment. Although the screw press, which is the predecessor of today's compression molding press, was patented in 1818 by McPherson Smith [4], the first documented *polymer processing* machinery dates back to 1820 when Thomas Hancock invented a rubber masticator. The primary use of this masticator, which consisted of a toothed rotor inside a toothed cylindrical cavity [5], was to reclaim rubber scraps which resulted from the manual manufacturing process of elastic straps[4]. In 1833, the development of the vulcanization[5] process by Charles Goodyear[6] [6] greatly enhanced

4 Perhaps the first plastics recycling program.

5 From the Greek *Vulcan*, god of fire.

6 In 1832, F. Lüdersdorff in Germany had already discovered that when incorporating sulfur into rubber it loses its tackiness if heated.

the properties of natural rubber, and in 1836, Edwin M. Chaffee invented the two roll steam-heated mill, the predecessor of the present day calender, for mixing additives into rubber for the continuous manufacturing of rubber coated textiles and leather. One of the first applications of vulcanized rubber was a set of rubber tires for Queen Victoria's carriage, but it took another 25 years for the veterinarian John Boyd Dunlop of Belfast, to develop the pneumatic tire to make his son's bicycle ride a little smoother.

As early as 1845, presses and dies were being used to mold buttons, jewelry, dominoes, and other novelties out of shellac and gutta-percha. *Gutta-percha* [7] or *gummi plasticum*[8], a gum found in trees similar to rubber, became the first wire insulation material and was used for ocean cable insulation for many years. A patent for cable coating was filed in 1846 for trans-gutta-percha and cis-hevea rubber and the first insulated wire was laid across the Hudson River for the Morse Telegraph Company in 1849. Charles Goodyear's brother, Nelson, patented hard rubber, or ebonite, in 1851 for the manufacturing of dental prostheses and combs. To allow the continuous coating of wires or textiles, the ram-type extruder was invented by Henry Bewley and Richard Brooman in 1845. The first *polymer processing* screw extruder[9], the most influential element in polymer processing, was patented by an Englishman named Mathew Gray in 1879 for the purpose of wire coating. Figure 2.4 [7] presents Mathew Gray's extruder as illustrated in his patent.

The demand for rubber grew rapidly in a world that was becoming more industrialized. In the late part of the 19th and early 20th Centuries, most of the world's rubber came from the Amazon region of South America. This not only led to a dependancy of the industrialized nations on a few local families and corporations in a relatively unstable part of the world, but also to horrible human rights abuses against the natives of the Amazon region by those entities that controlled the rubber cartel. To free themselves from the Latin-American rubber cartels, the British had planted rubber trees in Maylasia and other South East Asian colonies using seeds that had been smuggled out of South America by the 1920's. In Germany, to overcome its dependancy on rubber from other countries, and other countries' colonies, the German Kaiser, Wilhelm II, promised a 25,000 Reichsmark prize to anyone that would invent a synthetic replacement for natural rubber. By 1909, the German chemist Fritz Hoffmann, who was working at Bayer in Leverkusen had developed the first synthetic polyisoprene. The Kaiser's automobile was the first vehicle to be fitted with synthetic rubber tires (Fig 2.5[1]). When World War I broke out, the German government decided that since the war would not last long, it would drop the idea of building large polymerization plants to make synthetic rubber. Instead,

7 From the Malay *gettah* and *pertja* which means rubber and clump, respectively.

8 From the Latin plastic gum.

9 The screw pump is attributed to Archimedes, and the actual invention of the screw extruder by A.G. DeWolfe of the United States dates back to the early 1860s.

despite of the British embargo on Germany, the Germans smuggled natural rubber inside coffee cans from America through neutral countries. After 1916, the British blockade forced the Germans to produce synthetic rubber. In the next two years, Germany painstakingly produced 2,500 tons of synthetic rubber using polymerization processes that took several months.

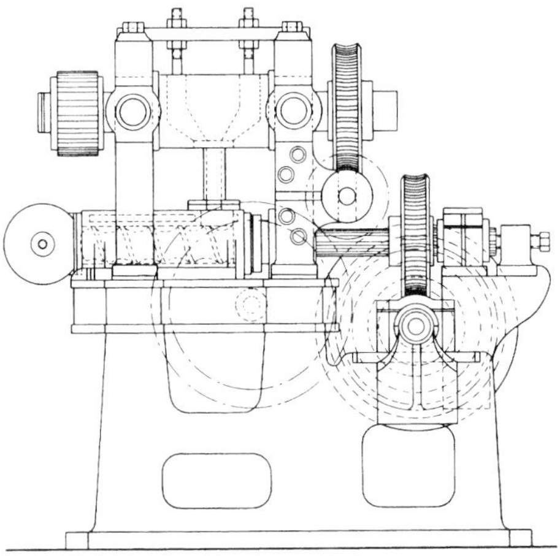

Figure 2.4 Mathew Gray's extruder as presented in his patent of 1879.

The low prices for rubber after the war -- 17¢ per pound in 1920 -- soon resulted in a loss of interest in continuing development of synthetic rubber. However, it did not take long for the rubber cartel to reorganize and as a result, prices of natural rubber skyrocketed again to a high of $1.21 per pound in 1925. The high rubber prices as well as being the home of the key polymer chemists of the time were the main incentives for the German government and industry to reinvest in their synthetic rubber research program.

Figure 2.5 The Kaiser (to left wearing a white coat) and his car fitted with methyl rubber tires (1912).

Eventually, work on developing synthetic rubber became an obsession on both sides of the Atlantic ocean. By the late 1920s, the Germans had found a way to speed up the polymerization process of the butadiene monomer by using sodium as a catalyst. Their new material was marketed under the tradename Buna, made up of the first syllabi of butadiene and natrium. In the United States, Wallace Carothers (see Section 2.6), of the DuPont Company, developed polychloroprene, which was first marketed as Duprene and finally, in 1932, as Neoprene. However, by 1932, the depression had caused the prices of natural rubber to drop to an all time low of 3.5¢ per pound. By now, Germany was already in its quest of becoming self-sufficient, and intense research in the field of synthetic rubber was encouraged and funded. This resulted in styrene-butadiene rubber, marketed as Buna S, and in butadiene-acrylonitrile copolymers, marketed as Buna N. The Nazi regime saw synthetic rubber as a strategic element, and so, in addition to stockpiling natural rubber, they invested enormously in their synthetic rubber program at the IG Farbenindustrie. IG Farben was without doubt the leader in Germany's effort to become self-sufficient, and therefore enjoyed the backing of the regime. They soon started building Buna rubber plants all over Germany. During the war, to keep up with the large demand for synthetic rubber, IG Farben built the largest synthetic rubber plant in Monowitz, Silecia, just outside of Auschwitz. This was not only a very strategic place, since the location was close to the Silician coal mines that were far from Allied bombers, but undoubtedly a place that could supply prisoners for forced labor. This plant became an integral part of the Auschwitz III concentration camp and it was referred to as the Buna sub-camp. Despite of its magnitude, the Auschwitz-Monowitz Buna plant never produced a pound of rubber [8], however, it remains forever closely linked to one of the darkest hours of humanity.

2.2 Cellulose and the $10,000 Idea

If we step back to the 19th Century, in addition to rubber, another natural polymer, cellulose, impacted everyday life. The invention of cellulose plastics, also known as Celluloid, Parkesine, Xylonite, or Ivoride has been attributed to three people: The Swiss professor Christian Schönbein, the English inventor Alexander Parkes, and the American entrepreneur John Wesley Hyatt.

Christian Friedrich Schönbein, a chemistry professor at the University of Basel, loved to perform chemistry experiments in the kitchen of his home, much to his wife's dismay. Early one morning in the spring of 1845, Schönbein spilled a mixture of nitric and sulfuric acids, part of that day's experiment, on the kitchen counter. He quickly took one of his wife's cotton aprons and wiped the mess up, then rinsing it with water before the acid would damage the cloth. As he hung the apron to dry over the hot stove, it went into the air in a loud puff and flame. It had exploded in front of his very eyes. After he recovered from the shock, Schönbein's curiosity led him to impregnate wads of cotton with the acid mixture. Every time, he was able to ignite the mass, leading to an enormous, uncontrollable explosion. He called his invention guncotton. He had invented cellulose nitrate. Guncotton was three times as powerful as gunpowder and did not leave a black cloud after the explosion. Schönbein sold his patent to the Austrian Empire's army, but found no buyers in Prussia, Russia, or France. Finally, he sold his patent to John Taylor, his English agent, who immediately began production of guncotton in England. The production ended when his factory exploded, killing 20 workers. Although there were no buyers, several laboratories did spring up across Europe to investigate guncotton; going up into the air faster than they were being built. In addition to its bellicose applications, Schönbein envisioned other uses for the nitrated cotton mass. He added a solvent or plasticizer made of ether and alcohol and found a way to nitrate the cellulose fibers into a less explosive material which he called kollodium, glue in Greek. He reported to his friend Michael Faraday that this mass "is capable of being shaped into all sorts of things and forms..." In the spring of 1846, after accidentally cutting himself on the hand, he covered the wound with a thin elastic translucent film made of kollodium. He sold his idea to the English, who for years supplied the world with the first adhesive bandages. In England, there was one person that took particular interest in the Swiss professor's inventions. His name was Alexander Parkes.

Alexander Parkes started playing around with cellulose nitrate in 1847, and spent the next 15 years in the laboratory perfecting the formulas and processes to manufacture cellulose nitrate. His final process took the nitrated cotton and added vegetable oils and organic solvents producing a "plastic mass" that was easily molded into any shape or form after it was softened under heat. He called his plastic mass Parkesine. The new applications for this versatile material, such as combs, knife handles, and decorations, made their debut at the 1862 World Exposition in London. In 1866, Parkes launched the Parkesine Company Ltd. Due to the low quality of its products, parkesine was not a success and the company was liquidated in 1868. The poor mixing of the additives and solvents caused Parkesine products to significantly warp only a few weeks after manufacture. In 1869, Parkes sold his patents to Daniel Spill, his chief engineer, who founded the Xylonite Company and renamed Parkesine, Xylonite. Parkes continued working on his material until his death in 1890 at the age of 77. Alexander Parkes, the inventor and engineer can be credited with improving on Schönbein's invention, paving the road for the future of the plastics industry. He is also credited with fathering a total of 20 children. A very busy man, to say the least.

At the same time as the plastics industry seemed to be going under in England, in the United States John Wesley Hyatt was launching an enterprise that finally made cellulose nitrate a success, under the name of celluloid. As the story goes, it all began when in 1865 the billiard ball manufacturer Phelan & Collendar placed an add that promised $10,000 to the person who would find a replacement for ivory in the manufacture of billiard balls. Elephants were being slaughtered at a rate of 70,000 per year, which would have led to the extinction of this great animal, exorbitant prices for the "white gold" from Africa, and reduced profits for the billiard ball industry. The $10,000 tag called the 28-year old Hyatt's attention. After returning home from his job as a printer, he worked on this project until eventually he stumbled upon nitrocellulose in 1869. After finding a way to better mix all the components as well as allowing the solvents to completely evaporate from the mass before solidification, he was soon manufacturing high quality billiard balls. Instead of cashing in on the $10,000 prize, John Hyatt founded the Albany Billiard Ball Company with his brother Isaiah, becoming a direct competitor to Phelan and Collendar. For the next thirty years, until Bakelite replaced celluloid on the billiard table, many guns were pulled in the Wild West when the volatile balls sometimes exploded upon collision. Another immediate application of celluloid were dentures (Fig. 2.6[9]), which up until then were made of hard rubber. In view of loosing a rather profitable business to plastics, the rubber industry started a propaganda campaign against cellulose in all major US newspapers. They falsely claimed that celluloid dentures could easily explode in ones mouth when coming in contact with hot food. This not only cheated people of a much prettier smile, but also started a rivalry between the two industries which has caused them to mantain as completely separate entities to this day. In fact, despite of the materials and processing similarities between plastics and rubber, the plastics industry and the rubber industry have completely separate societies and technical journals. A plastics engineer is likely to be found in meetings organized by the *Society of Plastics Engineers* (SPE) or the *Society of the Plastics Industry* (SPI), while a member of the rubber industry will attend meetings organized by their own society, the *Rubber Division of the American Chemical Society*.

With a new and versatile material, Hyatt and his coworkers needed equipment to mass-produce plastic products. Based on experience from metal injection molding, the Hyatt brothers built and patented the first injection molding machine in 1872, to mold cellulose materials [10], as well as the first blow molding machine, to manufacture hollow products. In the summer of 1869, Hyatt and Spill, respectively, filed for patents dealing with the manufacture of nitrocellulose materials. This started a lengthy and costly litigation that eventually ruled in Hyatt's favor in 1876. Spill died soon after, at age 55, of complications from diabetes. John Wesley Hyatt lived another 44 productive years in which he invented the injection molding and the blow molding machines with which he processed celluloid products. He can certainly be credited for being the first person to successfully mold a plastic mass into a useful, high quality final product. However, above all, we should credit him for saving the elephant on the road to a $10,000 prize he never claimed.

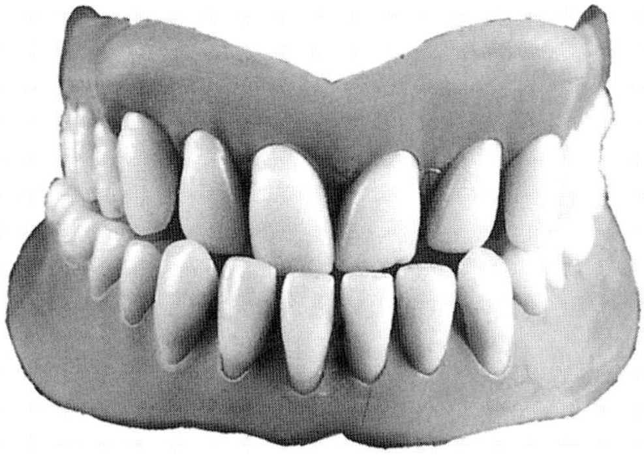

Figure 2.6 Celluloid dentures.

With the mass production of rubber, gutta-percha, cellulose, and shellac articles during the height of the industrial revolution, the polymer processing industry after 1870 saw the invention and development of internal kneading and mixing machines for the processing and preparation of raw materials [11]. One of the most notable inventions was the Banbury mixer, developed by Fernley Banbury in 1916. This mixer, with some modifications, is still in use today for rubber compounding.

2.3 Galalith -- The Milk Stone

Galalith, a popular plastic at the beginning of the 20th century, owes its discovery to a cat, a chemist, and an industrialist. Casein formaldehyde, a milk protein-based polymer is better known as casein plastic or by various trade names of which the most memorable ones are Galalith, Aladdinite, Erinoid, and Lactoid. As an industrial material it has come and gone, and all there is left are beautiful casein jewelry pieces, buttons, museum exhibits, and memories for plastic historians.

As the story goes, one night in 1897, the German chemist Adolf Spitteler's cat knocked over a small bottle of formaldehyde. The formaldehyde dripped from the chemists counter down to the floor into the cat's milk dish. The next morning, when Spitteler returned to his laboratory, he found that the formaldehyde that had dripped into the dish, had caused the milk to curdle and turn into a hard horn-like substance, much like celluloid. In fact, his cat had just invented the first semi-synthetic plastic since cellulose. Soon, Spitteler started experimenting with cheese curds (casein) and formaldehyde, and he found that the milk protein was rendered water-insoluble by letting it sit in a formaldehyde solution for extended periods of time.

Around the same time, Ernst W. Krische, who owned a small book binding and school supplies manufacturing company in Hanover, Germany, received a request from Turkey for washable, white writing boards. Pedagogists of the end of the 19th century believed that it would be better to teach children to write with black chalk on white boards rather than the other way around. However, the whiteboard had not been invented yet. Krische started experimenting by coating cardboard with casein. Casein, easily made by letting milk curdle, had been used for over 100 years as an adhesive, possibly also in Krische's book binding operation. In fact, the use of casein for non-food applications goes back even further, to at least 2 centuries B.C.E., in Egypt, where it was used as a binder for color pigments in the manufacture of paint. However, all these applications, which were well known at the end of the 19th century, did not require a water-insoluble material. In addition, there were many other hurdles that Krische had to overcome, such as the cracking and warpage which both resulted during the drying process of the wet casein.

Luckily, somehow Spitteler and Krische found each other. The chemist and the business man started playing with various materials that would soften or plasticize casein, such as glycerin and borax, and with various amounts and solutions of formaldehyde that would render the mass water-insoluble. Finally, in 1899 they were able to patent their invention. They were not only able to produce the washable whiteboards but soon their moldable plastic mass was marketed under the trade name of Galalith, from the Greek words for milk (gala) and stone (lithos). In Britain it was marketed under the trade name of Erinoid, derived from the Gaelic word for Ireland, the source of most of the British cheese curds required in the manufacture of the new plastic mass.

Beautiful products were produced using this new semi-synthetic material. Bright colored artifacts, such as fountain pens, buttons, clock cases, combs, faux jewelry, etc., were all made out of this polymer. In addition to pigments, fish scales were often blended with casein to produce a pearlised effect. By the late 1920s, the height of casein formaldehyde production, over 10,000 tons per year of dry casein were being absorbed by the plastics industry worldwide. At the same time, 35,000 tons of cellulose were being transformed into plastic products every year. But the end of casein plastics was around the corner. In addition to phenol formaldehyde, the advent of synthetic polymers caused the casein industry to slowly disappear. The new materials were less expensive and easier to process.

Today, the use of casein as well as other protein-based polymers is very small. However, it is still, for example, possible to find very exclusive buttons being made of casein formaldehyde. Although most people regard Galalith as a historic material, the fact that protein-based plastics are made of renewable resources, offering us an unlimited supply of raw materials, should not be ignored.

2.4 Leo Baekeland and the Plastics Industry

Leo Hendrik Baekeland is well known for inventing the first synthetic polymer and for being the father of today's plastics industry. Dr. Baekeland, the chemist and inventor, refined and made possible Bakelite, also known as phenolic. Leo Baekeland, the sometimes mischievous, boyish, Belgian immigrant of modest origins, was an entrepreneur who created and gave momentum to what is known today as the plastics industry.

Leo Hendrik Baekeland was born November 14, 1863, in Ghent, Belgium, to poor illiterate parents. He excelled at school and at age 17 enrolled as the youngest student at the University of Ghent, where he studied chemistry. After only two years he finished his Baccalaureate in chemistry, and at age 21 he received a doctorate *summa cum laude*. Shortly after receiving his doctorate, he taught chemistry at the University of Bruges, until he was offered a professorship at the University of Ghent. That same year he married his major professor's daughter, Céline, much to the old professor's dismay. Baekeland would recount years later: "The most important discovery that I ever made was Céline." In the end, a displeased senior faculty member, and a traveling fellowship to the United States was enough incentive for the young professor to embark for New York. During his visit, Professor Chandler at Columbia University encouraged Baekeland to stay in the United States. The 25-year-old Leo was now well aware that an academic career was not his calling. He was an entrepreneur at heart; even while an academic in Ghent he had started a small business that manufactured dry photographic plates, an enterprise that also satisfied his passion for photography. From New York, Baekeland telegraphed his resignation to the University of Ghent and soon after was offered a position as a chemist at A. & H.T. Anthony & Company, a large photographic supply house where he worked for two years. He then decided to try it on his own as a chemical consultant.

The next few years were very difficult for Baekeland. The country went into a deep recession, which adversely affected his consulting business. Worries about his finances and his future mounted, affecting Leo's health to the point that he was bedridden for an extended period of time. After recovering, he decided he would concentrate all his energies on one project where he knew he could strike it big. He returned to his old hobby, photography, and developed a photographic paper that was so light sensitive that it could be exposed using artificial light. Sunlight was the only light intense enough to expose the photographic plates of the late 1890s. *Velox*, the name he gave to his new product, was an instant success. It caught George Eastman's attention, who invited Baekeland to Rochester, NY, to negotiate the rights to this revolutionary photographic product. Baekeland had decided that he would ask Eastman for $50,000 for the rights of *Velox*, but he was willing go as low as $25,000. Before Baekeland was allowed to state his demands, the overpowering magnate took over the meeting and on the spot offered the young scientist $1 million for his invention. The year was 1899. Ten years after his arrival in the United States, the 35-year-old Leo Baekeland was wealthy beyond his wildest dreams.

Baekeland now had the necessary funds to embark on the research that would eventually lead to the development of Bakelite, the first synthetic plastic. Bakelite, also known as phenolic, results when phenol and formaldehyde undergo a condensation polymerization reaction. The reaction between these two readily available materials had been well known since 1872, when Adolf von Baeyer first reported it. Baekeland's contribution was converting the sticky useless material of the late 1800s into the chemical-resistant stiff product that could easily be shaped into any imaginable geometry during the 20th century.

Bakelite not only helped shape the streamlined pre-WWII years, but it also presented a material that made it possible to mass-produce items that made life easier in the home, office, street, field, and factory. The ability to mass-produce plastic products helped create the myth that plastics are a cheap replacement for materials of higher quality. The reality is that phenolic is a material of superior mechanical and electrical properties with higher chemical resistance than those materials it replaced. For example, tough, lightweight, and stiff phenolic-cloth composite propellers replaced wooden aircraft propellers that easily cracked, causing catastrophic failures in the early years of aviation. In another case, Bakelite replaced wood as the material of choice for radio housings (Fig. 2.7), not because it was a cheaper material, but because it was easier and faster to form into its finished shape. In addition, phenolic offered the freedom to experiment with design and esthetics, while at the same time made products, that were previously considered luxuries, available to everyone.

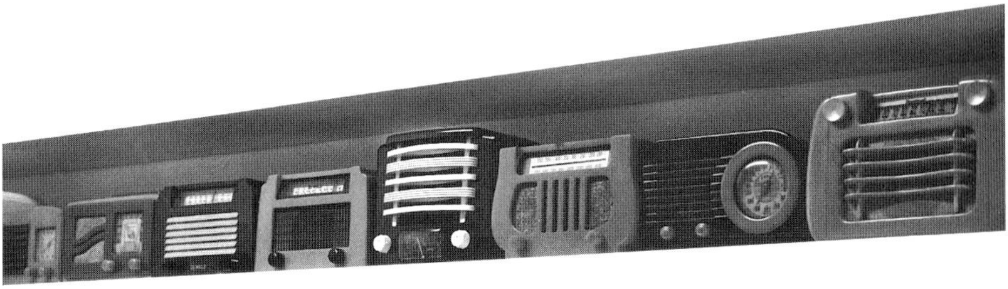

Figure 2.7 Bakelite radios. (The Kölsch Collection).

While president of the American Chemical Society, Baekeland once said: "If academics are given the opportunities to cut their teeth on some practical problems, they may grow to be of decidedly greater service to their science or its applications." Baekeland gave himself that opportunity and certainly grew greater than anyone before or after him in the plastics industry. He died at age 80 in Beacon, New York, on February 23, 1944.

2.5 Herman Mark and the American Polymer Education

Herman Francis Mark has been credited for being one of the founders of the field of polymer science and for being the father of polymer education in the United States. Over his long and productive lifetime, Herman Mark published 20 books and over 500 articles. His Institute of Polymer Science at the Brooklyn Polytechnic University in New York served as the incubator for some of the most notable polymer scientists of the second half of the twentieth century.

Herman Mark was born in Vienna, Austria, on May 3, 1895, to a Jewish father and a Lutheran mother. He spent his formative years in the Viennese fourth district where his father, a physician, and his mother welcomed many turn-of-the-century intellectuals into their home. One can only imagine the impact on young Mark when among his parent's friends were the musician Johan Strauss, Sigmund Freud and Theodor Herzl, who with Chaim Weizmann went on to found the Zionist movement. As a twelve-year-old, his life and world changed during a visit to a university laboratory. When he got home, the excited boy reported of "glass bottles and glass beakers, blue flames and Bunsen burners, bubbling liquids and those long rubber tubes through which vapors are diverted." Soon after, Herman built his own laboratory in his room, where he conducted his own analyses and syntheses, starting a love affair with chemistry that lasted over 80 years.

World War I forced him into adulthood (Fig.2.8[9]). He served as a combat soldier on various fronts for a period of five years. During this time he had two short stays in Vienna, the first to recover from a bullet wound. As usual, making the most of his time, during each stay he completed a semester of chemistry at the university. During the war, his many acts of bravery made Herman Mark a national hero. On one occasion, Mark single-handedly held a dozen Russian soldiers at bay until his whole commando reached safety. At the end of the war, after a short imprisonment in Italy, Mark, now the most decorated officer of the Austrian Army, continued his chemistry studies at the University of Vienna.

After Graduating *suma cum laude* with a Ph.D. in chemistry in the summer of 1921, he moved to Berlin with his advisor Wilhelm Schlenk who succeeded Professor Emil Fischer at the University of Berlin. The year 1922 was an eventful year for Mark: he married Marie Schramek (Mimi) and joined the Kaiser Wilhelm Institute for fiber chemistry (today the Max Planck Institute), directed by R. O. Herzog. In Berlin, Mark co-authored over 30 papers, which included his work on applying the field of X-ray diffraction when analyzing crystal structures. His stay in the postwar, inflation ridden Prussian capital gave him the opportunity to work and fraternize with people of various backgrounds and nationalities, broadening his intellect and interests. He always felt that a highlight of his years in Berlin was when he conducted experiments for Albert Einstein to confirm Einstein's light-quantum hypothesis.

In the summer of 1926, Professor Kurt H. Meyer, one of the research leaders at the I.G. Farbenindustrie chemical giant's BASF labs in Ludwigshafen, visited Mark in Berlin and proposed that the young scientist move to industry to apply his basic research abilities to industrial problems. Mark accepted Meyer's offer and in January of 1927, moved to Mannheim, where he and his family would live for the next five years. This dramatic change, along with Mark's association with Meyer, his immediate boss at I.G. Farben, and his continuing collaboration with Hermann Staudinger and other German academic scientists of the time, served as one of the catalysts that propelled the field of polymer science into the spotlight. In 1924, Hermann Staudinger[10] had proposed a model that described polymers as linear molecular chains. Once this model was accepted by other scientists, the concept used to synthesize new materials was realized. For his idea of giant molecules, Staudinger was ridiculed by his conservative peers, who at the time were still convinced that polymers such as natural rubber (polyisoprene) were made of colloids. It took until 1953 for Staudinger to receive the Nobel prize for his pioneering work on polymers. The young and open-minded Hermann Mark understood Staudinger's theories from the beginning. In 1928, Mark and Meyer proposed a crystal structure for cellulose, rubber, and silk, nearly identical to the ones still accepted today. Their theory was supportive of Staudinger's high molecular weight theory and refuted the commonly accepted colloidal association theory of the time. Mark's model for polymer molecules differed from Staudinger's in that the molecules were flexible, not rigid, as Staudinger suggested. Time would prove Mark right. In the relatively short time at I.G. Farben, Mark produced an impressive list of publications, which included over 80 papers, 17 patents and 3 books.

Figure 2.8 Hermann Mark in 1917.

10 The idea that polymers were formed by macromolecules was not well received by his peers at first. However, for this work Staudinger received the 1953 Nobel prize in chemistry.

With the political winds in Germany shifting to the right, and because of their Jewish ancestry, both Herman Mark's and Kurt Meyer's positions at I.G. Farben became uncertain in the summer of 1932. The advice that came from the top of the company was that Mark look for an academic position, preferably outside of Germany. However, since I.G. Farben did not want to loose a certain degree of control on Mark's research, he was promised financial support for his future academic work. And so, Herman Mark, his wife, Mimi, and two young children moved to Vienna in the fall of 1932, where he became director of the first Chemistry Institute at the University of Vienna. His friend and closest collaborator, Kurt Meyer, moved to Geneva that same year. With I.G. Farben funds, Mark was able to assemble an impressive team of scientists that included Eugene Guth, Robert Simha, and Fritz Eirich. In Vienna, Mark published many fundamental papers on polymer physics including groundbreaking theories on polymer relaxation, polymerization kinetics and molecular weight distribution. However, leaving Germany was not enough; during the thirties, the Austrian political airs also started turning brown, culminating March 15, 1938, when the Nazis marched into Vienna. Mark was arrested the following day, and released without his passport four days later. The next eight weeks were spent recovering his passport and securing a position in Canada, in order to get Canadian and Swiss visas. To finance his trip to North America, the resourceful Mark purchased over one kilogram of platinum wire that he carefully wrapped around coat hangers. With visas in-hand, and their precious currency, the Marks and their niece attached a Nazi flag on the hood of their car and placed rope and skis to the roof and headed up the Alps. They drove into Switzerland on May 10, 1938. After bribing their household possessions out of Austria, the Marks spent the summer in Switzerland, France, and England. Mark started his new position at the Canadian International Paper Company in September of 1938.

After almost two years in Canada, Mark moved to New York in the summer of 1940 where he founded The Institute of Polymer Science at the Brooklyn Polytechnic University in New York. This marks the official start of polymer science education in the United States. The Institute of Polymer Science was the first American institution to award the Ph.D. degree in polymer science, and thus, became the incubator of many American scientists and educators. He advised over 100 scientists that included Fritz Eirich, Edmund Immergut, Herbert Morawetz, and Charles Overberger. Some of his students went on as educators, multiplying the polymer science education all over the world. Eirich stayed at Brooklyn, Guth went to Notre Dame, Simha to Caltech and later to Case Western. Other pupils of Mark served as editors and writers such as Immergut, who had several editorial positions, including with Hanser Publishers between 1980 and 1997, which resulted in a vast polymer engineering and science book library, that has served as a reference and resource to many students, engineers, and scientists. During his years at

Brooklyn, Mark wrote hundreds of papers, many books and was the founder and editor of various journals including the Journal of Polymer Science. He also served as a consultant to DuPont, Polaroid, and others, and maintained relations with industry in his native Austria.

Herman Mark's lifetime achievements were recognized by many: he received 17 honorary doctorates, numerous medals and awards, and served as an honorary member to dozens of scientific academies and professional societies. Herman Mark, the war hero, scientist and teacher, died at age 96 in April of 1992.

2.6 Wallace Hume Carothers and Synthetic Polymers

Wallace Hume Carothers is perhaps the most influential American character in the history of polymers. Already as a child he was greatly admired by his grade school friends, who called him "Doc" for his ability to make, among other things, crystal radios. This was only the beginning of a short but productive life in a time when the field of polymer chemistry was still in its infancy. He went on to author over 50 technical papers, file over 100 U.S. and foreign patents, and above all, invent nylon, the world's first synthetic fiber. This specific contribution of Carothers opened the floodgates of invention in the field of thermoplastic polymers.

Born on April 27, 1896, in Burlington Iowa, Carothers was raised in Des Moines where his father taught at the Capital Cities Commercial College, which he attended after high school. He then went to Tarkio College in Missouri where he received a degree in Chemistry at age 24. From Tarkio he moved to the University of Illinois at Urbana-Champaign where he studied under Professor Roger Adams. In the summer of 1921 he received a science degree in chemistry and, after a one year teaching position at the University of South Dakota, defended his Ph.D. in the late spring of 1924. After two years as an instructor at Illinois, Carothers was recruited to teach at Harvard.

Carothers' success was also due in great part to the vision of Dr. Charles M. Stine, the Director of DuPont's Chemical Department. In 1927 Stine used his influence to accomplish something that was unheard of in a corporation of any size; he convinced the company's executive committee to fund research in pure science at the experimental station in Wilmington. With a $25,000 per month budget, Stine lured Carothers from Harvard were he had been teaching for about two years. With this practically unlimited budget and after assembling a team of extremely bright coworkers, including Paul Flory and Julian Hill, Carothers was ready to begin his 10-year tenure at DuPont. Building on his experience as a physical chemist and on the research of his contemporaries, such as Adams, Marvel, Staudinger and Mark, Carothers and his team[11] developed polychloroprene, polyester, and of course polyamide 6 and polyamide 66. During that time he coined the terms so widely used today: *addition polymerization* and *condensation polymerization*. Polychloroprene (Fig. 2.9[9]), an elastomer, was polymerized by addition polymerization, similar to the polymerization of polyisoprene in Fig. 2.2. Polyamide 66 is polymerized by condensation polymerization as shown in Fig. 2.10.

Figure 2.9 Wallace Carothers stretching a sheet of polychloroprene.

11 Paul Flory who received the 1974 Nobel prize in chemistry worked in Carother's group at DuPont's research laboratories.

$$n\ HOOC-(CH_2)_4-COOH \quad + \quad n\ H_2N-(CH_2)_6-NH_2$$

$$\text{Adipic acid} \qquad\qquad \text{Hexamethylendiamine}$$

$$\rightleftharpoons HO-\left[OC-(CH_2)_4-CO-NH-(CH_2)_6-NH\right]-H + (2n-1)H_2O\uparrow$$

$$\text{Polyamide 66} \qquad\qquad\qquad \text{Water}$$

Figure 2.10 Condensation polymerization of polyamide 66.

Unfortunately, Carothers never saw the dramatic impact of his inventions. He suffered from two illnesses that in the 1930s were still considered deadly: he was both a manic-depressive and an alcoholic. And so, in spite of his own efforts, and the constant care of friends, family, and doctors, the demons finally took over. Two days after his 41st birthday he was found dead in a Philadelphia hotel room. He had taken cyanide.

In 1938, a year after Carothers's death, nylon was introduced to the world, primarily as a replacement for silk in hose and stockings and as toothbrush bristles. As the story goes, nylon is not an acronym for New York and London, as many people often believe. It evolved from the suggested name *norun* (a descriptive word that was inaccurate, since fiber 66 stockings did run), to *nuron* (which sounded too brainy), to *nulon* (which would have been hard to market as *the new nulon*), to nilon (two possible pronunciations) to, finally, nylon. In the heated pre-WWII years, it was even suggested that the word nylon was an insulting acronym for the Japanese. Though this is doubtful, it is certain that its invention gravely affected the Japanese trade balance, and in consequence, the overall position of the Japanese industry in world markets at the threshold of World War II. The influence of this miracle fiber is indisputable. Allied use of nylon in parachutes during the invasion of Normandy may have played a decisive role in the war's military outcome. The most obvious influence may come from its impact on consumer consumption. A comical example of this came after the war, the day a department store in Washington offered to sell a pair of nylon stockings to the first 1000 callers. As a result the entire phone system in the D.C. area was flooded and collapsed. This phone system had always resisted collapse, including the day Pearl Harbor was bombed, the day the war ended, and every historically significant day since. This suggests the real power of Carothers' invention: a public whose lives have been changed forever with the advent of thermoplastic polymers.

As Wallace Hume Carothers once said, he had done enough for one lifetime.

2.7 Polyethylene - A Product of Brain and Brawn

By the 1930s Staudinger, Carothers and Mark had paved the road of synthetic polymers for the chemical industry. At the ICI laboratories in Winnington, England, in addition to knowledge in polymers and polymerization, it took muscle power and risk-taking youths, to invent polyethylene, today's highest volume commodity plastic.

Eric William Fawcett and Reginald Oswald Gibson are the two young British scientists who can be credited with polymerizing the first few grams of polyethylene at the Winnington ICI laboratories in England. Fawcett was a young Oxford graduate who was only twenty years old and had to have his father come to the Winnington laboratories to sign for him his employment contract with ICI. Gibson had just returned from Amsterdam, where he had worked with Professor Anton Michels, the world-leading expert on high-pressure vessel design. They both worked in Lab Z, a small ICI laboratory that housed a pressure vessel that everyone called "the bomb." It was no coincidence that the assistants that worked in Lab Z were all young men in great physical shape. It took significant muscle power to operate the hand pump that could pressurize the bomb to up to 2,500 atmospheres (almost 37,000 psi). After pressurizing various liquids and finding that they remained unfazed, they decided to switch to gases, which they hoped would lead to more exiting reactions.

Late on a Friday afternoon, March 24 of 1933, they filled the bomb with ethylene gas, raised the temperature to 170 ^{o}C and pressurized it to 1,900 atmospheres. They turned off the lights and went home for the weekend. Since they couldn't wait all weekend for their results, Fawcett and Gibson went back to the lab on Saturday morning and found that the pressure inside the vessel had dropped to 100 atmospheres, a sure sign for a chemical reaction having taken place. Again they pumped the bomb up to 1,900 atmospheres, and left for the remainder of the weekend. On Monday, they found the pressure dials marking 0 atmospheres. They found out that, in addition to the polymerization reaction that took place inside the vessel, the pressure drop was also due to a leak that sprang around an oil tube connection. After opening the vessel, they saw that its inside was covered with a waxy white powder, of which they were able to scrape out 0.4 grams; the first 0.4 grams of polyethylene. They quickly fixed the leak and repeated the experiment, recovering a total of 4 grams of polyethylene. They now had collected enough material that they were able to spin a couple of fibers and form a small thin film. In the third try, the bomb exploded. Even though the explosion did not harm anyone, the managers did not find these experiments as important as Fawcett and Gibson knew they were, and Lab Z was put to rest and the young men were given other tasks.

Two years later, Fawcett attended The Faraday Society of London polymer congress in Cambridge were Wallace Carothers, Herman Mark, Kurt Meyer, and Hermann Staudinger were the main guests and speakers. In the proceedings, Fawcett saw how the future Nobel Prize winner Staudinger claimed that it was not possible to polymerize ethylene. Fawcett got his courage together and went up to Staudinger's room at the University Arms Hotel to attempt to set him straight. There he told the prominent scientist that under certain conditions it was possible to polymerize ethylene. Staudinger ignored him and gave his talk as planned. After the talk Mark even supported Staudinger's findings. Undaunted by the prominence and belief of his opponents the young man rose and told the audience that ethylene polymerizes if maintained at 170 °C under very high pressures. This bold move impressed Carothers, then a well-known American scientist, who offered Fawcett a job and took him to the DuPont laboratories in Wilmington, Delaware.

Soon after, ICI resumed their high-pressure vessel work, reproducing Fawcett and Gibson's experiments and patenting polyethylene in February of 1936. ICI continued to produce polyethylene using their high-pressure polymerization process. Soon, their operation had been scaled-up sufficiently to produce several kilograms of polyethylene every hour. In the meantime they had also discovered that this new plastic had excellent dielectric properties, which made it ideal for insulating thin wires for telecommunications applications.

In 1940, the British were at war. The Germans were bombing major English cities, and although ground radars detected the bombers, it was usually too late. A major breakthrough came when the British were able to reduce the weight of a radar from several tons to just under 600 pounds with the use of polyethylene as a wire insulator. This allowed them to place radars inside their fighter planes, enabling them to detect Nazi planes, before they saw them, shooting them out of the air at the same rate as they were crossing the British Channel. This put the British at an advantage, saving their cities from destruction. Similarly, the radar rendered the German submarine fleet obsolete, protecting the life saving sea transports, from the U.S.A. to England. As a result, hardly any submarines returned to Germany after 1943.

Twenty years after Fawcett and Gibson produced polyethylene inside the bomb, the German professor and future Nobel Prize winner Karl Ziegler found a way to polymerize ethylene at low pressures using catalysts. His catalysts where metal based compounds that act as vehicles as they drive through a sea of ethylene, picking up each monomer in an orderly fashion resulting in long chains of polyethylene. One was now able to produce this versatile material at much higher rates, under safer conditions, and at lower energy costs. Ziegler sold his patent to several resin producers who immediately jumped into production. However, they soon found out that their new process produced polyethylene that cracked easily, rendering the first few thousand tons of high-density polyethylene useless. Luckily, in 1957 along came a small Californian toy manufacturer, The Wham-O Company, who proposed to use the inferior HDPE on their new toy: the Hula-Hoop (Fig. 2.11). While Wham-O was using up the world supply of imperfect polyethylene, and America was buying the hoops by the millions, the chemical companies were successfully working out the kinks in their polymerization processes. One of Ziegler's occasional co-workers, the Italian chemical engineer Giulio Natta, extended Ziegler's research on organometallic catalysts to discover polypropylene. Both, Ziegler and Natta received the Nobel Prize in chemistry in 1963, and today, Ziegler-Natta catalysts are still in use.

Figure 2.11 The Hula-Hoop.

Soon, polyethylene Hula-Hoops, Tupperware, and plastic bags became everyday items. As the years passed, catalysts not only produced other polymers but also allowed us to tailor an ultra-high molecular weight high-density polyethylene that can be spun into fibers that are stronger than steel. Today, UHMWHDPE is used to manufacture artificial hip joint replacements that have given many people a quality of life otherwise thought to be impossible.

Other thermoplastics were also developed during that time and in the years that followed. Polyvinyl chloride (PVC)[12] was made into a useful material in 1927 [12]. Due to its higher wear resistance, polyvinyl chloride replaced shellac as the material of choice for phonograph records in the early 1930s. Polyvinyl

12 The preparation of PVC was first recorded in 1835 by H.V. Regnault.

acetate, acrylic polymers, polystyrene (PS), polyurethanes, and polymethyl methacrylate (PMMA) were also developed in the 1930s [13], and polycarbonate (PC) in the 1950s. And the list goes on. Today, developing and synthesizing new polymeric materials has become increasingly expensive and difficult. Hence, developing new engineering materials by blending or mixing two or more polymers or by modifying existing ones with plasticizers has become a widely accepted procedure.

2.8 The Super Fiber and the Woman Who Invented It

When Stephanie Louise Kwolek (Fig.2.12) discovered a low viscosity and cloudy polymer in 1964, she did not discard it as a contaminated and useless polymer solution, but continued experimenting with it. Polymers in the melt state are usually not only highly viscous but also amorphous and therefore transparent or clear. Kwolek's curiosity and persistence soon showed her that she had discovered a polymer that was crystalline in the melt state; she had discovered liquid crystalline polymers or LCP's. This invention soon resulted in the discovery of Aramid fibers, probably better known by their trade name Kevlar fibers.

Kwolek, who was born in 1923 in New Kensington, Pennsylvania, had an early passion for fabrics and fashion design, which resulted in new clothes for her dolls. However, she never imagined that some day people's lives would be saved with bullet proof vests made with fabrics weaved out of her fibers. The super fibers that Stephanie Kwolek discovered were five times stronger than steel but were only a fifth of its density. Kwolek studied chemistry and biology at what is today known as Carnegie Mellon University. She wanted to continue her studies in medicine, but could not afford the high costs of medical school. To allow her to save some money, she decided to take a job at the DuPont Company. However, she soon became enamored with the work and shortly thereafter transferred to DuPont's Pioneering Research Laboratory of the Textile Fibers Department. She was in the perfect environment, working for a company that had discovered and made a name with synthetic polymer fibers. In 1964 she found herself working with a group of researchers whose main task was to develop an ultra strong fiber. The main application for this fiber would be the manufacture of lighter tires that would help reduce fuel consumption, motivated by early predictions of gasoline shortages. The main breakthrough came that same year when she discovered that extended-chain aromatic polyamides could be made into liquid crystalline solutions. These polymers could in turn be spun into fibers whose molecules were highly oriented, rendering a very stiff and strong final product.

Figure 2.12 Stephanie Louise Kwolek showing an Aramid molecule model.

After continued research refining her discovery, Kevlar was introduced to the market in 1971. Today, in addition to bullet proof vests, the super fiber can be found in brake pads of automobiles, trains, and aircraft as well as in safety helmets, skis, and numerous other applications.

2.9 One Last Word -- Plastics

The year 1967 was a significant year for plastics. One event, related to American pop culture and one that some of us tend to remember, is the advise given to the young graduate played by Dustin Hoffman in the movie *The Graduate*. The prophetic words told to the new graduate --*"I just want to say one word to you, Ben. Just one word -- plastics"*-- where a symbol of the times and a sign of things to come. The same year that the movie was showing in the theaters, the volume of plastics' production surpassed that of all metals combined. Today, almost forty years later, plastics production is six times higher than in 1967, while production of metals has barely doubled. However, to be fair, in the popularity contest between metals and plastics we can always present the data differently, namely by weight. This way, the tonnage of metals produced worldwide is over twice the tonnage production of polymers. Neverless, this is a book about polymers, and we like to point to the fact that today the volume of polymers produced is three times larger than that of metals.

In fact, the world's annual production of polymer resins has experienced a steady growth since the beginning of the century, with growth predicted way into the 21st century. This can be seen in Fig. 2.13 [14] which presents the world's annual polymer production in millions of tons. There has been a steady increase throughout the years, with slight dips during the oil crisis in the mid 1970s and during the recession in the early 1980s. In developed countries, the growth in annual polymer production has diminished somewhat in recent years. However, developing countries in South America and Asia are now starting to experience tremendous growth. With the exception of the recession years, the growth in US polymer production has been declining in the past 20 years to approximately 4% of annual growth rate. Since 1970, China has seen the highest annual growth in the world, ranging from a maximum approximately 50% between 1976 and 1977 to a low of 2% between 1980 and 1981.

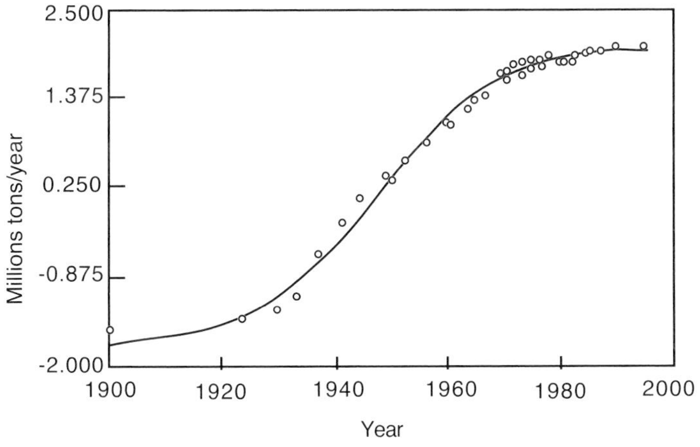

Figure 2.13 World annual plastics production since 1900.

In recent years, there has been a rapid increase in the demand for polymeric materials in developing countries such as Colombia. The demand for polymers in Colombia grew from 295,000 tons in 1992 to 389,000 tons in 1993. Although this number seems small when compared to the 34.4 million tons produced in the US in 1993, Colombia has a 32% annual growth compared to the US 4% growth. It should be pointed out that Colombia produces PE, PP, PS and PVC for its own consumption and for export to other Latin American countries, while it imports most of the needed engineering thermoplastics such as SAN, ABS, PC, etc. [15] from developed countries such as the US and Germany. Similar trends exist in other Latin American countries.

There are over 18,000 different grades of polymers, available today in the US alone[13]. As pointed out in Chapter 1, they can be divided into two general categories—thermosetting and thermoplastic polymers. In 1993, 90% of polymer produced in the United States were thermoplastics. However, in a 1995 worldwide projection, thermoplastics account for 83% of the total polymer production [16]. Figure 2.13[14] shows a percentage break down of US polymer production into major polymer categories. These categories include polyethylene, polypropylene, polystyrene, polyvinyl chloride, and thermosets. Polyethylene is by far the most widely used polymeric material, accounting for 41% of the US plastic production.

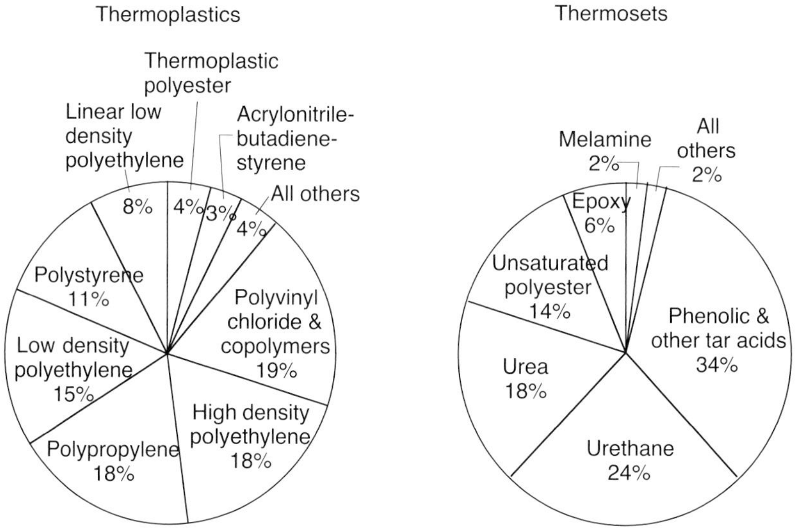

Figure 2.14 Break down of US polymer production into major polymer categories.

Packaging accounts for over one-third of the captive use of thermoplastics as graphically depicted in Fig. 2.15[15], whereas construction accounts for about

13 There are over 6,000 different polymer grades in Europe and over 10,000 in Japan.

14 Source: SPI Committee on Resin Statistics as compiled by Ernst & Young.

15 *Ibid.*

half that number, and transportation accounts for only 4% of the total captive use of thermoplastics. On the other hand, 69% of thermosets are used in building and construction, followed by 8% used in transportation. The transportation sector is one of the fastest growing areas of application for both thermoplastic and thermosetting resins. It should also be noted that 12% of the 1994 US polymer raw material production was exported.

It cannot be argued that, for some time now, polymers have become an indispensable material in everyday life. From sports to medicine, and from electronics to transportation, polymers are not only a material that is often used, but also the material that in many cases make it possible. One can sum it up with Hans Uwe Schenck's often quoted phrase -- *"Without natural polymers, there would be no life; without synthetic polymers, no standard of living."*

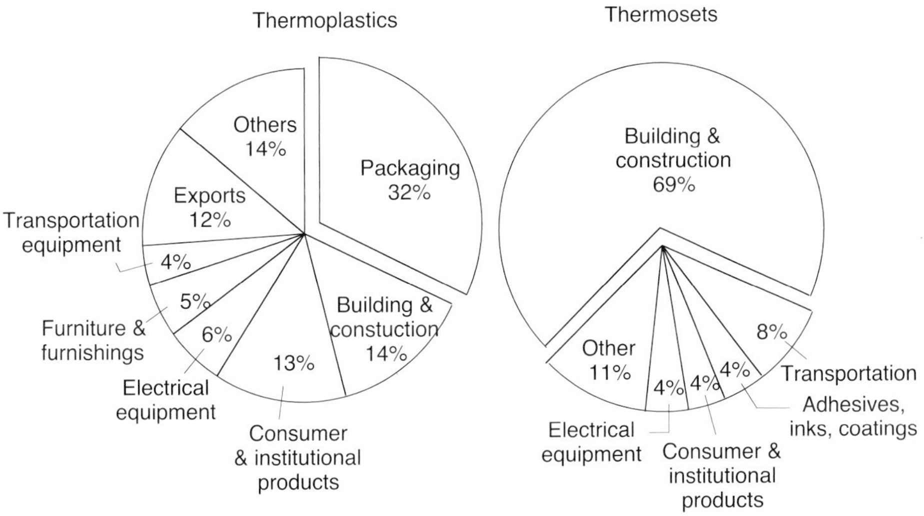

Figure 2.15 Break down of U.S. polymer production into major areas of application.

References

1. Giersch, U., and Kubisch, U., Gummi, *Die Elastische Faszination, Nicolai,* Berlin (1995).
2. Stern, H.J., *Rubber: Natural and Synthetic,* Maclaren and Sons LDT, London, (1967).
3. de la Condamine, C.M., Relation Abregee D'un Voyage Fait Dans l'interieur de l'Amerique Meridionale, Academie des Sciences, Paris, (1745).
4. DuBois, J.H., *Plastics History U.S.A.,* Cahners Publishing Co., Inc., Boston, (1972).
5. Tadmor, Z., and Gogos, C.G., *Principles of Polymer Processing,* John Wiley & Sons, New York, (1979).

6. McPherson, A.T., and Klemin, A., *Engineering Uses of Rubber*, Reinhold Publishing Corporation, New York, (1956).
7. Kaufman, M., The Chemistry and Industrial Production of Polyvinyl Chloride, Gordon and Beach, New York, (1969).
8. Levi, P., *If This is a Man*, Abacus, 1995.
9. Tschimmel, U., Die Zentausend-Dollar-Idee, ECON Verlag, Duesseldorf, (1989).
10. Sonntag, R., *Kunststoffe, 75*, 4, (1985).
11. Herrmann, H., *Kunststoffe, 75*, 2, (1985).
12. Regnault, H.V., Liebigs, A., 14, 22 (1835).
13. Ulrich, H., *Introduction to Industrial Polymers*, 2nd Ed., Hanser Publishers, Munich, (1993).
14. Utracki, L.A., *Polym. Eng. Sci.*, 35, 1, 2, (1995).
15. Noriega M. del P., ICIPC, Medellín, Colombia, Personal communication, (1994).
16. Progelhof, R.C., and Throne, J.L., *Polymer Engineering Principles*, Hanser Publishers, Munich, (1993).

3

Structure of Polymers

3.1 Macromolecular Structure of Polymers

Polymers are macromolecular structures that are generated synthetically or through natural processes. Cotton, silk, natural rubber, ivory, amber, and wood are a few materials that occur naturally with an organic macromolecular structure, whereas natural inorganic materials include quartz and glass. The other class of organic materials with a macromolecular structure are synthetic polymers, which are generated through addition or *chain growth polymerization*, and condensation or *radical initiated polymerization*.

In addition polymerization, monomers are added to each other by breaking the double-bonds that exist between carbon atoms, allowing them to link to neighboring carbon atoms to form long chains. The simplest example is the addition of ethylene monomers, schematically shown in Fig. 3.1, to form polyethylene molecules as shown in Fig. 3.2. The schematic shown in Fig. 3.2 can also be written symbolically as shown in Fig. 3.3. Here, the subscript n represents the number of repeat units which determines the molecular weight of a polymer. The number of repeat units is more commonly referred to as the degree of polymerization. Other examples of addition polymerization include polypropylene, polyvinyl chloride and polystyrene. The side groups, such as CH_3 for polypropylene and Cl for polyvinyl chloride, are sometimes referred to as the *X groups*.

Another technique to produce macromolecular materials is condensation polymerization. Condensation polymerization occurs when two components with end-groups that react with each other are mixed. When they are stoichiometric, these end-groups react, linking them to chains leaving a by-product such as water. A common polymer made by condensation polymerization is polyamide where diamine and diacid groups react to form polyamide and water, as explained n in Chapter 2.

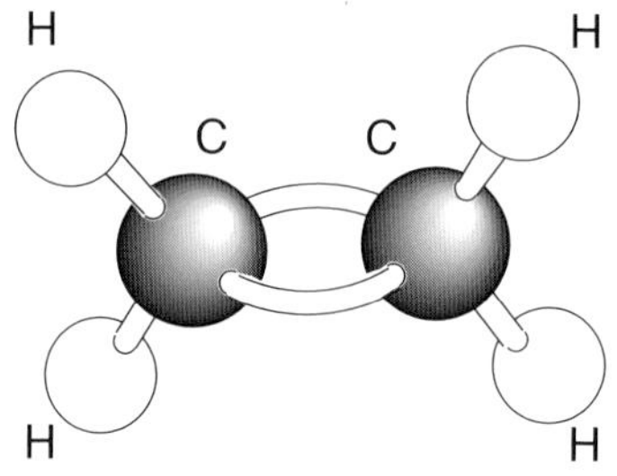

Figure 3.1 Schematic representation of an ethylene monomer.

At the molecular level, there are several forces that hold a polymeric material together. The most basic force present are covalent bonds which hold together the backbone of a polymer molecule, such as the -C-C bond. The energy holding together two carbon atoms is about 350 KJ/mol, which when translated would result in a polymer component strength between 1.4×10^4 and 1.9×10^4 MPa. However, as will be seen in Chapter 8, the strength of polymers only lies between 10 and 100 MPa.

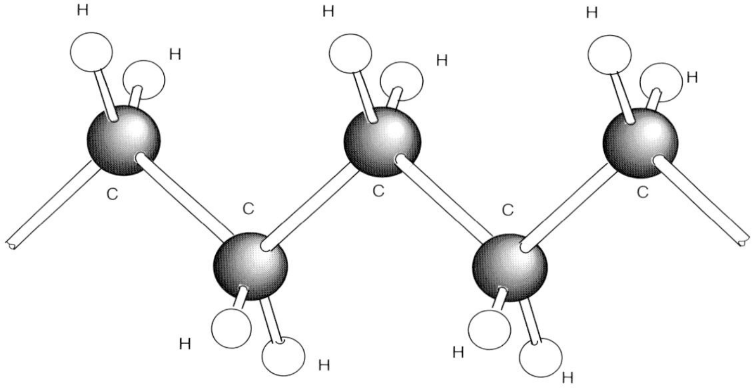

Figure 3.2 Schematic representation of a polyethylene molecule.

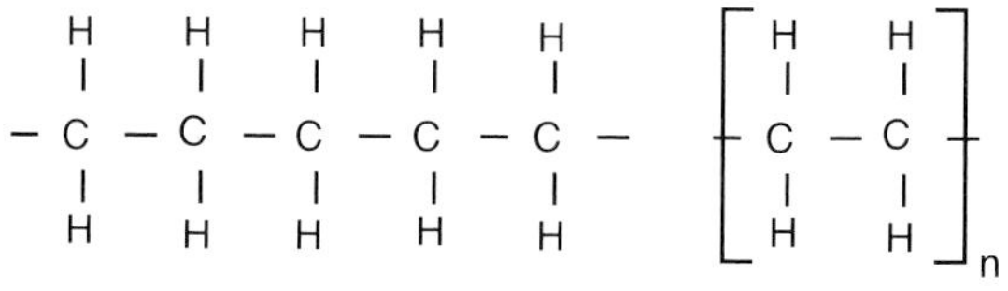

Figure 3.3 Symbolic representation of a polyethylene molecule.

3.2 Molecular Bonds and Inter-Molecular Attraction

Because of the comparatively low strength found in polymer components, it can be deduced that the forces holding a polymer component together do not come from the -C-C bonds but from intermolecular forces, or the so-called *Van-der-Waals forces*. The energy that generates the inter-molecular attraction between two polymer molecules increases, as the molecules come closer to each other by

$$F \sim \frac{1}{r^6} \tag{3.1}$$

where $F < 10\,\text{kJ/mol}$ and r is the distance between the molecules. Thus, it becomes clear that as a polymer sample is heated, the distance between the molecules increases as the vibration amplitude of the molecules increases. The vibration amplitude increase allows the molecules to move more freely, enabling the material to flow at the macroscopic level.

Another important point is that as solvents are introduced between the molecules, their inter-molecular separation increases which leads to a reduction in stiffness. This concept can be implemented by introducing *plasticizers* into the material, thus lowering the glass transition temperature below room temperature and bringing out rubber elastic material properties.

3.3 Molecular Weight

A polymeric material may consist of polymer chains of various lengths or repeat units. Hence, the molecular weight is determined by the average or mean molecular weight which is defined by

$$\overline{M} = \frac{W}{N} \tag{3.2}$$

where W is the weight of the sample and N the number of moles in the sample.

The properties of polymeric material are strongly linked to the molecular weight of the polymer as shown schematically in Fig. 3.4. A polymer such as polystyrene is stiff and brittle at room temperature with a degree of polymerization of 1,000. However, at a degree of polymerization of 10, polystyrene is sticky and soft at room temperature. Figure 3.5 shows the relation between molecular weight, temperature, and properties of a typical polymeric material. The stiffness properties reach an asymptotic maximum, whereas the flow temperature increases with molecular weight. On the other hand, the degradation temperature steadily decreases with increasing molecular weight. Hence, it is necessary to find the molecular weight that renders ideal material properties for the finished polymer product, while having flow properties that make it easy to shape the material during the manufacturing process. It is important to mention that the temperature scale in Fig. 3.5 corresponds to a specific time scale, e.g., time required for a polymer molecule to flow through an extrusion die. If the time scale is reduced (e.g., by increasing the extruder throughput) the molecules have more difficulty sliding past each other. This would require a somewhat higher temperature to assure flow. In fact, at a specific temperature, a polymer melt may behave as a solid if the time scale is reduced sufficiently. Hence, for this new time scale the stiffness properties and flow temperature curves must be shifted upward on the temperature scale. A limiting factor is that the thermal degradation curve remains fixed, limiting processing conditions to remain above certain time scales. This relation between time, or time scale, and temperature will be discussed in more detail later in this chapter.

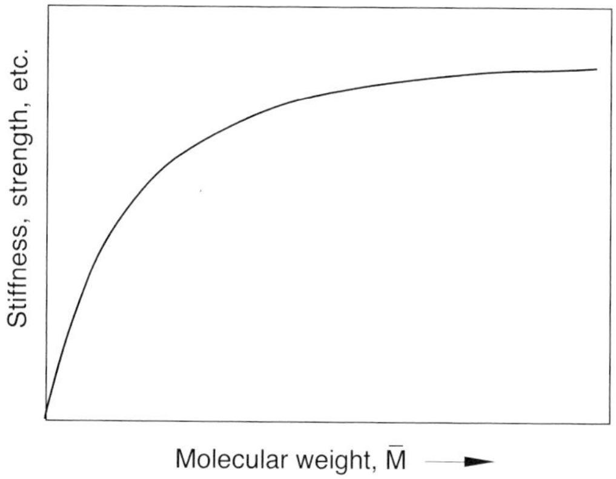

Figure 3.4 Influence of molecular weight on mechanical properties.

With the exception of maybe some naturally occurring polymers, most polymers have a molecular weight distribution such as shown in Fig. 3.6. For such a molecular weight distribution function, we can define a number average, weight average, and viscosity average[1]. The number average is the first moment and the weight average the second moment of the distribution function. In terms of mechanics this is equivalent to the center of gravity and

1 There are other definitions of molecular weight which depend on the type of measuring technique.

the radius of gyration as first and second moments, respectively. The number average is defined by

$$\overline{M}_n = \frac{\sum m_i}{\sum n_i} = \frac{\sum n_i M_i}{\sum n_i} \tag{3.3}$$

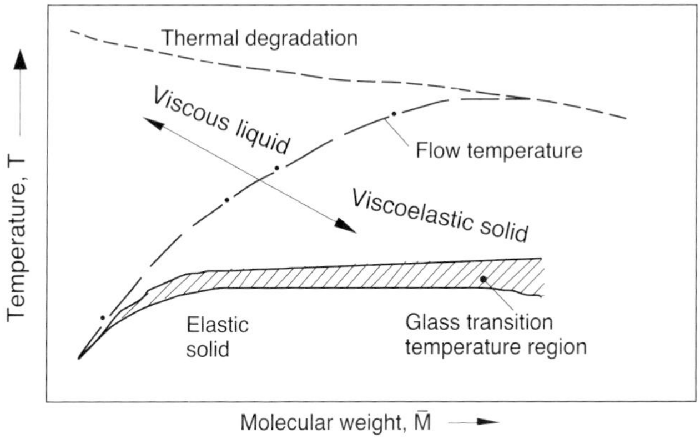

Figure 3.5 Diagram representing the relation between molecular weight, temperature, and properties of a typical thermoplastic.

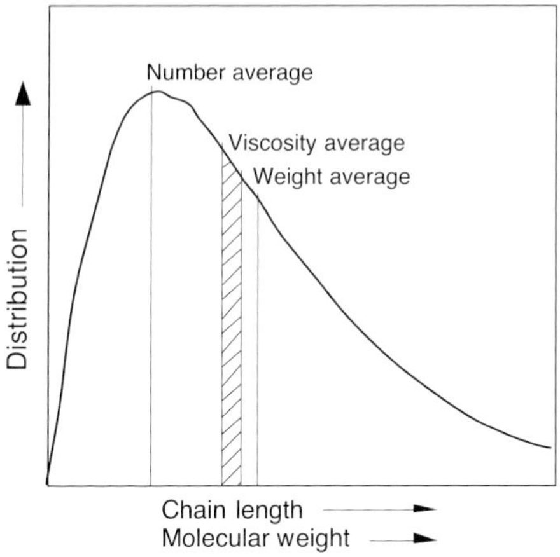

Figure 3.6 Molecular weight distribution of a typical thermoplastic.

where m_i is the weight, M_i the molecular weight and n_i the number of molecules with i repeat units. The weight average is calculated using

$$\overline{M}_W = \frac{\sum m_i M_i}{\sum m_i} = \frac{\sum n_i M_i^2}{\sum n_i M_i} \tag{3.4}$$

Another form of molecular weight average is the *viscosity average* which is calculated using

$$\overline{M}_V = \left(\frac{\sum m_i M_i^{\alpha+1}}{\sum m_i} \right)^{1/\alpha} \tag{3.5}$$

where α is a material dependent parameter which relates the *intrinsic viscosity*, $[\eta]$, to the molecular weight of the polymer. This relation is sometimes referred to as *Mark-Houwink relation* and is written as

$$[\eta] = k\overline{M}_v^{\alpha} \tag{3.6}$$

Figure 3.7 [1] presents the viscosity of various undiluted polymers as a function of molecular weight. The figure shows how for all these polymers the viscosity goes from a linear (a=1) to a power dependence (a=3.4) at some critical molecular weight. The linear relation is sometimes referred to as Staudinger's rule [2] and applies for a perfectly *monodispersed polymer* [2]. For monodispersed polymers we can write

$$\overline{M}_w = \overline{M}_n = \overline{M}_v \tag{3.7}$$

and for a polydispersed polymer [3]

$$\overline{M}_w > \overline{M}_v > \overline{M}_n \tag{3.8}$$

A measure of the broadness of a polymer's molecular weight distribution is the *polydispersity index* defined by

$$PI = \frac{\overline{M}_w}{\overline{M}_n} \tag{3.9}$$

2 A monodispersed polymer is composed of a single molecular weight species.

3 In most cases polymers are polydispersed systems which contain a range of molecular weights.

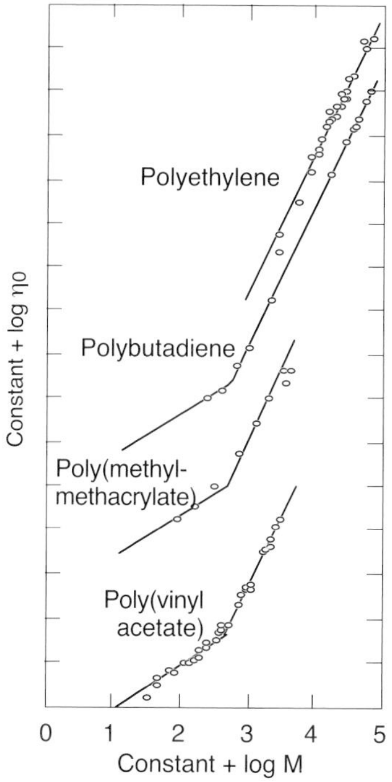

Figure 3.7 Zero shear rate viscosity for various polymers as a function of weight average molecular weight.

Figure 3.8 [3] presents a plot of flexural strength versus melt flow index[4] for polystyrene samples with three different polydispersity indices. The figure shows that low polydispersity index grade materials render higher strength properties and flowability, or processing ease, than high polydispersity index grades. Table 3.1 summarizes relations between polydispersity index, processing and material properties of a polymer. Figure 3.9 is a graphical representation of the relation between number average and weight average on properties of polymers. The arrows in the diagram indicate an increase in any specific property.

4 The melt flow index is the mass (grams) extruded through a capillary in a 10-min. period while applying a constant pressure. Increasing melt flow index signifies decreasing molecular weight. The melt flow index is discussed in more detail in Chapter 4.

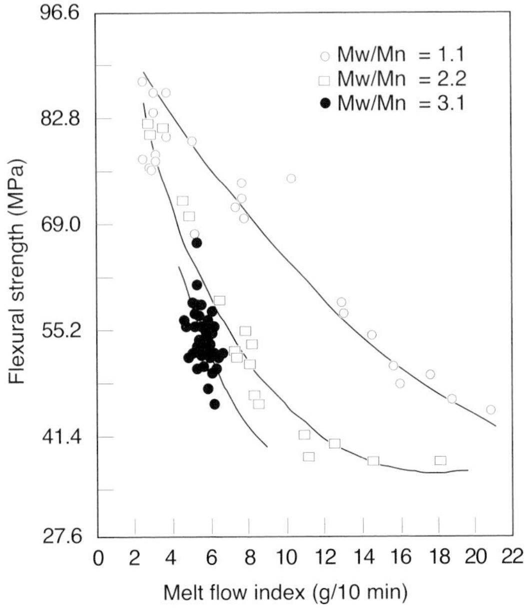

Figure 3.8 Effect of molecular weight on the strength-melt flow index interrelationship of polystyrene for three polydispersity indices.

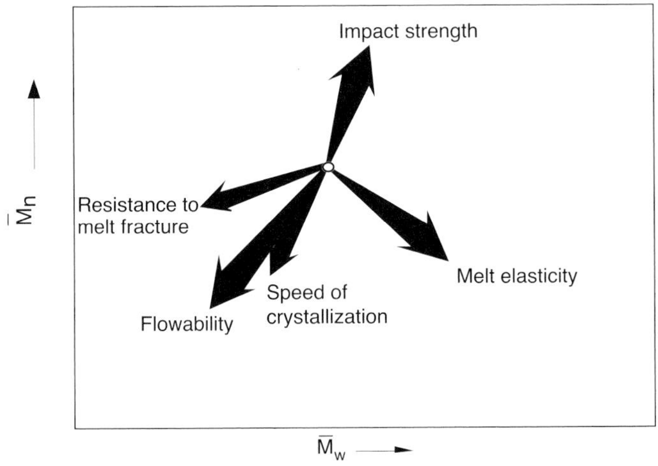

Figure 3.9 Influence of molecular weight number average and weight average on properties of thermoplastic polymers.

Table 3.1 Effect of Molecular Weight on Processing and Properties of Polymers.

$\overline{M}_w / \overline{M}_n > 1$ Processing	$\overline{M}_w / \overline{M}_n = 1$ Properties
Injection Molding:	Narrow distribution leads to higher impact strength.
Longer cycle times caused by slow crystallization and low thermal conductivity.	Short molecules lead to easy ripping.
Extrusion:	
Melt fracture is less likely to occur since short molecules act as lubricants.	
Higher degree of extrude swell caused by the long molecules.	

3.4 Conformation and Configuration of Polymer Molecules

The conformation and configuration of the polymer molecules have a great influence on the properties of the polymer component.

The conformation describes the preferential spatial positions of the atoms in a molecule, which is described by the polarity flexibility and regularity of the macromolecule. Typically, carbon atoms are tetravalent, which means that they are surrounded by four substituents in a symmetric tetrahedral geometry.

The most common example is methane, CH_4, schematically depicted in Fig. 3.10. As the figure demonstrates, the tetrahedral geometry sets the bond angle at 109.5^o. This angle is maintained between carbon atoms on the backbone of a polymer molecule, as shown in Fig. 3.11. As shown in the figure, each individual axis in the carbon backbone is free to rotate.

The configuration gives information about the distribution and spatial organization of the molecule. During polymerization it is possible to place the X groups on the carbon-carbon backbone in different directions. The order in which they are arranged is called the *tacticity*. The polymers with side groups that are placed in a random matter are called *atactic*. The polymers whose side groups are all on the same side are called *isotactic,* and those molecules with regularly alternating side groups are called *syndiotactic.*

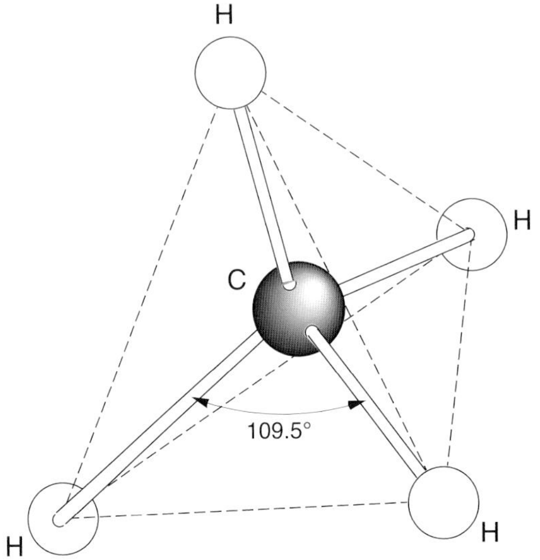

Figure 3.10 Schematic of tetrahedron formed by methane (CH_4).

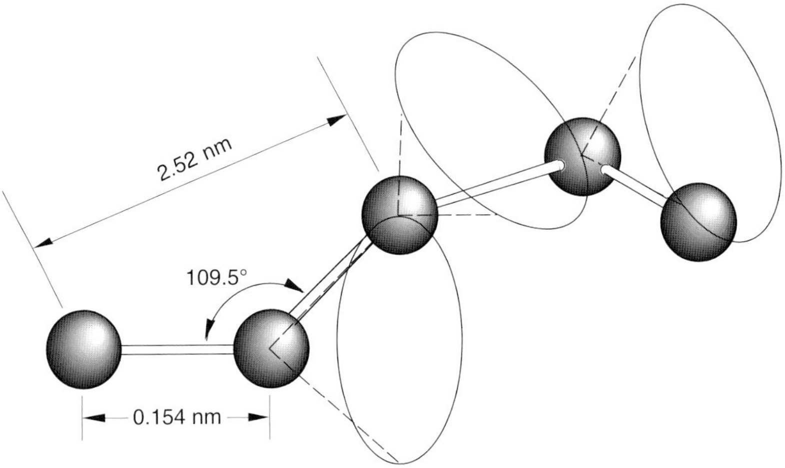

Figure 3.11 Random conformation of a polymer chain's carbon-carbon backbone.

Figure 3.12 shows the three different tacticity cases for polypropylene. The tacticity in a polymer determines the degree of crystallinity that a polymer can reach. For example, polypropylene with a high isotactic content will reach a high degree of crystallinity and as a result will be stiff, strong, and hard.

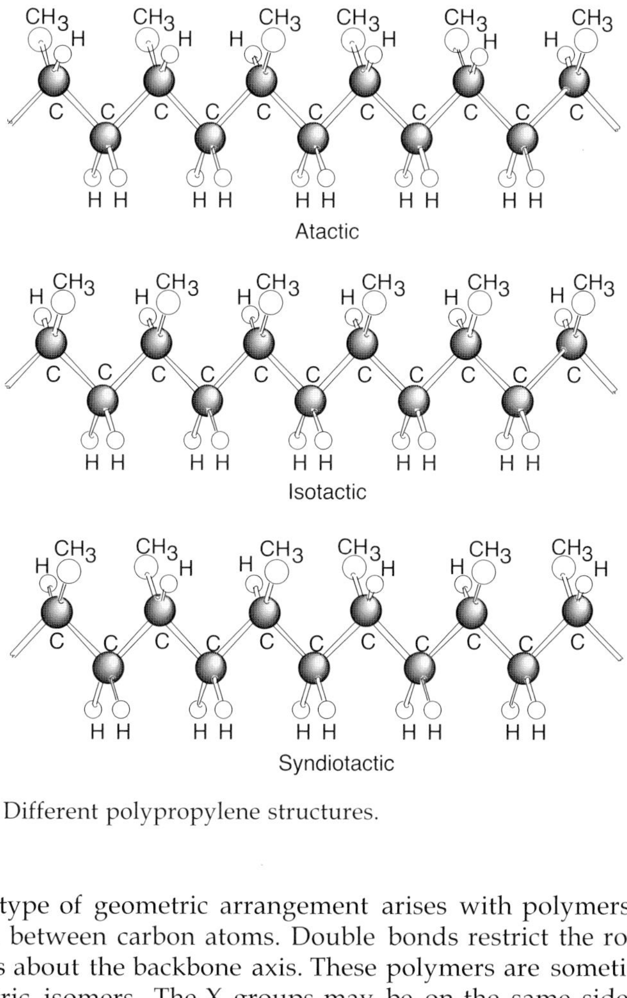

Figure 3.12 Different polypropylene structures.

Another type of geometric arrangement arises with polymers that have a double bond between carbon atoms. Double bonds restrict the rotation of the carbon atoms about the backbone axis. These polymers are sometimes referred to as geometric isomers. The X groups may be on the same side (*cis-*) or on opposite sides (*trans-*) of the chain as schematically shown for polybutadiene in Fig. 3.13. The arrangement in a cis-1,4- polybutadiene results in a very elastic rubbery material, whereas the structure of the trans-1,4- polybutadiene results in a leathery and tough material. A cis-1,4- polybutadiene can be used to manufacture the outer tread of an automotive tire. A trans-1,4- polybutadiene can be used to make the outer skin of a golf ball. The same geometric arrangement is found in natural rubber, polyisoprene. The cis-1,4- polyisoprene is the elastic natural rubber used for the body of a tire, and the latex used to manufacture "rubber" gloves and condoms. The trans-1,4- polyisoprene, or the so-called *gutta percha* or *ebony*, was used to make dentures, statues, and other decorative items in the 1800s.

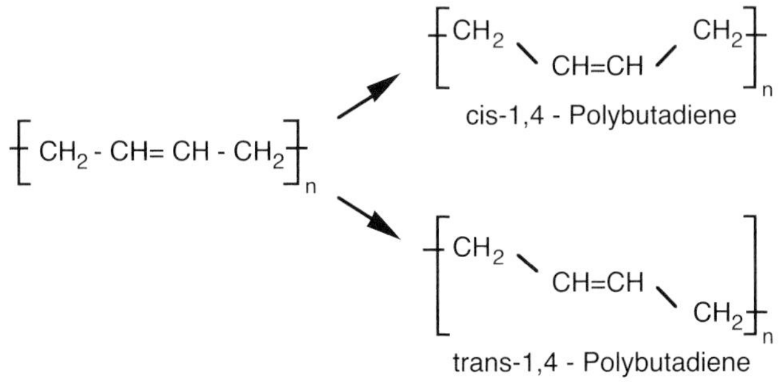

Figure 3.13 Symbolic representation of cis-1,4- and trans-1,4- polybutadiene molecules.

Branching of the polymer chains also influences the final structure, crystallinity, and properties of the polymeric material. Figure 3.14 shows the molecular architecture of high density, low density, and linear low density polyethylenes.

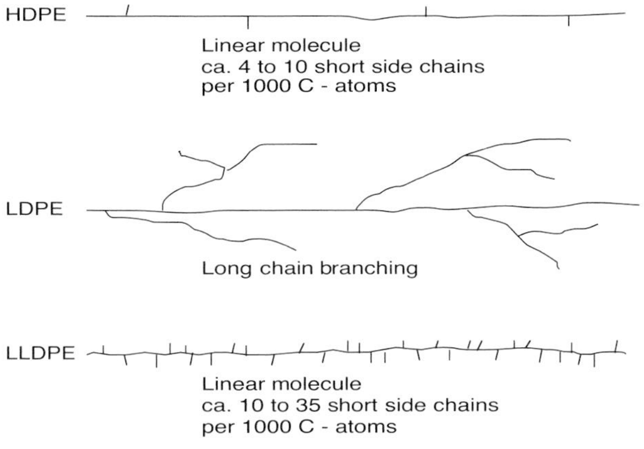

Figure 3.14 Schematic of the molecular structure of different polyethylenes.

The high density polyethylene has between 5 and 10 short branches every 1000 carbon atoms. The low density material has the same number of branches as HDPE; however, they are much longer and are themselves usually branched. The LLDPE has between 10 and 35 short chains every 1000 carbon atoms. Polymer chains with fewer and shorter branches can crystallize with more ease, resulting in higher density.

3.5 Arrangement of Polymer Molecules

As mentioned in Chapter 1, polymeric materials can be divided into two general categories: thermoplastics and thermosets. Thermoplastics are those that have the ability to remelt after they have solidified, and thermosets are those that solidify via a chemical reaction that causes polymer molecules to cross-link. These cross-linked materials cannot be remelted after solidification.

As thermoplastic polymers solidify, they take on two different types of structure: amorphous and semi-crystalline. Amorphous polymers are those where the molecules solidify in a random arrangement, whereas some of the molecules in semi-crystalline polymers align with their neighbors, forming regions with a three-dimensional order. In semi-crystalline polymers, the molecules that do not align into crystals, remain amorphous structures.

3.5.1 Thermoplastic Polymers

The formation of macromolecules from monomers occurs if there are unsaturated carbon atoms (carbon atoms connected with double bonds), or if there are monomers with reactive end-groups. The double bond, say in an ethylene monomer, is split, which frees two valences per monomer and leads to the formation of a macromolecule such as polyethylene. This process is often referred to as polymerization. Similarly, monomers (R) that possess two reactive end-groups (bifunctional) can react with other monomers (R') that also have two other reactive end-groups that can react with each other, also leading to the formation of a polymer chain. A list of typical reactive end-groups is given in Table 3.2.

Table 3.2 List of Selected Reactive End-Groups

Hydrogen in aromatic monomers	- H
Hydroxyl group in alcohols	- OH
Aldehyde group as in formaldehyde	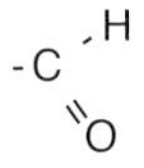
Carboxyl group in organic acids	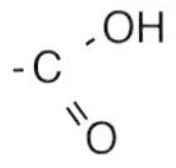
Isocyanate group in isocyanates	-N=C=O
Epoxy group in polyepoxys	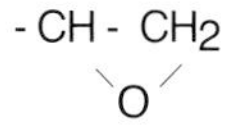
Amido groups in amides and polyamides	$-CO-NH_2$
Amino groups in amines	$-NH_2$

3.5.2 Amorphous Thermoplastics

Amorphous thermoplastics, with their randomly arranged molecular structure, are analogous to a bowl of spaghetti. Due to their random structure, the characteristic size of the largest ordered region is on the order of a carbon-carbon bond. This dimension is much smaller than the wavelength of visible light and so generally makes amorphous thermoplastics transparent.

Figure 3.15 [4] shows the shear modulus[5], G', versus temperature for polystyrene, one of the most common amorphous thermoplastics. The figure shows

5 The dynamic shear modulus, G', is obtained using the dynamic mechanical properties test described in Chapter 8.

two general regions: one where the modulus appears fairly constant[6], and one where the modulus drops significantly with increasing temperature. With decreasing temperatures, the material enters the glassy region where the slope of the modulus approaches zero. At high temperatures the modulus is negligible and the material is soft enough to flow. Although there is not a clear transition between "solid" and "liquid," the temperature that divides the two states in an amorphous thermoplastic is referred to as the *glass transition temperature*, T_g. For the polystyrene in Fig. 3.15 the glass transition temperature is approximately 110 °C. Although data are usually presented in the form shown in Fig. 3.15, it should be mentioned here that the curve shown in the figure was measured at a constant frequency. If the frequency of the test is increased—reducing the time scale—the curve is shifted to the right since higher temperatures are required to achieve movement of the molecules at the new frequency. Figure 3.16 [5] demonstrates this concept by displaying the elastic modulus as a function of temperature for polyvinyl chloride at various test frequencies. A similar effect is observed if the molecular weight of the material is increased. The longer molecules have more difficulty sliding past each other, thus requiring higher temperatures to achieve "flow."

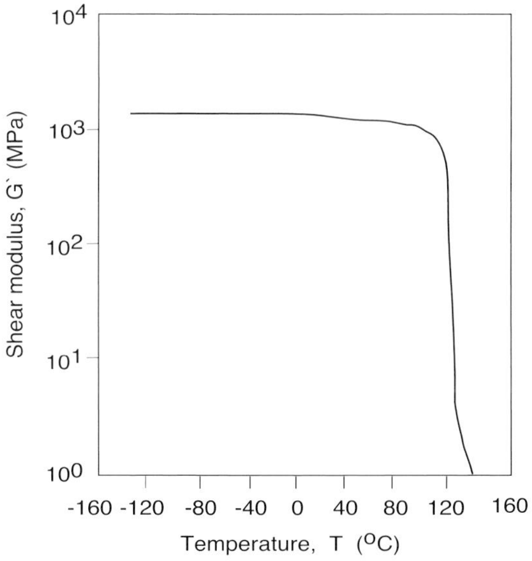

Figure 3.15 Shear modulus of polystyrene as a function of temperature.

6 When plotting G' versus temperature on a linear scale, a steady decrease of the modulus is observed.

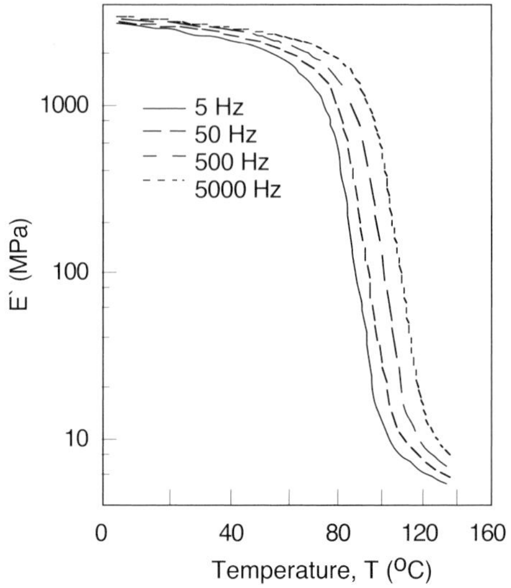

Figure 3.16 Modulus of polyvinyl chloride as a function of temperature at various test frequencies.

3.5.3 Semi-Crystalline Thermoplastics

Semi-crystalline thermoplastic polymers show more order than amorphous thermoplastics. The molecules align in an ordered crystalline form as shown for polyethylene in Fig. 3.17. The crystalline structure is part of a *lamellar crystal* which in turn forms the *spherulites*. The formation of spherulites during solidification of semi-crystalline thermoplastics is covered in Chapter 8. The schematic in Fig. 3.18 shows the general structure and hierarchical arrangement in semi-crystalline materials. The spherulitic structure is the largest domain with a specific order and has a characteristic size of 50 to 500 μm. This size is much larger than the wavelength of visible light, making semi-crystalline materials translucent and not transparent.

However, the crystalline regions are very small with molecular chains comprised of both crystalline and amorphous regions. The degree of crystallinity in a typical thermoplastic will vary from grade to grade. For example, in polyethylene the degree of crystallinity depends on the branching and the cooling rate. A low density polyethylene (LDPE) with its long branches (Fig. 3.16) can only crystallize to approximate 40-50%, whereas a high density polyethylene (HDPE) crystallizes to up to 80%. The density and strength of semi-crystalline thermoplastics increase with the degree of crystallinity as demonstrated in Table 3.3 [6], which compares low and high density polyethylenes. Figure 3.19 shows the different properties and molecular structure that may arise in polyethylene plotted as a function of degree of crystallinity and molecular weight.

Table 3.3 Influence of Crystallinity on Properties for Low and High Density Polyethylene

Property	Low density	High density
Density (g/cm^3)	0.91-0.925	0.941-0.965
% crystallinity	42-53	64-80
Melting temperature (°C)	110-120	130-136
Tensile modulus (MPa)	17-26	41-124
Tensile strength (MPa)	4.1-16	21-38

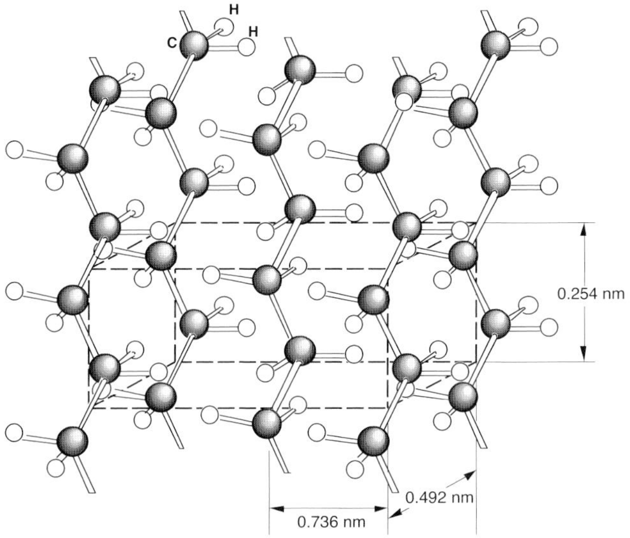

Figure 3.17 Schematic representation of the crystalline structure of polyethylene.

Figure 3.20 [4] shows the dynamic shear modulus versus temperature for a high density polyethylene, the most common semi-crystalline thermoplastic. Again, this curve presents data measured at one test frequency. The figure clearly shows two distinct transitions: one at about -110 °C, the *glass transition temperature*, and another near 140 °C, the *melting temperature*. Above the melting temperature, the shear modulus is negligible and the material will flow. Crystalline arrangement begins to develop as the temperature decreases below the melting point. Between the melting and glass transition temperatures, the material behaves as a leathery solid. As the temperature decreases below the glass transition temperature, the amorphous regions within the semi-crystalline structure solidify, forming a glassy, stiff, and in some cases brittle polymer.

To summarize, Table 3.4 presents the basic structure of several amorphous and semi-crystalline thermoplastics with their melting and/or glass transition temperatures.

Furthermore, Fig. 3.21 [7] summarizes the property behavior of amorphous, crystalline, and semi-crystalline materials using schematic diagrams of material properties plotted as functions of temperature.

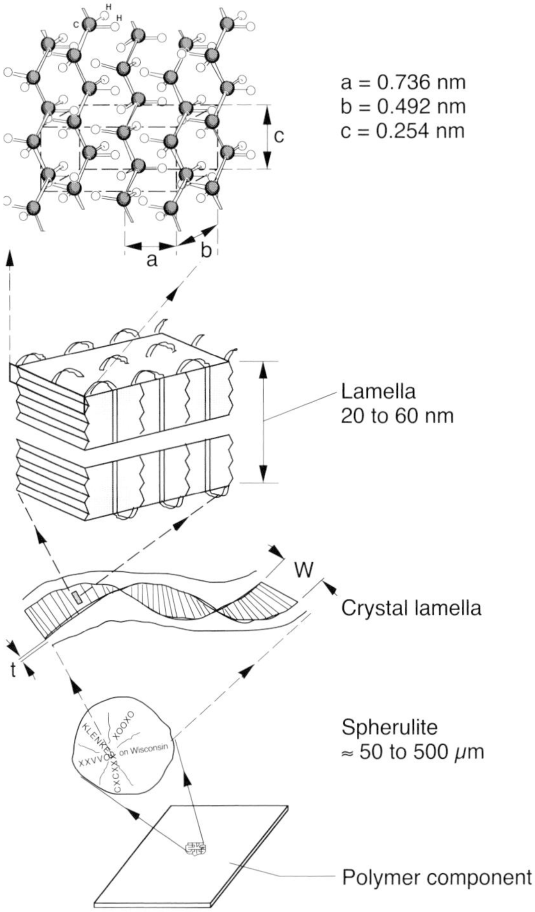

Figure 3.18 Schematic representation of the general molecular structure andarrangement of typical semi-crystalline materials.

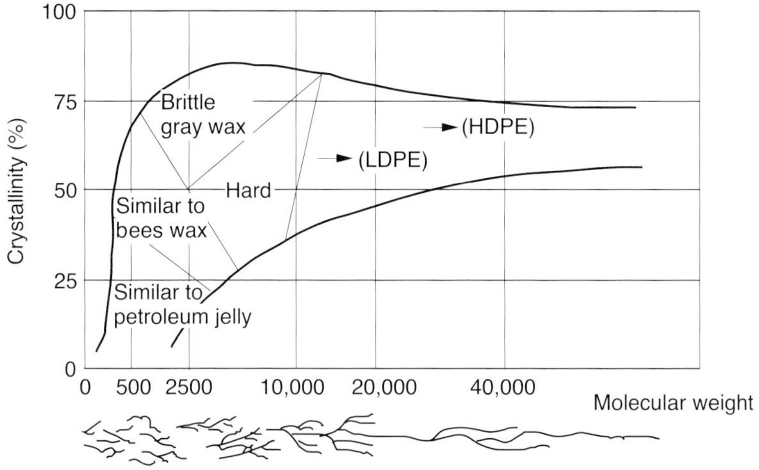

Figure 3.19 Influence of degree of crystallinity and molecular weight on different properties of polyethylene.

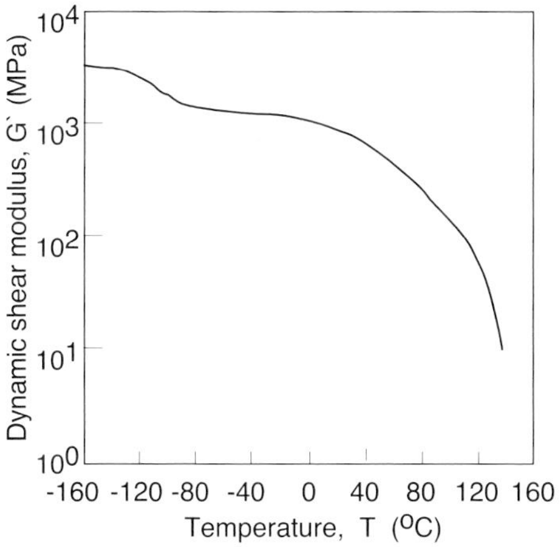

Figure 3.20 Shear modulus of a high density polyethylene as a function of temperature.

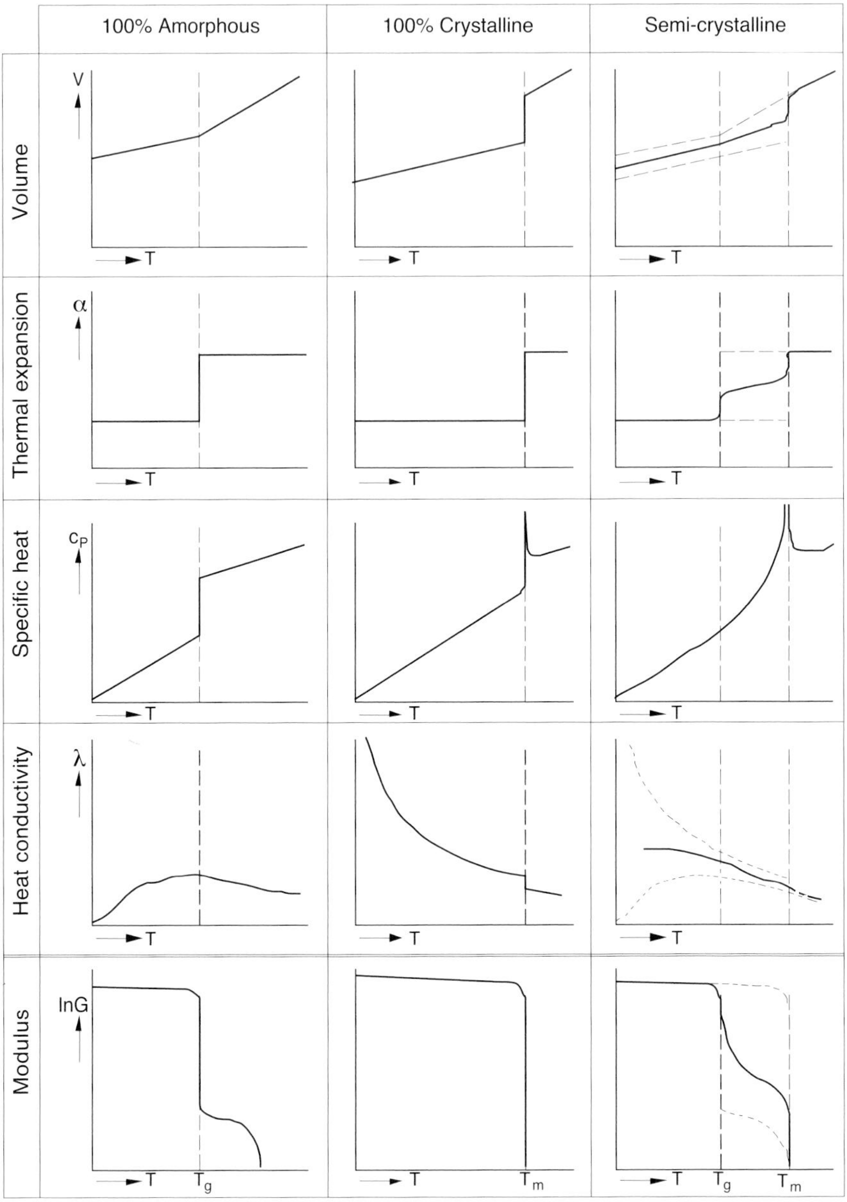

Figure 3.21 Schematic of the behavior of some polymer properties as a function of temperature for different thermoplastics.

Table 3.4 Structural Units for Selected Polymers with Glass Transition and Melting Temperatures.

Structural unit	Polymers	T_g(°C)	T_m(°C)
$-CH_2-CH_2-$	Linear polyethylene	-125	135
$-CH_2-CH-$ | CH_3	Isotactic polypropylene	-20	170
$-CH_2-CH-$ | C_2H_5	Isotactic Polybutene	-25	135
$-CH_2-CH-$ | $CH-CH_3$ | CH_3	Isotactic poly-3-methylbutene-1	50	310
$-CH_2-CH-$ | CH_2 | $CH-CH_3$ | CH_3	Isotactic poly-4-methylpentene -1	29	240
$-CH_2-CH-$ 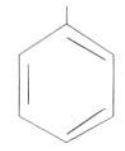	Isotactic polystyrene	100	240
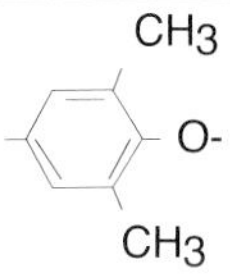	Polyphenylenether (PPE)	210	261
$-O-CH-$ | CH_3	Polyacetaldehyde	-30	165
	(continued)		

Table 3.4 Structural Units for Selected Polymers with Glass Transition and Melting Temperatures *(continued)*

Structural unit	Polymers	T_g(°C)	T_m(°C)
- O-CH_2-	Polyformaldehyde (polyacetal, polyoxymethylene)	-85	178-198
- O - CH_2-CH- | CH_3	Isotactic polypropyleneoxide	-75	75
CH_2- Cl | - O- CH_2 - C- CH_2- | CH_2 Cl	Poly-[2.2-bis-(chlormethyl)-trimethylene-oxide]	5	181
CH_3 | - CH_2- C- | CO_2 CH_3	Isotactic polymethyl-methacrylate	50	160
Cl F | | - C- C- | | F F	Polychlortri-fluoroethylene	45	220
- CF_2- CF_2	Polytetrafluoroethylene	-113,+127	330
Cl | - CH_2 - C - | Cl	Polyvinylidenechloride	-19	190
F | - CH_2 - C - | F	Polyvinylidenefluoride	-45	171
- CH_2 - CH | Cl	Polyvinylchloride (PVC) amorphous crystalline	80 80	- 212
	(continued)		

Table 3.4 Structural Units for Selected Polymers with Glass Transition and Melting Temperatures *(continued)*

Structural unit	Polymers	T_g(°C)	T_m(°C)
- CH_2- CH - F	Polyvinylfluoride (PVF)	-20	200
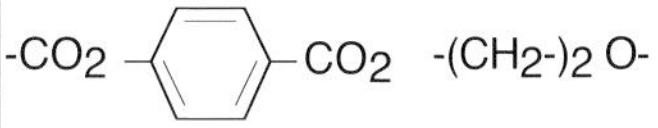	Polyethylene - terephthalate (PET) (linear polyester)	69	245
- CO- (CH_2-)$_4$ CO–NH- (CH_2)$_6$ NH–	Polyamide 66	57	265
- CO- (CH_2-)$_8$ CO–NH- (CH_2-)$_6$ NH–	Polyamide 610	50	228
- CO- (CH_2-)$_5$ NH–	Polycaprolactam, Polyamide 6	75	233
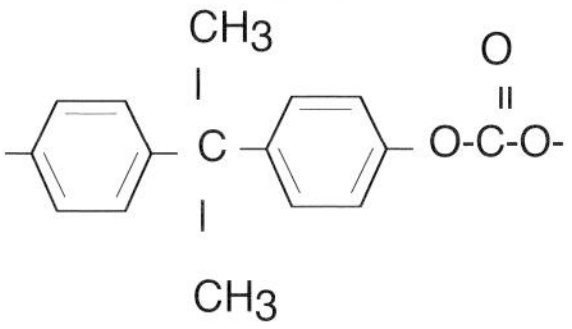	Polycarbonate (PC)	149	267
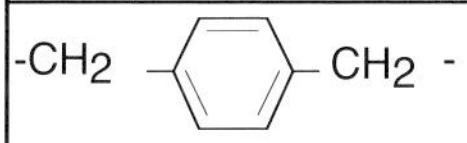	Poly-(p-xylene) (Parylene | | R)	-	400
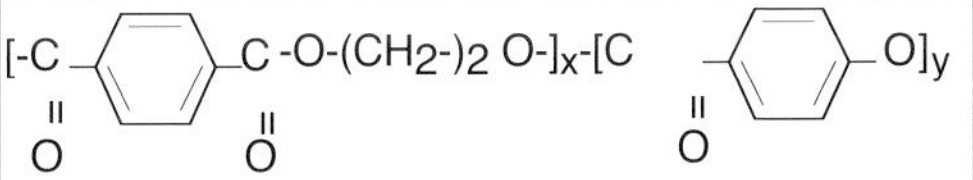 35% 65% | <—— PET ——> |<—— PHB ——>	Polyethylene terephthalate/p-Hydroxybenzoate-copolymers LC-PET, Polymers with flexible chains	75	280
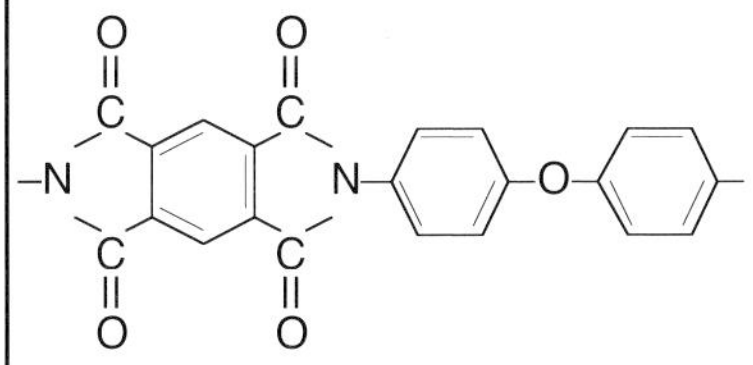	Polyimide (PI)	up to 400	

(continued)

Table 3.4 Structural Units for Selected Polymers with Glass Transition and Melting Temperatures *(continued)*

Structural unit	Polymers	T_g(°C)	T_m(°C)
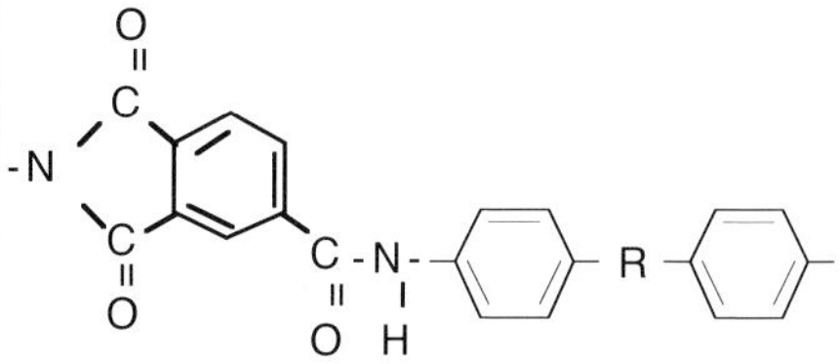	Polyamidimide (PAI)	≈ 260	
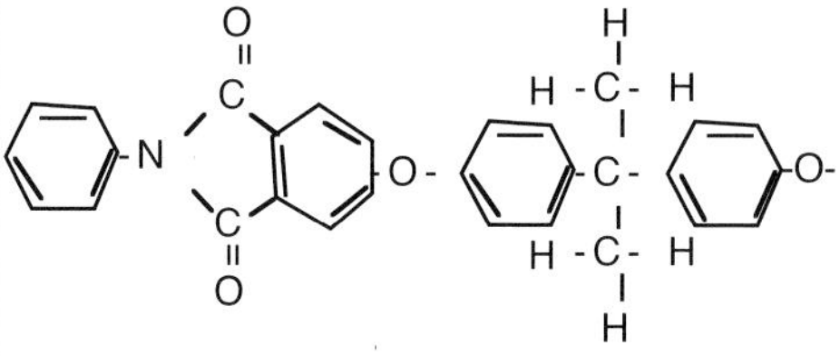	Polyetherimide (PEI)	≈ 215	
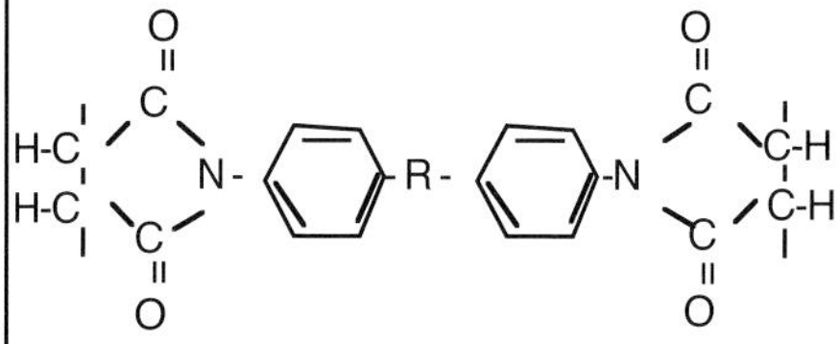	Polybismaleinimide (PBI)	≈ 250	
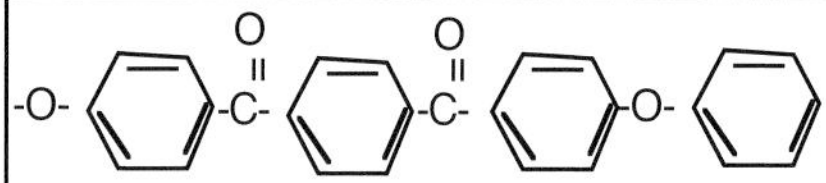	Polyoxybenzoate (POB)	≈ 290	
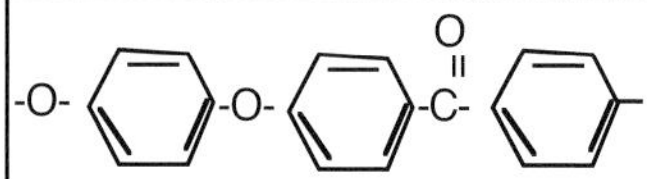	Polyetheretherketone (PEEK)	145	335
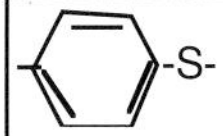	Polyphenylenesulfide (PPS)	~230	
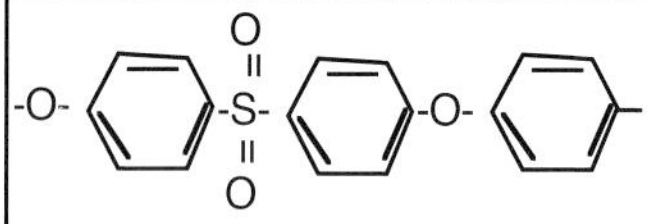	Polyethersulfone (PES)	85	280
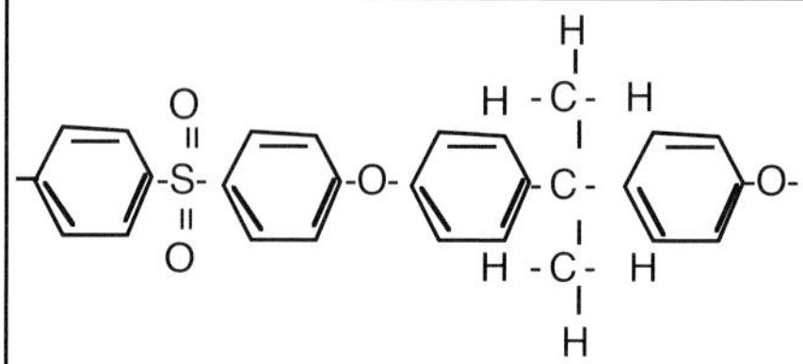	Polysulfone (PSU)	~ 180	

3.5.4 Thermosets and Cross-Linked Elastomers

Thermosets, and some elastomers, are polymeric materials that have the ability to cross-link. The cross-linking causes the material to become heat resistant after it has solidified. A more in-depth explanation of the cross-linking chemical reaction that occurs during solidification is given in Chapter 7.

The cross-linking usually is a result of the presence of double bonds that break, allowing the molecules to link with their neighbors. One of the oldest thermosetting polymers is phenol-formaldehyde, or phenolic. Figure 3.22 shows the chemical symbol representation of the reaction, and Fig. 3.23 shows a schematic of the reaction. The phenol molecules react with formaldehyde molecules to create a three-dimensional cross-linked network that is stiff and strong. The by-product of this chemical reaction is water.

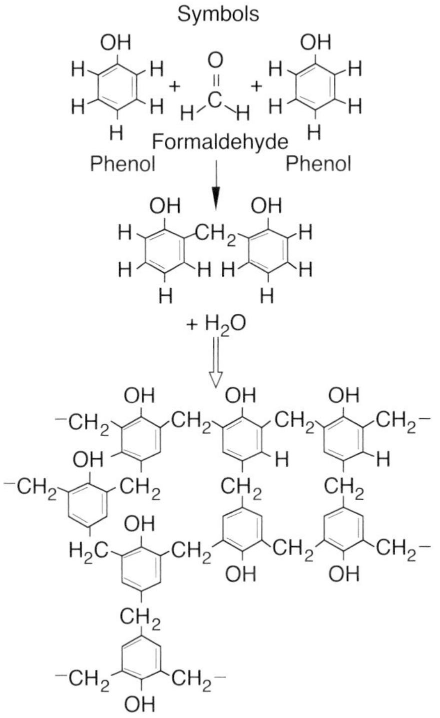

Figure 3.22 Symbolic representation of the condensation polymerization of phenol-formaldehyde resins.

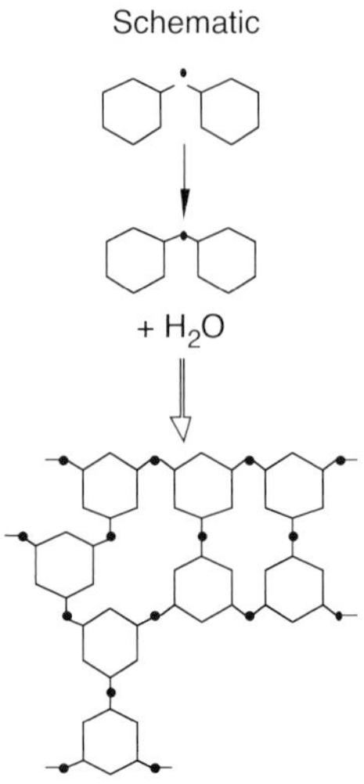

Figure 3.23 Schematic representation of the condensation polymerization of phenol-formaldehyde resins.

3.6 Copolymers and Polymer Blends

Copolymers are polymeric materials with two or more monomer types in the same chain. A copolymer that is composed of two monomer types is referred to as a *bipolymer*, and one that is formed by three different monomer groups is called a *terpolymer*. Depending on how the different monomers are arranged in the polymer chain, one distinguishes between *random*, alternating, block, or *graft* copolymers. The four types of copolymers are schematically represented in Fig. 3.24.

A common example of a copolymer is ethylene-propylene. Although both monomers would result in semi-crystalline polymers when polymerized individually, the melting temperature disappears in the randomly distributed copolymer with ratios between 35/65 and 65/35, resulting in an elastomeric material, as shown in Fig. 3.25. In fact EPDM[7] rubbers are continuously gaining acceptance in industry because of their resistance to weathering. On the other hand, the ethylene-propylene block copolymer maintains a melting temperature for all ethylene/propylene ratios, as shown in Fig. 3.26.

7 The D in EP(D)M stands for the added unsaturated diene component which results in a cross-linked elastomer.

Another widely used copolymer is high impact polystyrene (PS-HI), which is formed by grafting polystyrene to polybutadiene. Again, if styrene and butadiene are randomly copolymerized, the resulting material is an elastomer called styrene-butadiene-rubber (SBR). Another classic example of copolymerization is the terpolymer acrylonitrile-butadiene-styrene (ABS).

Polymer blends belong to another family of polymeric materials which are made by mixing or blending two or more polymers to enhance the physical properties of each individual component. Common polymer blends include PP-PC, PVC-ABS, PE-PTFE.

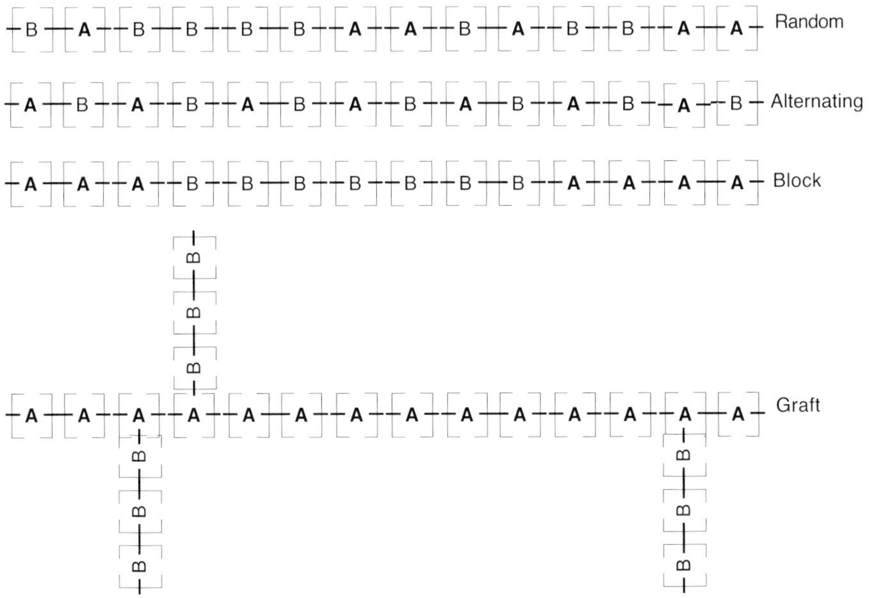

Figure 3.24 Schematic representation of different copolymers.

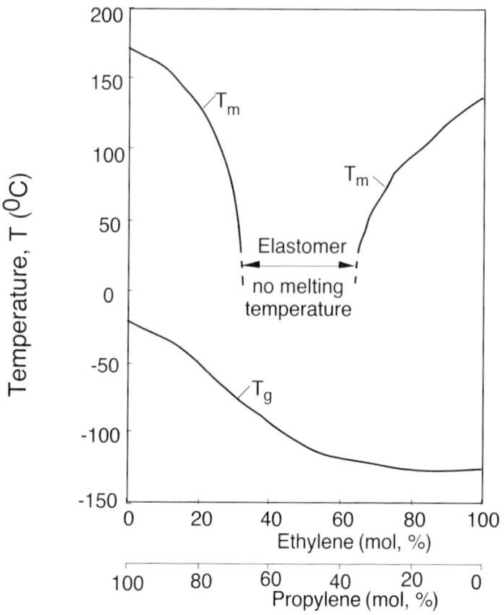

Figure 3.25 Melting and glass transition temperature for random ethylene-propylene copolymers.

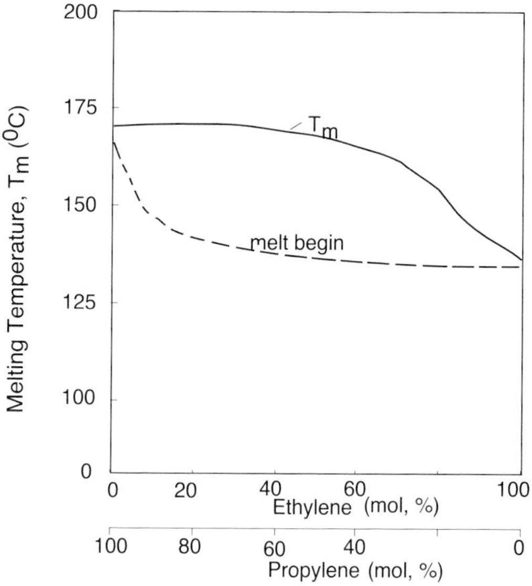

Figure 3.26 Melting temperature for ethylene-propylene block copolymers.

3.7 Polymer Additives

There are many polymer additives that are mixed into a polymer to improve the mechanical, optical, electrical, and acoustic—to name a few— performance of a component.

3.7.1 Flame Retardants

Since polymers are organic materials, most of them are flammable. The flammability of polymers has always been a serious technical problem. A parameter which can be used to assess the flammability of polymers is the *limiting oxygen index* (LOI), also known as the *critical oxygen index* (COI). This value defines the minimum volume percent of oxygen concentration, mixed with nitrogen, needed to support combustion of the polymer under the test conditions specified by ASTM D 2863. Since air contains 21% oxygen by volume, only those polymers with an LOI greater than 0.21 are considered self-extinguishing. In practice, an LOI value greater than 0.27 is recommended as the limiting self-extinguishing threshold. Table 3.5 presents LOI values for selected polymers.

It is impossible to make a polymer completely inflammable. However, some additives that contain halogens, such as bromine, chlorine, or phosphorous, reduce the possibility of either starting combustion within a polymer component or, once ignited, reduce the rate of flame spread. When rating the performance of flame retardants, bromine is more effective than chlorine.

In the *radical trap* theory of flame retardancy, it is believed that bromine or phosphorous containing additives compete in the reaction of a combustion process. To illustrate this we examine two examples of typical combustion reactions. These are:

$$CH_4 + OH \longrightarrow CH_3 + H_2O$$

and

$$CH_4 + O_2 \longrightarrow CH_3 + H + O_2$$

where OH and H are active chain carriers. With the presence of HBr the following reaction can take place

$$OH + HBr \longrightarrow H_2O + Br$$

Table 3.5 LOI Values for Selected Polymers

Polymer	LOI
Polyformaldehyde	0.15
Polyethyleneoxide	0.15
Polymethyl methacrylate	0.17
Polyacrylonitrile	0.18
Polyethylene	0.18
Polypropylene	0.18
Polyisoprene	0.185
Polybutadiene	0.185
Polystyrene	0.185
Cellulose	0.19
Polyethylene terephthalate	0.21
Polyvinyl alcohol	0.22
Polyamide 66	0.23
Epoxy	0.23
Polycarbonate	0.27
Aramide fibers	0.285
Polyphenylene oxide	0.29
Polysulfone	0.30
Phenolic resins	0.35
Polychloroprene	0.40
Polyvinyl chloride	0.42
Polyvinylidene fluoride	0.44
Polyvinylidene chloride	0.60
Carbon	0.60
Polytetrafluoroethylene	0.95

where the active chain carrier was replaced by the less active Br radical, helping in flame extinguishment. Similarly, with the presence of Br the following reaction can take place

$$CH_4 + Br \longrightarrow HBr + CH_3$$

Table 3.6 [8] lists selected polymers with one commonly used flame retardant.

Table 3.6 Selected Polymers with Typical Commercial Flame Retardants

Polymer	Flame retardants
Acrylonitrile butadiene styrene	Octabromodiphenyl oxide
High impact polystyrene	Decabromodiphenyl oxide
Polyamide	Dechlorane plus
Polycarbonate	Tetrabromobisphenol A carbonate oligomer
Polyethylene	Chlorinated paraffin
Polypropylene	Dechlorane plus
Polystyrene	Pentabromocyclododecane
Polyvinyl chloride	Phosphate ester

3.7.2 Stabilizers

The combination of heat and oxygen will bring about thermal degradation in a polymer. Heat or energy will produce free radicals which will react with oxygen to form carbonyl compounds, which give rise to yellow or brown discolorations in the final product.

Thermal degradation can be slowed by adding stabilizers such as antioxidants or peroxide decomposers. These additives do not eliminate thermal degradation but slow down the reaction process. Once the stabilizer has been consumed by the reaction with the free radicals, the protection of the polymer against thermal degradation ends. The time period over which the stabilizer renders protection against thermal degradation is called *induction time*. A test used to measure thermal stability of a polymer is the *Oxidative Induction Time* (OIT) by differential scanning calorimetry (DSC). The OIT test is defined as the time it takes for a polymer sample to thermally degrade in an oxygen environment at a set temperature above the polymer's transition temperature. The transitions must occur in a nitrogen environment. The standard test is described by ASTM D 3895. Another test used to measure the thermal stability of a polymer and its additives is the thermal gravimetric analysis (TGA) discussed in Chapter 3.

Polyvinyl chloride is probably the polymer most vulnerable to thermal degradation. In polyvinyl chloride, scission of the C-Cl bond occurs in the weakest point of the molecule. The chlorine radicals will react with their nearest CH group, forming HCl and creating new weak C-Cl bonds. A stabilizer must therefore neutralize HCl and stop the auto catalytic reaction, as well as preventing processing equipment corrosion.

3.7.3 Antistatic Agents

Since polymers have such low electric conductivity they can build-up electric charges quite easily. The amount of charge build-up is controlled by the rate at which the charge is generated compared to the charge decay. The rate of charge generation at the surface of the component can be reduced by reducing the intimacy of contact, whereas the rate of charge decay is increased through surface conductivity. Hence, a good antistatic agent should be an ionizable additive that allows charge migration to the surface, at the same time as creating bridges to the atmosphere through moisture in the surroundings. Typical antistatic agents are nitrogen compounds such as long chain amines and amides, polyhydric alcohols, etc.

3.7.4 Fillers

Fillers can be divided into two categories: those that reinforce the polymer and improve its mechanical performance, and those that are used to take up space and so reduce the amount of actual resin to produce a part—sometimes referred to as *extenders*. A third, less common, category of filled polymers are those where fillers are dispersed into the polymer to improve its electric conductivity.

Polymers that contain fillers that improve their mechanical performance are often referred to as *composites*. Composites can be divided into two further categories: composites with *high performance* reinforcements, and composites with *low performance* reinforcements. The high performance composites are those where the reinforcement is placed inside the polymer in such a way that optimal mechanical behavior is achieved, such as unidirectional glass fibers in an epoxy resin. High performance composites usually have 50–80% reinforcement by volume and the composites are usually laminates, tubular shapes containing braided reinforcements, etc. The low performance composites are those where the reinforcement is small enough that it can be dispersed well into the matrix, allowing to process these materials the same way their unreinforced counterparts are processed. The most common filler used to reinforce polymeric materials is glass fiber. However, wood fiber, which is commonly used as an extender, also increases the stiffness and mechanical performance of some thermoplastics. To improve the bonding between the polymer matrix and the reinforcement, *coupling agents* such as *silanes* and *titanates* are often added. Polymer composites and their performance are discussed in more detail in Chapters 8 and 9.

Extenders, used to reduce the cost of the component, often come in the form of particulate fillers. The most common particulate fillers are calcium carbonate, silica flour, clay, and wood flour or fiber. As mentioned earlier, some fillers also slightly reinforce the polymer matrix, such as clay, silica flour, and wood fiber. It should be pointed out that polymers with extenders often have significantly lower toughness than when unfilled. This concept is covered in more detail in Chapter 9.

3.7.5 Blowing Agents

The task of blowing or foaming agents is to produce cellular polymers, also referred to as expanded plastics. The cells can be completely enclosed (closed cell) or can be interconnected (open cell). Polymer foams are produced with densities ranging from 1.6 kg/m^3 to 960 kg/m^3. There are many reasons for using polymer foams such as their high strength/weight ratio, excellent insulating and acoustic properties, and high energy and vibration absorbing properties.

Polymer foams can be made by mechanically whipping gases into the polymer, or by either chemical or physical means. Some of the most commonly used foaming methods are [9]:

- Thermal decomposition of chemical blowing agents which generates nitrogen and/or carbon monoxide and dioxide. An example of such a foaming agent is *azodicarbonamide* which is the most widely used commercial polyolefin foaming agent.
- Heat induced volatilization of low-boiling liquids such as pentane and heptane in the production of polystyrene foams, and methylene chloride when producing flexible polyvinyl chloride and polyurethane foams.
- Volatilization by the exothermic reaction of gases produced during polymerization. This is common in the reaction of isocyanate with water to produce carbon dioxide.
- Expansion of the gas dissolved in a polymer upon reduction of the processing pressure.

The basic steps of the foaming process are nucleation of the cells, expansion or growth of the cells, and stabilization of the cells. The nucleation of a cell occurs when, at a given temperature and pressure, the solubility of a gas is reduced, leading to saturation, expelling the excess gas to form a bubble. Nucleating agents are used for initial formation of the bubbles. The bubble reaches an equilibrium shape when the pressure inside the bubble balances with the surface tension surrounding the cell.

3.8 Viscoelastic Behavior of Polymers

It was mentioned earlier in this chapter that a polymer, at a specific temperature and molecular weight, may behave as a liquid or a solid depending on the speed (time scale) at which its molecules are deformed. This behavior, which ranges between liquid and solid, is generally referred to as the viscoelastic behavior or material response. In this chapter we will limit the discussion to *linear viscoelasticity* which is valid for polymer systems that are undergoing *small deformations*. *Non-linear viscoelasticity*, required when modeling *large deformations* such as those encountered in flowing polymer melts, is covered in detail in Chapter 5.

In linear viscoelasticity the *stress relaxation test* is often used, along with the *time-temperature superposition principle* and the *Boltzmann superposition principle*, to explain the behavior of polymeric materials during deformation.

3.8.1 Stress Relaxation Test

In a stress relaxation test, a polymer test specimen is deformed a fixed amount, ε_0, and the stress required to hold that amount of deformation is recorded over time. This test is very cumbersome to perform, so the design engineer and the material scientist have tended to ignore it. In fact, several years ago the standard relaxation test ASTM D2991 was dropped by ASTM. Rheologists and scientists, however, have been consistently using the stress relaxation test to interpret the viscoelastic behavior of polymers.

Figure 3.27 [10] presents the stress relaxation modulus measured of polyisobutylene[8] at various temperatures. Here, the stress relaxation modulus is defined by

$$E_r(t) = \frac{\sigma(t)}{\varepsilon_0} \tag{3.10}$$

where ε_0 is the applied strain and $\sigma(t)$ is the stress being measured. From the test results it is clear that stress relaxation is time and temperature dependent, especially around the glass transition temperature where the slope of the curve is maximal. In the case of the polyisobutylene shown in Fig. 3.27, the glass

8 Better known as chewing gum.

transition temperature is about -70 °C. The measurements were completed in an experimental time window between a few seconds and one day. The tests performed at lower temperatures were used to record the initial relaxation while the tests performed at higher temperatures only captured the end of relaxation of the rapidly decaying stresses.

It is well known that high temperatures lead to small molecular relaxation times[9] and low temperatures lead to materials with large relaxation times. This is due to the fact that at low temperatures the free volume between the molecules is reduced, restricting or slowing down their movement. At high temperatures, the free volume is larger and the molecules can move with more ease. Hence, when changing temperature, the shape of creep[10] or relaxation test results remain the same except that they are horizontally shifted to the left or right, which represent lower or higher response times, respectively.

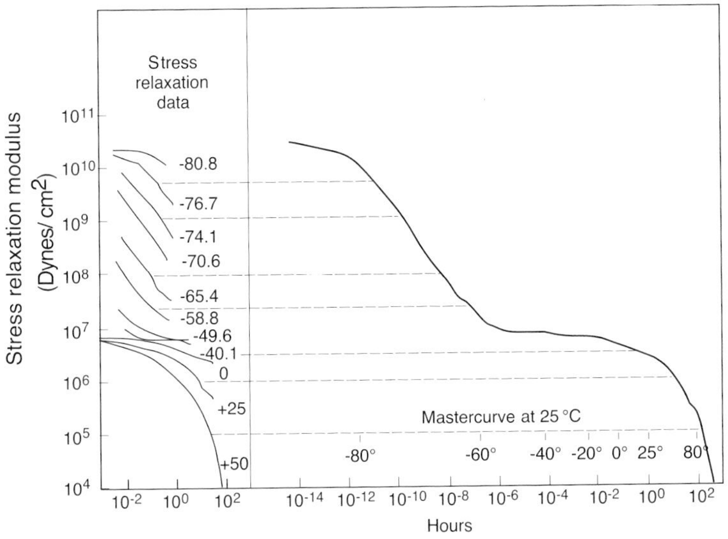

Figure 3.27 Relaxation modulus curves for polyisobutylene at various temperatures and corresponding master curve at 25 °C.

The same behavior is observed if the pressure is varied. As the pressure is increased, the free volume between the molecules is reduced, slowing down molecular movement. Here, an increase in pressure is equivalent to a decrease in temperature. In the melt state, the viscosity of a polymer increases with

9 By relaxation time one usually refers to the time it takes for applied stresses to relax within a material.

10 In a creep test the polymer specimen is loaded to a constant stress, and the strain response is recorded in time. The creep test is described in more detail in Chapter 9.

pressure. Figure 3.28 [11] is presented to illustrate the effect of pressure on stress relaxation.

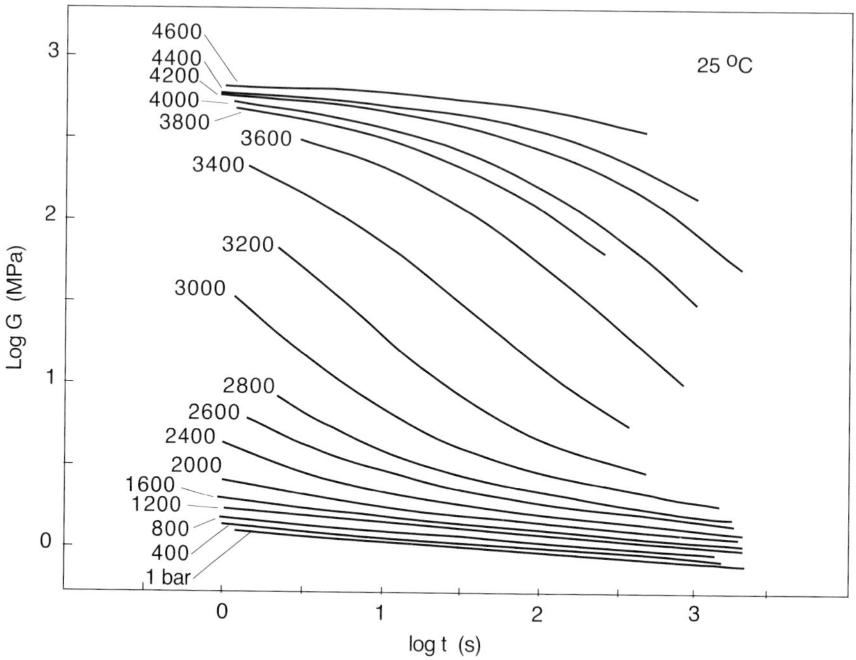

Figure 3.28 Shear relaxation modulus for a chlorosulfonated polyethylene at various pressures.

3.8.2 Time-Temperature Superposition (WLF-Equation)

The time-temperature equivalence seen in stress relaxation test results can be used to reduce data at various temperatures to one general *master curve* for a reference temperature, T_{ref}. To generate a master curve at the reference temperature, the curves shown in the left of Fig. 3.27 must be shifted horizontally, maintaining the reference curve stationary. Density changes are usually small and can be neglected, eliminating the need to perform tedious corrections. The master curve for the data in Fig. 3.27 is shown on the right side of the figure. Each curve was shifted horizontally until the ends of all the curves became superimposed. The amount that each curve was shifted can be plotted with respect to the temperature difference taken from the reference temperature. For the data in Fig. 3.27 the shift factor is shown in the plot in Fig. 3.29. The amounts the curves where shifted are represented by

$$\log t - \log t_{ref} = \log\left(\frac{t}{t_{ref}}\right) = \log a_T \tag{3.11}$$

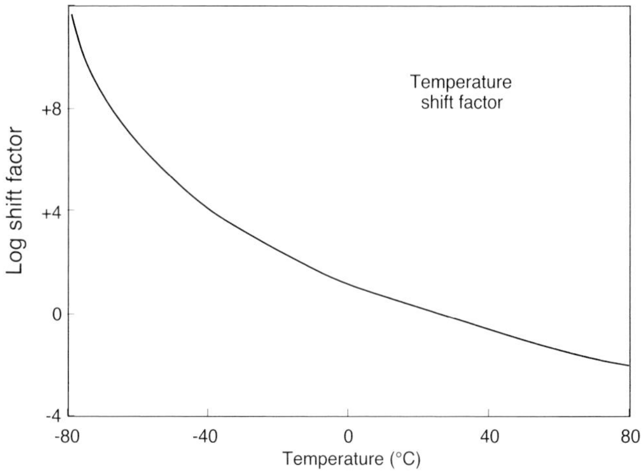

Figure 3.29 Plot of the shift factor as a function of temperature used to generate the master curve plotted in Fig. 3.27.

Although the results in Figure 3.29 where shifted to a reference temperature of 298 K (25 °C), Williams, Landel and Ferry [12] chose T_{ref} = 243 K for

$$\log a_T = \frac{-8.86\,(T - T_{ref})}{101.6 + T - T_{ref}} \tag{3.12}$$

which holds for nearly all polymers if the chosen reference temperature is 45 K above the glass transition temperature. In general, the horizontal shift, log a_T, between the relaxation responses at various temperatures to a reference temperature can be computed using the well known Williams-Landel-Ferry [12] (WLF) equation. The WLF equation is given by

$$\log a_T = -\frac{C_1\left(T - T_{ref}\right)}{C_2 + (T - T_{ref})} \tag{3.13}$$

where C_1 and C_2 are material dependent constants. It has been shown that with the assumption $C_1 = 17.44$ and $C_2 = 51.6$, Eq. 3.13 fits well a wide variety of polymers as long as the glass transition temperature is chosen as the reference temperature. These values for C_1 and C_2 are often referred to as universal constants. Often, the WLF equation must be adjusted until it fits the experimental data. Master curves of stress relaxation tests are important since the polymer's behavior can be traced over much greater periods of time than those determined experimentally.

3.8.3 The Boltzmann Superposition Principle

In addition to the *time-temperature superposition principle (WLF)*, the *Boltzmann superposition principle* is of extreme importance in the theory of linear viscoelasticity. The Boltzmann superposition principle states that the deformation of a polymer component is the sum or superposition of all strains that result from various loads acting on the part at different times. This means that the response of a material to a specific load is independent of already existing loads. Hence, we can compute the deformation of a polymer specimen upon which several loads act at different points in time by simply adding all strain responses. The Boltzmann superposition principle is schematically illustrated in Fig. 3.30. Mathematically, the Boltzmann superposition principle can be stated as follows

$$\varepsilon = \sigma_0 J(t - t_0) + (\sigma_1 - \sigma_0) J(t - t_1) + \ldots + (\sigma_i - \sigma_{i-1}) J(t - t_i) + \ldots \tag{3.14}$$

where J represents the material's compliance[11]. However, not all loadings and deformations consist of finite step changes, and Eq. 3.14 can be written in integral form as

$$\varepsilon(t) = \int_{\sigma(-\infty)}^{\sigma(t)} J(t - t')\, d\sigma(t') \tag{3.15}$$

which can also be written as

$$\varepsilon(t) = \int_{-\infty}^{t} J(t - t') \dot{\sigma}(t')\, dt' \tag{3.16}$$

Furthermore, one can invert Eq. 3.16 and write

$$\sigma(t) = \int_{-\infty}^{t} G(t - t') \dot{\varepsilon}\, dt' \tag{3.17}$$

The Boltzmann superposition principle holds as long as the polymer follows a linear viscoelastic behavior.

11 The compliance is the inverse of the stiffness, $J = 1/E$.

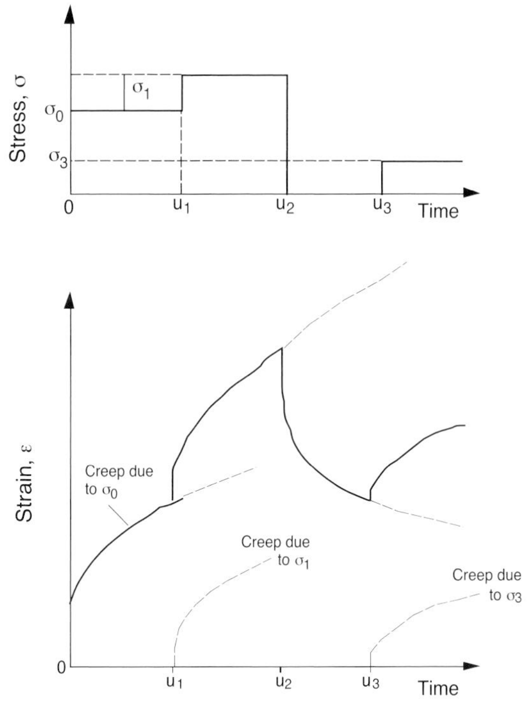

Figure 3.30 Schematic demonstration of Boltzmann's superposition principle. A, B, and C are sudden load changes.

3.8.4 Applying Linear Viscoelasticity to Describe the Behavior of Polymers

Most polymers exhibit a viscous as well as an elastic response to stress and strain. This puts them in the category of viscoelastic materials. Various combinations of elastic and viscous elements have been used to approximate the material behavior of polymeric melts. Most models are combinations of springs and dash pots— the most common one being the Maxwell model.

For clarity, let us first derive the stress-strain behavior for a Maxwell model shown in Fig. 3.31. The total strain, ε, in the model has an elastic, ε_e, and a viscous, ε_V, strain contribution and can be represented as follows:

$$\varepsilon = \varepsilon_e + \varepsilon_v \tag{3.18}$$

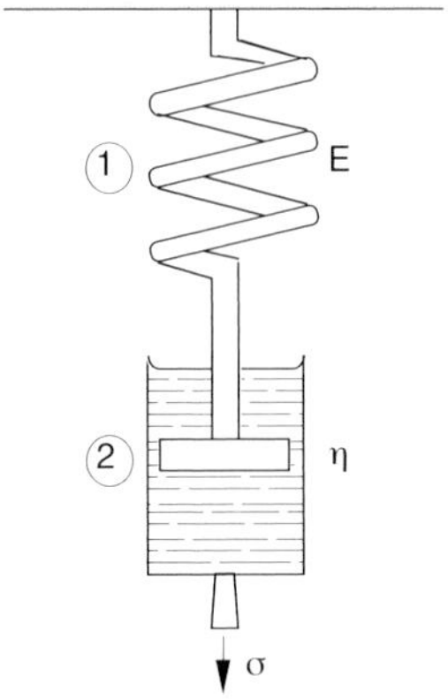

Figure 3.31 Schematic diagram of the Maxwell model.

Similarly, the strain rates are written as

$$\dot{\varepsilon} = \dot{\varepsilon}_e + \dot{\varepsilon}_v \tag{3.19}$$

Assuming the spring follows Hooke's law, the following relation holds

$$\dot{\varepsilon}_e = \frac{\dot{\sigma}}{E} \tag{3.20}$$

The viscous portion, represented by the dash pot, is written as follows

$$\dot{\varepsilon}_v = \frac{\sigma}{\eta} \tag{3.21}$$

Combining Eqs. 3.19–3.21 results in

$$\dot{\varepsilon} = \frac{\dot{\sigma}}{E} + \frac{\sigma}{\eta} \tag{3.22}$$

which can be rewritten as

$$\sigma + \frac{\eta}{E}\frac{d\sigma}{dt} = \eta\frac{d\varepsilon}{dt} \tag{3.23}$$

which is often referred to as the governing equation for Maxwell's model in *differential form.*

Maintaining the material at a constant deformation, such as in the relaxation test, the differential equation, Eq. 3.23, reduces to

$$\sigma + \frac{\eta}{E}\dot{\sigma} = 0 \tag{3.24}$$

Integrating Eq. 3.24 results in

$$\sigma = \sigma_0 e^{-t/\lambda} \tag{3.25}$$

where λ is known as the *relaxation time*. The relaxation response of the constant strain Maxwell model is depicted in Fig. 3.32. Using Eq. 3.25 one can show how after $t = \lambda$, the stresses relax to 37% of their initial value; $e^{-1} = 0.37$. When estimating relaxation of stresses, four relaxation times (4λ) are often used. Hence, by using $t = 4\lambda$ in Eq. 3.25 the stresses have relaxed to 1.8% of their original size. Using the Boltzmann superposition principle and Eq. 3.25 we can write the governing equation for the Maxwell model in *integral form* as

$$\sigma(t) = \int_{-\infty}^{t} E e^{-(t-t')/\lambda} \dot{\varepsilon}\, dt \tag{3.26}$$

For more accurate estimates or realistic analyses the Maxwell model is not sufficient. For a better fit with experimental data it is common to use several spring-dash pot models in parallel such as shown in Fig. 3.32 [13]. Such a configuration is often referred to as a *generalized Maxwell model*. The curve shown in the figure fits a four-parameter model with experimental relaxation and retardation data for a common polystyrene with a molecular weight of 260,000 g/mole. For this specific material, the relaxation behavior of the injected melt into a hot cavity, at a reference temperature of 113 $^{\circ}$C, is described by

$$\frac{\varepsilon}{\varepsilon_a} = 0.25\left(e^{-8.75\,t/\lambda} + e^{-1.0\,t/\lambda} + e^{-0.28\,t/\lambda} + e^{-0.0583\,t/\lambda}\right) \tag{3.27}$$

where ε_a is the strain after relaxation and is defined by

$$\varepsilon_a = \frac{l_a - l_0}{l_0} = \frac{S_0}{1 - S_0} \tag{3.28}$$

Here, l_a and l_0 represent the length of the stretched and relaxed sample, respectively, and S_0 represents the total shrinkage.

The terms $\frac{\lambda}{8.75}, \lambda, \frac{\lambda}{0.28}, \frac{\lambda}{0.0583}$, in Eq. 3.27 represent four individual relaxation times for this specific polystyrene, modeled using the four-parameter model. The relaxation time, λ, correlates with the time it takes for the initial strain to reduce, by relaxation, to one-half its initial value. This relaxation time is also temperature dependent as shown for various polymers in Fig. 3.33. Figure 3.33 shows how the shapes of the curves are all similar, only shifted by a certain temperature. It is important to note that the relaxation and retardation behavior of all amorphous thermoplastics is similar.

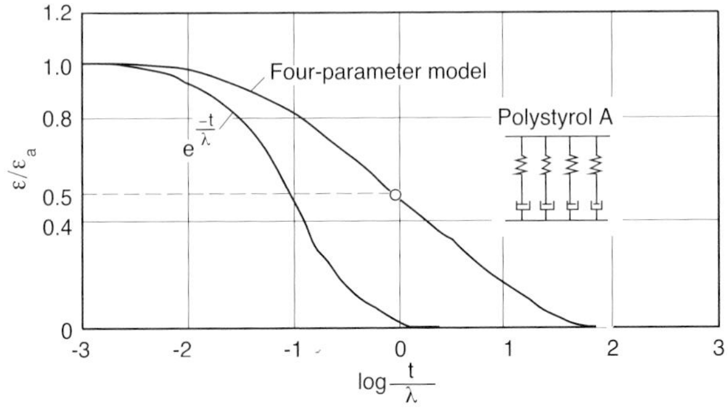

Figure 3.32 Relaxation response of a Maxwell model and a four-parameter Maxwell model.

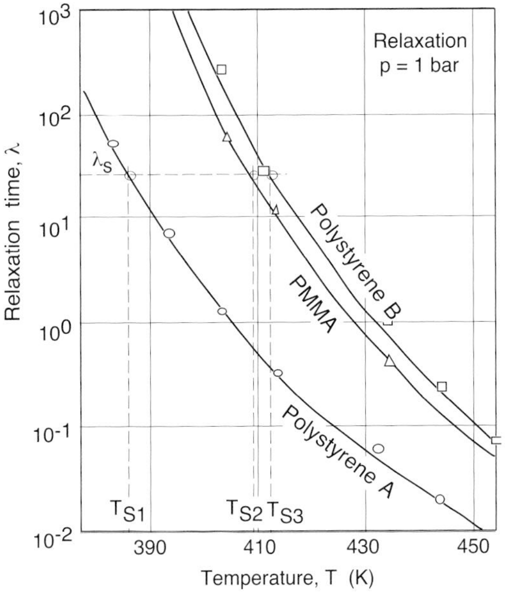

Figure 3.33 Relaxation time as a function of temperature for various thermoplastics.

Wübken [13] performed similar tests with different amorphous thermoplastics, and he found that, indeed, in all cases the measurements showed a correlation between time and temperature such as described by the WLF [12] equation. The data fit by the four-parameter model was generated via two different experiments: a relaxation test inside an injection mold between 100 and 180 °C, and a retardation test outside of the mold between 72 and 100 °C. The measured data are shown in Figs. 3.34 and 3.35 for the relaxation and retardation tests, respectively. The curves shown in both graphs were shifted horizontally to generate one master curve as shown in Fig. 3.36. The solid line in the figure is the four-parameter fit represented with Eq. 3.27.

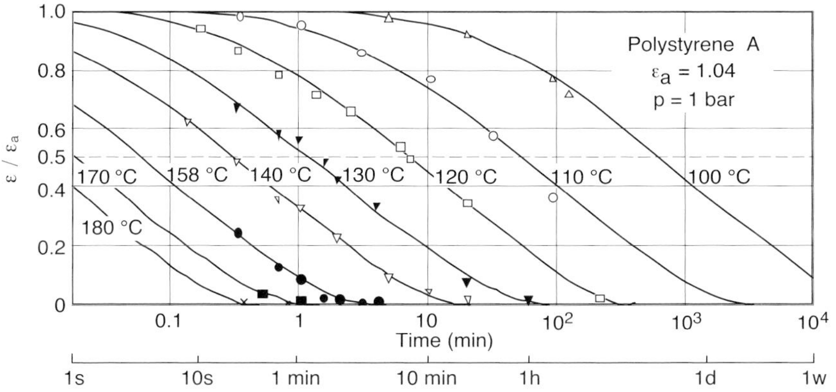

Figure 3.34 Relaxation response, inside an injection mold, of a polystyrene specimen at various temperatures.

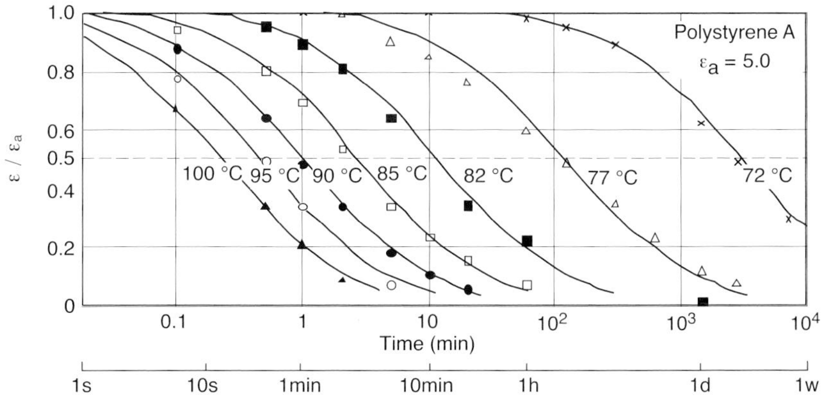

Figure 3.35 Recovery or retardation response after injection molding of a polystyrene specimen at various temperatures.

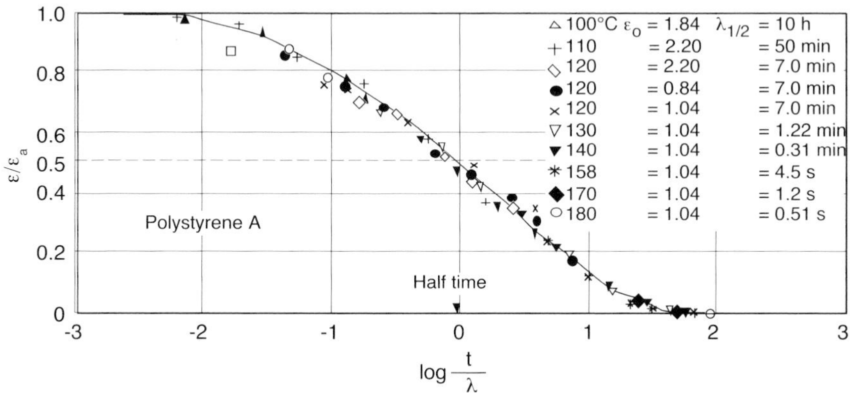

Figure 3.36 Master curve for the relaxation response, inside an injection mold, of a polystyrene specimen at various temperatures.

Hence, appropriate T_{ref} and l values must be found. However, the reference temperature is not quite independent of the relaxation behavior of the polymer but is related to the material properties. For the polystyrene A in Fig. 3.37, $T_{ref} = 113\,^{\circ}\mathrm{C}$, or about 48 °C above T_g. For example, for the polystyrene A of Fig. 3.37, the relaxation time, l, can be computed by

$$\text{Relaxation: } \log(\lambda) = \log(27) - \frac{8.86\,(T - T_s)}{101.6 + (T - T_s)} \tag{3.29}$$

$$\text{Creep: } \log(\lambda) = \log(0.0018) - \frac{8.86\,(T - T_s)}{101.6 + (T - T_s)} \tag{3.30}$$

where the constants 27 and 0.0018 are the relaxation times, λ, in minutes, at the reference temperature of 113 °C.

As discussed earlier, similar to the temperature induced shift, there is also a shift due to pressure[12]. If we refer to Fig. 3.38, which shows the influence of pressure on T_g, we can see that this effect can easily be included into the WLF equation i.e., there is approximately a 2 °C shift in the glass transition temperature of polystyrene for every 100 bar of pressure rise [14].

Models such as the Maxwell model described by Eq. 3.23 can also be used to simulate the dynamic response of polymers. In a dynamic test[13] the strain input is given by

$$\varepsilon = \varepsilon_0 \sin(\omega t) \tag{3.31}$$

12 This is discussed in Chapters 5 and 9 for shifts in viscosity and relaxation times, respectively.

13 The dynamic test is discussed in more detail in Chapter 8.

where ε_0 is the strain amplitude and ω the frequency in radians per second. Differentiating Eq. 3.31, combining with Eq. 3.23, and integrating results in

$$\sigma = \left(\frac{E\varepsilon_o \omega\lambda}{1+(\omega\lambda)^2} \right) \left(\omega\lambda \sin(\omega t) + \cos(\omega t) \right) \tag{3.32}$$

for a steady state response. Dividing Eq. 3.28 by the amplitude of the strain input results in a complex modulus, which is formed by an elastic component, that is in-phase with the strain input, and a viscous component. The elastic term is generally called the *storage modulus* and is defined by

$$E' = \left(\frac{E(\omega\lambda)^2}{1+(\omega\lambda)^2} \right) \tag{3.33}$$

and the viscous term, usually referred to as the *loss modulus*, is given by

$$E'' = \left(\frac{E\omega\lambda}{1+(\omega\lambda)^2} \right) \tag{3.34}$$

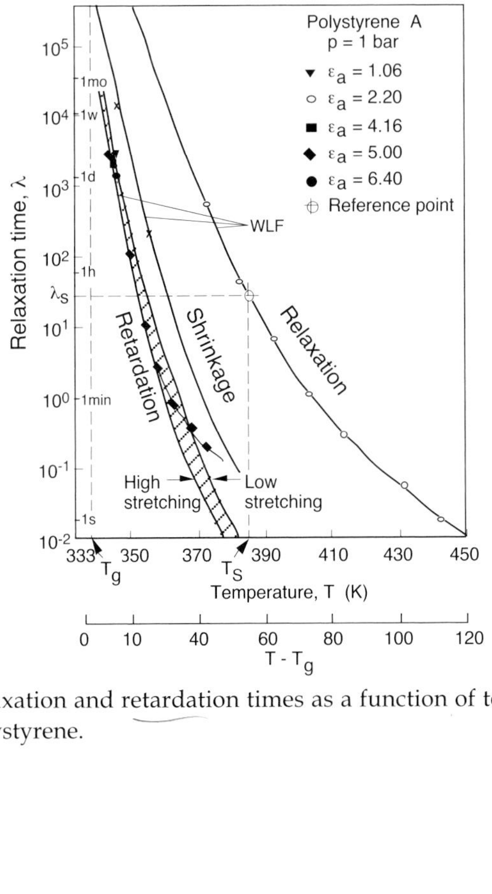

Figure 3.37 Relaxation and retardation times as a function of temperature for a polystyrene.

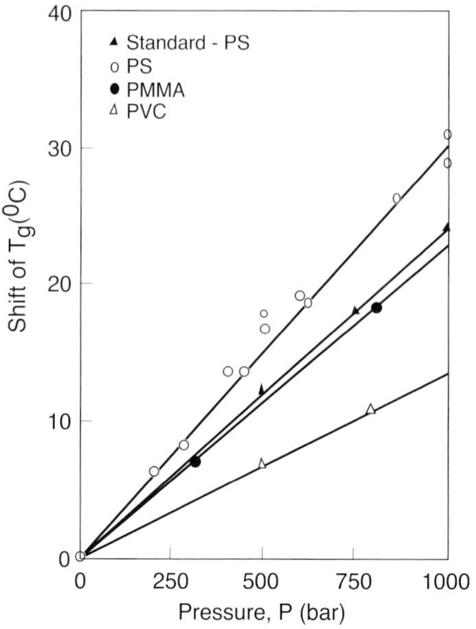

Figure 3.38 Influence of hydrostatic pressure on the glass transition temperature for various amorphous thermoplastics.

Examples

3.1 What is the maximum possible separation between the ends of a high density polyethylene molecule with an average molecular weight of 100,000.

Our first task is to estimate the number of repeat units, n, in the polyethylene chain. Each repeat unit has 2 carbons and 4 hydrogen atoms. The molecular weight of carbon is 12 and that of hydrogen 1. Hence,

MW/repeat unit = 2(12)+4(1)=28.

The number of repeat units is computed as

n = MW/(MW/repeat unit) = 100,000/28 = 3,571 units.

Using the diagram presented in Fig. 3.39 we can now estimate the length of the fully extended molecule using

$$\ell = 0.252\ \text{nm}(3571) = 890\ \text{nm} = 0.89\mu\text{m}$$

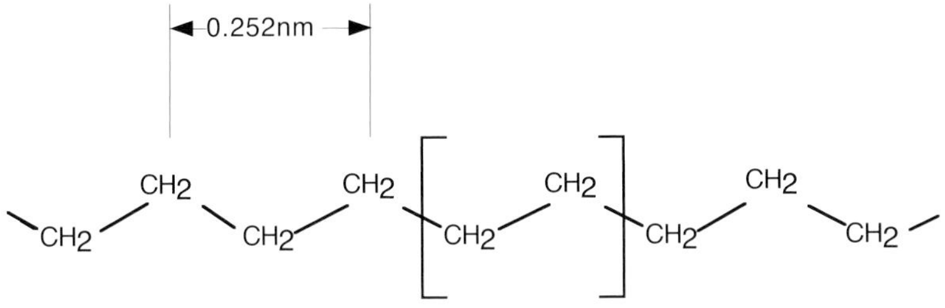

Figure 3.39 Schematic diagram of a polyethylene molecule.

As we know, even high density polyethylene molecules are branched and therefore our result is an overprediction.

3.2 For the poly-a-methylstyrene stress relaxation data in Fig. 3.40 [16], create a master creep curve at T_g (204 °C).

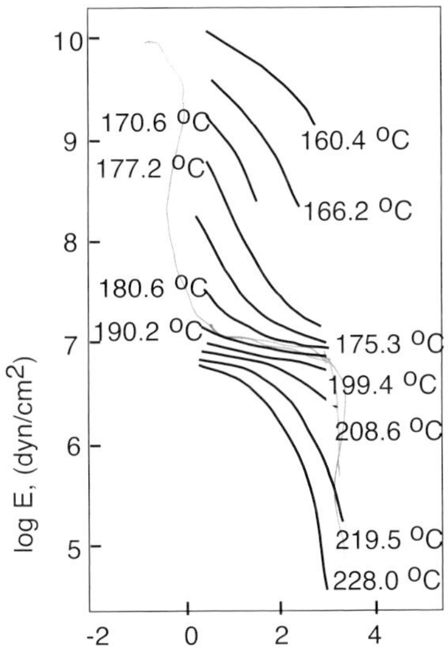

Figure 3.40 Stress relaxation data for poly-α-methylstyrene.

Identify the glassy, rubbery, viscous and viscoelastic regions of the master curve. Identify each region with a spring-dashpot diagram.

Develop a plot of the shift factor, $\log(a_T)$ versus T, used to create your master curve $\log(a_T)$ is the horizontal distance that the curve at temperature T was slid to coincide with the master curve.

What is the relaxation time of the polymer at the glass transition temperature?

The master creep curve for the above data is generated by sliding the individual relaxation curves horizontally until they match with their neighbors, using a fixed scale for a hypothetical curve at 204 °C. Since the curve does not exist for the desired temperature we can interpolate between 208.6 °C and 199.4 °C. The resulting master curve is presented in Fig. 3.41.

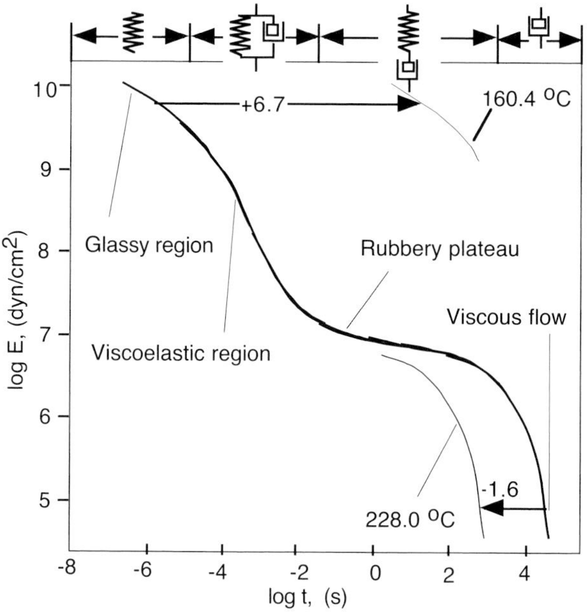

Figure 3.41 Master curve for poly-α-methylstyrene at 204 °C.

The amount each curve must be shifted from the master curve to their initial position is the shift factor, log a_T. The graph also shows the spring-dashpot models and the shift factor for a couple of temperatures. Figure 3.42 represents the shift factor versus temperature. The solid line indicates the shift factor predicted by the WLF equation.

The relaxation time for the poly-α-methylstyrene presented here is

$10^{4.5}$ s, or 31,623 s (8.8 h). The relaxation time for the remaining temperatures can be computed using the shift factor curve.

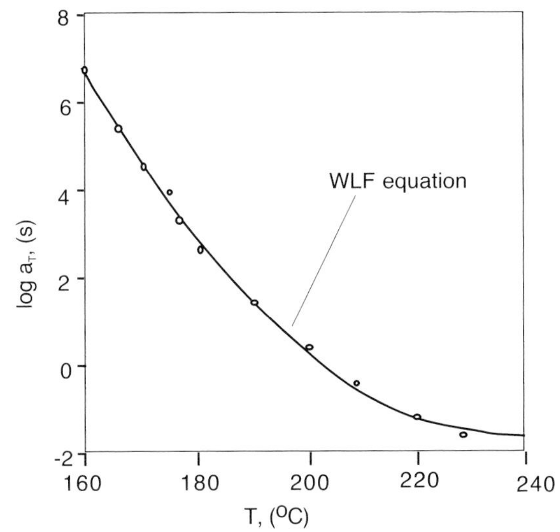

Figure 3.42 Shift factor and WLF curves for T_{ref}=204 °C.

Problems

3.1 Estimate the degree of polymerization of a polyethylene with an average molecular weight of 150,000. The molecular weight of an ethylene monomer is 28.

3.2 What is the maximum possible separation between the ends of a polystyrene molecule with a molecular weight of 160,000.

3.3 To enhance processability of a polymer why would you want to decrease its molecular weight?

3.4 Why would an uncrosslinked polybutadiene flow at room temperature?

3.5 Is it true that by decreasing the temperature of a polymer you can increase its relaxation time?

3.6 If you know the relaxation time of a polymer at one temperature, can you use the WLF equation to estimate the relaxation time of the same material at a different temperature? Explain.

3.7 What role does the cooling rate play in the morphological structure of semi-crystalline polymers?

3.8 Explain how cross-linking between the molecules affect the molecular mobility and elasticity of elastomers.

3.9 Increasing the molecular weight of a polymer increases its strength and stiffness, as well as its viscosity. Is too high of a viscosity a limiting factor when increasing the strength by increasing the molecular weight? Why?

3.10 Which broad class of thermoplastic polymers densifies the least during cooling and solidification from a melt state into a solid state? Why?

3.11 What class of polymers would you probably use to manufacture frying pan handles? Even though most polymers could not actually be used for this particular application, what single property do all polymers exhibit that would be considered advantageous in this particular application.

3.12 In terms of recycling, which material is easier to handle, thermosets or thermoplastics? Why?

3.13 You are to extrude a polystyrene tube at an average speed of is 0.1 m/s. The relaxation time, l, of the polystyrene, at the processing temperature, is 1 second. The die land length is 0.02 m. Will elasticity play a significant role in your process.

3.14 Figure 3.43 presents some creep compliance data for polystyrene at various temperatures [15]. Create a master curve at 109.8 ^{O}C by graphically sliding the curves at some temperatures horizontally until they line up.

Identify the glassy, rubbery, and viscoelastic regions of the master curve.

Develop a plot of the shift factor, $\log(a_T)$ versus T, used to create your master curve $\log(a_T)$ is the horizontal distance that the curve at temperature T was slid to coincide with the master curve. Compare your graphical result with the WLF equation.

Note: The WLF equation is for a master curve at Tg (85 ^{O}C for this PS), but your master curve is for 109.8 ^{O}C, so be sure you make a fair comparison.

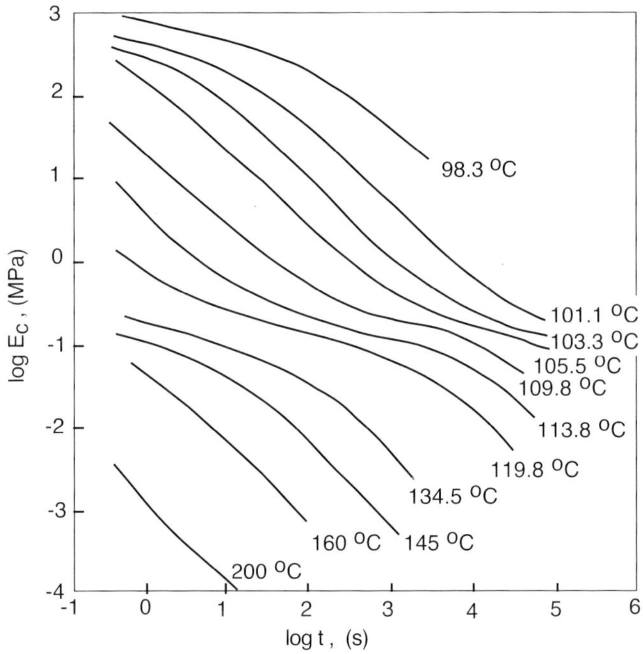

Figure 3.43 Creep modulus as a function of time for polystyrene.

3.15 Figure 3.44 presents relaxation data for polycarbonate at various temperatures [17]. Create a master curve at 25 °C by graphically sliding the curves at the various temperatures horizontally until they line up.

Identify the glassy, rubbery, and viscoelastic regions of the master curve.

Develop a plot of the shift factor, $\log(a_T)$ versus T, used to create your master curve. $\log(a_T)$ is the horizontal distance that the curve at temperature T was slid to coincide with the master curve.

Compare your graphical result with the WLF equation. Note that the resulting master curve is far from the glass transition temperature of polycarbonate.

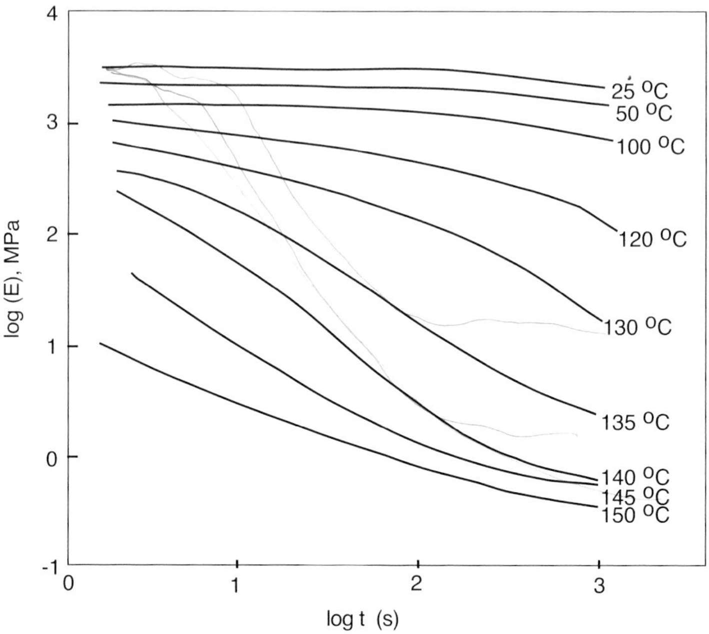

Figure 3.44 Relaxation modulus as a function of time for polycarbonate

3.16 Figure 3.28 presents shear relaxation data for a chlorosulfonated polyethylene at various pressures. Create a master curve at 1 bar by graphically sliding the curves at the various pressures horizontally until they line up. On the same graph draw the master curve at a pressure of 1200 bar, a high pressure encountered during injection molding.

References

1. Berry, G.C., and Fox, T.G., *Adv. Polymer Sci.*, 5, 261, (1968).
2. Staudinger, H., and Huer, W., *Ber. der Deutsch. Chem Gessel.*, 63, 222, (1930).
3. Crowder, M.L., Ogale, A.M., Moore, E.R., and Dalke, B.D.,*Polym. Eng. Sci., 34*, 19, 1497, (1994).
4. Domininghaus, H., *Plastics for Engineers*, Hanser Publishers, Munich, (1993).
5. Aklonis, J.J., and MacKnight, W.J., *Introduction to Polymer Viscoelasticity*, John Wiley and Sons, New York, (1983).
6. Rosen, S.L., *Fundamental Principles of Polymeric Materials*, John Wiley & Sons, Inc., New York, (1993).
7. Van Krevelen, D.W., and Hoftyzer, P.J., *Properties of Polymers*, 2nd. Ed., Elsevier, Amsterdam, (1976).

8. Green, J., *Thermoplastic Polymer Additives*, Chapter 4, J.T. Lutz, Jr., Ed., Marcel Dekker, Inc., New York, (1989).
9. Klempner, D., and Frisch, K.C. , *Handbook of Polymeric Foams and Foam Technology*, Hanser Publishers, Munich, (1991).
10. Castiff, E. and Tobolsky, A.V.J., *Colloid Sci., 10*, 375, (1955).
11. Fillers, R.W., and Tschoegl, N.W., *Trans. Soc. Rheol, 21*, 51 (1977).
12. Williams, M.L., Landel, R.F., and Ferry, J.D., *J. Amer. Chem. Soc., 77*, 3701, (1955).
13. Wübken, G., Ph.D. Thesis, IKV, RWTH-Aachen, (1974).
14. Münstedt, H., *Rheol. Acta*, 14, 1077, (1975).
15. Pazek, D.J., *J. Polym. Sci.*, A-2, (6), 621, (1968).
16. Fujimoto, T., Ozaki, M., and Nagasawa, M., *J. Polymer Sci.,2*, 6, (1968).
17. Carreau, P.J., De Kee, D.C.R., and Chhabra, R.P., *Rheology of Polymeric Systems*, Hanser Publishers, Munich, (1997).

4

Thermal Properties of Polymers

The heat flow through a material can be defined by Fourier's law of heat conduction. Fourier's law can be expressed as[1]

$$q_x = -k_x \frac{\partial T}{\partial x} \tag{4.1}$$

where q_x is the energy transport per unit area in the x direction, k_x the thermal conductivity and the temperature gradient. At the onset of heating the polymer responds solely as a heat sink, and the amount of energy per unit volume, Q, stored in the material before reaching steady state conditions can be approximated by

$$Q = \rho C_p \Delta T \tag{4.2}$$

where ρ is the density of the material, C_p the specific heat, and ΔT the change in temperature.

Using the notation found in Fig. 4.1 and balancing the heat flow through the element via conduction, including the transient, convective, and viscous heating effects, the energy balance can be written as

$$\rho C_p \frac{\partial T}{\partial t} + \rho C_p \underline{v} \cdot \nabla T = (\nabla \cdot \underline{k} \cdot \nabla T) + \frac{1}{2}\mu\left(\dot{\underline{\underline{\gamma}}} : \dot{\underline{\underline{\gamma}}}\right) + \dot{Q} \tag{4.3}$$

where on the left are the transient and convective terms, and on the right the conductive, viscous heating, and an arbitrary heat generation term. The full form of the energy equation is found along with the continuity equation and momentum balance in Tables IV and V of the appendix of this book.

1 For more detailed information on transport phenomena the reader can refer to: Bird, R.B., Steward, W.E. and Lightfoot, E.N., *Transport Phenomena*, John Wiley & Sons, (1960).

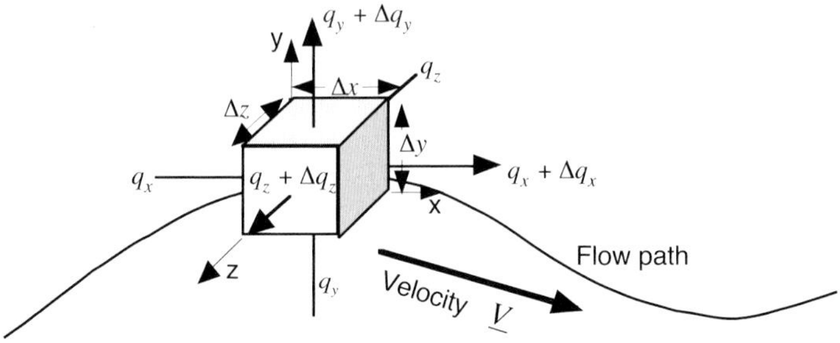

Figure 4.1 Heat flow through a differential material element.

The material properties found in Eqs. 4.1–4.3 are often written as one single property, namely the thermal diffusivity, a, which for an isotropic material is defined by

$$\alpha = \frac{k}{\rho C_P} \tag{4.4}$$

Typical values of thermal properties for selected polymers are shown in Table 4.1 [1]. For comparison, the properties for stainless steel are also shown at the end of the list.

It should be pointed out that the material properties of polymers are not constant and may vary with temperature, pressure or phase changes. This chapter will discuss each of these properties individually and present examples of some of the most widely used polymers and measurement techniques. For a more in-depth study of thermal properties of polymers the reader is encouraged to consult the literature [2–4].

Table 4.1 Thermal Properties for Selected Polymeric Materials

Polymer	Specific gravity	Specific heat (kJ/kg°C)	Thermal conductivity (W/m/K)	Coeff. of therm. exp. (μm/m/K)	Thermal diffusivity $(m^2/s)x10^{-7}$	Max temperature (°C)
ABS	1.04	1.47	0.3	90	1.7	70
Acetal (homo-pol.)	1.42	1.47	0.2	80	0.7	85
Acetal (Co-pol.)	1.41	1.47	0.2	95	0.72	90
Acrylic	1.18	1.47	0.2	70	1.09	50
Cellulose acetate	1.28	1.50	0.15	100	1.04	60
Epoxy	1.90	-	0.23	70	-	130
Mod. PPO	1.06	-	0.22	60	-	120
PA66	1.14	1.67	0.24	90	1.01	90
PA66 +30%GF	1.38	1.26	0.52	30	1.33	100
PET	1.37	1.05	0.24	90	-	110
PET +30%GF	1.63	-	-	40	-	150
Phenolic	1.40	1.30	0.35	22	1.92	185
PC	1.15	1.26	0.2	65	1.47	125
u-Polyester	1.20	1.20	0.2	100	-	-
PP	0.905	1.93	0.24	100	0.65	100
PS	1.05	1.34	0.15	80	0.6	50
LDPE	0.92	2.30	0.33	200	1.17	50
HDPE	0.95	2.30	0.63	120	1.57	55
PTFE	2.10	1.00	0.25	140	0.7	50
u-PVC	1.40	1.00	0.16	70	1.16	50
p-PVC	1.30	1.67	0.14	140	0.7	50
SAN	1.08	1.38	0.17	70	0.81	60
PS-foam	0.032	-	0.032	-	-	-
Steel	7.854	0.434	60.00	-	14.1	800

4.1 Material Properties

4.1.1 Thermal Conductivity

When analyzing thermal processes, the thermal conductivity, *k*, is the most commonly used property that helps quantify the transport of heat through a material. By definition, energy is transported proportionally to the speed of sound. Accordingly, thermal conductivity follows the relation

$$k \approx C_p \rho u l \tag{4.5}$$

where u is the speed of sound and l the molecular separation. Amorphous polymers show an increase in thermal conductivity with increasing temperature, up to the glass transition temperature, T_g. Above T_g, the thermal conductivity decreases with increasing temperature. Figure 4.2 [2] presents the thermal conductivity, below the glass transition temperature, for various amorphous thermoplastics as a function of temperature.

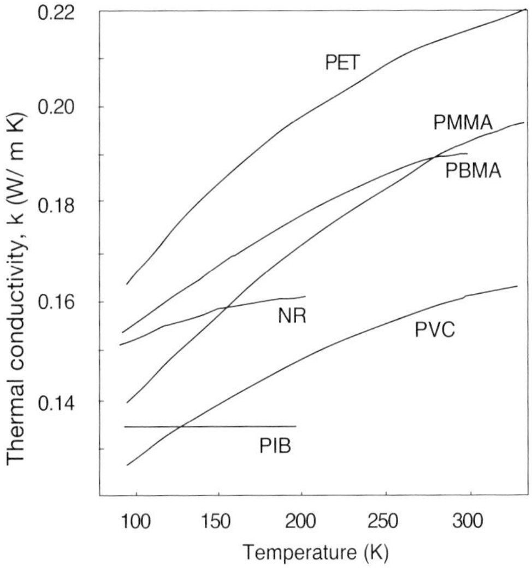

Figure 4.2 Thermal conductivity of various materials.

Due to the increase in density upon solidification of semi-crystalline thermoplastics, the thermal conductivity is higher in the solid state than in the melt. In the melt state, however, the thermal conductivity of semi-crystalline polymers reduces to that of amorphous polymers as can be seen in Fig. 4.3 [5]. Furthermore, it is not surprising that the thermal conductivity of melts increases with hydrostatic pressure. This effect is clearly shown in Fig. 4.4 [6].

As long as thermosets are unfilled, their thermal conductivity is very similar to amorphous thermoplastics.

Anisotropy in thermoplastic polymers also plays a significant role in the thermal conductivity. Highly drawn semi-crystalline polymer samples can have a much higher thermal conductivity as a result of the orientation of the polymer chains in the direction of the draw. For amorphous polymers, the increase in thermal conductivity in the direction of the draw is usually not higher than two. Figure 4.5 [2] presents the thermal conductivity in the directions parallel and perpendicular to the draw for high density polyethylene, polypropylene, and polymethyl methacrylate.

A simple relation exists between the anisotropic and the isotropic thermal conductivity [7]. This relation is written as

$$\frac{1}{k_{\parallel}} + \frac{2}{k_{\perp}} = \frac{3}{k} \tag{4.6}$$

where the subscripts $||$ and $\perp$ represent the directions parallel and perpendicular to the draw, respectively.

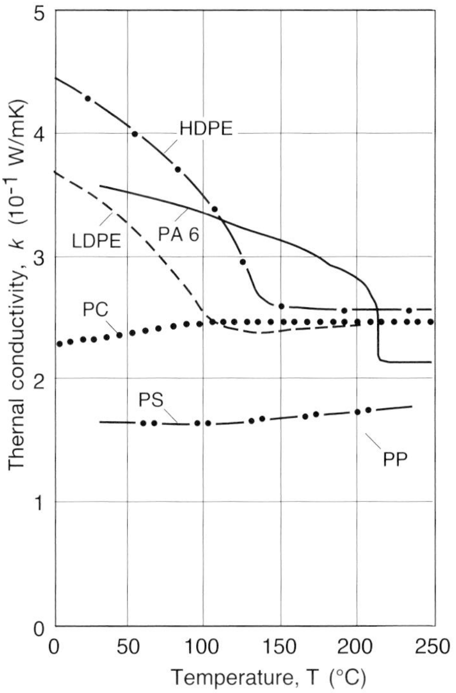

Figure 4.3 Thermal conductivity of various thermoplastics.

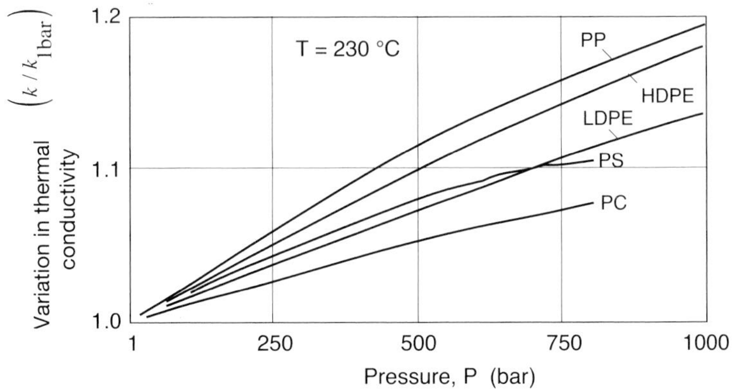

Figure 4.4 Influence of pressure on thermal conductivity of various thermoplastics.

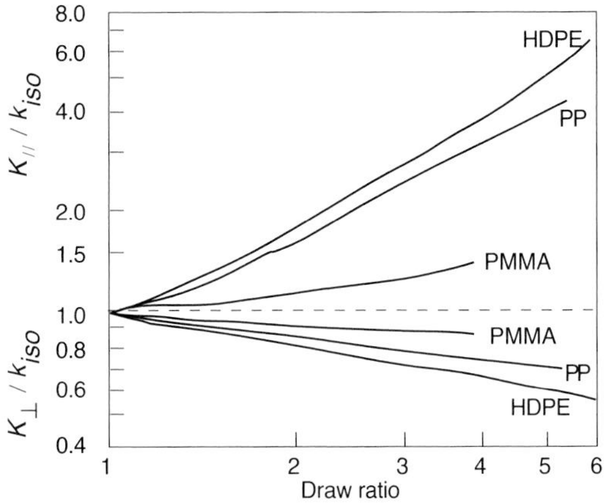

Figure 4.5 Thermal conductivity as a function of draw ratio in the directions perpendicular and parallel to the stretch for various oriented thermoplastics.

The higher thermal conductivity of inorganic fillers increases the thermal conductivity of filled polymers. Nevertheless, a sharp decrease in thermal conductivity around the melting temperature of crystalline polymers can still be seen with filled materials. The effect of filler on thermal conductivity of various thermoplastics is shown in Figs. 4.6 to 4.9. Figure 4.6 [8] shows the effect of fiber orientation as well as the effect of quartz powder on the thermal conductivity of low density polyethylene. Figures 4.7 to 4.9 show the effect of various volume fractions of glass fiber in polyamide 6, polycarbonate, and ABS, respectively. Figure 4.10 demonstrates the influence of gas content on expanded or foamed polymers, and the influence of mineral content on filled polymers. There are various models available to compute the thermal conductivity of foamed or filled plastics [7, 9, 10]. A rule of mixtures, suggested by Knappe [7], commonly used to compute thermal conductivity of composite materials is written as

$$k_c = \frac{2k_m + k_f - 2\phi_f (k_m - k_f)}{2k_m + k_f + \phi_f (k_m - k_f)} k_m \tag{4.7}$$

where, ϕ_f is the volume fraction of filler, and k_m, k_f and k_c are the thermal conductivity of the matrix, filler and composite, respectively. Figure 4.11 compares Eq. 4.7 with experimental data [11] for an epoxy filled with copper particles of various diameters. The figure also compares the data to the classic model given by Maxwell [9] which is written as

$$k_c = \left(1 + 3\phi_f \left(\frac{k_f / k_m - 1}{k_f / k_m + 2}\right)\right) k_m \tag{4.8}$$

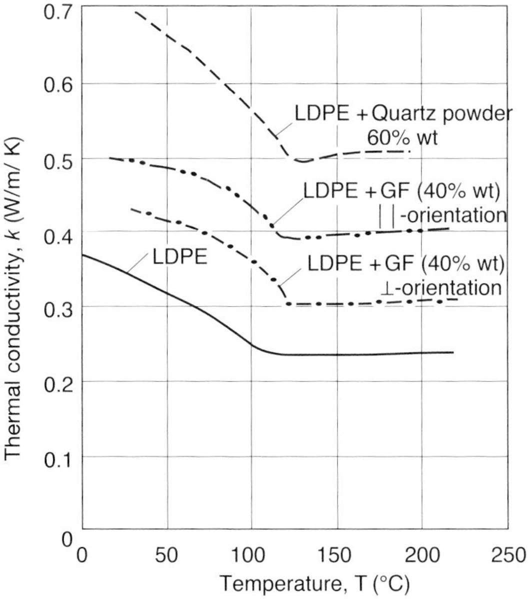

Figure 4.6 Influence of filler on the thermal conductivity of LDPE.

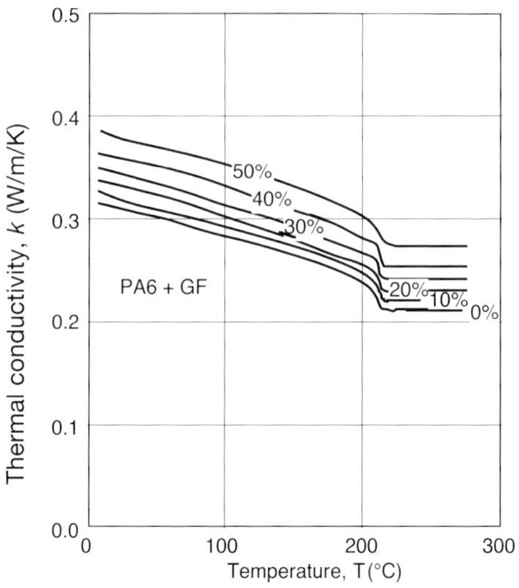

Figure 4.7 Influence of glass fiber on the thermal conductivity of polyamide 6. Courtesy of Bayer AG, Germany.

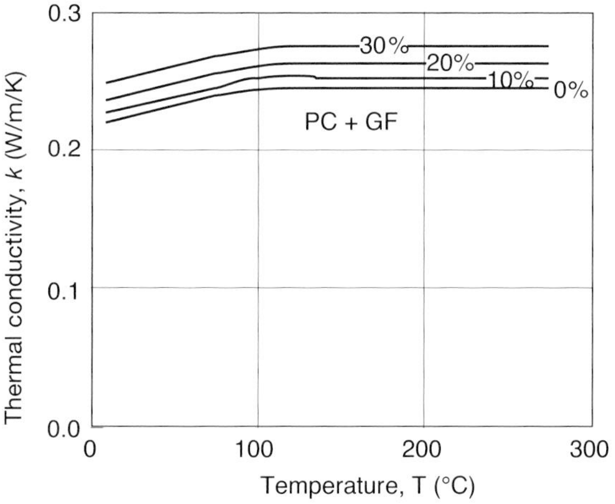

Figure 4.8 Influence of glass fiber on the thermal conductivity of polycarbonate. Courtesy of Bayer AG, Germany.

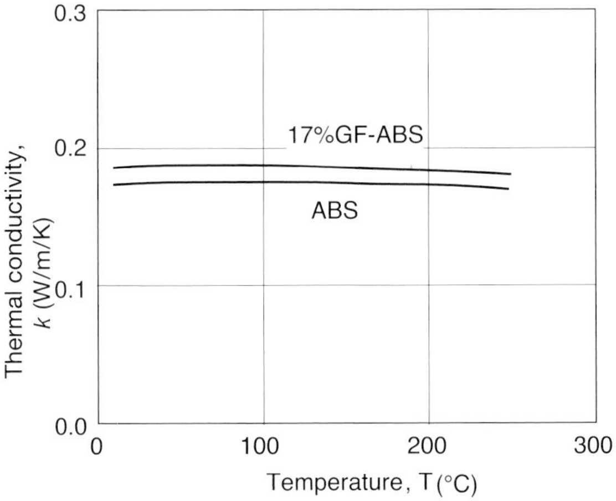

Figure 4.9 Influence of glass fiber on the thermal conductivity of ABS. Courtesy of Bayer AG, Germany.

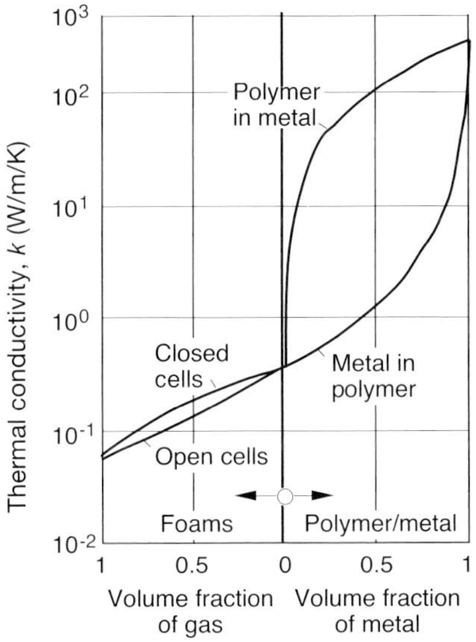

Figure 4.10 Thermal conductivity of plastics filled with glass or metal.

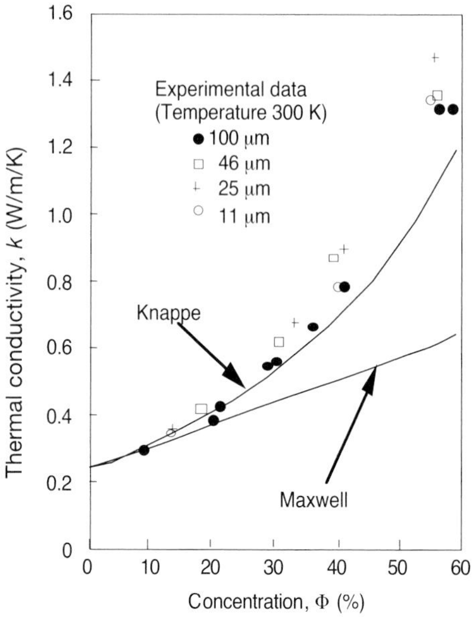

Figure 4.11 Thermal conductivity versus volume concentration of metallic particles of an epoxy resin. Solid lines represent predictions using Maxwell and Knappe models.

In addition, a model derived by Meredith and Tobias [10] applies to a cubic array of spheres inside a matrix. Consequently, it cannot be used for volumetric concentration above 52% since the spheres will touch at that point. However, their model predicts the thermal conductivity very well up to 40% by volume of particle concentration.

When mixing several materials the following variation of Knappe's model applies

$$k_c = \frac{1 - \sum_{i=1}^{n} 2\phi_i \dfrac{k_m - k_i}{2k_m + k_i}}{1 - \sum_{i=1}^{n} \phi_i \dfrac{k_m - k_i}{2k_m - k_i}} k_m \tag{4.9}$$

where k_i is the thermal conductivity of the filler and ϕ_i its volume fraction. This relation is useful for glass fiber reinforced composites (FRC) with glass concentrations up to 50% by volume. This is also valid for FRC with unidirectional reinforcement. However, one must differentiate between the direction longitudinal to the fibers and that transverse to them. For high fiber content one can approximate the thermal conductivity of the composite by the thermal conductivity of the fiber.

The thermal conductivity can be measured using the standard tests ASTM C177 and DIN 52612. A new method is currently being balloted, ASTM D20.30, which is preferred by most people today.

4.1.2 Specific Heat

The specific heat, C, represents the energy required to change a unit mass of material by one degree in temperature. It can be measured at either constant pressure, C_p, or constant volume, C_v. Since the specific heat at constant pressure includes the effect of volumetric change, it is larger than the specific heat at constant volume. However, the volume changes of a polymer with changing temperatures have a negligible effect on the specific heat. Hence, one can usually assume that the specific heat at constant volume or constant pressure are the same. It is usually true that specific heat only changes modestly in the range of practical processing and design temperatures of polymers. However, semi-crystalline thermoplastics display a discontinuity in the specific heat at the melting point of the crystallites. This jump or discontinuity in specific heat includes the heat that is required to melt the crystallites which is usually called the *heat of fusion*. Hence, specific heat is dependent on the degree of crystallinity. Values of heat of fusion for typical semi-crystalline polymers are shown in Table 4.2.

Table 4.2[2] Heat of Fusion of Various Thermoplastic Polymers

Polymer	λ (kJ/kg)	T_m (°C)
Polyethylene	268- 300	141
Polypropylene	209- 259	183
Polyvinyl-chloride	181	285
Polybutadiene	170- 187	148
Polyamide 6	193- 208	223
Polyamide 66	205	265

The chemical reaction that takes place during solidification of thermosets also leads to considerable thermal effects. In a hardened state, their thermal data are similar to the ones of amorphous thermoplastics. Figure 4.12 shows the specific heat graphs for the three polymer categories.

2 The values for heat of fusion were computed using data taken from: van Krevelen, D.W., and Hoftyzer, P.J. , *Properties of Polymers*, Elsevier Scientific Publishing Company, Amsterdam, (1976).

For filled polymer systems with inorganic and powdery fillers a rule of mixtures[3] can be written as

$$C_P(T) = (1 - \psi_f)C_{PM}(T) + \psi_f C_{Pf}(T) \tag{4.10}$$

where ψ_f represents the weight fraction of the filler and C_{PM} and C_{Pf} the specific heat of the polymer matrix and the filler, respectively. As an example of using Eq. 4.10, Fig. 4.13 shows a specific heat curve of an unfilled polycarbonate and its corresponding computed specific heat curves for 10%, 20%, and 30% glass fiber content. In most cases temperature dependence of C_P of inorganic fillers is minimal and need not be taken into consideration.

The specific heat of copolymers can be calculated using the mole fraction of the polymer components.

$$C_{Pcopolymer} = \sigma_1 C_{P1} + \sigma_2 C_{P2} \tag{4.11}$$

where σ_1 and σ_2 are the mole fractions of the comonomer components and C_{P1} and C_{P2} the corresponding specific heats.

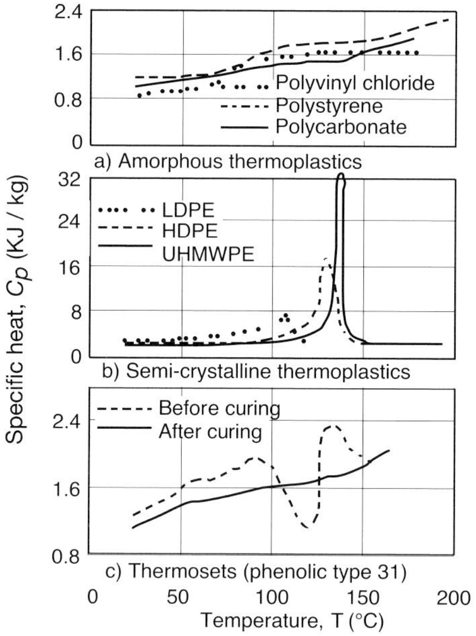

Figure 4.12 Specific heat curves for selected polymers of the three general polymer categories.

3 Valid up to 65% filler content by volume.

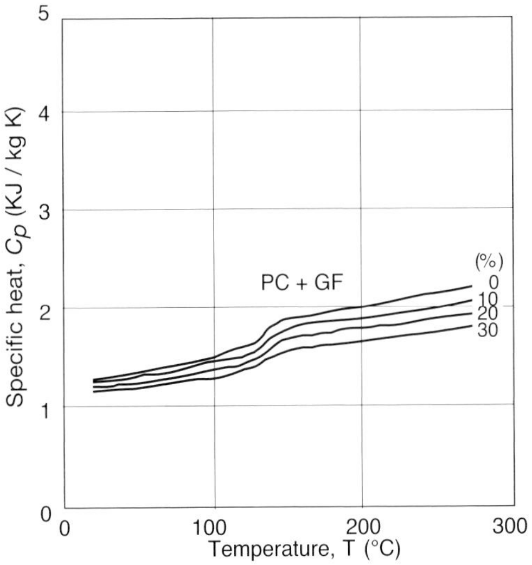

Figure 4.13 Generated specific heat curves for a filled and unfilled polycarbonate. Courtesy of Bayer AG, Germany.

4.1.3 Density

The density or its reciprocal, the specific volume, is a commonly used property for polymeric materials. The specific volume is often plotted as a function of pressure and temperature in what is known as a p-v-T diagram. A typical p-v-T diagram for an unfilled and filled amorphous polymer is shown, using polycarbonate as an example, in Figs. 4.14 to 4.16. The two slopes in the curves represent the specific volume of the melt and of the glassy amorphous polycarbonate, separated by the glass transition temperature. Figure 4.17 presents the p-v-T diagram for polyamide 66 as an example of a typical semi-crystalline polymer. Figure 4.18 shows the p-v-T diagram for polyamide 66 filled with 30% glass fiber. The curves clearly show the melting temperature (i.e., $T_m \approx 250\,^{\circ}\mathrm{C}$ for the unfilled PA66 cooled at 1 bar, which marks the beginning of crystallization as the material cools). It should also come as no surprise that the glass transition temperatures are the same for the filled and unfilled materials.

When carrying out die flow calculations, the temperature dependence of the specific volume must often be dealt with analytically. At constant pressures, the density of pure polymers can be approximated by

$$\rho(T) = \rho_0 \frac{1}{1+\alpha_t (T - T_\circ)} \tag{4.12}$$

where ρ_o is the density at reference temperature, T_o, and α_t is the linear coefficient of thermal expansion.

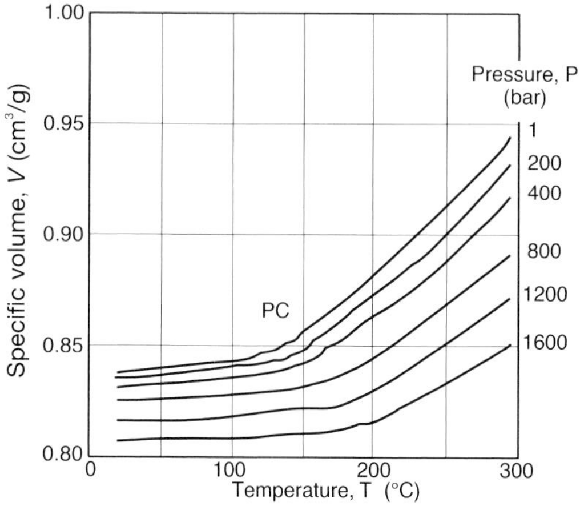

Figure 4.14 p-v-T diagram for a polycarbonate. Courtesy of Bayer AG, Germany.

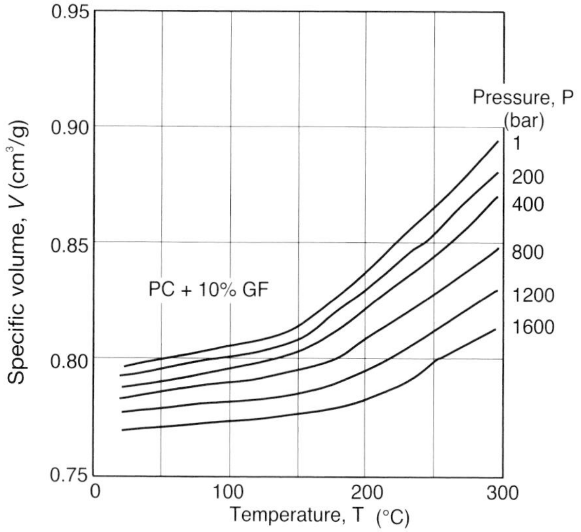

Figure 4.15 p-v-T diagram for a polycarbonate filled with 10% glass fiber. Courtesy of Bayer AG, Germany.

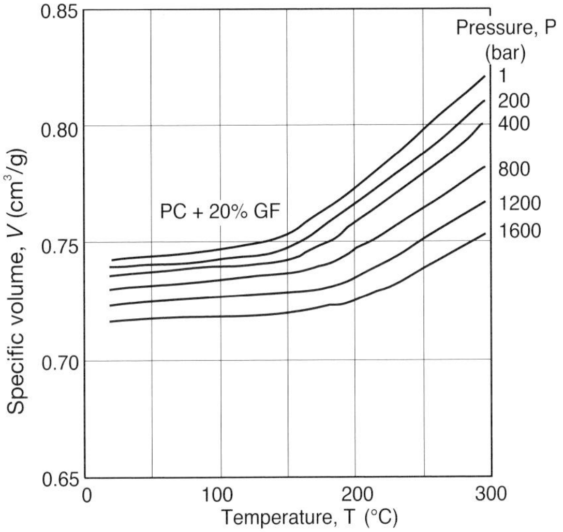

Figure 4.16 p-v-T diagram for a polycarbonate filled with 20% glass fiber. Courtesy of Bayer AG, Germany.

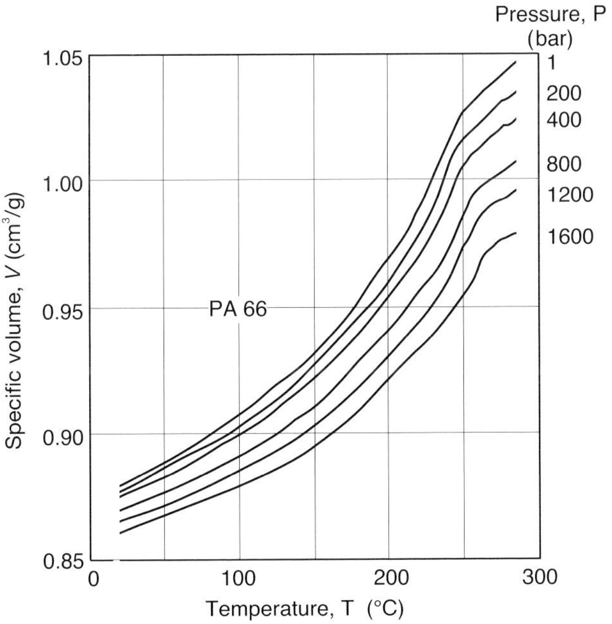

Figure 4.17 p-v-T diagram for a polyamide 66. Courtesy of Bayer AG, Germany.

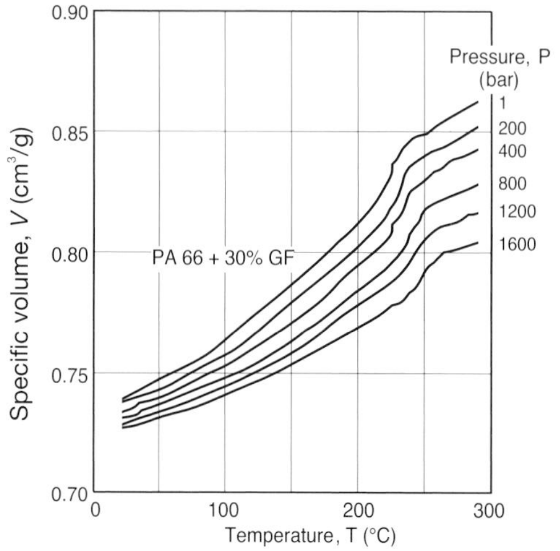

Figure 4.18 p-v-T diagram for a polyamide 66 filled with 30% glass fiber. Courtesy of Bayer AG, Germany.

For amorphous polymers, Eq. 4.12 is valid only for the linear segments (i.e., below or above T_g), and for semi-crystalline polymers it is only valid for temperatures above T_m.

The density of polymers filled with inorganic materials can be computed at any temperature using the following rule of mixtures

$$\rho_c(T) = \frac{\rho_m(T)\rho_f}{\psi\rho_m(T) + (1-\psi)\rho_f} \tag{4.13}$$

where ρ_c, ρ_m and ρ_f are the densities of the composite, polymer and filler, respectively, and ψ the weight fraction of filler.

4.1.4 Thermal Diffusivity

Thermal diffusivity, defined in Eq. 4.4, is the material property that governs the process of thermal diffusion over time. The thermal diffusivity in amorphous thermoplastics decreases with temperature. A small jump is observed around the glass transition temperature due to the decrease in heat capacity at T_g. Figure 4.19 [2] presents the thermal diffusivity for selected amorphous thermoplastics.

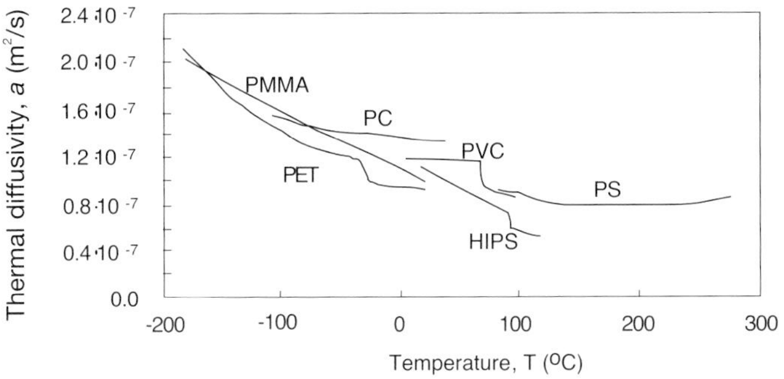

Figure 4.19 Thermal diffusivity as a function of temperature for various amorphous thermoplastics.

A decrease in thermal diffusivity, with increasing temperature, is also observed in semi-crystalline thermoplastics. These materials show a minimum at the melting temperature as demonstrated in Fig. 4.20 [2] for a selected number of semi-crystalline thermoplastics. It has also been observed that the thermal diffusivity increases with increasing degree of crystallinity and that it depends on the rate of crystalline growth, hence, on the cooling speed.

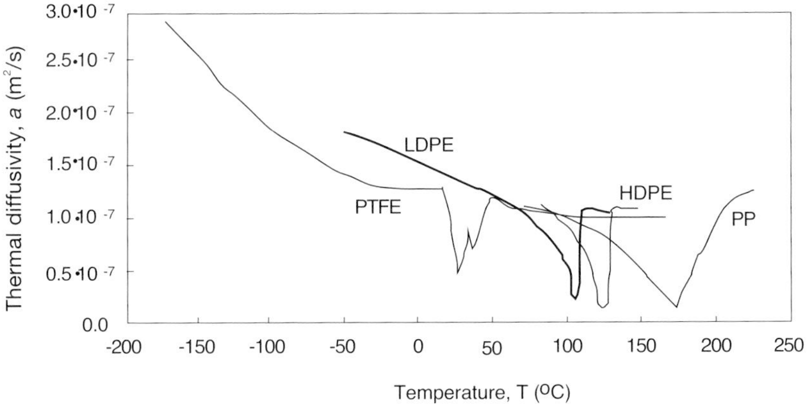

Figure 4.20 Thermal diffusivity as a function of temperature for various semi-crystalline thermoplastics.

4.1.5 Linear Coefficient of Thermal Expansion

The linear coefficient of thermal expansion is related to volume changes that occur in a polymer due to temperature variations and is well represented in the p-v-T diagram. For many materials, thermal expansion is related to the melting temperature of that material, clearly demonstrated for some important polymers in Fig. 4.21. Although the linear coefficient of thermal expansion varies with temperature, it can be considered constant within typical design and processing conditions. It is especially high for polyolefins where it ranges from 1.5 x 10^{-4}/K to 2 x 10^{-4}/K, however fibers and other fillers significantly reduce thermal expansion. A rule of mixtures is sufficient to calculate the thermal expansion coefficient of polymers that are filled with powdery and small particles as well as with short fibers. For this case the rule of mixtures is written as

$$\alpha_c = \alpha_p(1-\phi_f) + \alpha_f \phi_f \tag{4.14}$$

where ϕ_f is the volume fraction of the filler, and α_c, α_p and α_f are coefficients for the composite, the polymer and the filler, respectively. In case of continuous fiber reinforcement, the rule of mixtures presented in Eq. 4.14 applies for the coefficient perpendicular to the reinforcing fibers. In the fiber direction, however, the thermal expansion of the fibers determines the linear coefficient of thermal expansion of the composite. Extensive calculations are necessary to determine coefficients in layered laminated composites.

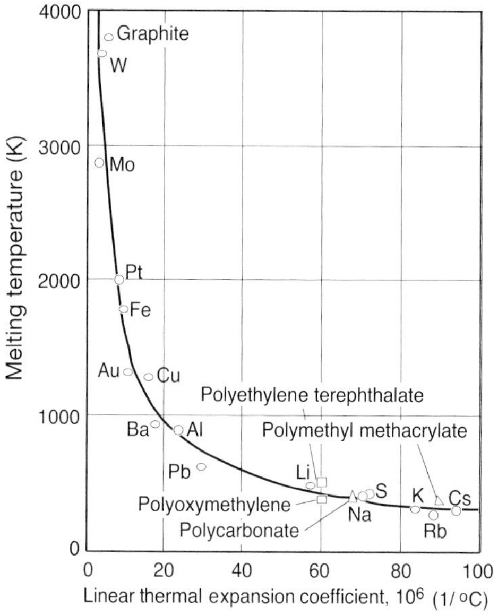

Figure 4.21 Relation between thermal expansion of some metals and plastics at 2 °C and their melting temperature.

4.1.6 Thermal Penetration

In addition to thermal diffusivity, the thermal penetration number is of considerable practical interest. It is given by

$$b = \sqrt{k\, C_p\, \rho} \tag{4.15}$$

If the thermal penetration number is known, the contact temperature T_C which results when two bodies A and B, which are at different temperatures, touch can easily be computed using

$$T_C = \frac{b_A T_A + b_B T_B}{b_A + b_B} \tag{4.16}$$

where T_A and T_B are the temperatures of the touching bodies and b_A and b_B are the thermal penetrations for both materials. The contact temperature is very important for many objects in daily use (e.g., from the handles of heated objects or drinking cups made of plastic, to the heat insulation of space crafts). It is also very important for the calculation of temperatures in tools and molds during polymer processing. The constants used to compute temperature dependent thermal penetration numbers for common thermoplastics are given in Table 4.3 [12].

Table 4.3 Thermal Penetration of Some Plastics

Plastic	**Coefficients to calculate the thermal penetration** $b = a_b T + b_b$ $(\mathrm{W/s^{1/2}/m^2K})$	
	a_b	b_b
HDPE	1.41	441.7
LDPE	0.0836	615.1
PMMA	0.891	286.4
POM	0.674	699.6
PP	0.846	366.8
PS	0.909	188.9
PVC	0.649	257.8

4.1.7 Glass Transition Temperature

The glass transition temperature, T_g, is closely related to the secondary forces. Typical values for the glass transition temperature of common thermoplastics are listed in Chapter 2. If a polymer is mixed with a solvent, the glass transition temperature can be lowered and we can compute it with the following rule of mixtures[4]

$$T_{gM} = \frac{\phi_p \alpha_{tp} T_{gp} - (1-\phi_p) T_{gs}}{\alpha_{tp}\phi_p + (1-\phi_p)\alpha_{ts}} \tag{4.17}$$

where α_{tp} and α_{ts} are the linear coefficients of thermal expansion for the polymer and the solvent, respectively; T_{gM}, T_{gp} and T_{gs} are the glass transition temperatures of the mixture, polymer and solvent, respectively; and ϕ_p is the volume fraction of polymer in the mixture. Stabilizers, plasticizers, and similar auxiliary processing agents work the same way. Usually, the rule of thumb applies that 1% by volume of plasticizer reduces the glass transition temperature by 2 K .

When mixing two incompatible polymers, two glass transition temperatures result, which are visible when measuring the elastic or loss modulus of the polymer blend.

4.1.8 Melting Temperature

The melting temperature, T_m is the highest temperature at which crystalline structures can exist. Above this temperature the polymer can be considered a viscous or viscoelastic liquid, depending on the molecular weight of the polymer and the time scale associated with its deformation. An interesting observation made is the relation between the melting and the glass transition temperatures. This relation can be written as

$$\frac{T_g}{T_m} \approx \frac{2}{3} \tag{4.18}$$

where the temperatures are expressed in degrees Kelvin.

4.2 Measuring Thermal Data

Thanks to modern analytical instruments it is possible to measure thermal data with a high degree of accuracy. These data allow a good insight into chemical and manufacturing processes. Accurate thermal data or properties are necessary for everyday calculations and computer simulations of thermal processes. Such

4 This rule of mixtures only applies for compatible or miscible materials.

analyses are used to design polymer processing installations and to determine and optimize processing conditions.

In the last twenty years several physical thermal measuring devices have been developed to determine thermal data used to analyze processing and polymer component behavior.

4.2.1 Differential Thermal Analysis (DTA)

The differential thermal analysis test serves to examine transitions and reactions which occur on the order between seconds and minutes, and involve a measurable energy differential of less than 0.04 J/g. Usually the measuring is done dynamically (i.e., with linear temperature variations in time). However, in some cases isothermal measurements are also done. The DTA is mainly used to determine the transition temperatures. The principle is shown schematically in Fig. 4.22. Here, the sample, S, and an inert substance, I, are placed in an oven that has the ability to raise its temperature linearly. Two thermocouples that monitor the samples are connected opposite to one another such that no voltage is measured as long as S and I are at the same temperature:

$$\Delta T = T_S - T_I = 0 \tag{4.19}$$

However, if a transition or a reaction occurs in the sample at a temperature, T_c, then heat is consumed or released, in which case $\Delta T \neq 0$. This thermal disturbance in time can be recorded and used to interpret possible information about the reaction temperature, T_C, the heat of transition or reaction, ΔH, or simply about the existence of a transition or reaction.

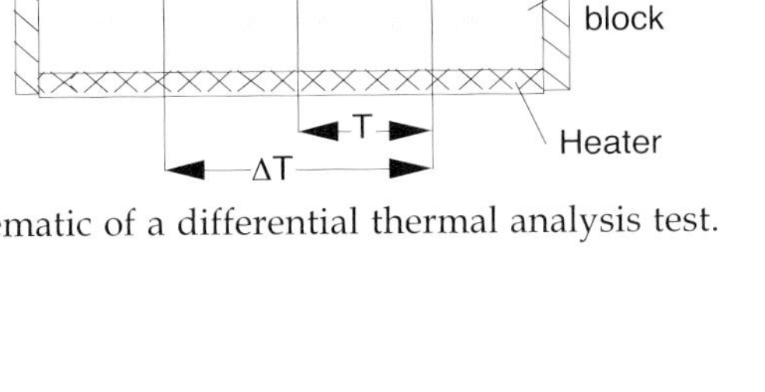

Figure 4.22 Schematic of a differential thermal analysis test.

Figure 4.23 presents the temperature history in a sample with an endothermic melting point (i.e., such as the one that occurs during melting of semi-crystalline polymers). The figure also shows the functions $\Delta T(T_I)$ and $\Delta T(T_S)$ which result from such a test. A comparison between Figs. 4.23 (b) and (c) demonstrates that it is very important to record the sample temperature, T_S, to determine a transition temperature such as the melting or glass transition temperature.

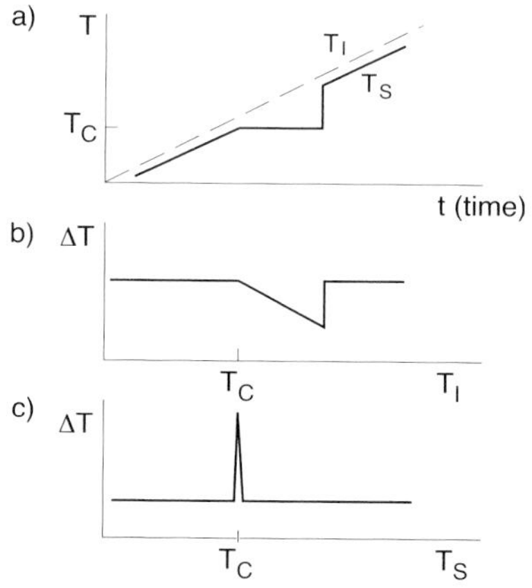

Figure 4.23 Temperature and temperature differences measured during melting of a semi-crystalline polymer sample.

4.2.2 Differential Scanning Calorimeter (DSC)

The differential scanning calorimeter permits us to determine thermal transitions of polymers in a range of temperatures between -180 and +600 °C. Unlike the DTA cell in the DSC device, thermocouples are not placed directly inside the sample or the reference substance. Instead, they are embedded in the specimen holder or stage on which the sample and reference pans are placed; the thermocouples make contact with the containers from the outside. A schematic diagram of a differential scanning calorimeter is very similar to the one shown in Fig. 4.22. Materials that do not show or undergo transition or react in the measuring range (e.g., air, glass powder, etc.) are placed inside the reference container. For standardization one generally uses mercury, tin, or zinc, whose properties are exactly known. In contrast to the DTA test, where samples larger than 10 g are needed, the DSC test requires samples that are in the mg range (< 10 mg). Although DSC tests are less sensitive than the DTA tests, they are the most widely used tests for thermal analysis. In fact, DTA tests are rarely used in the polymer industry.

Figure 4.24 [4] shows a typical DSC curve measured using a partly crystalline polymer sample. In the figure, the area which is enclosed between the trend line (bold) and the base line is a direct measurement for the amount of heat, ΔH, needed for transition. In this case, the transition is melting and the area corresponds to the *heat of fusion*.

The degree of crystallinity, x, is determined from the ratio of the heat of fusion of a polymer sample, ΔH_{SC}, and the enthalpy of fusion of a 100% crystalline sample ΔH_C.

$$\chi = \frac{\Delta H_{SC}}{\Delta H_C} \tag{4.20}$$

In a DSC analysis of a semi-crystalline polymer, a jump in the specific heat curve, as shown in Fig. 4.24, becomes visible; a phenomenon which can easily be traced with a DSC. The glass transition temperature, T_g, is determined at the inflection point of the specific heat curve. The release of residual stresses as a material's temperature is raised above the glass transition temperature is often observed in a DSC analysis.

Specific heat, C_p, is one of the many material properties that can be measured with the DSC. During a DSC temperature sweep, the sample pan and the reference pan are maintained at the same temperature. This allows the measurement of the differential energy required to maintain identical temperatures. The sample with the higher heat capacity will absorb a larger amount of heat, which is proportional to the difference between the heat capacity of the measuring sample and the reference sample.

It is also possible to determine the purity of a polymer sample when additional peaks or curve shifts are detected in a DSC measurement.

Thermal degradation is generally accompanied by an exothermic reaction which may result from oxidation. Such a reaction can easily be detected in a DSC output. By further warming of the test sample, cross-linking may take place and, finally, chain breakage, as shown in Fig. 4.24.

An important aspect in DSC data interpretation is the finite heat flow resistance between the sample pan and the furnace surface. Recent studies by Janeschitz-Kriegl, Eder and co-workers [13, 14] have demonstrated that the heat transfer coefficient between the sample pan and furnace is of finite value, and cannot be disregarded while interpreting the data. In fact, with materials that have a low thermal conductivity, such as polymers, the finite heat transfer coefficient will significantly influence the temperature profiles of the samples.

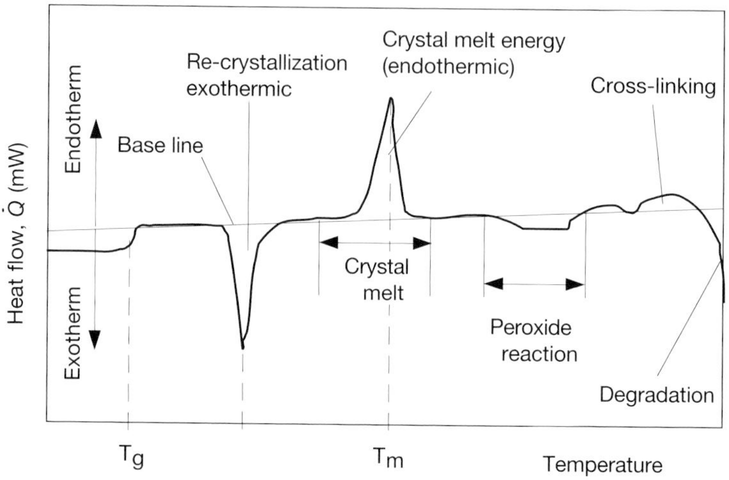

Figure 4.24 Typical DSC heat flow for a semi-crystalline polymer.

4.2.3 Thermomechanical Analysis (TMA)

The thermomechanical analysis (TMA) measures shape stability of a material at elevated temperatures by physically penetrating it with a metal rod. A schematic diagram of TMA equipment is shown in Fig. 4.25. In TMA, the test specimen's temperature is raised at a constant rate, the sample is placed inside the measuring device, and a rod with a specified weight is placed on top of it. To allow for measurements at low temperatures, the sample, oven, and rod can be cooled with liquid nitrogen. The TMA also allows one to measure the Vicat temperature described in Chapter 8.

Most instruments are so precise that they can be used to measure the melting temperature of the material and, by using linear dilatometry, to measure the thermal expansion coefficients. The thermal expansion coefficient can be measured using

$$\alpha_t = \frac{1}{L_0}\frac{\Delta L}{\Delta T} \tag{4.21}$$

where L_0 is the initial dimension of the test specimen, ΔL the change in size and ΔT the temperature difference. For isotropic materials a common relation between the linear and the volumetric thermal expansion coefficient can be used:

$$\gamma = 3\alpha_t \tag{4.22}$$

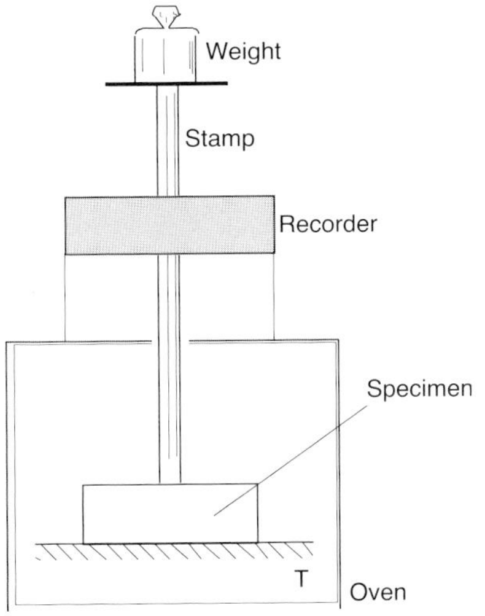

Figure 4.25 Schematic diagram of the thermomechanical analysis (TMA) device.

4.2.4 Thermogravimetry (TGA)

A thermogravimetric analyzer can measure weight changes of less than 10 μg as a function of temperature and time. This measurement technique, typically used for thermal stability, works on the principle of a beam balance. The testing chamber can be heated (up to approximate 1200 °C) and rinsed with gases (inert or reactive). Measurements are performed on isothermal reactions or at temperatures sweeps of less than 100 K/min. The maximum sample weight used in thermogravimetric analyses is 500 mg. Thermogravimetry is often used to identify the components in a blend or a compound based on the thermal stability of each component. Figure 4.26[5] shows results from a TGA analysis on a PVC fabric. The figure clearly shows the transitions at which the various components of the compound decompose. The percent of the original sample weight is recorded along with the change of the weight with respect to temperature. Five transitions representing (1) the decomposition of volatile components, (2) decomposition of the DOP plasticizer, (3) formation of HCl, (4) carbon-carbon scission, and (5) the forming of CO_2, are clearly visible.

5 Courtesy of the ICIPC, Medellín, Colombia.

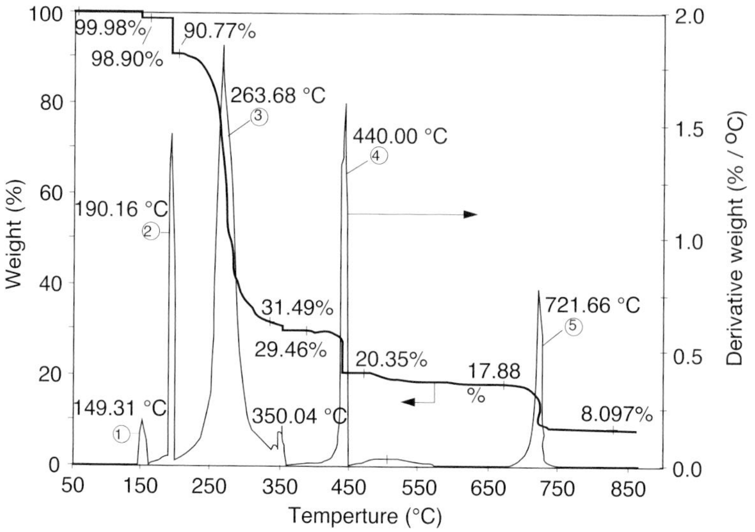

Figure 4.26 TGA analysis on a PVC fabric. (1) volatiles: humidity, monomers, solvents etc., (2) DOP plasticizer, (3) HCl formation, (4) carbon-carbon scission, and (5) CO_2 formation.

4.2.5 Density Measurements

One simple form of calculating the density of a polymer sample is to first weigh the sample immersed in water. Assuming the density of water to be 1.0 g/cm^3 we can use the relation

$$\rho = \frac{m}{(m - m_i)\left(1\,cm^3/g\right)} \tag{4.23}$$

where m is mass of the specimen, m_i is the mass of the immersed specimen and $(m - m_i)$ is the mass of the displaced body of water.

Some common ways of determining density of polymeric materials are described by the ASTM D792, ISO 1183, and DIN 53 479 test methods. Another common way of measuring density is the "through flow density meter." Here, the density of water is changed to that of the polymer by adding ethanol until the plastic shavings are suspended in the solution. The density of the solution is then measured in a device which pumps the liquid through a U-pipe, where it is measured using ultrasound techniques. A density gradient technique is described by the standard ASTM D1505 test method.

Examples

4.1 A differential scanning calorimetry (DSC) test was performed on a 25 mg polyethylene terephthalate (PET) sample taken from the screw-top of a soda bottle. The test was performed using a heating rate of 5K/minute (5K rise every minute). The DSC output is presented in Fig. 4.27.

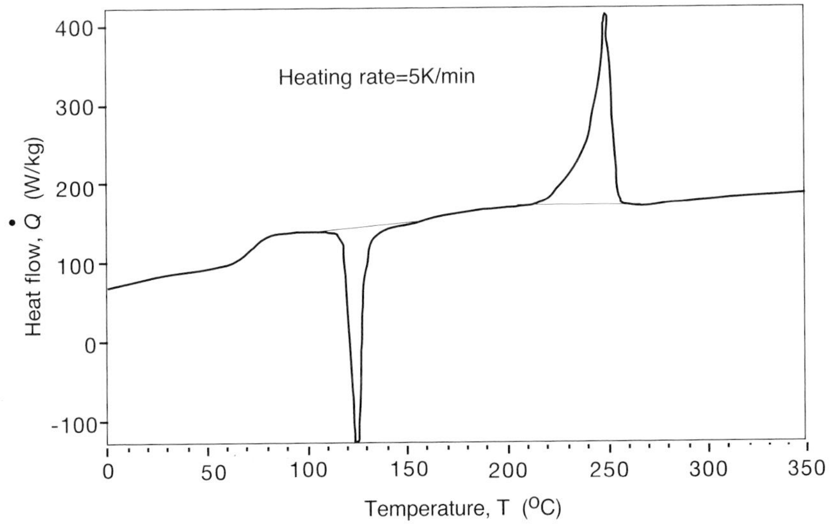

Figure 4.27 DSC scan of a PET bottle screw-top sample

From the curve estimate the glass transition temperature, T_g, the melting temperature, T_m, the crystallization temperature, T_c and the heat of fusion, λ, for this specific PET sample during the temperature ramp-up.

If the heat of fusion for a hypothetically 100% crystalline PET sample is 137 kJ/kg [15], what was the degree of crystallinity in the PET bottle screw-top?

From the curve presented in Fig. 4.27 we can deduce that the glass transition temperature is around 72 °C, the crystallization temperature at 125 °C and the melting temperature at 250 °C. Note that in all three cases there is range of temperatures at which the transition occurs. To compute the heat of fusion during the ramp-up we need to find the area between $\dot{Q}$ and the base-line for the endothermic deviation around the melting point, between 210 and 260 °C. To do this we must first transform the temperature scale to time by dividing it by the heating rate as

$$t = \frac{T}{5\text{K/min}} \frac{60s}{1\text{min}}$$

Hence, 50 °C becomes 600 s, 100 °C becomes 1200 s, etc. The integral equals 37.8 kJ/kg, which represents the heat of fusion of the sample during the temperature ramp-up. However, one must consider that this melting energy includes the extra crystallization that occurs between 108 and 155 °C. The exothermic energy is computed by integrating the curve between those two temperatures in a transformed time scale. The integral equals 22.9 kJ/kg. We can also find the area under the curve by transforming the heat flow, $\dot{Q}$, to heat capacity, C_p, and integrating using the temperature scale instead of a time scale. Heat capacity can be computed using

$$C_p = \frac{\dot{Q}}{5\text{K/min}} \frac{60s}{\text{min}}$$

The degree of crystallinity of the initial PET bottle screw-top can now easily be computed using

$$\chi = \frac{37.8\text{kJ/kg} - 22.9\text{kJ/kg}}{137\text{kJ/kg}} = 0.109 \quad \text{or} \quad 10.9\%$$

Problems

4.1 Does the coefficient of linear expansion of a polymer increase or decrease upon the addition of glass fibers?

4.2 Plot T_g versus T_m for several polymers. What trend or relation do you observe?

4.3 In a soda bottle, how does the degree of crystallinity in the screw-top region compare to the degree of crystallinity in the wall? Explain.

4.4 A 5K/min heating and 5K/min cooling differential scanning calorimetry (DSC) test (Fig. 4.28) was performed on a 10.8 mg sample of PE-LD. What is the specific heat of the PE-LD just after melting during heating and just before crystallization during cooling. What is the degree of crystallinity of the initial and the final samples.

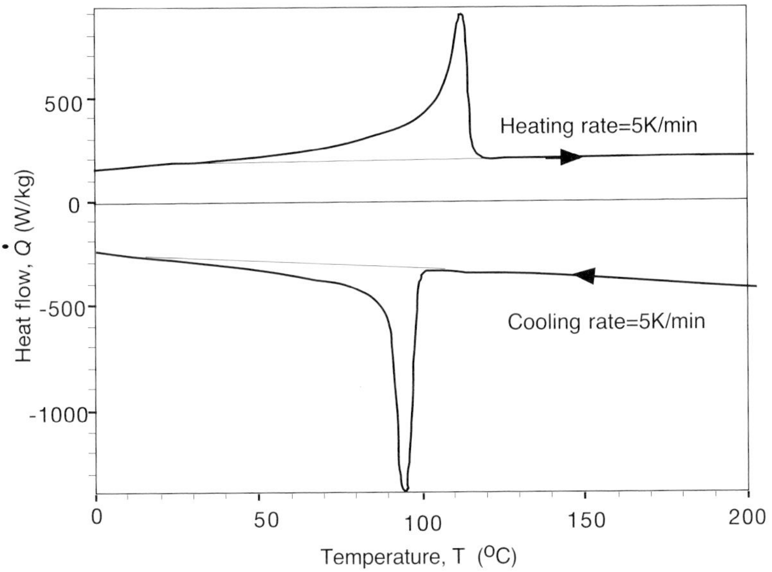

Figure 4.28 Heating and cooling DSC scans of a PE-LD sample

4.5 A differential scanning calorimetry (DSC) test was performed on an 11.4 mg polyethylene terephthalate (PET) sample using the standard ASTM D 3417 test method. The ASTM test calls for a temperature heating rate of 20 °C/minute (20 °C rise every minute). The DSC output is presented below[6].

From the curve in Fig. 4.29 estimate the glass transition temperature, T_g, the melting temperature, T_m, the crystallization temperature, T_c and the heat of fusion, λ, for this specific PET sample during the temperature ramp-up. Note that the heat flow scale has already been transformed to heat capacity. How do T_g and T_m compare to the "book values" ?

If the heat of fusion for a hypothetically 100% crystalline PET is 137 kJ/kg, what was the degree of crystallinity of the original PET sample.

On the same graph below sketch a hypothetical DSC output for the original PET sample with a temperature heating rate that is too fast to allow any additional crystallization during heating.

6 Courtesy of ICIPC, Medellín, Colombia.

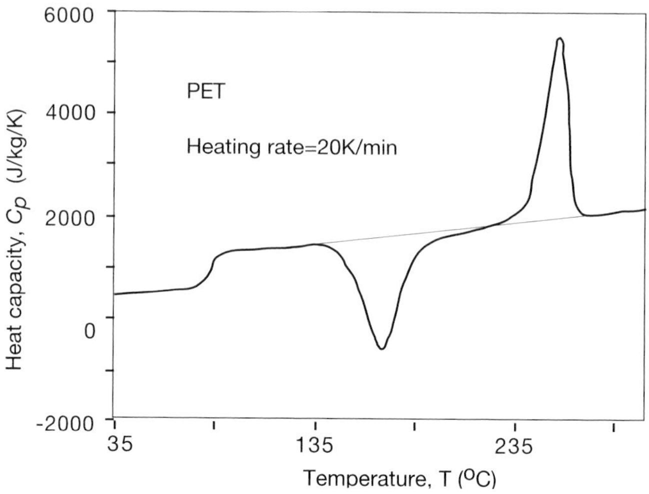

Figure 4.29 DSC scan of a PET sample.

4.6 A typical injection pack/hold pressure during injection molding of polyamide 66 components is 1000 bar and the injection temperature is 280 °C. The gate freezes shut when the average temperature inside the mold reaches 225 °C.

Draw the process on the pvT diagram given in Fig. 4.17.

What volume shrinkage should be taken into account when designing the mold? Note that the shrinkage is mostly taken up by a thickness reduction.

4.7 Isothermal differential scanning calorimetry (DSC) tests were performed on three unsaturated polyester (UPE) samples at three different temperatures (100 °C, 110 °C, and 120 °C). The output for the three DSC tests are presented in the figure below[7]. On the graph, label which curve is associated with which test temperature. From the curves in Fig. 4.30 estimate the total heat of reaction, Q_T.

7 Courtesy of GenCorp Research, Akron, OH.

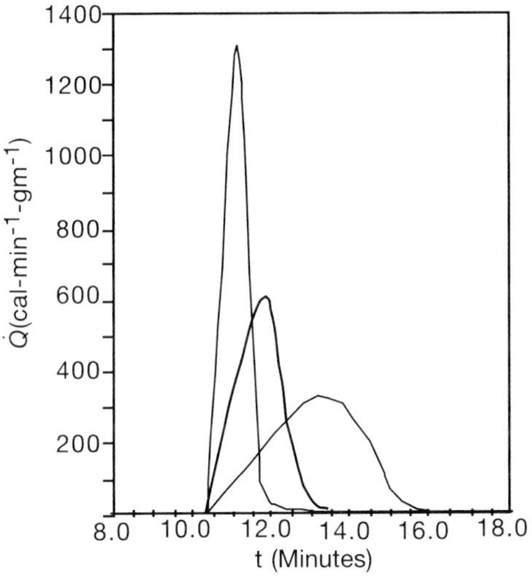

Figure 4.30 Isothermal DSC measurements of UPE samples.

4.8 A differential scanning calorimetry (DSC) test (Fig. 4.31) was performed on an 18.3 mg sample of polystyrene. What is the glass transition temperature of the sample. Determine the specific heat of this PS just before the glass transition temperature has been reached. What is C_p just after T_g? Why is the heat larger as the temperature increases?

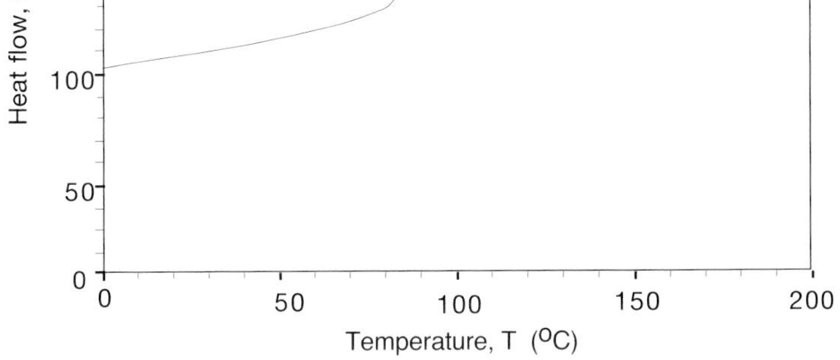

Figure 4.31 DSC scan of a PS sample.

4.9 Sketch the pvT diagrams for a semi-crystalline polymer with a high and a low cooling rate.

References

1. Crawford, R.J., *Plastics Engineering*, 2nd Ed., Pergamon Press, Oxford, (1987).
2. Godovsky, Y.K., *Thermophysical Properties of Polymers*, Springer-Verlag, Berlin, (1992)
3. Mathot, V.B.F., *Calorimetry and Thermal Analysis of Polymers*, Hanser Publishers, Munich, (1994).
4. van Krevelen, D.W., and Hoftyzer, P.J., *Properties of Polymers*, 2nd. Ed., Elsevier, Amsterdam, (1976).
5. Knappe, W., *Kunststoffe, 66*, 5, 297, (1976).
6. Dietz, W., *Kunststoffe, 66,* 3, 161, (1976)
7. Knappe, W., *Adv. Polym. Sci., 7,* 477, (1971).
8. Fischer, F., *Gummi-Asbest-Kunststoffe,* 32, 12, 922, (1979).
9. Maxwell, J.C., *Electricity and Magnetism*, Clarendon Press, Oxford, (1873).
10. Meredith, R.E., and Tobias, C.W., *J. Appl. Phys. ,31*, 1270, (1960).
11. Araujo, F.F.T., and Rosenberg, H.M., *J. Phys. D:Appl. Phys., 9,* 665, (1976).
12. Catic, I., Ph.D. Thesis, IKV, RWTH-Aachen, Germany, (1972).
13. Janeschitz-Kriegl, H., Wippel, H., Paulik, Ch., and Eder, G., *Colloid Polym. Sci.*, 271, 1107, (1993).
14. Wu, C.H., Eder, G., and Janeschitz-Kriegl, H.,*Colloid Polym. Sci.*, 271, 1116, (1993).
15. Tadmor, Z., and Gogos, C.G., *Principles of Polymer Processing,* John Wiley & Sons, New York, (1979).

5

Rheology of Polymer Melts

5.1 Introduction

Rheology is the field of science that studies fluid behavior during flow-induced deformation. From the variety of materials that rheologists study, polymers have been found to be the most interesting and complex. Polymer melts are shear thinning, viscoelastic, and their flow properties are temperature dependent. Viscosity is the most widely used material parameter when determining the behavior of polymers during processing. Since the majority of polymer processes are shear rate dominated, the viscosity of the melt is commonly measured using shear deformation measurement devices. However, there are polymer processes, such as blow molding, thermoforming, and fiber spinning, which are dominated by either elongational deformation or by a combination of shear and elongational deformation. In addition, some polymer melts exhibit significant elastic effects during deformation. This chapter will concentrate on shear deformation models but, for completeness, elongational flows, concentrated suspensions, and viscoelastic fluids will also be covered. Modeling and simulation of polymer flows will be briefly discussed. For further reading on rheology of polymer melts, the reader should consult the literature [1–6]. For more detail on polymer flow and processing simulation the literature [7, 8] should also be reviewed.

5.1.1 Continuum Mechanics

When analyzing the flow or deformation of polymers during processing, a balance of energy, mass and forces must be preserved.

Using the notation found in Fig. 5.1 and performing a mass balance on the differential element, for an incompressible liquid[1] one can write

1 It is clear that a polymer melt is not incompressible. However, assuming incompressibility simplifies the analysis significantly without losing accuracy in flow predictions.

$$\nabla \cdot \underline{\mathbf{v}} = 0 \tag{5.1}$$

where Eq. 5.1 represents the divergence of the velocity vector and is presented in Appendix A in expanded form for various coordinate systems.

Using a similar differential element that is moving with the fluid, as shown in Fig. 5.2, a force balance results in

$$\rho \frac{D\underline{\mathbf{v}}}{Dt} = \left[\nabla \cdot \underline{\underline{\sigma}}\right] + \rho \underline{g} \tag{5.2}$$

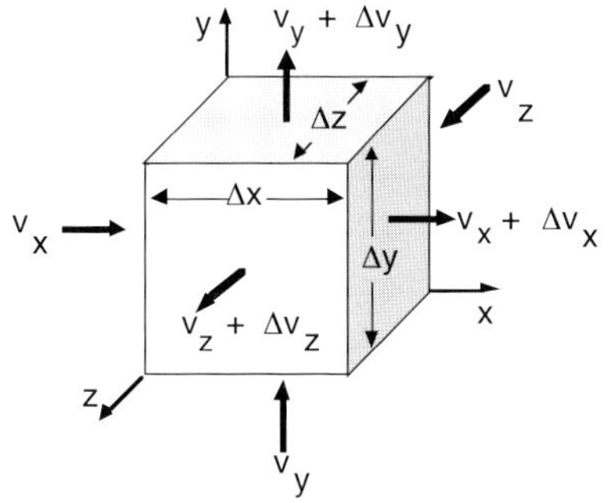

Figure 5.1 Differential element used for mass balance.

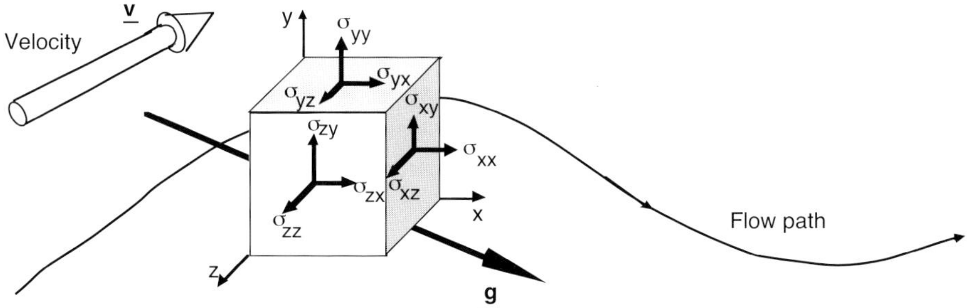

Figure 5.2 Differential element used for force balance.

where the term on the left side represents the inertia of the polymer melt, the first term on the right is the divergence of the stress tensor and represents viscous and elastic forces, and the second term on the right represents the gravitational effects. The operator D/Dt is the *substantial* or *material derivative* defined by

$$\frac{D}{Dt} = \frac{\partial}{\partial t} + \underline{\mathbf{v}} \cdot \nabla \tag{5.3}$$

which represents a convective or embedded frame of reference that moves with a material particle.

The stress, $\underline{\underline{\sigma}}$, sometimes referred to as the *total stress,* can be split into *hydrostatic pressure,* $\underline{\underline{\sigma}}_H$, and *deviatoric stress,* $\underline{\underline{\tau}}$. The hydrostatic pressure is a stress vector that acts only in the normal direction to the surfaces and can be represented by $-p\underline{\underline{\delta}}$, where p is the pressure and $\underline{\underline{\delta}}$ is a unit tensor. The momentum balance can now be written as

$$\rho\frac{D\underline{\mathbf{v}}}{Dt} = -\nabla p + \left[\nabla \cdot \underline{\underline{\tau}}\right] + \rho g \tag{5.4}$$

The full form of Eq. 5.4 is presented in Table VI of the appendix.

5.1.2 The Generalized Newtonian Fluid

The viscosity of most polymer melts is *shear thinning* and temperature dependent. The shear thinning effect is the reduction in viscosity at high rates of deformation. This phenomenon occurs because at high rates of deformation the molecules are stretched out, enabling them to slide past each other with more ease, hence, lowering the bulk viscosity of the melt. Figure 5.3 clearly shows the shear thinning behavior and temperature dependance of the viscosity a specific general purpose polystyrene. To take into consideration these *non-Newtonian effects,* it is common to use a viscosity which is a function of the strain rate and temperature to calculate the stress tensor[2] in Eq. 5.4:

$$\underline{\underline{\tau}} = \eta(\dot{\gamma}, T)\dot{\gamma} \tag{5.5}$$

where η is the viscosity and $\underline{\underline{\dot{\gamma}}}$ the *strain rate* or *rate of deformation tensor* defined by

$$\underline{\underline{\dot{\gamma}}} = \nabla\underline{\mathbf{v}} + \nabla\underline{\mathbf{v}}^t \tag{5.6}$$

where $\nabla\underline{\mathbf{v}}$ represents the velocity gradient tensor. This model describes the *Generalized Newtonian Fluid.* In Eq. 5.5, $\dot{\gamma}$ is the magnitude of the strain rate tensor and can be written as

$$\dot{\gamma} = \sqrt{\frac{1}{2}II} \tag{5.7}$$

where II is the second invariant of the strain rate tensor defined by

$$II = \sum_i \sum_j \dot{\gamma}_{ij}\dot{\gamma}_{ji} \tag{5.8}$$

2 As will be shown later, this is only true when the elastic effects are negligible during deformation of the polymeric material.

The strain rate tensor components in Eq. 5.8 are defined by

$$\dot{\gamma}_{ij} = \frac{\partial v_i}{\partial x_j} + \frac{\partial v_j}{\partial x_i} \tag{5.9}$$

The temperature dependence of the polymer's viscosity is normally factored out as

$$\eta(T,\dot{\gamma}) = f(T)\eta(\dot{\gamma}) \tag{5.10}$$

where for small variations in temperature $f(T)$ can be approximated using an exponential function such as

$$f(T) = \exp\left(-a\left(T - T_0\right)\right). \tag{5.11}$$

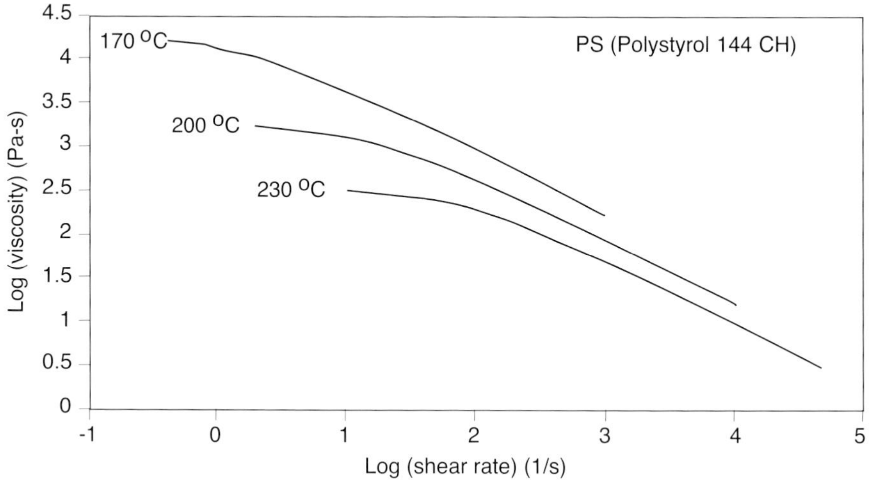

Figure 5.3 Viscosity curves for a polystyrene.

However, as mentioned in Chapter 2, a variation in temperature corresponds to a shift in the time scale. A shift commonly used for semi-crystalline polymers is the Arrhenius shift, which is written as

$$a_T(T) = \frac{\eta_0(T)}{\eta_0(T_0)} = \exp\left(\frac{E_0}{R}\left(\frac{1}{T} - \frac{1}{T_0}\right)\right) \tag{5.12}$$

where E_0 is the activation energy, T_0 a reference temperature, and R the gas constant. Using this shift, one can translate viscosity curves measured at different temperatures to generate a master curve at a specific temperature. Figure 5.4 [9] presents the viscosity of a low density polyethylene with measured values shifted to a reference temperature of 150 °C. For the shift in Fig. 5.4, an activation energy E_0 = 54 kJ/mol was used.

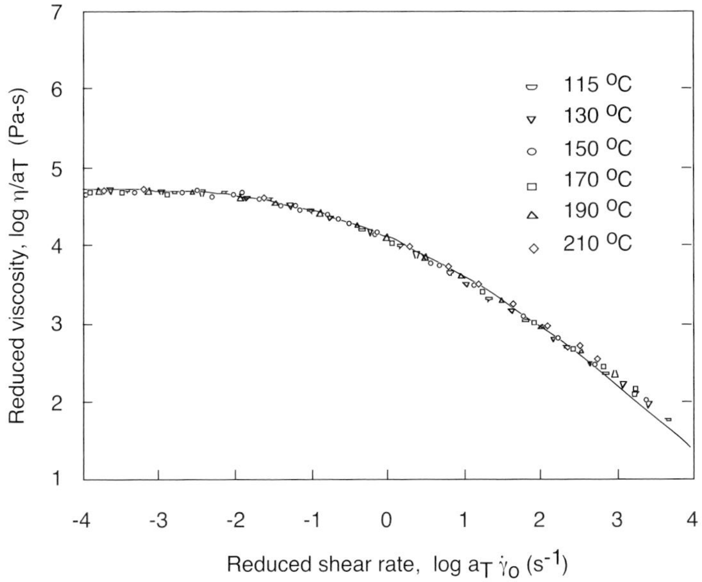

Figure 5.4 Reduced viscosity curve for a low density polyethylene at a reference temperature of 150 °C.

Several models that are used to represent the strain rate dependence of polymer melts are presented later in this chapter.

5.1.3 Normal Stresses in Shear Flow

The tendency of polymer molecules to "curl-up" while they are being stretched in shear flow results in normal stresses in the fluid. For example, shear flows exhibit a deviatoric stress defined by

$$\tau_{xy} = \eta(\dot{\gamma})\dot{\gamma}_{xy} \tag{5.13}$$

Measurable normal stress differences, $N_1 = \tau_{xx} - \tau_{yy}$ and $N_2 = \tau_{yy} - \tau_{zz}$ are referred to as the *first* and *second normal stress differences*. The first and second normal stress differences are material dependent and are defined by

$$N_1 = \tau_{xx} - \tau_{yy} = -\psi_1(\dot{\gamma}, T)\dot{\gamma}_{xy}^2 \tag{5.14}$$

$$N_2 = \tau_{yy} - \tau_{zz} = -\psi_2(\dot{\gamma},T)\dot{\gamma}_{xy}^2 \tag{5.15}$$

The material functions, ψ_1 and ψ_2, are called the primary and secondary normal stress coefficients, and are also functions of the magnitude of the strain rate tensor and temperature. The first and second normal stress differences do not change in sign when the direction of the strain rate changes. This is reflected in Eqs. 5.14 and 5.15. Figure 5.5 [9] presents the first normal stress difference coefficient for the low density polyethylene melt of Fig. 5.4 at a reference temperature of 150 °C. The second normal stress difference is difficult to measure and is often approximated by

$$\psi_2(\dot{\gamma}) \approx -0.1\psi_1(\dot{\gamma}) \tag{5.16}$$

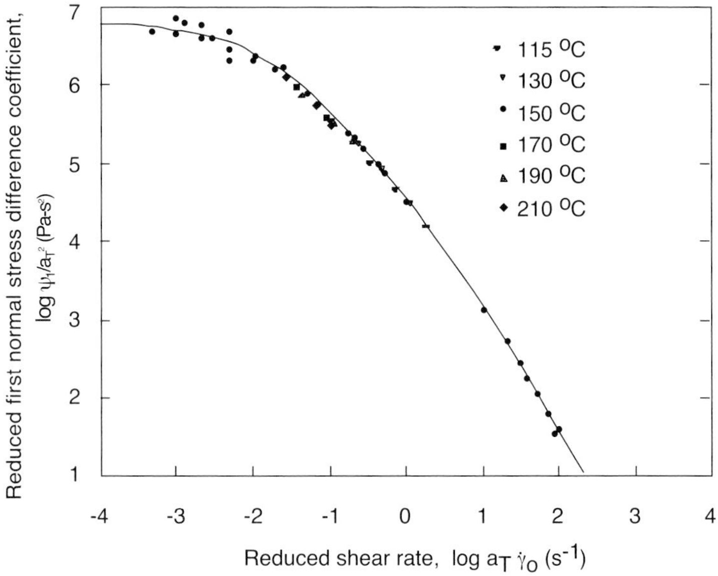

Figure 5.5 Reduced first normal stress difference coefficient for a low density polyethylene melt at a reference temperature of 150 °C.

5.1.4 Deborah Number

A useful parameter often used to estimate the elastic effects during flow is the Deborah number[3], De. The Deborah number is defined by

3 From the Song of Deborah, Judges 5:5—"The mountains flowed before the Lord." M. Reiner is credited for naming the Deborah number; *Physics Today*, (January 1964).

$$De = \frac{\lambda}{t_p} \tag{5.17}$$

where λ is the relaxation time of the polymer and t_p is a characteristic process time. The characteristic process time can be defined by the ratio of characteristic die dimension and average speed through the die. A Deborah number of zero represents a viscous fluid and a Deborah number of ∞ an elastic solid. As the Deborah number becomes > 1, the polymer does not have enough time to relax during the process, resulting in possible extrudate dimension deviations or irregularities such as *extrudate swell* [4], *shark skin*, or even *melt fracture*.

Although many factors affect the amount of extrudate swell, fluid "memory" and normal stress effects are the most significant ones. However, abrupt changes in boundary conditions, such as the separation point of the extrudate from the die, also play a role in the swelling or cross section reduction of the extrudate. In practice, the fluid memory contribution to die swell can be mitigated by lengthening the land length of the die. This is schematically depicted in Fig. 5.6. A long die land separates the polymer from the manifold for enough time to allow it to "forget" its past shape.

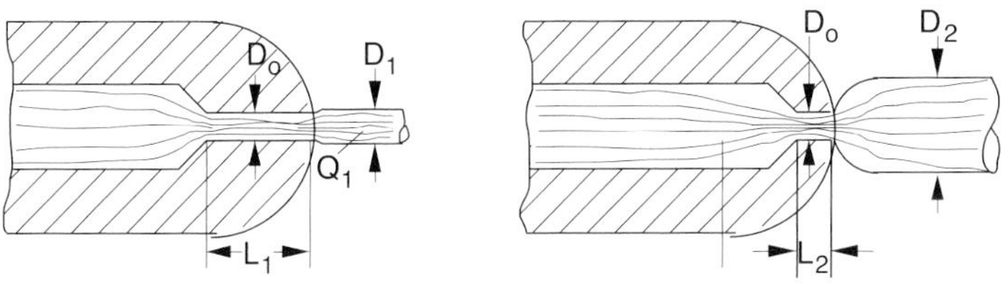

Figure 5.6 Schematic diagram of extrudate swell during extrusion.

Waves in the extrudate may also appear as a result of high speeds during extrusion, where the polymer is not allowed to relax. This phenomenon is generally referred to as *shark skin* and is shown for a high density polyethylene in Fig. 5.7a [10]. It is possible to extrude at such high speeds that an intermittent separation of melt and inner die walls occurs as shown in Fig. 5.7b. This phenomenon is often referred to as the *stick-slip effect* or *spurt flow* and is attributed to high shear stresses between the polymer and the die wall. This phenomena occurs when the shear stress is near the critical value of 0.1 MPa [11–13]. If the speed is further increased, a helical geometry is extruded as

4 It should be pointed out that Newtonian fluids, which do not experience elastic or normal stress effects, also show some extrudate swell or reduction. A Newtonian fluid that is being extruded at high shear rates reduces its cross-section to 87% of the diameter of the die, whereas if extruded at very low shear rate it swells to 113% of the diameter of the die. This swell is due to inertia effects caused by the change from the parabolic velocity distribution inside the die to the flat velocity distribution of the extrudate.

shown for a polypropylene extrudate in Fig. 5.7c. Eventually, the speeds are so high that a chaotic pattern develops such as the one shown in Fig. 5.7d. This well known phenomenon is called *melt fracture*. The shark skin effect is frequently absent and spurt flow seems to occur only with linear polymers.

The critical shear stress has been reported to be independent of the melt temperature but to be inversely proportional to the weight average molecular weight [14, 12]. However, Vinogradov et al. [15] presented results where the critical stress was independent of molecular weight except at low molecular weights. Dealy and co-workers [16], and Denn [17] give an extensive overview of various melt fracture phenomena which is recommended reading.

To summarize, the Deborah number and the size of the deformation imposed upon the material during processing determine how the system can most accurately be modeled. Figure 5.8 [1] helps visualize the relation between time scale, deformation and applicable model. At small Deborah numbers, the polymer can be modeled as a Newtonian fluid, and at very high Deborah numbers the material can be modeled as a Hookean solid. In between, the viscoelastic region is divided in two: the linear viscoelastic region for small deformations, and the non-linear viscoelastic region for large deformations. Linear viscoelasticity was briefly discussed in Chapter 2.

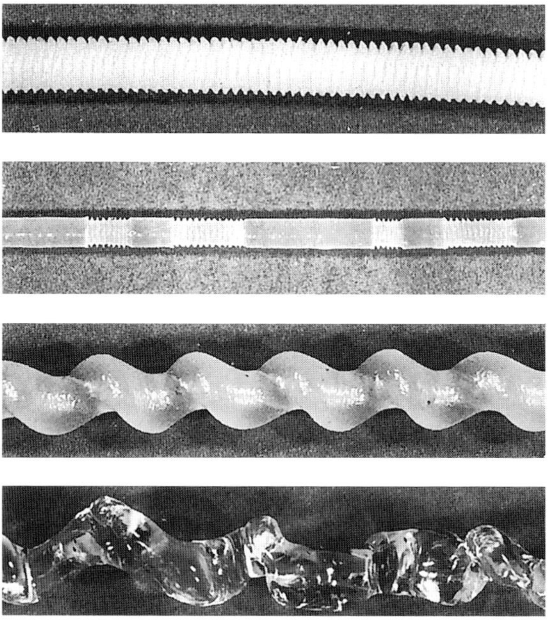

(a)

(b)

(c)

(d)

Figure 5.7 Various shapes of extrudates under melt fracture.

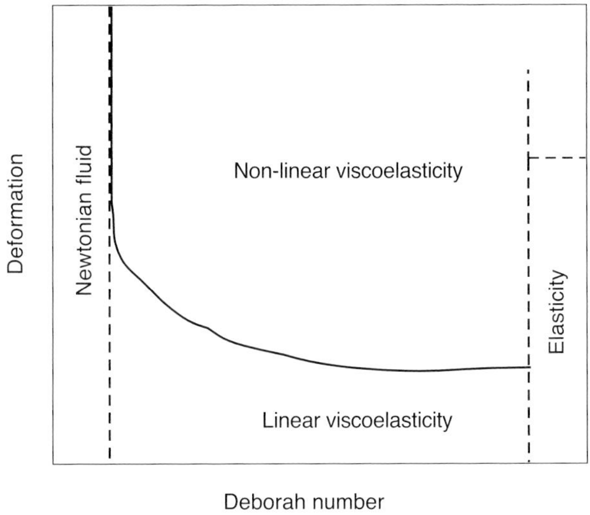

Figure 5.8 Schematic of Newtonian, elastic, linear, and non-linear viscoelastic regimes as a function of deformation and Deborah number during deformation of polymeric materials.

5.2 Viscous Flow Models

Strictly speaking, the viscosity η, measured with shear deformation viscometers, should not be used to represent the elongational terms located on the diagonal of the stress and strain rate tensors. Elongational flows are briefly discussed later in this chapter. A rheologist's task is to find the models that best fit the data for the viscosity represented in Eq. 5.5. Some of the models used by polymer processors on a day-to-day basis to represent the viscosity of industrial polymers are presented in the next section.

5.2.1 The Power Law Model

The power law model proposed by Ostwald [18] and de Waale [19] is a simple model that accurately represents the shear thinning region in the viscosity versus strain rate curve but neglects the Newtonian plateau present at small strain rates. The power law model can be written as follows:

$$\eta = m(T)\dot{\gamma}^{n-1} \tag{5.18}$$

where m is referred to as the *consistency index* and n the *power law index*. The consistency index may include the temperature dependence of the viscosity such as represented in Eq. 5.11, and the power law index represents the shear thinning behavior of the polymer melt. Figure 5.9 presents normalized velocity distributions inside a tube for a fluid with various power law indices calculated using the power law model. It should be noted that the limits of this model are

$$\eta \rightarrow 0 \; as \; \dot{\gamma} \rightarrow \infty \text{ and}$$

$$\eta \rightarrow \infty \; as \; \dot{\gamma} \rightarrow 0$$

The infinite viscosity at zero strain rates leads to an erroneous result in problems where there is a region of zero shear rate, such as at the center of a tube. This results in a predicted velocity distribution that is flatter at the center than the experimental profile. In computer simulation of polymer flows, this problem is often overcome by using a truncated model such as

$$\eta = m_0(T)\dot{\gamma}^{n-1} \quad \text{for} \quad \dot{\gamma} > \dot{\gamma}_0 \text{ and} \tag{5.19a}$$

$$\eta = m_0(T) \quad \text{for} \quad \dot{\gamma} \le \dot{\gamma}_0 \tag{5.19b}$$

where m_0 represents a zero shear rate $(\dot{\gamma}_0)$ viscosity. Table 5.1 presents a list of typical power law and consistency indices for common thermoplastics.

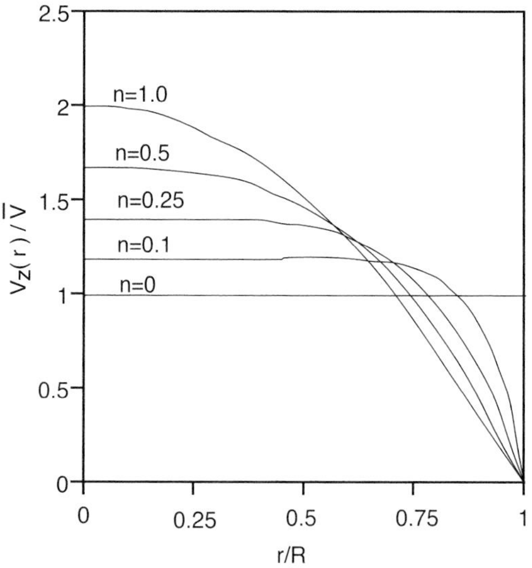

Figure 5.9 Pressure flow velocity distributions inside a tube for fluids with various power law indices.

Table 5.1 Power Law and Consistency Indices for Common Thermoplastics

Polymer	m(Pa-s^n)	n	T(^{o}C)
Polystyrene	2.80 x10^4	0.28	170
High density polyethylene	2.00 x10^4	0.41	180
Low density polyethylene	6.00 x10^3	0.39	160
Polypropylene	7.50 x10^3	0.38	200
Polyamide 66	6.00 x10^2	0.66	290
Polycarbonate	6.00 x10^2	0.98	300
Polyvinyl chloride	1.70 x10^4	0.26	180

5.2.2 The Bird–Carreau–Yasuda Model

A model that fits the whole range of strain rates was developed by Bird and Carreau [20] and Yasuda [21] and contains five parameters:

$$\frac{\eta - \eta_0}{\eta_0 - \eta_\infty} = \left[1 + |\lambda\dot{\gamma}|^a\right]^{(n-1)/a} \tag{5.20}$$

where η_0 is the zero shear rate viscosity, η_∞ is an infinite shear rate viscosity, λ is a time constant and n is the power law index. In the original Bird–Carreau model, the constant a = 2. In many cases, the infinite shear rate viscosity is negligible, reducing Eq. 5.20 to a three parameter model. Equation 5.20 was modified by Menges, Wortberg and Michaeli [22] to include a temperature dependence using a WLF relation. The modified model, which is used in commercial polymer data banks, is written as follows:

$$\eta = \frac{k_1 a_T}{\left[1 + k_2 \dot{\gamma} a_T\right]^{k_3}} \tag{5.21}$$

where the shift a_T applies well for amorphous thermoplastics and is written as

$$\ln a_T = \frac{8.86(k_4 - k_5)}{101.6 + k_4 - k_5} - \frac{8.86(T - k_5)}{101.6 + T - k_5} \tag{5.22}$$

Table 5.2 presents constants for Carreau–WLF (amorphous) and Carreau–Arrhenius models (semi-crystalline) for various common thermoplastics. In addition to the temperature shift, Menges, Wortberg and Michaeli [22] measured a pressure dependence of the viscosity and proposed the following model, which includes both temperature and pressure viscosity shifts:

$$\log\eta(T,p)=\log\eta_0+\frac{8.86\left(T^*-T_s\right)}{101.6+T^*-T_s}-\frac{8.86\left(T^*-T_s+0.02p\right)}{101.6+\left(T^*-T_s+0.02p\right)} \quad (5.23)$$

where p is in bar, and the constant 0.02 represents a 2 °C shift per bar.

Table 5.2 Constants for Carreau–WLF (Amorphous) and Carreau-Arrhenius (Semi-Crystalline) Models for Various Common Thermoplastic

Polymer	K1 (Pa-s)	K2 (s)	K3	K4 (°C)	K5 (°C)	T0 (°C)	E0 (J/mol)
Polystyrene	1777	0.064	0.73	200	123	-	-
High density polyethylene	24198	1.38	0.60	-	-	200	22272
Low density polyethylene	317	0.015	0.61	-	-	189	43694
Polypropylene	1386	0.091	0.68	-	-	220	427198
Polyamide 66	44	0.00059	0.40	-	-	300	123058
Polycarbonate	305	0.00046	0.48	320	153	-	-
Polyvinyl chloride	1786	0.054	0.73	185	88	-	-

5.2.3 The Bingham Fluid

The Bingham fluid is an empirical model that represents the rheological behavior of materials that exhibit a "no flow" region below certain yield stresses, τ_Y, such as polymer emulsions and slurries. Since the material flows like a Newtonian liquid above the yield stress, the Bingham model can be represented by

$$\eta=\infty \quad \text{or} \quad \dot{\gamma}=0 \quad \tau\le\tau_Y \quad (5.24a)$$

$$\eta=\mu_0+\frac{\tau_y}{\dot{\gamma}} \quad \tau\ge\tau_Y \quad (5.24b)$$

Here, τ is the magnitude of the deviatoric stress tensor and is computed in the same way as in Eq. 5.7.

5.2.4 Elongational Viscosity

In polymer processes such as fiber spinning, blow molding, thermoforming, foaming, certain extrusion die flows, and compression molding with specific processing conditions, the major mode of deformation is elongational.

To illustrate elongational flows, consider the fiber spinning process shown in Fig. 5.10.

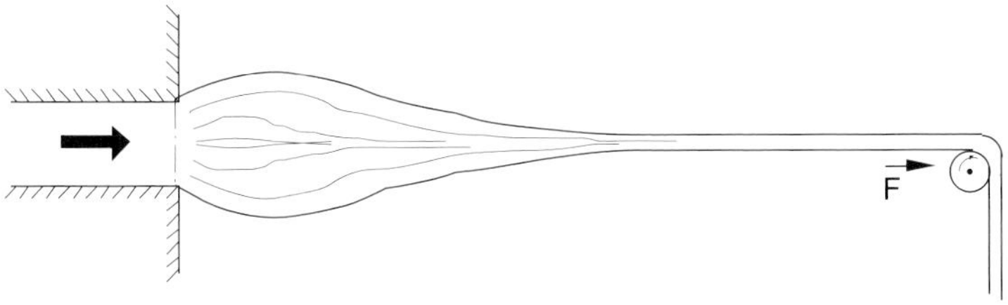

Figure 5.10 Schematic diagram of a fiber spinning process.

A simple elongational flow is developed as the filament is stretched with the following components of the rate of deformation:

$$\dot{\gamma}_{11} = -\dot{\varepsilon} \tag{5.25a}$$

$$\dot{\gamma}_{22} = -\dot{\varepsilon} \tag{5.25b}$$

$$\dot{\gamma}_{33} = -2\dot{\varepsilon} \tag{5.25c}$$

where $\dot{\varepsilon}$ is the elongation rate, and the off-diagonal terms of $\dot{\gamma}_{ij}$ are all zero. The diagonal terms of the total stress tensor can be written as

$$\sigma_{11} = -p - \eta\dot{\varepsilon} \tag{5.26a}$$

$$\sigma_{22} = -p - \eta\dot{\varepsilon} \tag{5.26b}$$

and

$$\sigma_{33} = -p + 2\eta\dot{\varepsilon} \tag{5.26c}$$

Since the only outside forces acting on the fiber are in the axial or 3 direction, for the Newtonian case, σ_{11} and σ_{12} must be zero. Hence,

$$p = -\eta\dot{\varepsilon} \quad \text{and} \tag{5.27}$$

$$\sigma_{33} = 3\eta\dot{\varepsilon} = \overline{\eta}\,\dot{\varepsilon} \tag{5.28}$$

which is known as *elongational viscosity or Trouton viscosity* [23]. This is analogous to elasticity where the following relation between elastic modulus, E, and shear modulus, G, can be written

$$\frac{E}{G} = 2(1 + v) \tag{5.29}$$

where v is Poisson's ratio. For the incompressibility case, where $v = 0.5$, Eq. 5.29 reduces to

$$\frac{E}{G} = 3 \tag{5.30}$$

Figure 5.11 [24] shows shear and elongational viscosities for two types of polystyrene. In the region of the Newtonian plateau, the limit of 3, shown in Eq. 5.28, is quite clear. Figure 5.12 presents plots of elongational viscosities as a function of stress for various thermoplastics at common processing conditions. It should be emphasized that measuring elongational or extensional viscosity is an extremely difficult task. For example, in order to maintain a constant strain rate, the specimen must be deformed uniformly exponentially. In addition, a molten polymer must be tested completely submerged in a heated neutrally buoyant liquid at constant temperature.

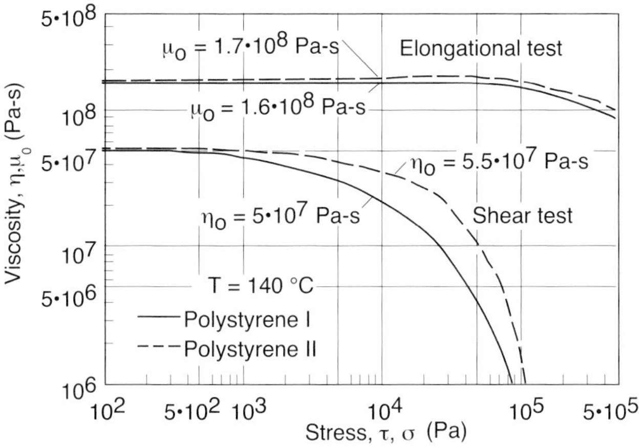

Figure 5.11 Shear and elongational viscosity curves for two types of polystyrene.

5.2.5 Rheology of Curing Thermosets

A curing thermoset polymer has a conversion or cure dependent viscosity that increases as the molecular weight of the reacting polymer increases. For the vinyl ester whose curing history[5] is shown in Fig. 5.13 [25], the viscosity behaves as shown in Fig. 5.14 [25]. Hence, a complete model for viscosity of a reacting polymer must contain the effects of strain rate, $\dot{\gamma}$, temperature, T, and degree of cure, c, such as

$$\eta = \eta(\dot{\gamma}, T, c) \tag{5.31}$$

[5] A more in-depth view of curing and solidification processes of thermosetting polymers is given in Chapter 7.

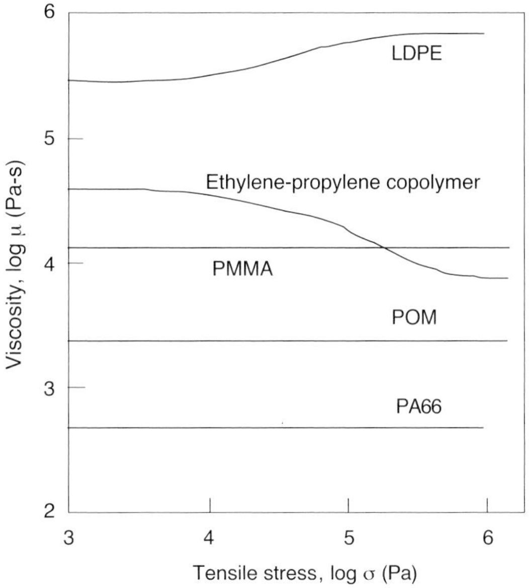

Figure 5.12 Elongational viscosity curves as a function of tensile stress for several thermoplastics.

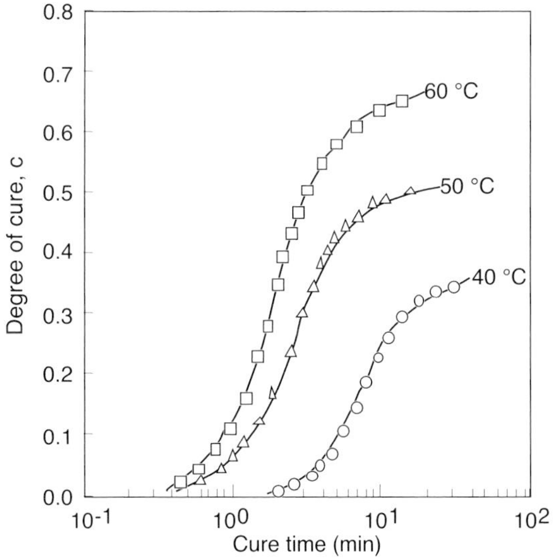

Figure 5.13 Degree of cure as a function of time for a vinyl ester at various isothermal cure temperatures.

There are no generalized models that include all these variables for thermosetting polymers. However, extensive work has been done on the viscosity of polyurethanes [26, 27] used in the reaction injection molding process. An empirical relation which models the viscosity of these mixing-activated polymers, given as a function of temperature and degree of cure, is written as

$$\eta = \eta_0 e^{E/RT} \left(\frac{c_g}{c_g - c} \right)^{C_1 + C_2 c} \tag{5.32}$$

where E is the activation energy of the polymer, R is the ideal gas constant, T is the temperature, c_g is the gel point[6], c the degree of cure, and C_1 and C_2 are constants that fit the experimental data. Figure 5.15 shows the viscosity as a function of time and temperature for a 47% MDI-BDO P(PO-EO) polyurethane, and Fig. 5.16 shows the viscosity as a function of degree of cure.

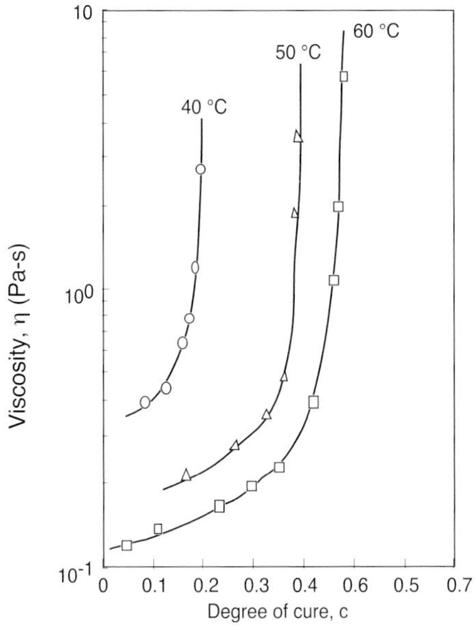

Figure 5.14 Viscosity as a function of degree of cure for a vinyl ester at various isothermal cure temperatures.

6 At the gel point the change of the molecular weight with respect to the degree of cure goes to infinity. Hence, it can be said that at this point all the molecules are interconnected.

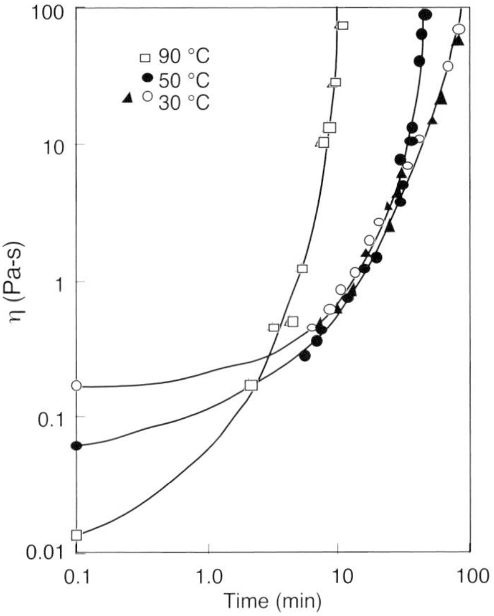

Figure 5.15 Viscosity as a function of time for a 47% MDI-BDO P(PO-EO) polyurethane at various isothermal cure temperatures.

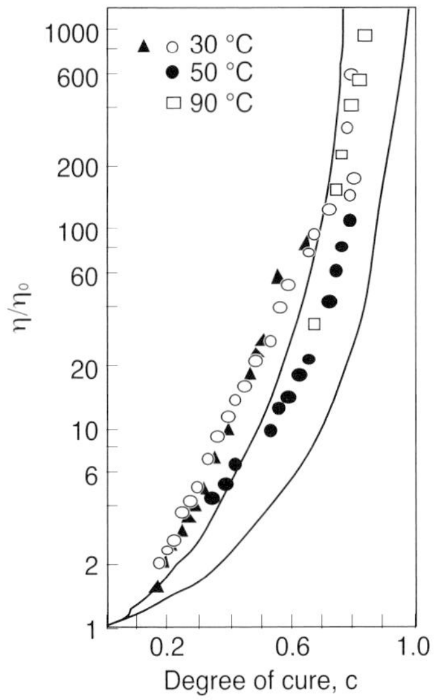

Figure 5.16 Viscosity as a function of degree of cure for a 47% MDI-BDO P(PO-EO) polyurethane at various isothermal cure temperatures.

5.2.6 Suspension Rheology

Particles suspended in a material, such as in filled or reinforced polymers, have a direct effect on the properties of the final article and on the viscosity during processing. Numerous models have been proposed to estimate the viscosity of filled liquids [28–32]. Most models proposed are a power series of the form[7]

$$\frac{\eta_f}{\eta_0} = 1 + a_1\phi + a_2\phi^2 + a_3\phi^3 + \dots \tag{5.33}$$

The linear term in Eq. 5.33 represents the narrowing of the flow passage caused by the filler that is passively entrained by the fluid and sustains no deformation as shown in Fig. 5.17. For instance, Einstein's model, which only includes the linear term with $a_1 = 2.5$, was derived based on a viscous dissipation balance. The quadratic term in the equation represents the first-order effects of interaction between the filler particles. Geisbüsch suggested a model with a yield stress and, where the strain rate of the melt increases by a factor κ as

$$\eta_f = \frac{\tau_0}{\dot{\gamma}} + \kappa\eta_0(\kappa\dot{\gamma}) \tag{5.34}$$

For high deformation stresses, which are typical in polymer processing, the yield stress in the filled polymer melt can be neglected. Figure 5.18 compares Geisbüsch's experimental data to Eq. 5.33 using the coefficients derived by Guth [30]. The data and Guth's model seem to agree well. A comprehensive survey on particulate suspensions was recently given by Gupta [33], and on short-fiber suspensions by Milliken and Powell [34].

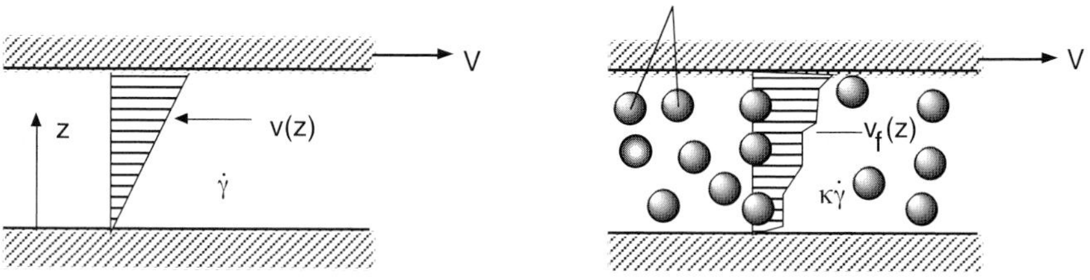

Figure 5.17 Schematic diagram of strain rate increase in a filled system.

---

7 The model which best fits experimental data is the one given by Guth (1938): $\frac{\eta_f}{\eta_0} = 1 + 2.5\phi + 14.1\phi^2$

However, a full analysis of the first-order particle interactions gives an analytical value for the quadratic term of 6.96.

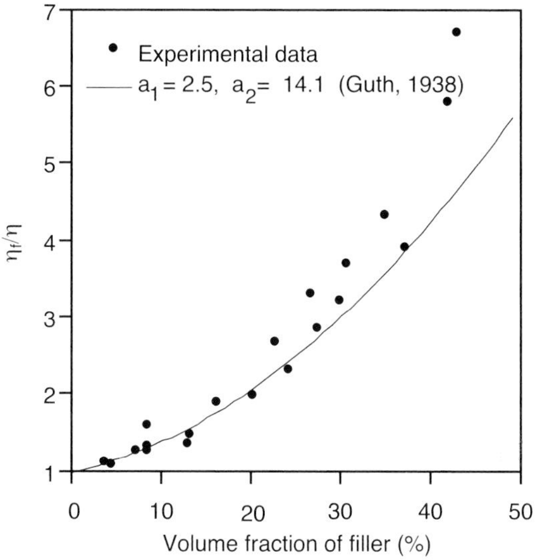

Figure 5.18 Viscosity increase as a function of volume fraction of filler for polystyrene and low density polyethylene containing spherical glass particles with diameters ranging between 36 μm and 99.8 μm.

5.3 Simplified Flow Models Common in Polymer Processing

Many polymer processing operations can be modeled using simplified geometries and material models. This section presents several isothermal flow models in simple geometries using a Newtonian viscosity and a power-law viscosity as described in Eq. 5.18. Although it is very common to simplify analyses of polymer processes by assuming isothermal conditions, most operations are non-isothermal since they include melting and are often influenced by viscous dissipation. Hence, the temperature of the polymer melt lies between T_g (amorphous polymers) or T_m (semi-crystalline polymers) and the heater temperature, T_w[8], and often exceeds T_w due to viscous dissipation. An estimate of the maximum temperature rise due to viscous heating is

$$\Delta T_{max} = \frac{\eta v_0^2}{8k} \tag{5.35}$$

where v_0 represents a characteristic velocity in the flow system, such as plate velocity in a simple shear flow, and k represents the thermal conductivity. To

8 The subscript w is often used for "wall."

estimate the importance of viscous dissipation, the Brinkman number, Br, is often computed

$$Br = \frac{\eta v_0^2}{k\left(T_w - T_{g\ or\ m}\right)} \tag{5.36}$$

The Brinkman number is the ratio of the heat generated via viscous dissipation and the heat conducted out of the system via conduction. A Brinkman number > 1 implies that the temperature field will be affected by viscous dissipation. The choice of temperatures in the denominator of the equation depends on the type of material and problem being analyzed.

5.3.1 Simple Shear Flow

Simple shear flows as represented in Fig. 5.19 are very common in polymer processing, such as inside extruders as well as in certain coating flows. The flow field in simple shear is the same for all fluids and is described by

$$v_z = v_0 \frac{y}{h} \tag{5.37}$$

for the velocity distribution and

$$Q = \frac{v_0 h W}{2} \tag{5.38}$$

for the volumetric throughput, where W represents the width of the plates.

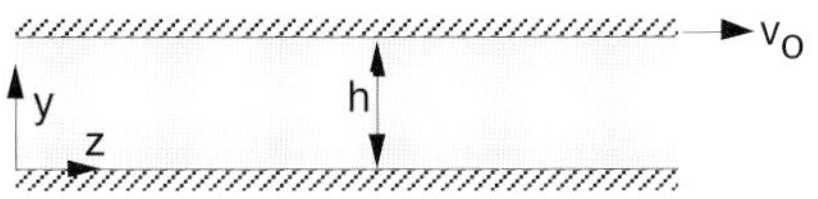

Figure 5.19 Schematic diagram of a simple shear flow.

5.3.2 Pressure Flow Through a Slit

The pressure flow through a slit, as depicted in Fig. 5.20, is encountered in flows through extrusion dies or inside injection molds. The flow field for a Newtonian fluid is described by

$$v_z(y) = \left(\frac{h^2 \Delta p}{8 \mu L}\right)\left[1 - \left(\frac{2y}{h}\right)^2\right] \tag{5.39}$$

and

$$Q = \frac{W h^3 \Delta p}{12 \mu L} \tag{5.40}$$

for the net volumetric throughput. When using the power-law model the flow field is described by

$$v_z(y) = \left(\frac{h}{2(s+1)}\right)\left(\frac{h \Delta p}{2mL}\right)^s \left[1 - \left(\frac{2y}{h}\right)^{s+1}\right] \tag{5.41}$$

and

$$Q = \frac{W h^2}{2(s+2)}\left(\frac{h \Delta p}{2mL}\right)^s \tag{5.42}$$

for the throughput, where $s = 1/n$, and n the power-law index.

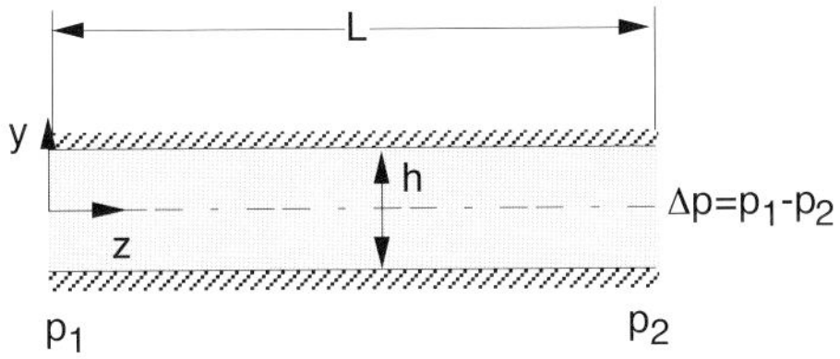

Figure 5.20 Schematic diagram of pressure flow through a slit.

5.3.3 Pressure Flow through a Tube - Hagen-Poiseuille Flow

Pressure flow through a tube (Fig. 5.21), or Hagen-Poiseuille flow, is the type that exists inside the runner system in injection molds, as well as extrusion dies and inside the capillary viscometer. For a Newtonian fluid, the flow field inside a tube is described by

$$v_z(r) = \frac{R^2 \Delta p}{4 \mu L}\left[1 - \left(\frac{r}{R}\right)^2\right] \tag{5.43}$$

and

$$Q = \frac{\pi R^4 \Delta p}{8 \mu L} \tag{5.44}$$

for the throughput.

Using the power law model, the flow though a tube is described by

$$v_z(r) = \frac{R}{1+s}\left(\frac{R\Delta p}{2mL}\right)^s\left[1-\left(\frac{r}{R}\right)^{s+1}\right] \tag{5.45}$$

and

$$Q = \left(\frac{\pi R^3}{s+3}\right)\left(\frac{R\Delta p}{2mL}\right)^s \tag{5.46}$$

for the throughput.

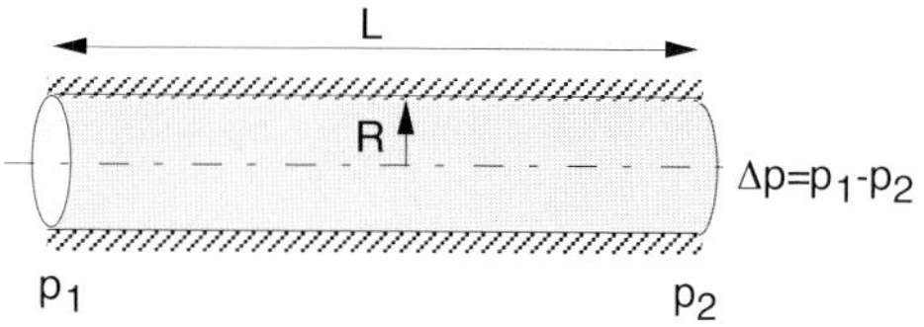

Figure 5.21 Schematic diagram of pressure flow through a tube.

5.3.4 Couette Flow

The Couette device, as depicted in Fig. 5.22, is encountered in bearings and in certain types of rheometers. It is also used as a simplified flow model for mixers and extruders. The Newtonian flow field in a Couette device is described by

$$v_\theta(r) = \frac{\Omega}{\kappa^2 - 1}\left(\frac{R_0^2 - r^2}{r}\right) \tag{5.47}$$

where $\kappa = R_0/R_i$. Using the power law model, the flow field inside a Couette device is described by

$$v_\theta(r) = \frac{\Omega}{\kappa^{2s} - 1}\left(\frac{R_0^{2s} - r^{2s}}{r^{2s-1}}\right) \tag{5.48}$$

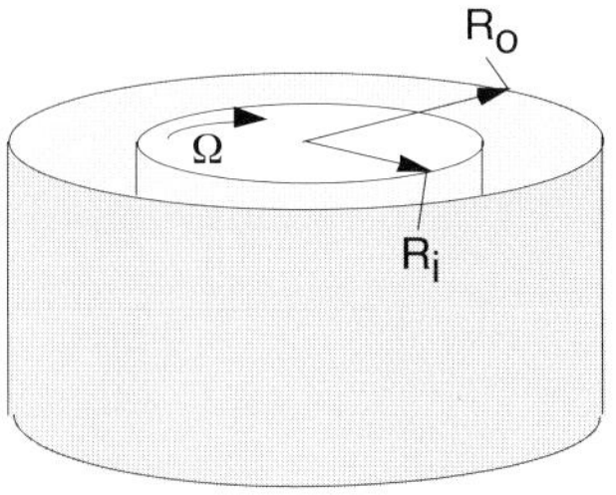

Figure 5.22 Schematic diagram of a Couette device

5.4 Viscoelastic Flow Models

Viscoelasticity has already been introduced in Chapter 3, based on linear viscoelasticity. However, in polymer processing large deformations are imposed on the material, requiring the use of non-linear viscoelastic models. There are two types of general non-linear viscoelastic flow models: the differential type and the integral type.

5.4.1 Differential Viscoelastic Models

Differential models have traditionally been the choice for describing the viscoelastic behavior of polymers when simulating complex flow systems. Many differential viscoelastic models can be described with the general form

$$\mathbf{Y}\underline{\underline{\tau}} + \lambda_1 \underline{\underline{\tau}}_{(1)} + \lambda_2\left(\underline{\underline{\dot{\gamma}}} \cdot \underline{\underline{\tau}} + \underline{\underline{\tau}} \cdot \underline{\underline{\dot{\gamma}}}\right) + \lambda_3\left(\underline{\underline{\tau}} \cdot \underline{\underline{\tau}}\right) = \eta_0 \underline{\underline{\dot{\gamma}}} \tag{5.49}$$

where $\underline{\underline{\tau}}_{(1)}$ is the *first contravariant convected time derivative* of the deviatoric stress tensor and represents rates of change with respect to a convected coordinate system that moves and deforms with the fluid. The *convected derivative* of the deviatoric stress tensor is defined as

$$\underline{\underline{\tau}}_{(1)} = \frac{D\underline{\underline{\tau}}}{D_t} - \left(\nabla \underline{v}^t \cdot \underline{\underline{\tau}} + \underline{\underline{\tau}} \cdot \nabla \underline{v}\right) \tag{5.50}$$

The constants in Eq. 5.49 are defined in Table 5.3 for various viscoelastic models commonly used to simulate polymer flows. A recent review by Bird and Wiest [6] gives a more complete list of existing viscoelastic models.

The *upper convective model* and the *White-Metzner model* are very similar with the exception that the White-Metzner model incorporates the strain rate effects of the relaxation time and the viscosity. Both models provide a first order approximation to flows in which shear rate dependence and memory effects are important. However, both models predict zero second normal stress coefficients. The *Giesekus model* is molecular-based, non-linear in nature and describes the power law region for viscosity and both normal stress coefficients. The *Phan-Thien Tanner models* are based on network theory and give non-linear stresses. Both the Giesekus and Phan-Thien Tanner models have been successfully used to model complex flows.

Table 5.3 Definition of Constants in Eq. 5.49

Constitutive model	Y	λ_1	λ_2	λ_2
Generalized Newtonian	1	0	0	0
Upper convected Maxwell	1	λ_1	0	0
White–Metzner	1	$\lambda_1(\dot{\gamma})$	0	0
Phan-Thien Tanner-1	$\exp\left(-\varepsilon\left(\lambda / \eta_0\right) tr\underline{\underline{\tau}}\right)$	λ	$\frac{1}{2}\xi\lambda$	0
Phan-Thien Tanner-2	$1-\varepsilon\left(\lambda / \eta_0\right) tr\underline{\underline{\tau}}$	λ	$\frac{1}{2}\xi\lambda$	0
Giesekus	1	λ_1	0	$-(\alpha\lambda_1 / \eta_0)$

An overview of numerical simulation of viscoelastic flow systems and an extensive literature review on the subject was given by Keunings [35], and detail on numerical implementation of viscoelastic models are given by Crochet et al. [36] and Debbaut et al. [37]. As an example of the application of differential models to predict flow of polymeric liquids, it is worth mentioning recent work by Dietsche and Dooley [38], who evaluated the White–Metzner, the Phan-Thien Tanner-1, and the Giesekus models by comparing finite element[9] and experimental results of the flow inside multi-layered coextrusion dies. Figure 5.23 [39] presents the progression of a matrix of colored circular polystyrene strands flowing in an identical polystyrene matrix down a channel with a square cross section of 0.95 x 0.95 cm. The cuts in the figure are shown at intervals of 7.6 cm.

The circulation pattern caused by the secondary normal stress differences inside non-circular dies were captured well by the Phan-Thien Tanner and Giesekus models but, as expected, not by the White–Metzner model. Figure 5.24 presents flow patterns predicted by the Phan-Thien Tanner model along with the experimental rearrangement of 165 initially horizontal layers of polystyrene in square, rectangular, and tear-drop shaped dies[10]. In all three

9 For their simulation they used the commercially available code POLYFLOW.

10 These geometries are typical for distribution manifolds used in sheeting dies.

cases, the shape of the circulation patterns were predicted accurately. The flow simulation of the square die predicted a velocity on the order of 10^{-5} m/s along the diagonal of the cross section, which was in agreement with the experimental results. Also worth mentioning is work recently done by Baaijens [40], who evaluated the Phan-Thien Tanner models 1 and 2, and the Giesekus models. He compared finite element results to measured isochromatic birefringence patterns using complex experiments with polymer melts and solutions. His simulation results predicted the general shape of the measured birefringence patterns. He found that at high Deborah numbers, the Phan-Thien Tanner models converged much more easily than the Giesekus model.

Figure 5.23 Polystyrene strand profile progression in a square die.

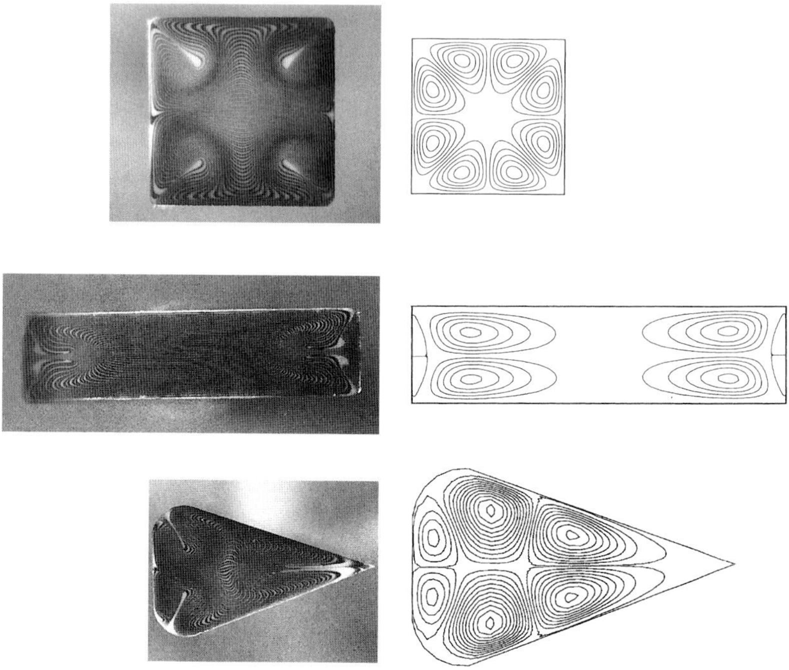

Figure 5.24 Comparison between experimental and predicted flow patterns of polystyrene in square, rectangular, and tear-drop shaped dies.

5.4.2 Integral Viscoelastic Models

Integral models with a memory function have been widely used to describe the viscoelastic behavior of polymers and to interpret their rheological measurements [9,41,42]. In general one can write the single integral model as

$$\underline{\underline{\tau}} = \int_{-\infty}^{t} M(t-t')\underline{\underline{S}}(t')dt' \tag{5.51}$$

where $M(t-t')$ is a memory function and a deformation dependent tensor defined by

$$\underline{\underline{S}}(t') = \phi_1(I_{1,}I_2)\underline{\underline{\gamma}}_{[0]} + \phi_2(I_{1,}I_2)\underline{\underline{\gamma}}^{[0]} \tag{5.52}$$

where I_1 and I_2 are the first invariant of the Cauchy and Finger strain tensors respectively.

Table 5.4 [43–47] defines the constants ϕ_1 and ϕ_2 for various models. In Eq.

5.52, $\underline{\underline{\gamma}}_{[0]}$ and $\underline{\underline{\gamma}}^{[0]}$ are the *finite strain tensors* given by

$$\underline{\underline{\gamma}}_{[0]} = \underline{\underline{\Delta}}^{t} \cdot \underline{\underline{\Delta}} - \underline{\underline{\delta}} \quad \text{and} \tag{5.53}$$

$$\underline{\underline{\gamma}}^{[0]} = \underline{\underline{\delta}} - \underline{\underline{E}} \cdot \underline{\underline{E}}^{t} \tag{5.54}$$

Table 5.4. Definition of Constants in Eq. 5.52

Constitutive model	ϕ_1	ϕ_2
Lodge rubber-like liquid	1	0
K-BKZ*	$\dfrac{\partial W}{\partial I_1}$	$\dfrac{\partial W}{\partial I_2}$
Wagner**	$\exp(-\beta)\sqrt{\alpha I_1 + (1-\alpha) I_2 - 3}$	0
Papanastasiou–Scriven–Macosko***	$\dfrac{\alpha}{(\alpha-3) + \beta I_1 + (1-\beta) I_2}$	0

* W(I1,I2) represents a potential function which can be derived from empiricisms or molecular theory.

** Wagner's model is a special form of the K-BKZ model

*** The Papanastasiou-Scriven Macosko model is also a special form of the K-BKZ model.

The terms and are displacement gradient tensors[11] defined by

$$\Delta_{ij} = \frac{\partial x_i'(x,t,t')}{\partial x_j} \quad \text{and} \tag{5.55}$$

$$E_{ij} = \frac{\partial x_i(x',t',t)}{\partial x_j'} \tag{5.56}$$

where the components Δ_{ij} measure the displacement of a particle at past time t' relative to its position at present time t, and the terms E_{ij} measure the material displacements at time t relative to the positions at time t'.

A memory function $M(t-t')$ which is often applied and which leads to commonly used constitutive equations is written as

$$M(t-t') = \sum_{k=1}^{n} \frac{\eta_k}{\lambda_k^{2}} \exp\left(-\frac{t-t'}{\lambda_k}\right) \tag{5.57}$$

11 Another combination of the displacement gradient tensors which are often used are the *Cauchy strain tensor* and the *Finger strain tensor* defined by $\underline{\underline{B}}^{-1} = \underline{\underline{\Delta}}^{t} \cdot \underline{\underline{\Delta}}$ and $\underline{\underline{B}} = \underline{\underline{E}} \cdot \underline{\underline{E}}^{t}$ respectively.

where λ_k and η_k are relaxation times and viscosity coefficients at the reference temperature T_{ref}, respectively.

Once a memory function has been specified one can calculate several material functions using [4]

$$\eta(\dot{\gamma}) = \int_0^\infty M(s)s(\phi_1 + \phi_2)ds \tag{5.58}$$

$$\psi_1(\dot{\gamma}) = \int_0^\infty M(s)s^2(\phi_1 + \phi_2)ds \quad \text{and} \tag{5.59}$$

$$\psi_2(\dot{\gamma}) = \int_0^\infty M(s)s^2(\phi_2)ds \tag{5.60}$$

For example, Figs. 5.25 and 5.26 present the measured [48] viscosity and first normal stress difference data, respectively, for three blow molding grade high density polyethylenes along with a fit obtained from the Papanastasiou–Scriven–Macosko [47] form of the K-BKZ equation. A memory function with a relaxation spectrum of 8 relaxation times was used. The coefficients used to fit the data are summarized in Table 5.5 [41]. The viscosity and first normal stress coefficient data presented in Figs. 5.4 and 5.5 where fitted with the Wagner [9] form of the K-BKZ equation [9]. Luo and Mitsoulis used the K-BKZ model with the data in Table 5.5 to simulate the flow of HDPE through annular dies. Figure 5.27 [41] shows simulation results for a converging, a straight, and a diverging die geometry. The results shown in Fig. 5.27 were in good agreement with experimental results[12].

Table 5.5 Material Parameter Values in Eq. 5.57 for Fitting Data of High Density Polyethylene Melts at 170 °C

k	λ_k (s)	η_k (Pa-s)
1	0.0001	52
2	0.001	148
3	0.01	916
4	0.1	4210
5	1.0	8800
6	10.0	21,200
7	100.0	21,000
8	1000.0	600

12 The quality of the agreement between experiment and simulation varied between the resins.

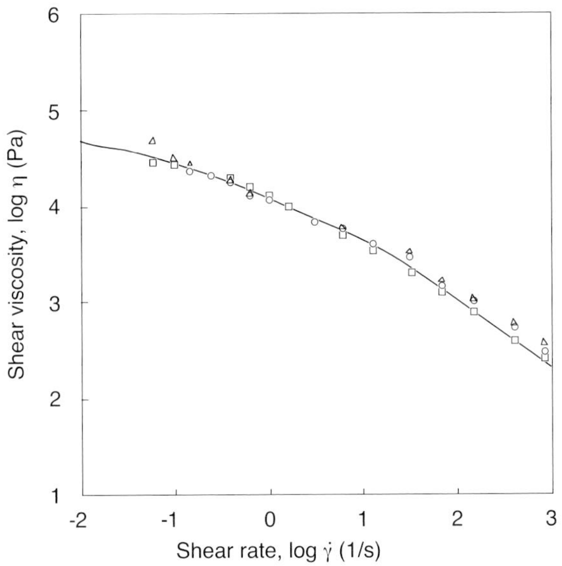

Figure 5.25 Measured and predicted shear viscosity for various high density polyethylene resins at 170 °C.

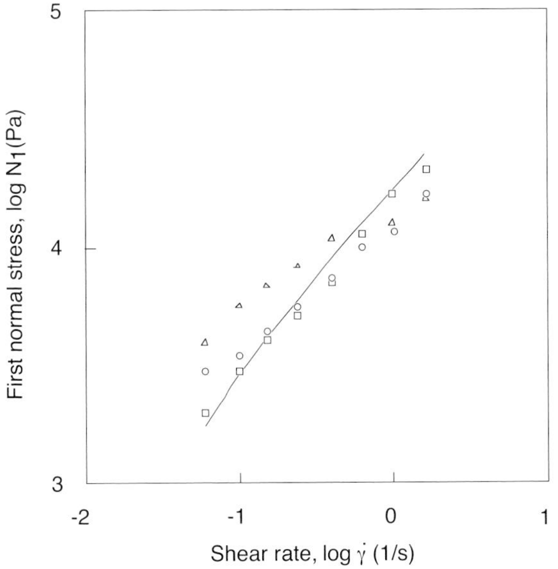

Figure 5.26 Measured and predicted first normal stress difference for various high density polyethylene resins at 170 °C.

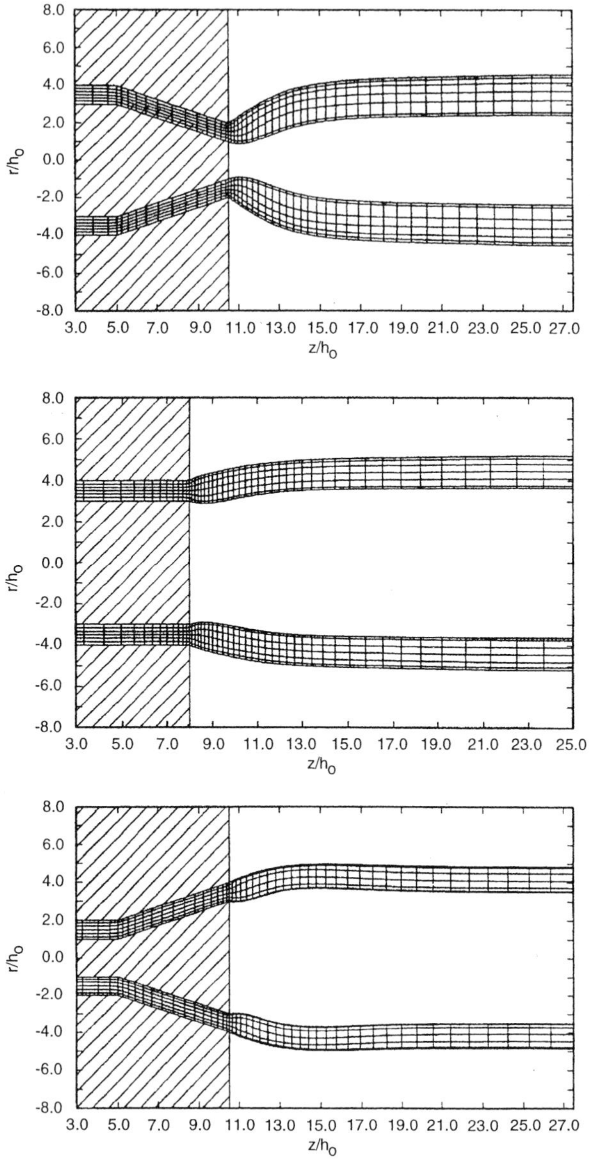

Figure 5.27 Predicted extrudate geometry for: (a) converging, (b) straight, and (c) diverging annular dies.

5.5 Rheometry

In industry there are various ways to qualify and quantify the properties of the polymer melt. The techniques range from simple analyses for checking the consistency of the material at certain conditions, to more complex measurements to evaluate viscosity, and normal stress differences. This section includes three such techniques, to give the reader a general idea of current measuring techniques.

5.5.1 The Melt Flow Indexer

The melt flow indexer is often used in industry to characterize a polymer melt and as a simple and quick means of quality control. It takes a single point measurement using standard testing conditions specific to each polymer class on a ram type extruder or extrusion plastometer as shown in Fig. 5.28. The standard procedure for testing the flow rate of thermoplastics using a extrusion plastometer is described in the ASTM D1238 test [49]. During the test, a sample is heated in the barrel and extruded from a short cylindrical die using a piston actuated by a weight. The weight of the polymer in grams extruded during the 10-minute test is the melt flow index (MFI) of the polymer.

5.5.2 The Capillary Viscometer

The most common and simplest device for measuring viscosity is the capillary viscometer. Its main component is a straight tube or capillary, and it was first used to measure the viscosity of water by Hagen [50] and Poiseuille [51]. A capillary rheometer has a pressure driven flow for which the velocity gradient or strain rate and also the shear rate will be maximum at the wall and zero at the center of the flow, making it a non-homogeneous flow.

Since pressure driven viscometers employ non-homogeneous flows, they can only measure steady shear functions such as viscosity, $\eta(\dot{\gamma})$. However, they are widely used because they are relatively inexpensive to build and simple to operate. Despite their simplicity, long capillary viscometers give the most accurate viscosity data available. Another major advantage is that the capillary rheometers has no free surfaces in the test region, unlike other types of rheometers such as the cone and plate rheometers, which we will discuss in the next section. When the strain rate dependent viscosity of polymer melts is measured, capillary rheometers may be the only satisfactory method of obtaining such data at shear rates > 10 s^{-1}. This is important for processes with higher rates of deformation like mixing, extrusion, and injection molding. Because its design is basic and it only needs a pressure head at its entrance, capillary rheometers can easily be attached to the end of a screw- or ram-type extruder for on-line measurements. This makes the capillary viscometer an efficient tool for industry.

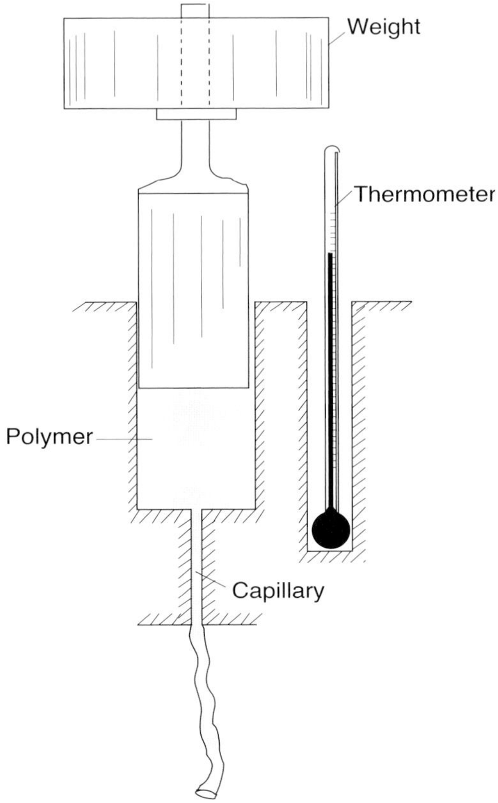

Figure 5.28 Schematic diagram of an extrusion plastometer used to measure melt flow index.

The basic features of the capillary rheometers are shown in Fig. 5.29. A capillary tube of radius R and length L is connected to the bottom of a reservoir. Pressure drop and flow rate through this tube are used to determine the viscosity.

To derive the viscosity relation, the following assumptions are made:

- no velocity in the r and θ directions,
- the polymer is incompressible, and
- fully developed, steady, isothermal, laminar flow.

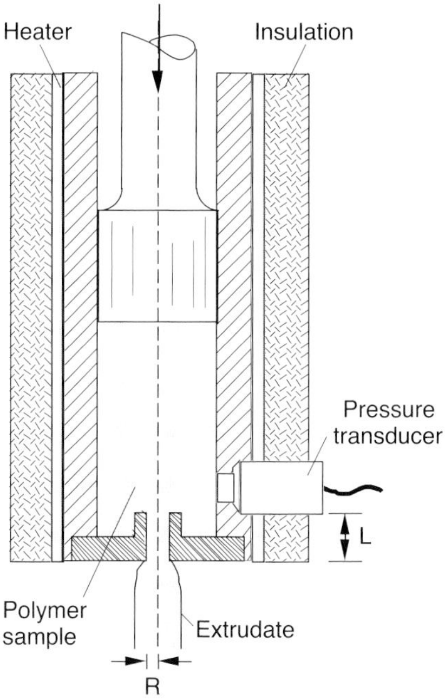

Figure 5.29 Schematic diagram of a capillary viscometer.

The capillary viscometer can be modeled using the z-component of the equation of motion in terms of stress, τ, as

$$0 = \frac{dp}{dz} + \frac{1}{r}\frac{d}{dr}(r\tau_{rz}) \tag{5.61}$$

where,

$$\frac{dp}{dz} = \frac{p_0 - p_L}{L} \tag{5.62}$$

Integrating for the shear stress term gives:

$$\tau_{rz} = \frac{(p_0 - p_L)r}{2L} + \frac{C_1}{r} \tag{5.63}$$

The constant C_1 is taken to be zero since the stress can not be infinite at the tube axis.

5.5.3 Computing Viscosity Using the Bagley and Weissenberg-Rabinowitsch Equations

At the wall the shear stress is:

$$\tau_{r=R} = \tau_w = \frac{R}{2}\frac{(p_0 - p_L)}{L} = \frac{R}{2}\frac{\Delta p}{L} \tag{5.64}$$

Equation 5.64 requires that the capillary be sufficiently long to assure a fully developed flow where end effects are insignificant. However, due to end effects the actual pressure profile along the length of the capillary exhibits a curvature. The effect is shown schematically in Fig. 5.30 [52] and was corrected by Bagley [53] using the end correction e as,

$$\tau_w = \frac{1}{2}\frac{(p_0 - p_L)}{\left(L/R + e\right)} \tag{5.65}$$

The correction factor e at a specific shear rate can be found by plotting pressure drop for various capillary L/D ratios as shown in Fig. 5.31 [52].

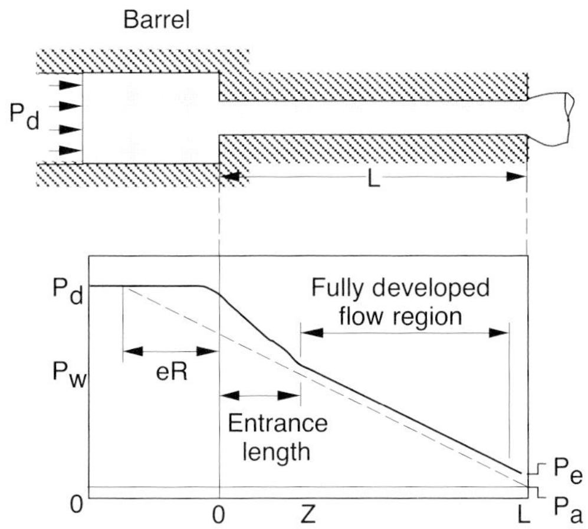

Figure 5.30 Entrance effects in a typical capillary viscometer.

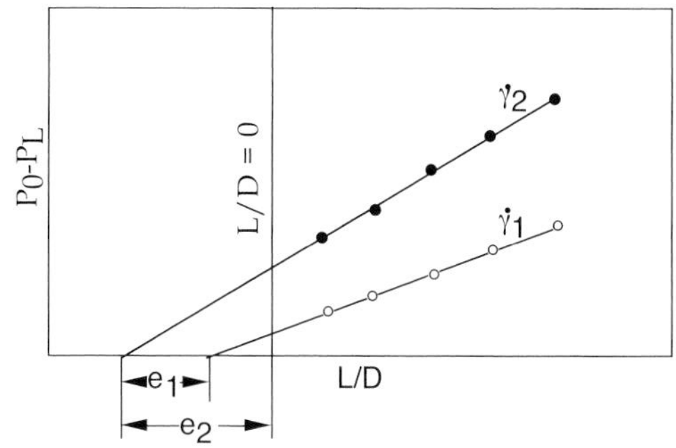

Figure 5.31 Bagley plots for two shear rates.

The equation for shear stress can now be written as

$$\tau_{rz} = \frac{r}{R}\tau_w \tag{5.66}$$

To obtain the shear rate at the wall, $\frac{dv_z}{dr}$, the Weissenberg–Rabinowitsch [54] equation can be used

$$-\frac{dv_z}{dr} = \dot{\gamma}_w = \frac{1}{4}\dot{\gamma}_{aw}\left(3 + \frac{d(\ln Q)}{d(\ln \tau)}\right) \tag{5.67}$$

where, $\dot{\gamma}_{aw}$ is the apparent or Newtonian shear rate at the wall and is written as

$$\dot{\gamma}_{aw} = \frac{4Q}{\pi R^3} \tag{5.68}$$

The shear rate and shear stress at the wall are now known. Therefore, using the measured values of the flow rate, Q, and the pressure drop, $p_0 - p_L$, the viscosity can be calculated using

$$\eta = \frac{\tau_w}{\dot{\gamma}_w} \tag{5.69}$$

5.5.4 Viscosity Approximation Using the Representative Viscosity Method

A simplified method to compute viscosity, developed by Schümmer and Worthoff [55], takes advantage of the fact that the Newtonian and the shear thinning materials have a common streamline at which the strain rate is the same. This is schematically represented in Fig. 5.32 where the common streamline is located at r_S. The position of that streamline is related to the power law index and varies between 0.7715*R* and 0.8298*R* for power law indices between 1.4 and 0.25. A close approximation is given by[13]

$$r_s \approx \frac{\pi}{4} R = 0.7854R \tag{5.70}$$

and the strain rate at that point is given by

$$\bar{\dot{\gamma}} = \frac{4}{\pi} \frac{Q}{R^4} r_s \approx \frac{Q}{R^3} \tag{5.71}$$

The shear stress at the location r_S can be calculated using

$$\bar{\tau} = \left(\frac{p_0 - p_L}{L} \right) \frac{r_s}{2} \approx \frac{\pi}{8} R \left(\frac{p_0 - p_L}{L} \right) \tag{5.72}$$

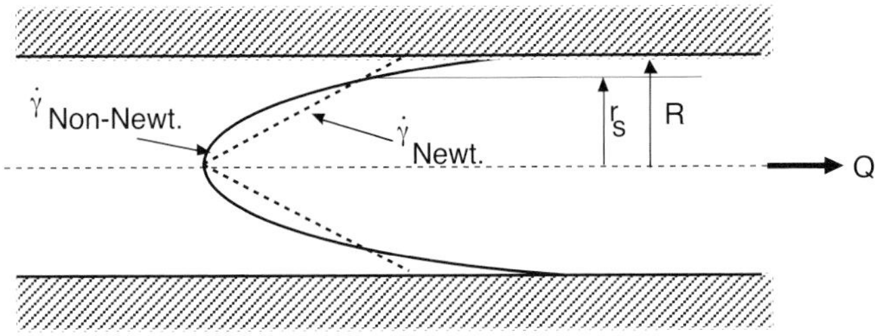

Figure 5.32 Strain rate distribution in Newtonian and non-Newtonian fluids flowing through a capillary.

5.5.5 The Cone-Plate Rheometer

The cone-plate rheometer is often used when measuring the viscosity and the primary and secondary normal stress coefficient functions as a function of shear rate and temperature. The geometry of a cone-plate rheometer is shown in Fig. 5.33. Since the angle θ_0 is very small, typically < 5°, the shear rate can be considered to be constant and is given by

$$\dot{\gamma}_{\theta\phi} = \frac{\Omega}{\theta_0} \tag{5.73}$$

where Ω is the angular velocity of the cone. The shear stress can also be considered to be constant and can be related to the measured torque, *T*,

13 The value $\pi/4$ was not mathematically derived but offers a significant simplification to the equations with a final error in viscosity of less than 5%.

$$\tau_{\theta\phi} = \frac{3T}{2\pi R^3} \tag{5.74}$$

The viscosity function can now be obtained from

$$\eta(\dot{\gamma}_{\theta\phi}) = \frac{\tau_{\theta\phi}}{\dot{\gamma}_{\theta\phi}} \tag{5.75}$$

The primary normal stress coefficient function, ψ_1, can be calculated by measuring the force, *F*, required to maintain the cone in place and can be computed using

$$\psi_1 = \frac{2F}{\pi R^2 \dot{\gamma}^2} \tag{5.76}$$

Although it is also possible to determine the secondary stress coefficient function from the normal stress distribution across the plate, it is very difficult to get accurate data.

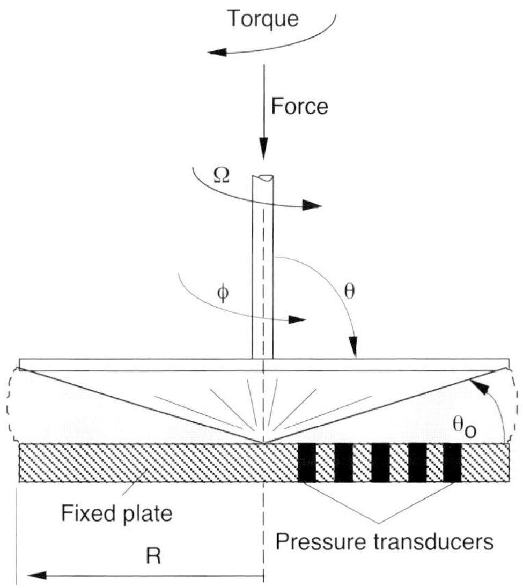

Figure 5.33 Schematic diagram of a cone-plate rheometer.

5.5.6 The Couette Rheometer

Another rheometer commonly used in industry is the concentric cylinder or Couette flow rheometer schematically depicted in Fig. 5.34. The torque, T, and rotational speed, Ω, can easily be measured. The torque is related to the shear stress that acts on the inner cylinder wall and can be computed as follows:

$$\tau_i = \frac{T}{2\pi r_i^2 L} \tag{5.77}$$

If we consider a power-law fluid confined between the outer and inner cylinder walls of a Couette device, the shear rate at the inner wall can be computed using

$$\dot{\gamma}_i = \frac{2\Omega}{n\left(1-\left(r_i/r_0\right)^{2/n}\right)} \tag{5.78}$$

The power-law index can be determined with experimental data using

$$n = \frac{d\log\tau_i}{d\log\Omega} \tag{5.79}$$

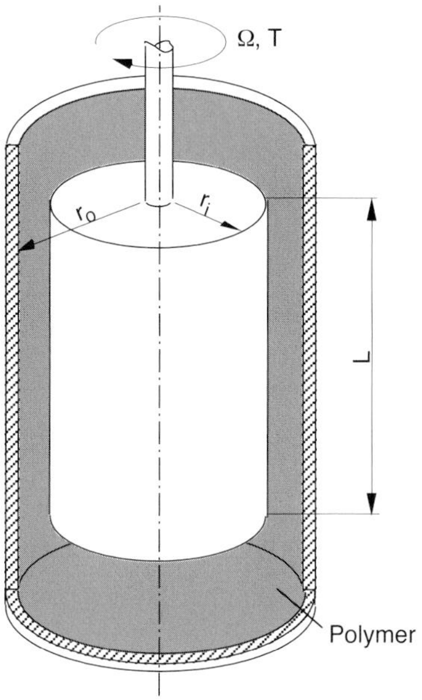

Figure 5.34 Schematic diagram of a Couette rheometer.

Once the shear strain rate and stress are known the viscosity can be computed using

$$\eta = \frac{\tau_i}{\dot{\gamma}} \tag{5.80}$$

The major sources of error in a concentric cylinder rheometer are the end-effects. One way of minimizing these effects is by providing a large gap between the inner cylinder end and the bottom of the closed end of the outer cylinder.

5.5.7 Extensional Rheometry

It should be emphasized that the shear behavior of polymers measured with the equipment described in the previous sections cannot be used to deduce the extensional behavior of polymer melts. Extensional rheometry is the least understood field of rheology. The simplest way to measure extensional viscosities is to stretch a polymer rod held at elevated temperatures at a speed that maintains a constant strain rate as the rod reduces its cross-sectional area. The viscosity can easily be computed as the ratio of instantaneous axial stress to elongational strain rate. The biggest problem when trying to perform this measurement is to grab the rod at its ends as it is pulled apart. The most common way to grab the specimen is with toothed rotary clamps to maintain a constant specimen length [56]. A schematic of Meissner's extensional rheometer incorporating rotary clamps is shown in Fig. 5.35 [56].

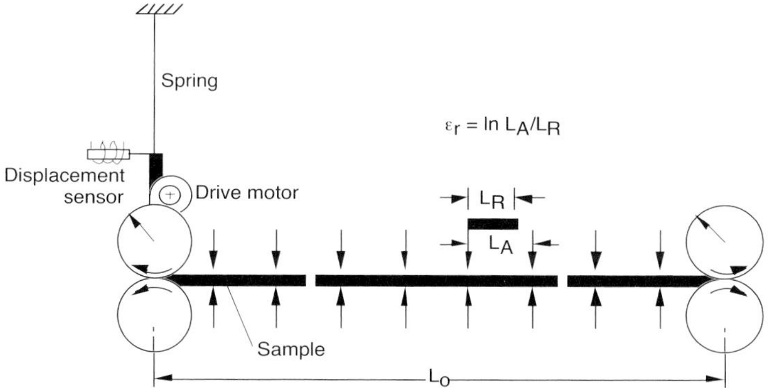

Figure 5.35 Schematic diagram of an extensional rheometer.

Another set-up that can be used to measure extensional properties without clamping problems and without generating orientation during the measurement is the lubricating squeezing flow [57], which generates an equibiaxial deformation. A schematic of this apparatus is shown in Fig. 5.36.

It is clear from the apparatus description on Fig. 5.35 that carrying out tests to measure extensional rheometry is a very difficult task. One of the major problems arises because of the fact that, unlike shear tests, it is not possible to achieve steady state condition with elongational rheometry tests. This is simply because the cross-sectional area of the test specimen is constantly diminishing. Figure 5.37 [56] shows this effect by comparing shear and elongational rheometry data on polyethylene.

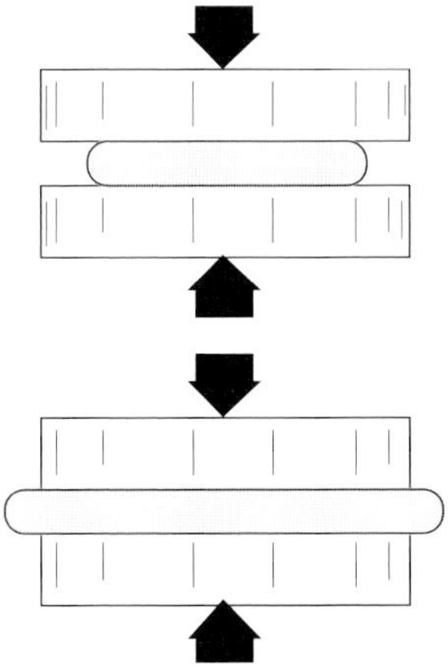

Figure 5.36 Schematic diagram of squeezing flow.

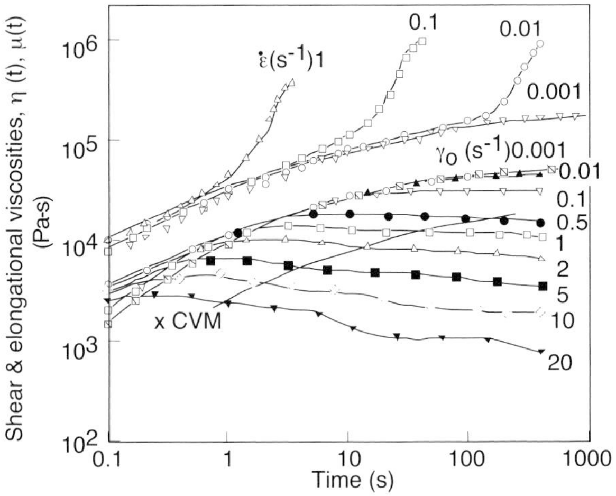

Figure 5.37 Development of elongational and shear viscosities during deformation for polyethylene samples.

Finally, another equibiaxial deformation test is carried out by blowing a bubble and measuring the pressure required to blow the bubble and the size of the bubble during the test. This test has been successfully used to measure extensional properties of polymer membranes for blow molding and thermoforming applications. Here, a sheet is clamped between two plates with circular holes and a pressure differential is introduced to deform it. The pressure applied and deformation of the sheet are monitored with time and related to extensional properties of the material. Assuming an incompressible material, the instantaneous thickness of the sheet can be computed using the notation shown in Fig. 5.38:

$$t = t_0\left(\frac{D^2}{8Rh}\right) \tag{5.81}$$

The instantaneous radius of curvature of the sheet is related to bubble height by

$$R = \frac{D^2}{8h} + \frac{h}{2} \tag{5.82}$$

The biaxial strain can be computed using

$$\varepsilon_B = \ln\left(\frac{2\alpha R}{D}\right) \tag{5.83}$$

and the biaxial stress can be calculated using

$$\sigma_B = \frac{R\Delta P}{2t} \tag{5.84}$$

For more detail on extensional rheometry, beyond the scope of this book, the reader should refer to the literature [58].

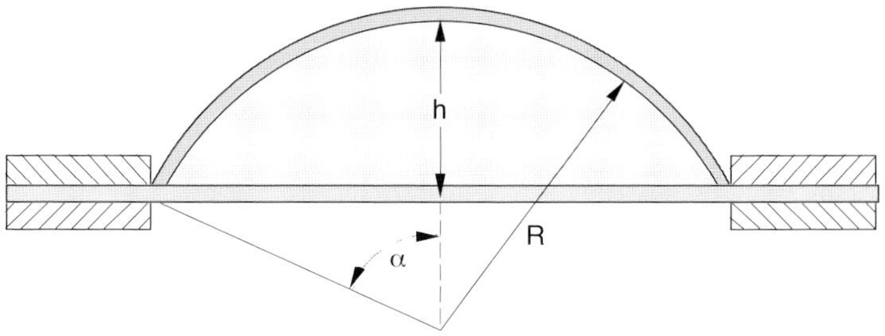

Figure 5.38 Schematic diagram of sheet inflation.

5.6 Surface Tension

Although surface tension is generally not included in rheology chapters, it does play a significant role in the deformation of polymers during flow, especially in dispersive mixing of polymer blends. Surface tension, σ_S, between two materials appears as a result of different intermolecular interactions. In a liquid-liquid system, surface tension manifests itself as a force that tends to maintain the surface between the two materials to a minimum. Thus, the equilibrium shape of a droplet inside a matrix which is at rest is a sphere. When three phases touch, such as liquid, gas, and solid, we get different contact angles depending on the surface tension between the three phases. Figure 5.45 schematically depicts three different cases. In case 1, the liquid perfectly wets the surface with a continuous spread, leading to a wetting angle of zero. Case 2, with moderate surface tension effects, shows a liquid that has a tendency to flow over the surface with a contact angle between zero and $\pi/2$. Case 3, with a high surface tension effect, is where the liquid does not wet the surface which results in a contact angle greater than $\pi/2$. In Fig. 5.39, σ_S denotes the surface tension between the gas and the solid, σ_l the surface tension between the liquid and the gas, and σ_{sl} the surface tension between the solid and liquid. Using geometry one can write

$$\cos\theta = \frac{\sigma_s - \sigma_{sl}}{\sigma_l} \tag{5.85}$$

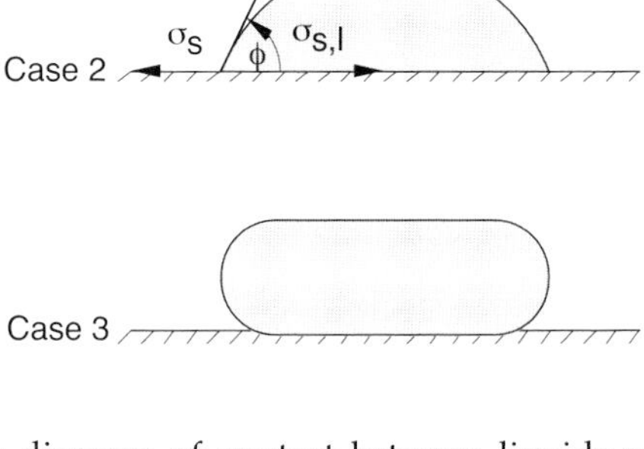

Figure 5.39 Schematic diagram of contact between liquids and solids with various surface tension effects.

The wetting angle can be measured using simple techniques such as a projector, as shown schematically in Fig. 5.40. This technique, originally developed by Zisman [59], can be used in the ASTM D2578 [60] standard test. Here, droplets of known surface tension, σ_l are applied to a film. The measured values of cos ϕ are plotted as a function of surface tension, σ_l, as shown in Fig. 5.41, and extrapolated to find the *critical surface tension* , σ_C, required for wetting.

For liquids of low viscosity, a useful measurement technique is the tensiometer, schematically represented in Fig. 5.42. Here, the surface tension is related to the force it takes to pull a platinum ring from a solution. Surface tension for selected polymers are listed in Table 5.6 [61], for some solvents in Table 5.7 [62] and between polymer-polymer systems in Table 5.8 [61].

Furthermore, Hildebrand and Scott [63] found a relationship between the solubility parameter[14], δ, and surface tension, σ_S for polar and non-polar liquids. Their relationship can be written as [64]

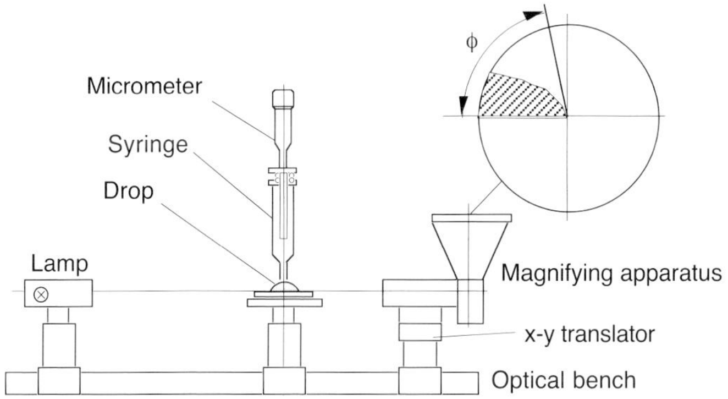

Figure 5.40 Schematic diagram of apparatus to measure contact angle between liquids and solids.

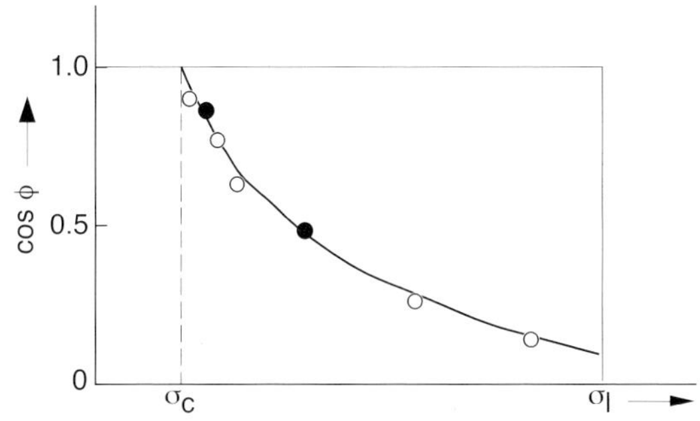

Figure 5.41 Contact angle as a function of surface tension.

14 Solubility parameter is defined in Chapter 6.

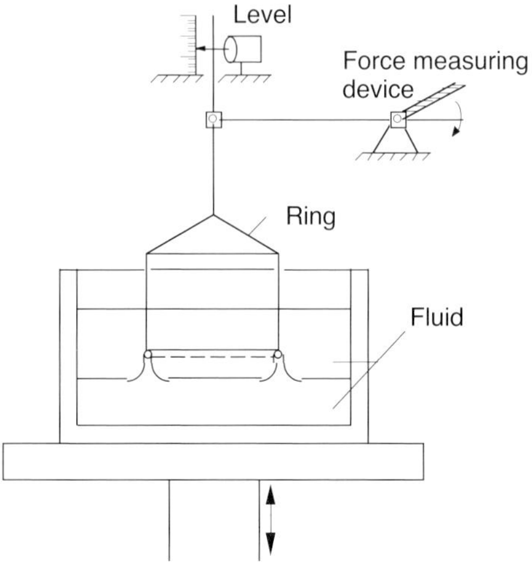

Figure 5.42 Schematic diagram of a tensiometer used to measure surface tension of liquids.

$$\sigma_s = 0.24\,\delta^{2.33}\,V^{0.33} \tag{5.86}$$

where V is the molar volume of the material. The molar volume is defined by

$$V = \frac{M}{\rho} \tag{5.87}$$

where M is the molar weight. It should be noted that the values in Eqs. 5.86 and 5.87 must be expressed in cgs units.

There are many areas in polymer processing and in engineering design with polymers where surface tension plays a significant role. These are mixing of polymer blends, which is discussed in detail in Chapter 6 of this book, adhesion, treatment of surfaces to make them non-adhesive and sintering.

During manufacturing it is often necessary to coat and crosslink a surface with a liquid adhesive or bonding material. To enhance adhesion it is often necessary to raise surface tension by oxidizing the surface, by creating COOH groups, using flames, etching or releasing electrical discharges. This is also the case when enhancing the adhesion properties of a surface before painting.

On the other hand it is often necessary to reduce adhesiveness of a surface such as required when releasing a product from the mold cavity or when coating a pan to give it nonstick properties. A material that is often used for this purpose is polytetrafluoroethylene (PTFE), mostly known by its tradename of teflon.

Table 5.6 Typical Surface Tension Values of Selected Polymers at 180 ^oC

Polymer	σ_S (N/m)	$\partial\sigma_S/\partial T$ (N/m/^oC)
Polyamide resins (290 ^oC)	0.0290	–
Polyethylene (linear)	0.0265	-5.7 x 10^{-5}
Polyethylene terephthalate (290 ^oC)	0.027	–
Polyisobutylene	0.0234	-6.6 x 10^{-5}
Polymethyl methacrylate	0.0289	-7.6 x 10^{-5}
Polypropylene	0.0208	-5.8 x 10^{-5}
Polystyrene	0.0292	-7.2 x 10^{-5}
Polytetrafluoroethylene	0.0094	-6.2 x 10^{-5}
Polyvinyl acetate	0.0259	-6.6 x 10^{-5}

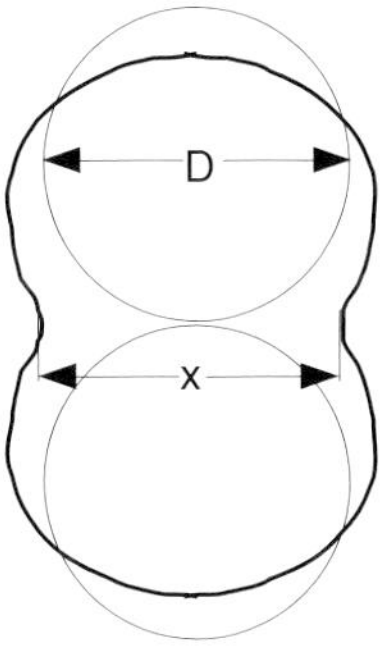

Figure 5.43 Schematic of the sintering process.

Finally, during the sintering process, surface tension plays a key role, especially when bringing together pairs of particles at high temperatures without applying pressure. Using the notation for the sintering process presented in Fig. 5.43, Menges and Reichstein [65] develop the relation

$$x^2 \approx \frac{\sigma_s D t}{\eta} \tag{5.88}$$

where, σ_s is the surface tension, t is time and η the viscosity.

Examples

5.1 Derive the equation that describes the pressure driven velocity field inside a tube of constant diameter (Eq. 5.43).

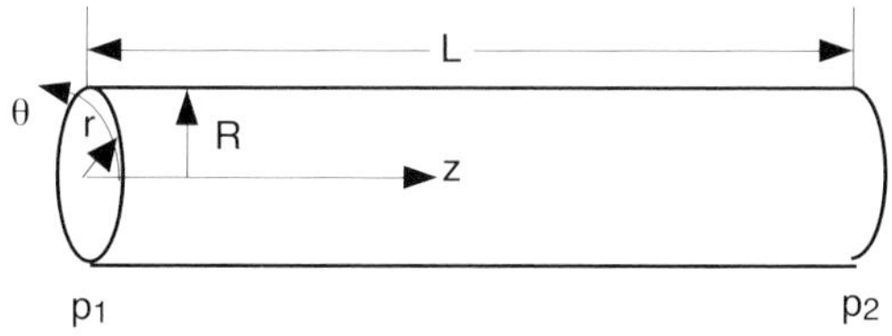

Figure 5.44 Tube geometry and coordinate system.

In our solution scheme, we must first choose a coordinate system:

Next we must state our assumptions:

- no variations in the θ direction: $\frac{\partial}{\partial \theta} = v_\theta = 0$,
- steady state: $\frac{\partial}{\partial t} = 0$,
- no velocity in the radial direction: $v_r = 0$.

The continuity equation (Appendix 1) reduces to

$$\frac{\partial v_z}{\partial z} = 0$$

Using the reduced continuity equation and the above assumptions, only the z-component of the equation of motion (Appendix 1) is needed

$$0 = -\frac{\partial p}{\partial z} + \mu \left[\frac{1}{r} \frac{\partial}{\partial r} \left(r \frac{\partial v_z}{\partial r} \right) \right]$$

Letting $\frac{\partial p}{\partial z} = \frac{\Delta p}{L}$ and integrating gives

$$r \frac{\partial v_z}{\partial r} = \frac{r^2}{2\mu} \left(\frac{\Delta p}{L} \right) + C_1$$

Since $\frac{\partial v_z}{\partial r} = 0$ at $r = 0$, $C_1 = 0$. Integrating again, we get

$$v_z = \frac{r^2}{4\mu}\left(\frac{\Delta p}{L}\right) + C_2$$

Since $v_z = 0$ at $r = R$, the above equation becomes

$$v_z = \frac{R^2 \Delta p}{4\mu L}\left[1 - \left(\frac{r}{R}\right)^2\right]$$

5.2 Derive the equation that describes the annular flow in a wire coating process. The wire travels through the annulus at a speed, V, and we assume that the pressure inside the die equals atmospheric pressure. Use the notation presented in Fig. 5.45. Assume a power law viscosity.

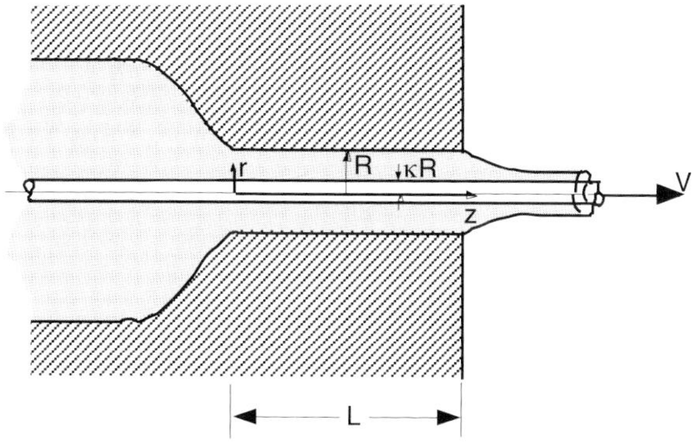

Figure 5.45 Wire coating process geometry and coordinate system.

When solving this problem we make similar assumptions as in Example 5.1. This results in $\frac{\partial}{\partial\theta} = v_\theta = 0$, $\frac{\partial}{\partial t} = 0$ and $v_r = 0$. Here too, the continuity equation reduces to

$$\frac{\partial v_z}{\partial z} = 0$$

However, the pressure gradient, $\frac{\partial p}{\partial z}$, can be neglected. Since this problem is non-Newtonian we need to use the general equation of motion (Appendix 1). The z-component reduces to

$$\frac{\partial}{\partial r}\left(r\tau_{rz}\right) = 0$$

Next, we define the deviatoric stress as a function of rate of deformation using

$$\tau_{rz} = \eta(T,\dot{\gamma})\dot{\gamma}_{rz} = m_0\dot{\gamma}^{n-1}\dot{\gamma}_{rz}$$

Since the only deformation that occurs is the shear deformation in the rz-plane the magnitude of the rate of deformation tensor is given by

$$\dot{\gamma} = \dot{\gamma}_{rz} = \frac{\partial v_z}{\partial r}$$

The equation of motion, can now be written as

$$\frac{\partial}{\partial r}\left(rm_0\left[\frac{\partial v_z}{\partial r}\right]^n\right) = 0$$

Integrating this equation twice, and applying two boundary conditions, $v_z = V$ at $r = \kappa R$, and $v_z = 0$ at $r = R$ we get

$$v_z = V\frac{\left[\frac{r}{R}\right]^{1-\frac{1}{n}} - 1}{\kappa^{1-\frac{1}{n}} - 1}$$

5.3 Formulate a design equation for the end-fed sheeting die shown in Fig. 5.46 so that the extruded sheet is of uniform thickness. Your job is to specify the length of the approach zone or die land as a function of the manifold direction to achieve uniform flow. The manifold diameter is constant and the flow can be assumed isothermal with a Newtonian viscosity of η.

After assigning a coordinate system, and assuming the notation presented in Fig. 5.46, the flow of the manifold can be represented using the Hagen-Poiseuille equation as

$$Q = \frac{\pi R^4}{8\eta}\left(-\frac{dp}{dz}\right)$$

and the flow in the die land (per unit width) using slit flow

$$q = \frac{h^3}{12\eta}\left(-\frac{dp}{dl}\right) = \frac{h^3}{12\eta}\left(\frac{p}{L_l}\right)$$

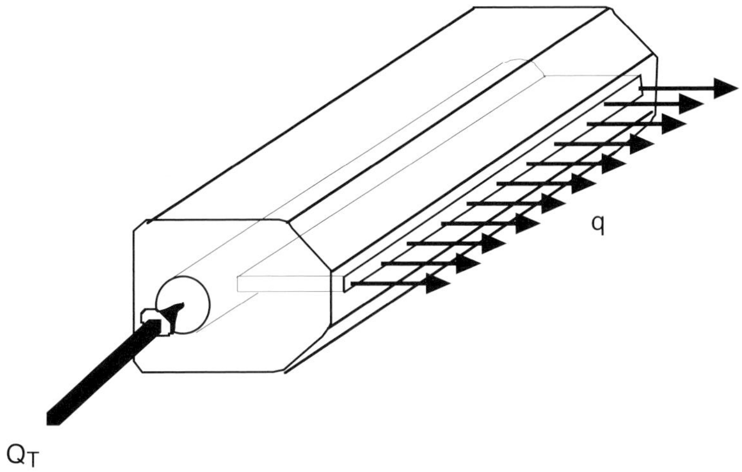

Figure 5.46 Schematic of an end-fed sheeting die.

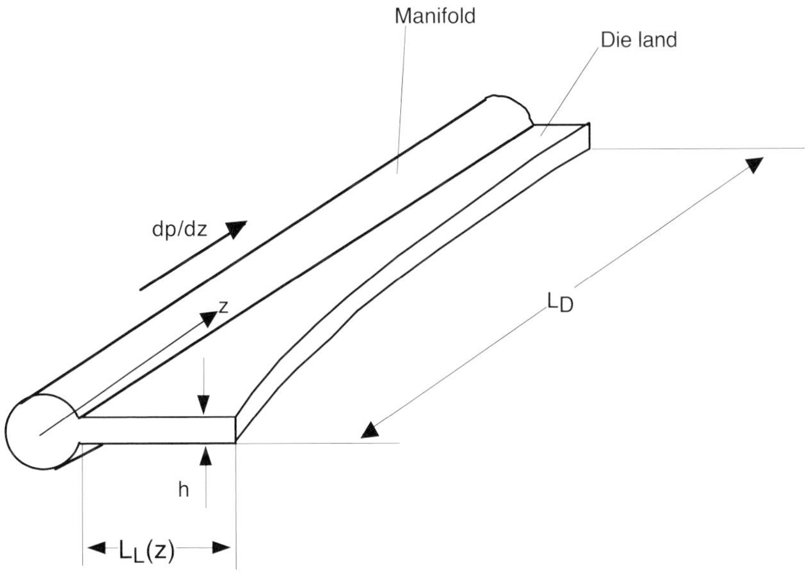

Figure 5.47 End-fed sheeting die coordinates and dimensions.

A manifold that leads to a uniform sheet must deliver a constant throughput along the die land. Performing a flow balance within the differential element shown in Fig. 5.48 results in

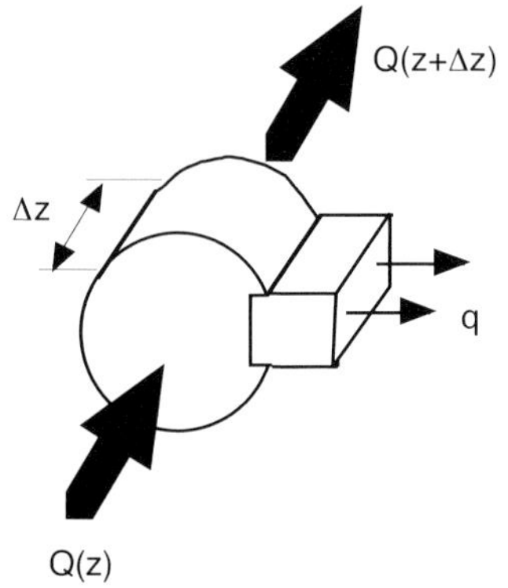

Figure 5.48 Differential element of an end-fed sheeting die.

$$Q(z+\Delta z) - Q(z) - q(z)\Delta z = 0$$

and letting $\Delta z \to 0$ results in

$$\frac{dQ}{dz} = -q = \text{constant}$$

If we integrate and let $Q = Q_T$ at $z = 0$ and $Q = 0$ at $z = L_D$ we get

$$Q(z) = Q_T\left(1 - \frac{z}{L_D}\right)$$

Hence,

$$\frac{dQ}{dz} = -\frac{Q_T}{L_D} = -\frac{h^3}{12\eta}\frac{p(z)}{L_l(z)}$$

which results in

$$L_l(z) = \frac{h^3}{12\eta}\frac{L_D}{Q_T}p(z)$$

where $p(z)$ is unknown. We can now rewrite the manifold equation as

$$\frac{dp}{dz} = -\frac{8\eta}{\pi R^4}Q_T\left(1 - \frac{z}{L_D}\right)$$

and then, letting $p = p_0$ at $z = 0$, we get

$$p = p_0 - \frac{8\eta Q_T L_D}{\pi R^4}\left[\left(\frac{z}{L_D}\right) - \frac{1}{2}\left(\frac{z}{L_D}\right)^2\right]$$

which leads to a land length of

$$L_l = \frac{h^3 L_D p_0}{12\eta Q_T} - \frac{2h^3 L_D^2}{3\pi R^4}\left[\left(\frac{z}{L_D}\right) - \frac{1}{2}\left(\frac{z}{L_D}\right)^2\right]$$

Problems

5.1 During the filling stage of the injection molding cycle, would the pressure requirements for a PP with a MFI = 10 be larger or smaller, relative to a PP with a MFI = 4? Why?

5.2 Model the viscosity of the PE-LD shown in Fig. 5.4 using the power law model and the Bird–Carreau model.

5.3 Derive the equation that predicts the volumetric throughput during pressure flow through a slit, assuming a Newtonian viscosity (Eq. 5.40).

5.4 Derive the Hagen-Poiseuille equation for a power law fluid (Eq. 5.46).

5.5 You are to extrude a polystyrene sheet through a die with a land length of 0.1 m. What is the maximum speed you can extrude the sheet if you are extruding a polystyrene with the viscoelastic properties presented in Fig. 3.44 of Chapter 3.

5.6 The metering section of a single screw extruder can be modeled using a combination of shear flow and pressure flow between parallel plates. Derive the relation between volumetric throughput versus pressure build-up, Δp, in a melt extruder of a Newtonian fluid. Hint: Use the notation given in Chapter 6.

5.7 In Example 5.2 obtain an expressions for the flowrate, coating thickness, and the axial force required to pull the wire through the die.

5.8 Formulate the design equation for an end-fed sheeting die such as the one in Example 3. Unlike the example, the new die should have a variable radius manifold as in Fig. 5.48. In this case, the axial distance from the manifold center to the die exit must be constant.

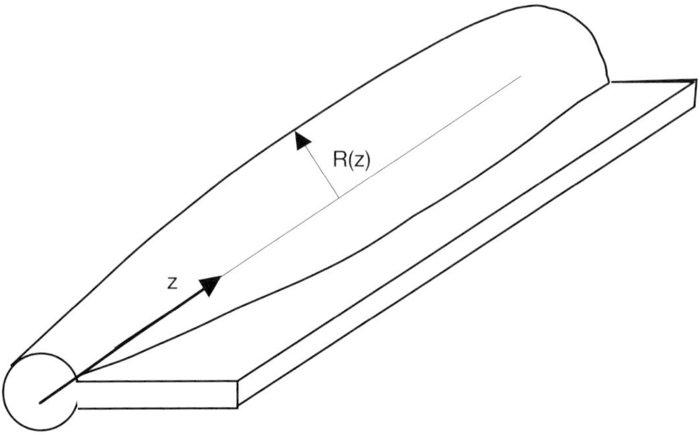

Figure 5.49 End-fed variable diameter manifold sheeting die coordinates and dimensions.

References

1. Macosko, C.W., Rheology: Principles, Measurements and Applications, VCH, (1994).
2. Dealy, J.M., and Wissbrun, K.F., *Melt Rheology and Its Role in Plastics Processing*, Van Nostrand, New York, (1990).
3. Carreau, P.J., De Kee, D.C.R., and Chhabra, R.P., *Rheology of Polymeric Systems*, Hanser Publishers, Munich, (1997).
4. Bird, R.B., Armstrong, R.C., and Hassager, O., *Dynamics of Polymeric Liquids*, 2nd Ed., Vol.1, John Wiley & Sons, New York, (1987).
5. Gordon, G.V., and Shaw, M.T., *Computer Programs for Rheologists*, Hanser Publishers, Munich, (1994).
6. Bird, R.B., and Wiest, J.M., *Annu. Rev. Fluid Mech., 27*, 169, (1995).
7. Tucker III, C.L., *Fundamentals of Computer Modeling for Polymer Processing*, Hanser Publishers, Munich, (1989).
8. Isayev, A.I., *Modeling of Polymer Processing*, Hanser Publishers, Munich, (1991).
9. Laun, H.M., *Rheol. Acta, 17*,1, (1978).
10. Agassant, J.-F., Avenas, P., J.-Ph. Sergent, and P.J. Carreau, *Polymer Processing: Principles and Modeling*, Hanser Publishers, Munich, (1991).
11. Vinogradov, G.V., Malkin, A.Y., Yanovskii, Y.G., Borisenkova, E.K., Yarlykov, B.V., and Berezhnaya, G.V., *J. Polym. Sci. Part A-2, 10*, 1061, (1972).
12. Vlachopoulos, J., and Alam, M., *Polym. Eng. Sci., 12*, 184, (1972).
13. Hatzikiriakos, S.G., and Dealy, J.M., *ANTEC Tech. Papers, 37*, 2311 (1991).
14. Spencer, R.S., and Dillon, R.D., *J. Colloid Inter. Sci., 3*, 163, (1947).
15. Vinogradov, G.V., Malkin, A.Y., Yanovskii, Y.G., Borisenkova, E.K., Yarlykov, B.V., and Berezhnaya, G.V., *J. Polym. Sci. Part A-2, 10*, 1061, (1972).

16. Hatzikiriakos, S.G., and Dealy, J.M., *J. Rheol., 36*, 845, (1992).
17. Denn, M.M., *Annu. Rev. Fluid Mech.*, 22, 13, (1990).
18. Ostwald, W., *Kolloid-Z.*, 36, 99, (1925).
19. de Waale, A., *Oil and Color Chem. Assoc. Journal*, 6, 33, (1923).
20. Carreau, P.J., Ph.D. Thesis, University of Wisconsin-Madison, USA, (1968).
21. Yasuda, K., Armstrong, R.C., and Cohen, R.E., *Rhel. Acta, 20*, 163, (1981).
22. Menges, G., Wortberg, F., and Michaeli, W., *Kunststoffe, 68*, 71, (1978).
23. Trouton, F.T., *Proc. Roy. Soc., A77*, (1906).
24. Münstedt, H., *Rheologica Acta, 14*, 1077, 92, (1975).
25. Han, C.D. and Lem, K.W., *J. Appl. Polym. Sci.*, 29, 1879, (1984).
26. Castro, J.M. and Macosko, C.W., *AIChe J., 28*, 250, (1982).
27. Castro, J.M., Perry, S.J., and Macosko, C.W., *Polymer Comm., 25*, 82, (1984).
28. Einstein, A., *Ann. Physik, 19*, 549, (1906).
29. Guth, E., and Simha, R., *Kolloid-Zeitschrift, 74*, 266, (1936).
30. Guth, E., *Proceedings of the American Physical Society*, (1937); *Physical Review, 53*, 321, (1938).
31. Batchelor, G.K., *Annu. Rev. Fluid Mech.*, 6, 227, (1974).
32. Geisbüsch, P., Ph.D. Thesis, IKV, RWTH-Aachen, Germany, (1980).
33. Gupta, R.K., *Flow and Rheology in Polymer Composites Manufacturing*, Ed. S.G. Advani, Elsevier, Amsterdam, (1994).
34. Milliken, W.J., and Powell, R.L., *Flow and Rheology in Polymer Composites Manufacturing*, Ed. S.G. Advani, Elsevier, Amsterdam, (1994).
35. Keunings, R., *Simulation of Viscoelastic Fluid Flow*, in *Computer Modeling for Polymer Processing*, Ed. C.L. Tucker III, Hanser Publishers, Munich, (1989).
36. Crocket, MJ., Davies, A.R., and Walters, K., *Numerical Simulation of Non-Newtonian Flow*, Elsevier, Amsterdam, (1984).
37. Debbaut, B., Marchal, J.M., and Crochet, M.J., *J. Non-Newtonian Fluid Mech., 29*, 119, (1988).
38. Dietsche, L., and Dooley, J., *SPE ANTEC, 53*, 188, (1995).
39. Dooley, J., and Hughes, K., *SPE ANTEC, 53*, 69, (1995).
40. Baaijens, J.P.W., *Evaluation of Constitutive Equations for Polymer Melts and Solutions in Complex Flows*, Ph.D. Thesis, Eindhoven University of Technology, Eindhoven, The Netherlands, (1994).
41. Luo, X.-L., and Mitsoulis, E., *J. Rheol., 33*, 1307, (1989).
42. Kiriakidis, D.G., and Mitsoulis, E., *Adv. Polym. Techn., 12*, 107, (1993).
43. Lodge, A.S., *Elastic Liquids*, Academic Press, London, (1960).
44. Kaye, A., *Non-Newtonian Flow in Incompressible Fluids*, CoA Note No.134, The College of Aeronautics, Cranfield, (1962).
45. Bernstein, B., Kearsley, E., and Zapas, L., *Trans. Soc. Rheol., 7*, 391, (1963).
46. Wagner, M.H., *Rheol. Acta*, 18, 33, (1979).
47. Papanastasiou, A.C., Scriven, L.E., and Macosko, C.W., *J. Rheol.*, 27, 387, (1983).
48. Orbey, N., and Dealy, J.M., *Polym. Eng. Sci., 24*, 511, (1984).
49. ASTM, 8.01, Plastics (I), ASTM, Philadelphia, (1994).
50. Hagen, G.H.L., *Annalen der Physik, 46*, 423, (1839).
51. Poiseuille, L.J., *Comptes Rendus 11*, 961, (1840).
52. Dealy, J.M., *Rheometers for Molten Plastics*, Van Nostrand Reinhold Company, New York, (1982).

53. Bagley, E.B., *J. Appl. Phys., 28*, 624, (1957).
54. Rabinowitsch, B., *Z. Phys. Chem., 145*, 1, (1929).
55. Schümmer, P., and Worthoff, R.H., *Chem. Eng. Sci., 38*, 759, (1978).
56. Meissner, J., *Rheol. Acta, 10* , 230, (1971)..
57. Chatrei, Sh., Macosko, C.W., and Winter, H.H., *J. Rheol., 25* , 433, (1981).
58. Macosko, C.W., *Rheology: Principles, Measurements and Applications*, VCH, (1994).
59. Zisman, W.A., *Ind. Eng. Chem., 55*, 19, (1963).
60. ASTM, 8.02, Plastics (II), ASTM, Philadelphia, (1994).
61. Wu, S., *J. Macromol. Sci. - Revs. Macromol. Chem., C10*, 1, (1974).
62. Owens, D.K. and Wendt, R.C., *J. Appl. Polymer Sci.*, 13, 1741, (1969).
63. Hildebrand, J. and Scott, R.L., *The Solubility of Non-Electrolytes*, 3rd Ed., Reinhold Publishing Co., New York, (1949).
64. Van Krevelen, D.W., and Hoftyzer, P.J., *Properties of Polymers*, Elsevier, Amsterdam, (1976).
65. Reichstein, H., Ph.D. thesin, IKV, RWTH-Aachen, (1982).

Part II

Influence of Processing Properties

6

Introduction to Processing

The mechanical properties and the performance of a finished product are always the result of a sequence of events. Manufacturing of a plastic part begins with material choice in the early stages of part design. Processing follows this, at which time properties of the final part are made and frozen into place. During design and manufacturing of any plastic product one must always be aware that material, processing and design properties all go hand-in-hand and cannot be decoupled. This approach is often referred to as the four P's: polymer, processing, product and performance. This chapter presents the most important polymer processing techniques available today[1]. Extrusion[2] is covered first, followed by mixing processes and injection molding[3]. Secondary shaping shaping operations are discussed next. At the end of the chapter other processes such as calendering, coating, compression molding, and rotational molding are presented.

6.1 Extrusion

During extrusion, a polymer melt is pumped through a shaping die and formed into a profile. This profile can be a plate, a film, a tube, or have any shape for its cross section. Ram-type extruders were first built by J. Bramah in 1797 to extrude seamless lead pipes. The first ram-type extruders for rubber were built by Henry Bewley and Richard Brooman in 1845. In 1846, a patent for cable coating was filed for trans-gutta-percha and cis-hevea rubber and the first insulated wire was laid across the Hudson River for the Morse Telegraph Company in 1849. The first screw extruder was patented by Mathew Gray in 1879 for the purpose of wire coating. However, the screw pump can be attributed to Archimedes, and the actual invention of the screw extruder in

1 An in-depth view of polymer processing is given by Tadmor and Gogos [1].

2 For further reading in area of extrusion we recommend Rauwendaal's book [2].

3 For further reading in the area of injection molding we recommend Osswald, Turng and Gramann [3]

polymer processing by A.G. DeWolfe of the United States dates to the early 1860s. The first extrusion of thermoplastic polymers was done at the Paul Troester Maschinenfabrik in Hannover, Germany in 1935.

Although ram and screw extruders are both used to pump highly viscous polymer melts through passages to generate specified profiles, they are based on different principles. The schematic in Fig. 6.1 shows under what principles ram extruders, screw extruders, and other pumping systems work.

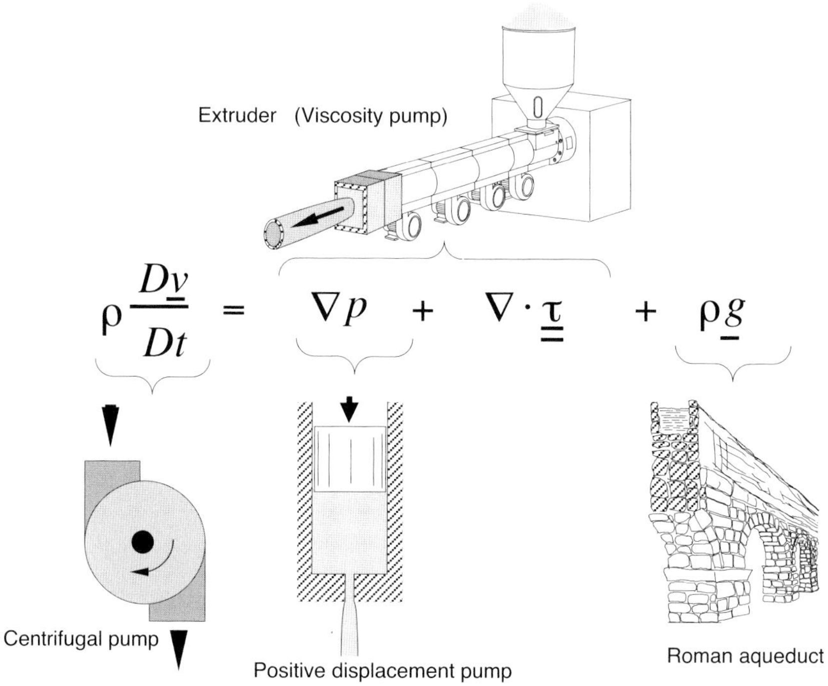

Figure 6.1 Schematic of pumping principles.

The ram extruder is a positive displacement pump based on the pressure gradient term of the equation of motion. Here, as the volume is reduced, the fluid is displaced from one point to the other, resulting in a pressure rise. The gear pump, widely used in the polymer processing industry, also works on this principle. On the other hand, a screw extruder is a viscosity pump that works based on the pressure gradient term and the deformation of the fluid, represented as the divergence of the deviatoric stress tensor in Fig. 6.1. The centrifugal pump, based on the fluid inertia, and the Roman aqueduct, based on the potential energy of the fluid, are also represented in the figure and are typical of low viscosity liquids.

In today's polymer industry, the most commonly used extruder is the single screw extruder, schematically depicted in Fig. 6.2. The single screw extruder can either have a smooth inside barrel surface, called a conventional single screw extruder, or a grooved feed zone, called a grooved feed extruder. In some cases, an extruder can have a degasing zone, required to extract moisture, volatiles, and other gases that form during the extrusion process.

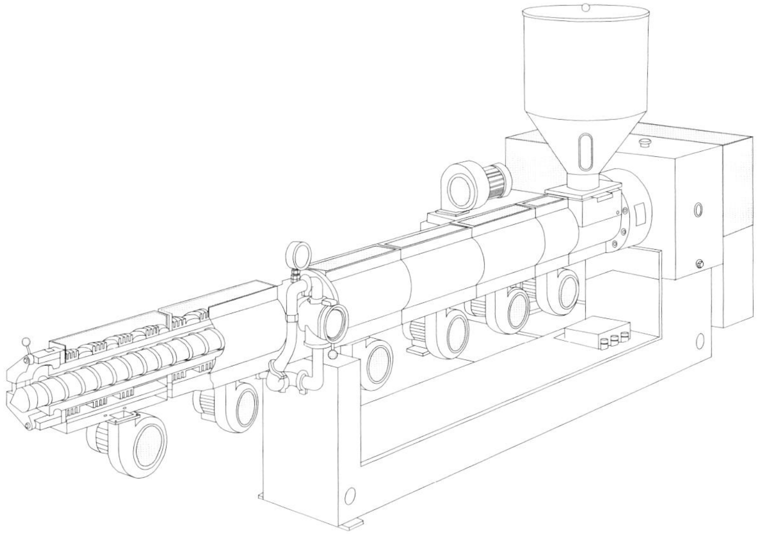

Figure 6.2 Schematic of a single screw extruder (Reifenhäuser).

Another important class of extruders are the twin screw extruders, schematically depicted in Fig. 6.3. Twin screw extruders can have co-rotating or counter-rotating screws, and the screws can be intermeshing or non-intermeshing. Twin screw extruders are primarily employed as mixing and compounding devices, as well as polymerization reactors. The mixing aspects of single and twin screw extruders are detailed later in this chapter.

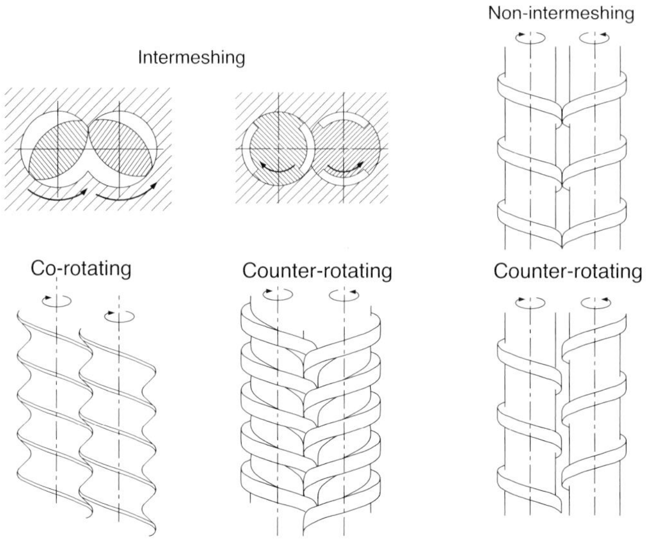

Figure 6.3 Schematic of different twin screw extruders.

6.1.1 The Plasticating Extruder

The plasticating single screw extruder is the most common equipment in the polymer industry. It can be part of an injection molding unit and found in numerous other extrusion processes, including blow molding, film blowing, and wire coating. A schematic of a plasticating or three-zone, single screw extruder, with its most important elements is given in Fig. 6.4. Table 6.1 presents typical extruder dimensions and relationships common in single screw extruders.

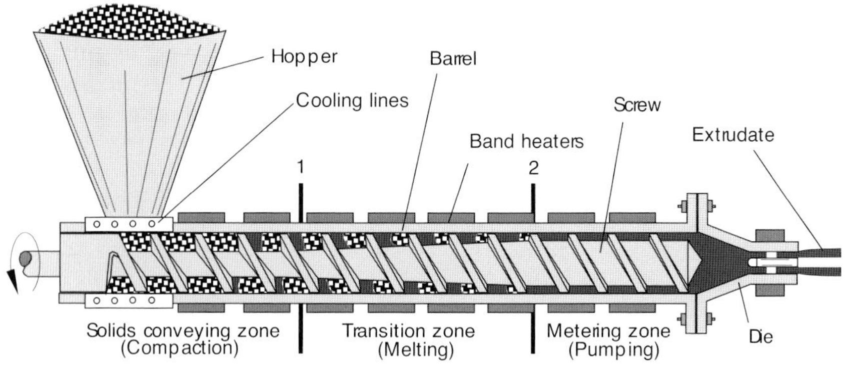

Figure 6.4 Schematic of a plasticating single screw extruder.

Table 6.1 Typical Extruder Dimensions and Relationships (the Notation for Table 6.1 is Defined in Fig. 6.5.)

L/D	Length to Diameter ratio 20 or less for feeding or melt extruders 25 for blow molding, film blowing, and injection molding 30 or higher for vented extruders or high output extruders
D	Standard diameter
US (inches)	0.75, 1.0, 1.5, 2, 2.5, 3.5, 4.5, 6, 8, 10, 12, 14, 16, 18, 20, and 24
Europe (mm)	20, 25, 30, 35, 40, 50, 60, 90, 120, 150, 200, 250, 300, 350, 400, 450, 500, and 600
ϕ	Helix angle 17.65° for a square pitch screw where L_s=D New trend: $0.8 < L_s/D < 1.2$
h	Channel depth in the metering section (0.05-0.07)D for D<30 mm (0.02-0.05)D for D>30 mm
β	Compression ratio: $h_{feed} = \beta h$ 2 to 4
δ	Clearance between the screw flight and the barrel 0.1 mm for D<30 mm 0.15 mm for D>30 mm
N	Screw speed 1-2 rev/s (60-120 rpm) for large extruders 1-5 rev/s (60-300 rpm) for small extruders
V_b	Barrel velocity (relative to screw speed) = πDN 0.5 m/s for most polymers 0.2 m/s for unplasticized PVC 1.0 m/s for LDPE

The plasticating extruder can be divided into three main zones:

- The solids conveying zone
- The melting or transition zone
- The metering or pumping zone

The tasks of a plasticating extruder are to:

- Transport the solid pellets or powder from the hopper to the screw channel
- Compact the pellets and move them down the channel
- Melt the pellets
- Mix the polymer into a homogeneous melt
- Pump the melt through the die

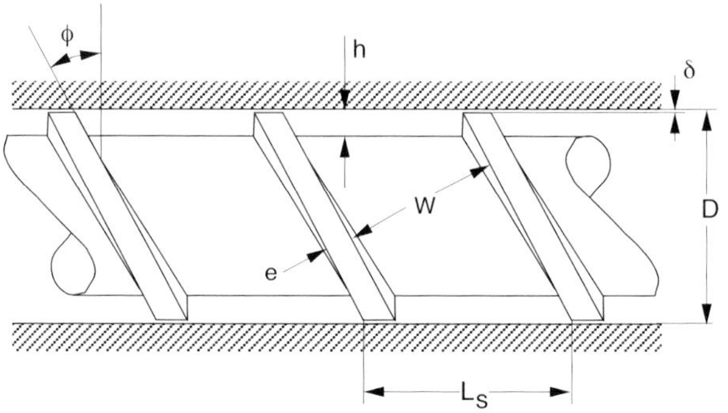

Figure 6.5 Schematic diagram of a screw section.

The pumping capability and characteristic of an extruder can be represented with sets of die and screw characteristic curves. Figure 6.6 presents such curves for a conventional (smooth barrel) single screw extruder.

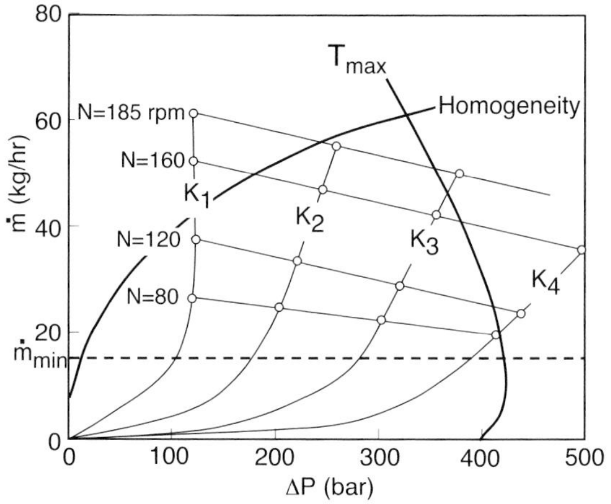

Figure 6.6 Screw and die characteristic curves for a 45 mm diameter extruder with an LDPE.

The die characteristic curves are Landel K1, K2, K3, and K4 in ascending order of die restriction. Here, K1 represents a low resistance die such as for a thick plate, and K4 represents a restrictive die, such as is used for film. The different screw characteristic curves represent different screw rotational speeds. In a screw characteristic curve the point of maximum throughput and no pressure build-up is called the point of open discharge. This occurs when there is no die. The point of maximum pressure build-up and no throughput is called the point of closed discharge. This occurs when the extruder is plugged.

Shown in Fig. 6.6 are also lines that represent critical aspects encountered during extrusion. The curve labeled Tmax represents the conditions at which excessive temperatures are reached as a result of viscous heating. The feasibility line ($\dot{m}_{min}$) represents the throughput required to have an economically feasible system. The processing conditions to the right of the homogeneity line render a thermally and physically heterogeneous polymer melt.

6.1.1.1 The Solids Conveying Zone

The task of the solids conveying zone is to move the polymer pellets or powders from the hopper to the screw channel. Once the material is in the screw channel, it is compacted and transported down the channel. The process to compact the pellets and to move them can only be accomplished if the friction at the barrel surface exceeds the friction at the screw surface. This can be visualized if one assumes the material inside the screw channel to be a nut sitting on a screw. As we rotate the screw without applying outside friction, the nut (polymer pellets) rotates with the screw without moving in the axial direction. As we apply outside forces (barrel friction), the rotational speed of the nut is less than the speed of the screw, causing it to slide in the axial direction. Virtually, the solid polymer is then "unscrewed" from the screw.

To maintain a high coefficient of friction between the barrel and the polymer, the feed section of the barrel must be cooled, usually with cold water cooling lines. The frictional forces also result in a pressure rise in the feed section. This pressure compresses the solids bed which continues to travel down the channel as it melts in the transition zone. Figure 6.7 presents the pressure build-up in a conventional, smooth barrel extruder. In these extruders, most of the pressure required for pumping and mixing is generated in the metering section.

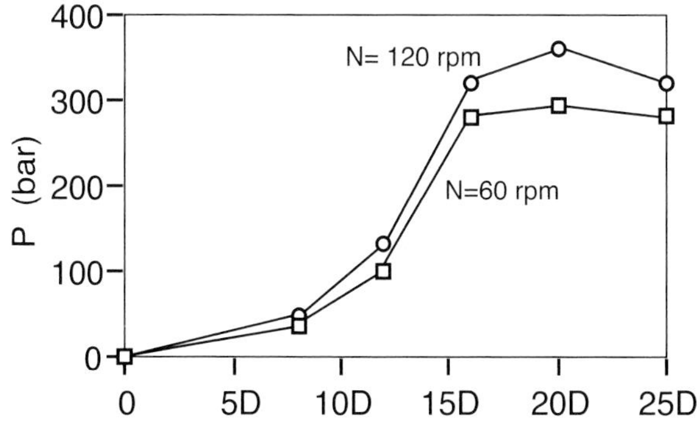

Figure 6.7 Conventional extruder pressure distribution.

The simplest mechanism for ensuring high friction between the polymer and the barrel surface is grooving its surface in the axial direction [4, 5]. Extruders with a grooved feed section where developed by Menges and Predöhl [4, 5] in 1969, and are called grooved feed extruders. To avoid excessive pressures that can lead to barrel or screw failure, the length of the grooved barrel section must not exceed 3.5D. A schematic diagram of the grooved section in a single screw extruder is presented in Fig. 6.8. The key factors that propelled the development and refinement of the grooved feed extruder were the processing problems, excessive melt temperature, and reduced productivity posed by high viscosity and low coefficients of friction typical of high molecular weight polyethylenes and polypropylenes.

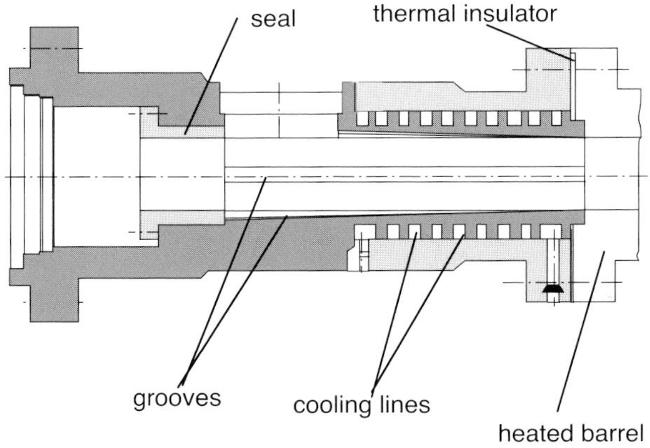

Figure 6.8 Schematic diagram of the grooved feed section of a single screw extruder.

In a grooved feed extruder, the conveying and pressure build-up tasks are assigned to the feed section. Figure 6.9 presents the pressure build-up in a single screw extruder with a grooved feed section. The high pressures in the feed section lead to the main advantages over conventional systems. With grooved feed systems, there is a higher productivity and a higher melt flow stability and pressure invariance. This is demonstrated with the screw characteristic curves in Fig. 6.10, which presents screw characteristic curves for a 45 mm diameter grooved feed extruder with comparable mixing sections and die openings as shown in Fig. 6.6.

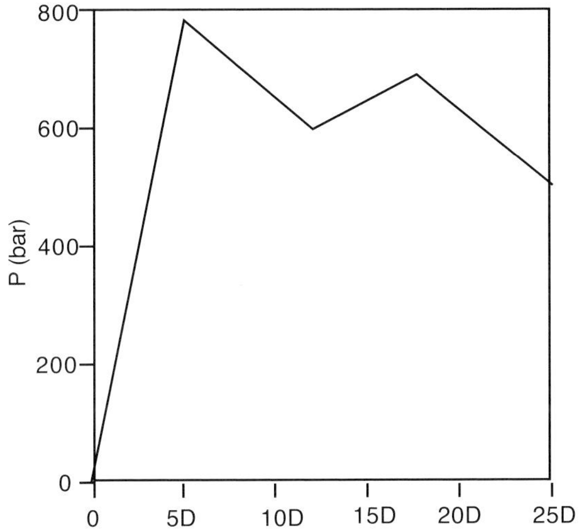

Figure 6.9 Grooved feed extruder pressure distribution.

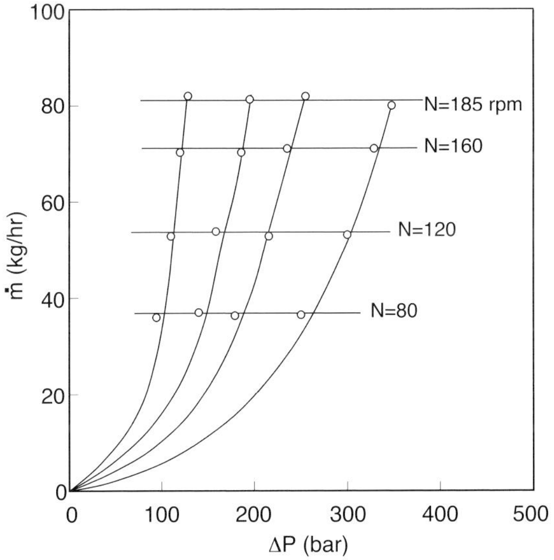

Figure 6.10 Screw and die characteristic curves for a grooved feed 45 mm diameter extruder with an LDPE.

The behavior of the two extruders in Figs. 6.6 and 6.10 are best compared if the throughput and the pressure build-up are non-dimensionalized. The dimensionless throughput is

$$\hat{m} = \frac{\dot{m}}{\rho N D^3} \tag{6.1}$$

and the dimensionless pressure build-up is

$$\Delta\hat{p} = \frac{\Delta p D}{m N^n L_c} \tag{6.2}$$

where L represents the total channel length and for a $25L/D$ extruder is

$$L_c = \frac{25D}{\sin(\phi)} \tag{6.3}$$

where f is assumed to be 17.65º (square pitch). Figure 6.11 presents the results shown in Figs. 6.6 and 6.10 after having been non-dimensionalized using Eqs. 6.2 and 6.3. The figure clearly shows the higher productivity of the grooved feed extruder where the throughput is at least 50% more than that observed with the conventional system for a comparable application. Used with care, Fig. 6.11 can also be used for scale-up.

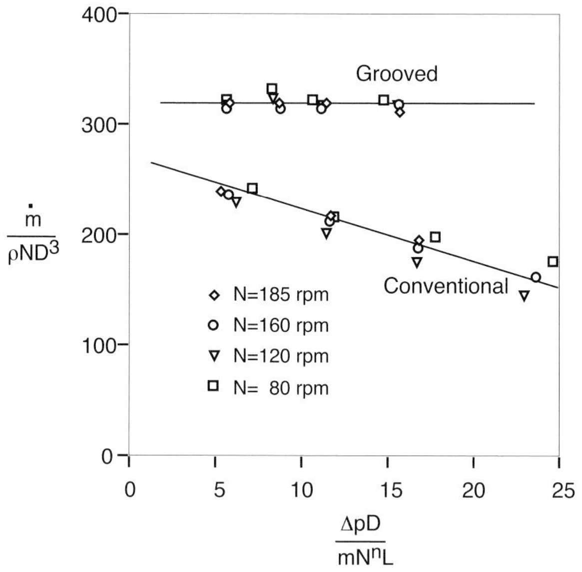

Figure 6.11 Dimensionless screw characteristic curves for conventional and grooved feed extruders.

6.1.1.2 The Melting Zone

The melting or transition zone is the portion of the extruder were the material melts. The length of this zone is a function of the material properties, screw geometry, and processing conditions. During melting, the size of the solid bed shrinks as a melt pool forms at its side, as depicted in Fig. 6.12a which shows the polymer unwrapped from the screw channel.

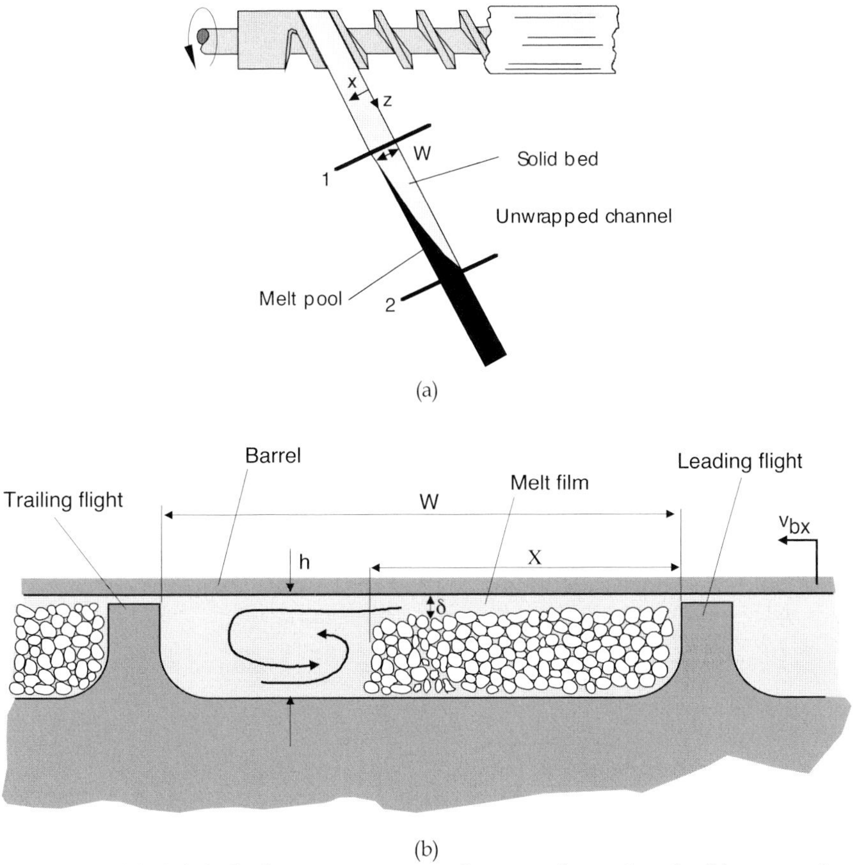

Figure 6.12 (a) Solids bed in an unwrapped screw channel and (b) screw channel cross section.

Figure 6.12b presents a cross section of the screw channel in the melting zone. The solid bed is pushed against the leading flight of the screw as freshly molten polymer is wiped from the melt film into the melt pool by the relative motion between the solids bed and the barrel surface.

Knowing where the melt starts and ends is important when designing a screw for a specific application. The most widely used model to predict melting in a plasticating single screw extruder is the well known Tadmor Model [5]. Using the Tadmor Model, one can accurately predict the solid bed profile in the single screw extruder. Figure 6.13 presents the experimental and predicted solids bed profile of an LDPE in a single screw extruder. The material properties and processing conditions used in the calculations are given in Table 6.2.

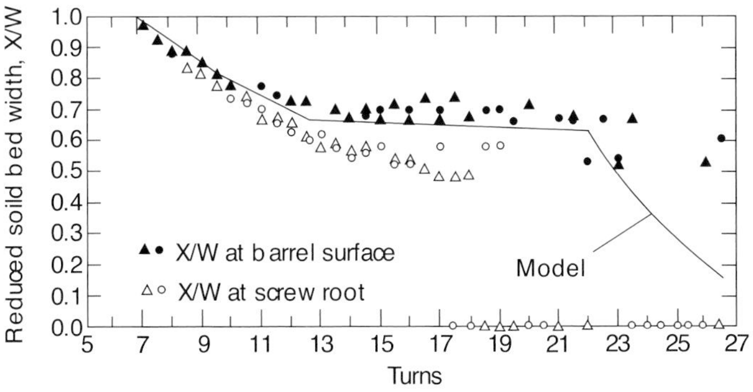

Figure 6.13 Predicted (Tadmor Model) and experimental solids bed profile.

Table 6.2 Extruder Parameters, Processing Conditions, and Material Properties for the Solids Bed Profile Results in Fig. 6.13.

Extruder Geometry:

Square pitch screw, D=63.5 mm, L/D=26.5, W=54.16 mm

Feed zone	- 12.5 turns	h_1=9.4 mm	
Transition zone	- 9.5 turns	h_1=9.4 mm	h_2=3.23 mm
Metering zone	- 4.5 turns	h_2=3.23 mm	

Processing Conditions:

T_0=24 °C T_b=149 °C N=60 rpm Dp=204 bar =61.8 kg/hr

Material properties (LDPE):

Viscosity: n=0.345 a=0.01 $^oC^{-1}$ $m_0=5.6\times10^4$ Pa-s^n T_m=110 °C

Thermal:

k_m=0.1817 W/m°C C_m=2.596 kJ/kg°C C_s=2.763 kJ/kg°C

ρ_{bulk}=595 kg/m^3 ρ_s=915.1 kg/m^3 $\rho_m = 852.7 + 5.018\times10^{-7}p - 0.4756T$

λ=129.8 kJ/kg

From experiment to experiment there are always large variations in the experimental solids bed profiles. The variations in this section of the extruder are caused by slight variations in processing conditions and by the uncontrolled solids bed break-up towards the end of melting. This effect can be eliminated by introducing a screw with a barrier flight that separates the solids bed from the melt pool. The Maillefer screw and barrier screw in Fig. 6.14 are commonly used for high quality and reproducibility. The Maillefer screw maintains a constant solids bed width, using most effectively the melting with melt-removal mechanism, while the barrier screw uses a constant channel depth with a gradually decreasing solids bed width.

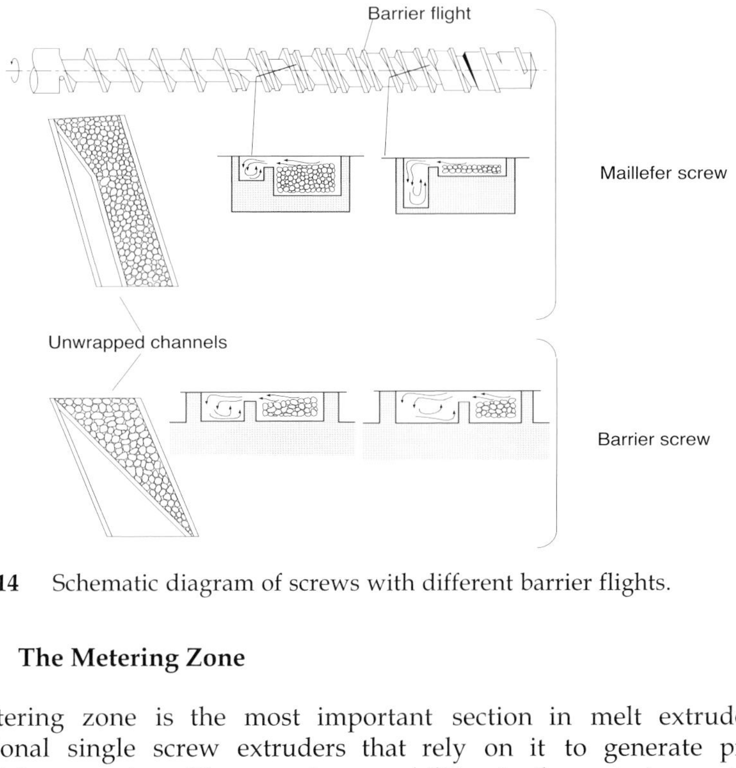

Figure 6.14 Schematic diagram of screws with different barrier flights.

6.1.1.3 The Metering Zone

The metering zone is the most important section in melt extruders and conventional single screw extruders that rely on it to generate pressures sufficient for pumping. The pumping capabilities in the metering section of a single screw extruder can be estimated by solving the equation of motion with appropriate constitutive laws. For a Newtonian fluid in an extruder with a constant channel depth, the screw and die characteristic curves for different cases are represented in Fig. 6.15. The figure shows the influence of the channel depth on the screw characteristic curves. A restrictive extrusion die would clearly work best with a shallow channel screw, and a less restrictive die would render the highest productivity with a deep channel screw.

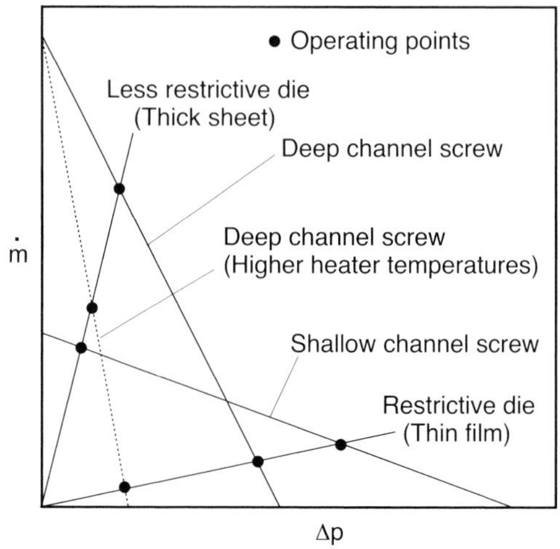

Figure 6.15 Screw characteristic curves (Newtonian).

In both the grooved barrel and the conventional extruder, the diameter of the screw determines the metering or pumping capacity of the extruder. Figure 6.16 presents typical normalized mass throughput as a function of screw diameter for both systems.

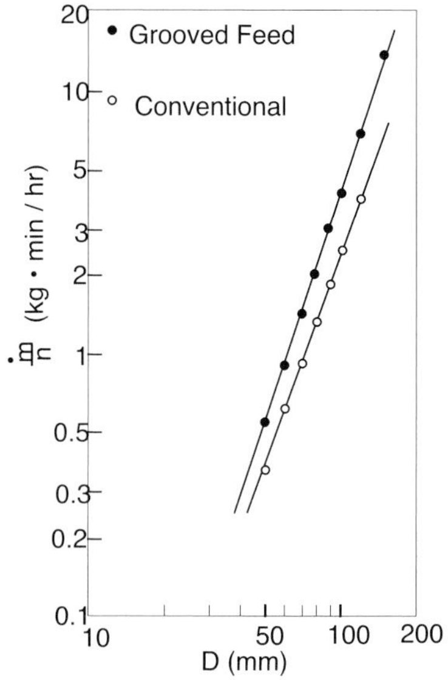

Figure 6.16 Throughput for conventional and grooved feed extruders.

6.1.2 Extrusion Dies

The extrusion die shapes the polymer melt into its final profile. The extrusion die is located at the end of the extruder and its used to extrude

- Flat films and sheets
- Pipes and tubular films for bags
- Filaments and strands
- Hollow profiles for window frames
- Open profiles

As shown in Fig. 6.17, depending on the functional needs of the product, several rules of thumb can be followed when designing an extruded plastic profile. These are:

- Avoid thick sections. Thick sections add to the material cost and increase sink marks caused by shrinkage.
- Minimize the number of hollow sections. Hollow sections add to die cost and make the die more difficult to clean.
- Generate profiles with constant wall thickness. Constant wall thickness in a profile makes it easier to control the thickness of the final profile and results in a more even crystallinity distribution in semi-crystalline profiles.

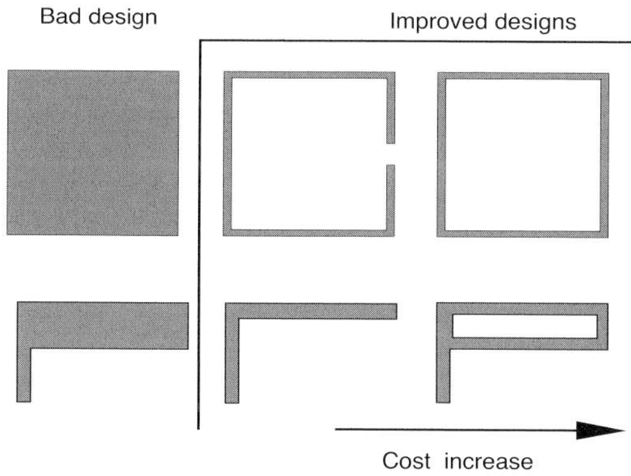

Figure 6.17 Extrusion profile designs.

6.1.2.1 Sheeting Dies

One of the most widely used extrusion dies is the coat-hanger sheeting die. A sheeting die, such as depicted in Fig. 6.18, is formed by the following elements:

- Manifold: evenly distributes the melt to the approach or land region
- Approach or land: carries the melt from the manifold to the die lips
- Die lips: perform the final shaping of the melt
- Flex lips: for fine tuning when generating a uniform profile

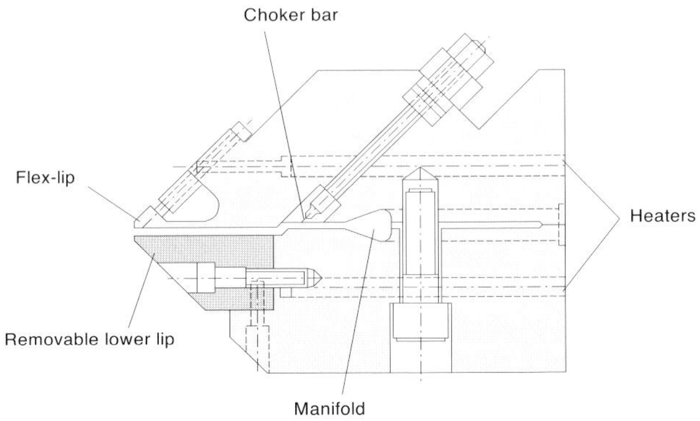

Figure 6.18 Cross section of a coat-hanger die.

To generate a uniform extrudate geometry at the die lips, the geometry of the manifold must be specified appropriately. Figure 6.19 presents the schematic of a coat-hanger die with a pressure distribution that corresponds to a die that renders a uniform extrudate. It is important to mention that the flow though the manifold and the approach zone depend on the non-Newtonian properties of the polymer extruded. So the design of the die depends on the shear thinning behavior of the polymer. Hence, a die designed for one material does not necessarily work for another.

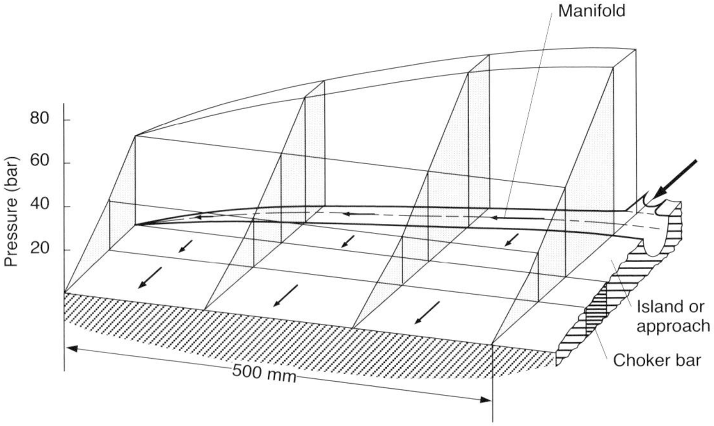

Figure 6.19 Pressure distribution in a coat-hanger die.

6.1.2.2 Tubular Dies

In a tubular die, the polymer melt exits through an annulus. These dies are used to extrude plastic pipes and tubular film. The film blowing operation is discussed in more detail later in this chapter.

The simplest tubing die is the spider die, depicted in Fig. 6.20. Here, a symmetric mandrel is attached to the body of the die by several legs. The polymer must flow around the spider legs causing weld lines along the pipe or film. These weld lines, visible streaks along the extruded tube, are weaker regions.

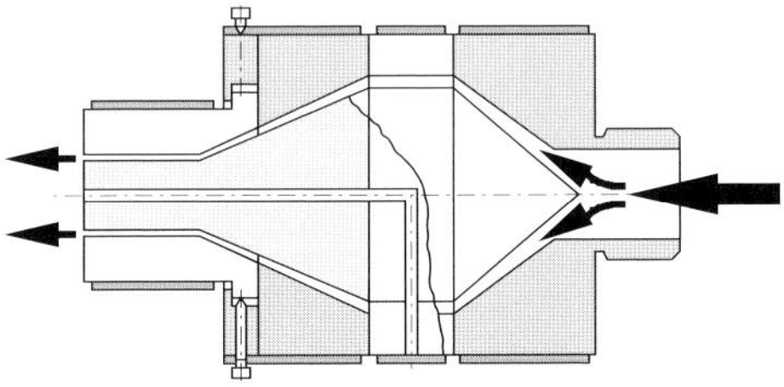

Figure 6.20 Schematic of a spider leg tubing die.

To overcome weld line problems, the cross-head tubing die is often used. Here, the die design is similar to that of the coat-hanger die, but wrapped around a cylinder. This die is depicted in Fig. 6.21. Since the polymer melt must flow around the mandrel, the extruded tube exhibits one weld line. In addition, although the eccentricity of a mandrel can be controlled using adjustment screws, there is no flexibility to perform fine tuning such as in the coat-hanger die. This can result in tubes with uneven thickness distributions.

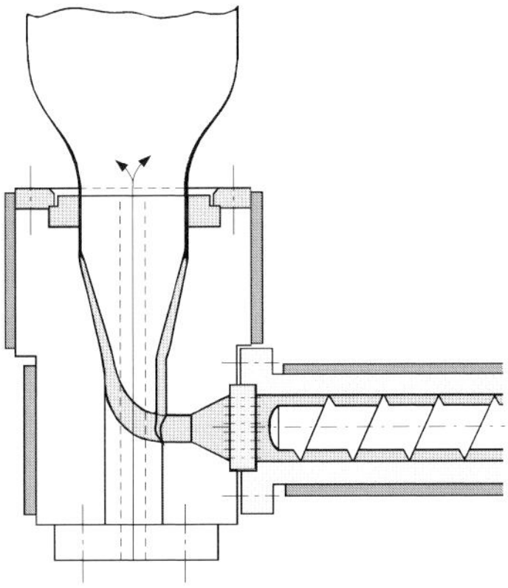

Figure 6.21 Schematic of a cross-head tubing die used in film blowing.

The spiral die, commonly used to extrude tubular blown films, eliminates weld line effects and produces a thermally and geometrically homogeneous extrudate. The polymer melt in a spiral die flows through several feed ports into independent spiral channels wrapped around the circumference of the mandrel. This type of die is schematically depicted in Fig. 6.22.

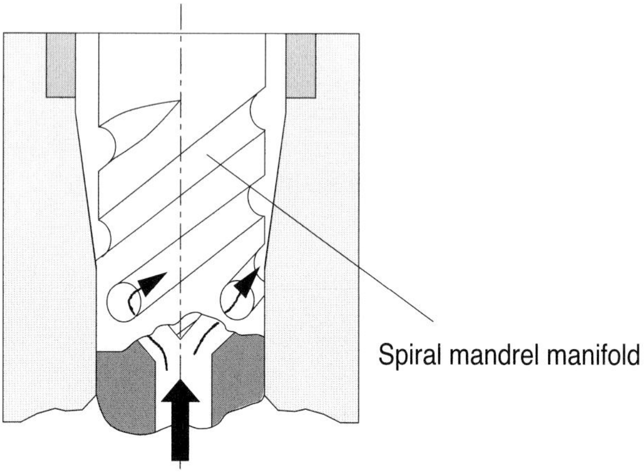

Figure 6.22 Schematic of a spiral die.

6.2 Mixing Processes

Today, most processes involve some form of mixing. As discussed in the previous section, an integral part of a screw extruder is a mixing zone. In fact, most twin screw extruders are primarily used as mixing devices. Similarly, the plasticating unit of an injection molding machine often has a mixing zone. This is important because the quality of the finished product in almost all polymer processes depends in part on how well the material was mixed. Both the material properties and the formability of the compound into shaped parts are highly influenced by the mixing quality. Hence, a better understanding of the mixing process helps to optimize processing conditions and increase part quality.

The process of polymer blending or mixing is accomplished by distributing or dispersing a minor or secondary component within a major component serving as a matrix. The major component can be thought of as the continuous phase, and the minor components as distributed or dispersed phases in the form of droplets, filaments, or agglomerates.

When creating a polymer blend, one must always keep in mind that the blend will probably be remelted in subsequent processing or shaping processes. For example, a rapidly cooled system, frozen as a homogenous mixture, can separate into phases because of coalescence when re-heated. For all practical purposes, such a blend is not processable. To avoid this problem, compatibilizers, which are macromolecules used to ensure compatibility in the boundary layers between the two phases, are common [6].

The mixing can be distributive or dispersive. The morphology development of polymer blends is determined by three competing mechanisms: distributive mixing, dispersive mixing, and coalescence. Figure 6.23 presents a model, proposed by Macosko and co-workers [6], that helps us visualize the mechanisms governing morphology development in polymer blends. The process begins when a thin tape of polymer is melted away from the pellet. As the tape is stretched, surface tension causes it to rip and to form into threads. These threads stretch and reduce in radius, until surface tension becomes significant enough which leads to Rayleigh distrurbances. These cause the threads to break down into small droplets.

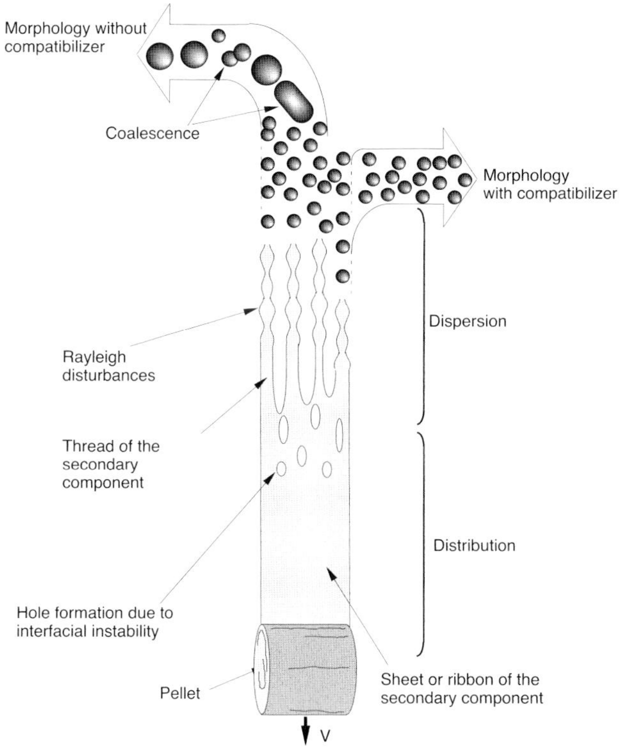

Figure 6.23 Mechanism for morphology development in polymer blends.

There are three general categories of mixtures that can be created:

- Homogeneous mixtures of compatible polymers,
- Single phase mixtures of partly incompatible polymers, and
- Multi-phase mixtures of incompatible polymers.

Table 6.3 lists examples of compatible, partially incompatible, and incompatible polymer blends.

Table 6.3 Common Polymer Blends

Compatible polymer blends	
	Natural rubber and polybutadiene
	Polyamides (e.g., PA 6 and PA 66)
	Polyphenylene ether (PPE) and polystyrene
Partially incompatible polymer blends	
	Polyethylene and polyisobutylene
	Polyethylene and polypropylene (5% PE in PP)
	Polycarbonate and polybutylene terephthalate
Incompatible polymers blends	
	Polystyrene/polyethylene blends
	Polyamide/polyethylene blends
	Polypropylene/polystyrene blends

6.2.1 Distributive Mixing

Distributive mixing or laminar mixing of compatible liquids is usually characterized by the distribution of the droplet or secondary phase within the matrix. This distribution is achieved by imposing large strains on the system such that the interfacial area between the two or more phases increases and the local dimensions, or striation thicknesses, of the secondary phases decrease. This concept is shown schematically in Fig. 6.24 [7]. The figure shows a Couette flow device with the secondary component having an initial striation thickness of δ_0. As the inner cylinder rotates, the secondary component is distributed through the systems with constantly decreasing striation thickness; striation thickness depends on the strain rate of deformation which makes it a function of position. The total strain that a droplet or secondary phase undergoes is defined by

$$\gamma(\tau) = \int_0^{\tau} \dot{\gamma}(t)dt \tag{6.4}$$

where $\dot{\gamma}$ is the magnitude of the strain rate of deform tion defined by Eqs. 4.7

and 4.8, and τ is an arbitrary point in time. For a sphere, which is deformed into an ellipsoid, the total strain can be related to the striation thickness using

$$\delta = 2R\left(1+\gamma^2\right)^{-0.25} \tag{6.5}$$

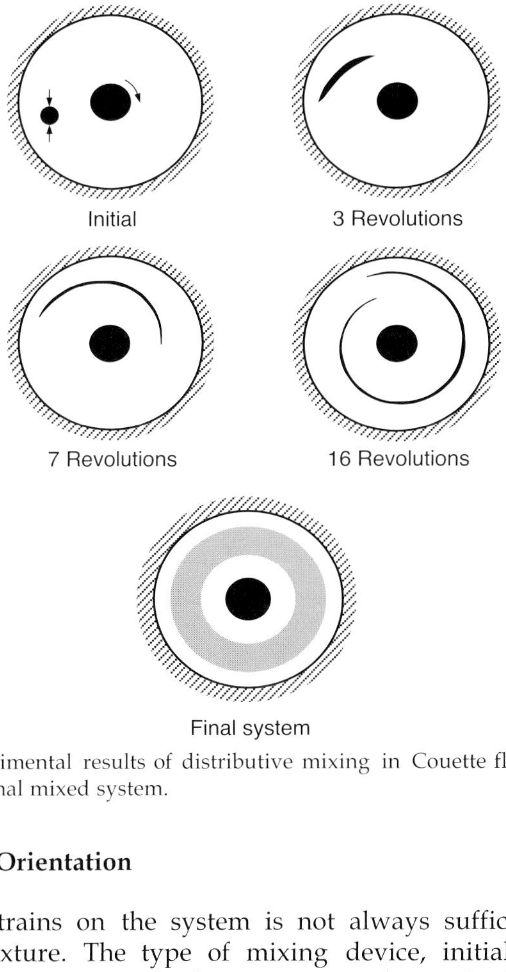

Figure 6.24 Experimental results of distributive mixing in Couette flow, and schematic of the final mixed system.

6.2.1.1 Effect of Orientation

Imposing large strains on the system is not always sufficient to achieve a homogeneous mixture. The type of mixing device, initial orientation and position of the two or more fluid components play a significant role in the quality of the mixture. For example the mixing problem shown in Fig. 6.24 homogeneously distributes the melt within the region contained by the streamlines cut across by the initial secondary component. The final mixed system is shown in Fig. 6.24. Figure 6.25 [8] shows another variation of initial orientation and arrangement of the secondary component. Here, the secondary phase cuts across all streamlines, which leads to a homogeneous mixture throughout the Couette device, under appropriate conditions.

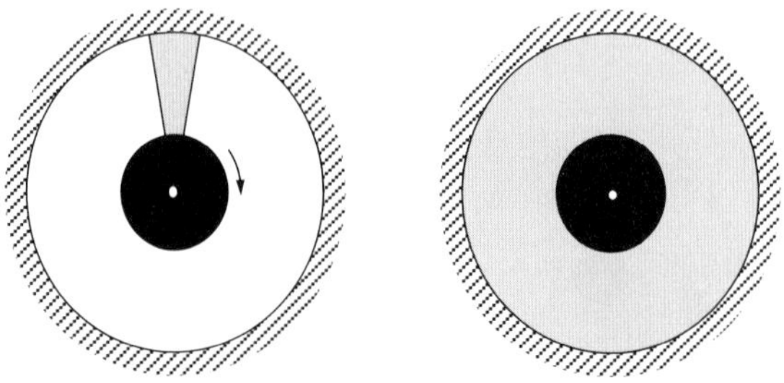

Figure 6.25 Schematic of distributive mixing in Couette flow.

A common way of quantifying mixing is by following the growth of the interface between the primary and secondary fluids. In a simple shear flow, a simple expression exists that relates the growth of the interface, the strain, and the orientation of the area of the secondary fluid with respect to the flow direction [9]:

$$\frac{A}{A_0} = \gamma \cos\alpha \qquad (6.6)$$

where A_0 is the initial interface area, A is the final interface area, γ is the total strain and α the angle that defines the orientation of the surface, or normal vector, with respect to the direction of flow. Figure 6.26 [2] demonstrates this concept. Here, both cases (a) and (b) start with equal initial areas, A_0, and undergo the same amount of strain, $\gamma = 10$. The circular secondary component in (a) has a surface that is randomly oriented, between 0 and 2π, whereas most of the surface of the elongated secondary component in (b) is oriented at $\pi/2$ leading to negligible growth of the interface area. An ideal case would have been a long slender secondary component with a surface oriented in the direction of flow or vertically between the parallel plates. Hence, the maximum interface growth inside a simple shear mixer can be achieved if the direction of the interface is maintained in an optimal orientation ($\cos\alpha = 1$). In a simple shear flow this would require a special stirring mechanism that would maintain the interface between the primary and secondary fluid components in a vertical position. Erwin [10] and Ng [11] demonstrated this in an experimental study that involved placing black and white polyethylene blocks in a Couette device (Fig. 6.27a). Figure 6.27b shows that after applying a small amount of shear, the surfaces that were originally oriented in the radial direction have stretched a certain amount and have changed their orientation. It is clear from the photograph that the same surface tends to align with the planes of shear, reducing the mixing efficiency of the process. Hence, in order to increase the effectiveness of the mixer Ng [11] took the Couette content and cut it into new blocks, that were placed inside the Couette device rotated by 90º(Fig. 6.27c). This changed the orientation of the surfaces back to a position where they can more effectively feel the effects of deformation. By repeating this procedure several times, Erwin and co-workers [10, 11] were able to demonstrate that the area growth is also a function of the number of re-orientation that occur during the mixing process. If N is the total number of shearing stages, separated by a re-orientation, the area growth can be computed using

$$\frac{A}{A_0} = \left(\frac{\gamma_{Total}}{N}\right)^N \tag{6.7}$$

where γ_{Total} is the total strain applied during the process. Using this concept, Erwin [10] demonstrated that the upper bound for the ideal mixer is found in a mixer that applies a plane strain extensional flow or pure shear flow to the fluid and where the surfaces are maintained ideally oriented during the whole process; this occurs when $N = \infty$ and each time an infinitesimal amount of shear is applied. In such a system the growth of the interfacial areas follows the relation given by

$$\frac{A}{A_0} = e^{\gamma_{Total}/2} \tag{6.8}$$

In Erwin's ideal mixer the amount of mixing increases in an exponential fashion, compared to a linear increase if the orientation of the fluids' interfaces remain undisturbed.

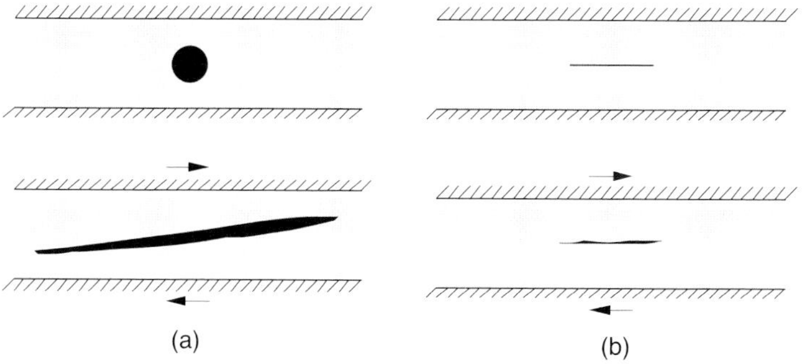

Figure 6.26 Effect of initial surface orientation on distributive mixing.

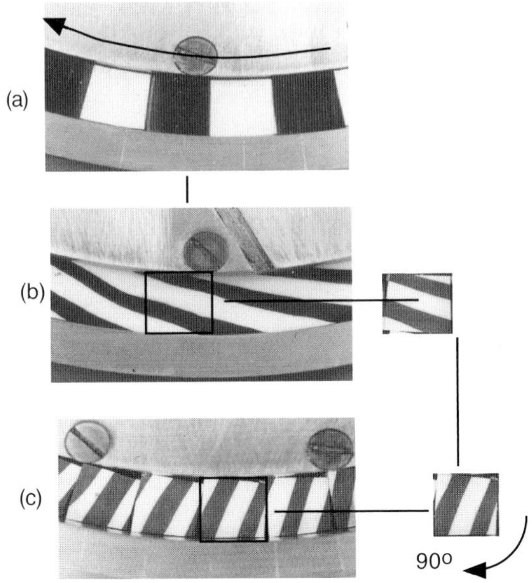

Figure 6.27 Couette deformation and reorientation scheme.

6.2.2 Dispersive Mixing

Dispersive mixing in polymer processing involves breaking a secondary immiscible fluid or an agglomerate of solid particles and dispersing them throughout the matrix. Here, the imposed strain is not as important as the imposed stress which causes the system to break-up. Hence, the type of flow inside a mixer plays a significant role on the break-up of solid particle clumps or fluid droplets when dispersing them throughout the matrix.

6.2.2.1 Break-Up of Particulate Agglomerates

The most common example of dispersive mixing of particulate solid agglomerates is the dispersion and mixing of carbon black into a rubber compound. The dispersion of such a system is schematically represented in Fig. 6.28. However, the break-up of particulate agglomerates is best explained using an ideal system of two small spherical particles that need to be separated and dispersed during a mixing process.

If the mixing device generates a simple shear flow, as shown in Fig. 6.29, the maximum separation forces that act on the particles as they travel on their streamline occur when they are oriented in a 45° position as they continuously rotate during flow. The magnitude of the force trying to separate the "agglomerate" is given by [12]

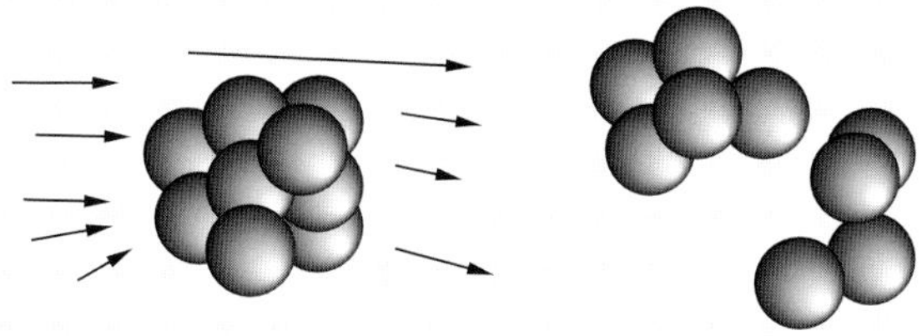

Figure 6.28 Break-up of particulate agglomerates during flow.

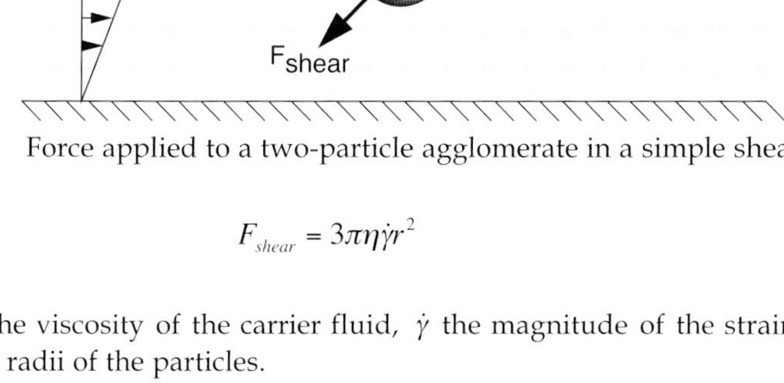

Figure 6.29 Force applied to a two-particle agglomerate in a simple shear flow.

$$F_{shear} = 3\pi\eta\dot{\gamma}r^2 \tag{6.9}$$

where η is the viscosity of the carrier fluid, $\dot{\gamma}$ the magnitude of the strain rate tensor, and r are the radii of the particles.

However, if the flow field generated by the mixing device is a pure elongational flow, such as shown in Fig. 6.30, the particles will always be oriented at 0°; the position of maximum force. The magnitude of the force for this system is given by

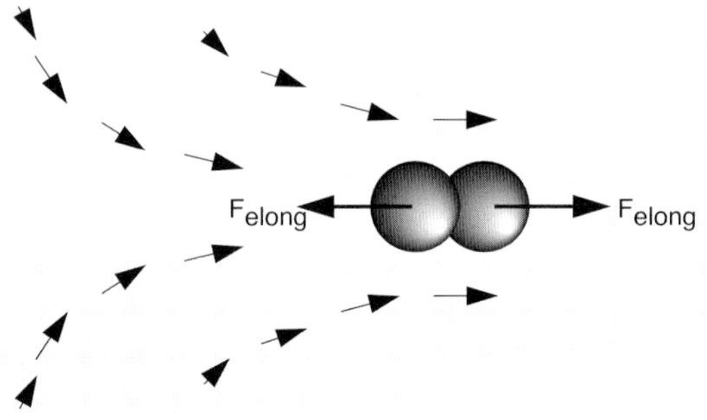

Figure 6.30 Force applied to a two-particle agglomerate in an elongational flow.

$$F_{shear} = 6\pi\eta\dot{\gamma}r^2 \tag{6.10}$$

which is twice as large as the maximum force generated by the system that produces a simple shear flow. In addition, in an elongational flow, the agglomerate is always oriented in the direction of maximum force generation, whereas in simple shear flow the agglomerate tumbles quickly through the position of maximum force[4].

The above analysis makes it clear that for mixing processes which require break-up and dispersion of agglomerates, elongation is the preferred mode of deformation. This is only valid if the magnitude of the rate of deformation tensor can be kept the same in elongation as in shear. Hence, when optimizing mixing devices it is important to know which mode of deformation is dominant. This can be accomplished by computing a flow number [13], defined by

$$\lambda = \frac{\dot{\gamma}}{\dot{\gamma} - \omega} \tag{6.11}$$

where $\dot{\gamma}$ is the magnitude of the rate of deformation tensor and ω the magnitude of the vorticity tensor. A flow number of 0 implies pure rotational flow, a value of 0.5 represents simple shear flow, and pure elongational flow is implied when l is 1.

4 A full description of the relation between flow field and rotation of fibers and agglomerates is given in Chapter 7.

6.2.2.2 Break-Up of Fluid Droplets

In general, droplets inside an incompatible matrix tend to stay or become spherical due to the natural tendencies of the drop trying to maintain the lowest possible surface to volume ratio. However, a flow field within the mixer applies a stress on the droplets, causing them to deform. If this stress is high enough, it will eventually cause the drops to disperse. The droplets will disperse when the surface tension can no longer maintain their shape in the flow field and the filaments break-up into smaller droplets. This phenomenon of dispersion and distribution continues to repeat itself until the deviatoric stresses of the flow field can no longer overcome the surface tension of the new droplets formed.

As can be seen, the mechanism of fluid agglomerate break-up is similar in nature to solid agglomerate break-up in the sense that both rely on forces to disperse the particulates. Hence, elongation is also the preferred mode of deformation when breaking up fluid droplets and threads, making the flow number, λ, an indispensable quantity when quantifying mixing processes that deal with such systems.

A parameter commonly used to determine whether a droplet will disperse is the capillary number defined by

$$Ca = \frac{\tau R}{\sigma_s} \tag{6.12}$$

where τ is the flow induced or deviatoric stress, R the characteristic dimension of the droplet and σ_S the surface tension that acts on the drop. The capillary number is the ratio of flow stresses to droplet surface stresses. Droplet break-up occurs when a critical capillary number, Ca_{crit}, is reached. This break-up is clearly shown in Fig. 6.31 [1], which shows the disintegration of a Newtonian thread in a Newtonian matrix. Because of the continuously decreasing thread radius, the critical capillary number will be reached at some specific point in time. Due to the competing deviatoric stresses and surface forces, the cylindrical shape becomes unstable and small disturbances at the surface lead to a growth of capillary waves. These waves are commonly referred to as Rayleigh disturbances. Disturbances with various wavelengths form on the cylinder surface, but only those with a wavelength greater than the circumference ($2\pi R_0$) of the thread lead to a monotonic decrease of the interfacial area.

Figure 6.32 [14] shows the critical capillary number as a function of viscosity ratio, ϕ, and flow type, described by the mixing parameter λ. For a viscosity ratio of 1 the critical capillary number is of order 1 [1]. Distributive mixing is implied when Ca is much greater than Ca_{crit} since the

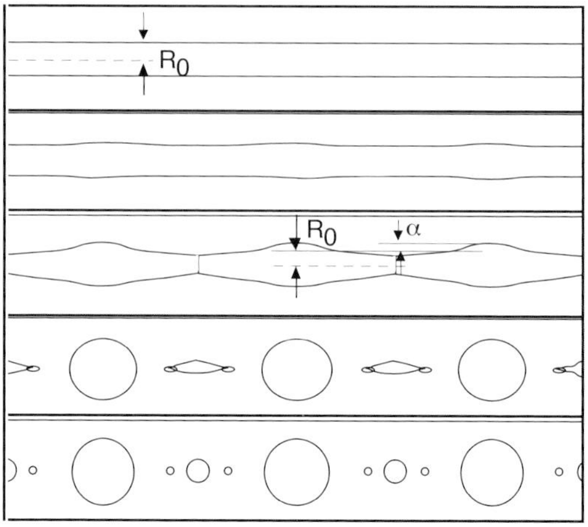

Figure 6.31 Disintegration of a Newtonian 0.35 mm diameter castor oil thread in a Newtonian silicon oil matrix. Redrawn from photographs taken every second.

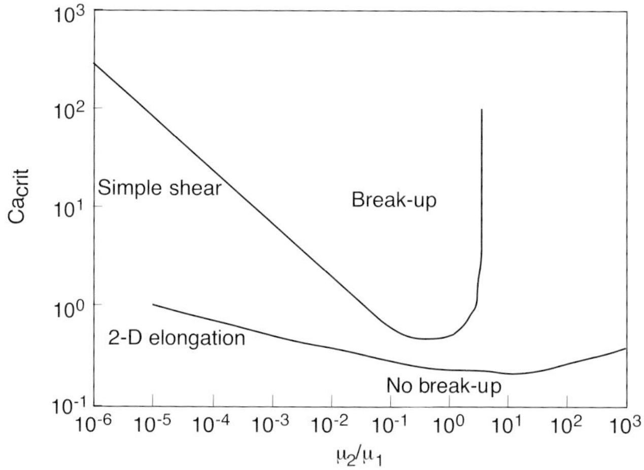

Figure 6.32 Critical capillary number for drop break-up as a function of viscosity ratio in a simple shear and a 2-D elongational flow.

interfacial stress is much smaller than shear stresses. For such a case the capillary waves which would cause droplet break-up would not develop. Dispersive mixing is implied when *Ca* is close to the value of the critical *Ca* or when interfacial stresses are almost equal to the deviatoric stresses causing droplet break-up. In addition, break-up can only occur if enough time is given for this to happen. The disturbance amplitude, α, is assumed to grow exponentially as

$$\alpha = \alpha_0 e^{qt} \tag{6.13}$$

where α_0 is the initial disturbance amplitude, sometimes assumed to be 0.3% of the thread radius, and the growth rate q defined by

$$q = \frac{\sigma_s \Omega}{2\eta_1 R_0} \tag{6.14}$$

In the above equation R_0 represents the initial radius of the thread and Ω a dimensionless growth rate presented in Fig. 6.33 as a function of viscosity ratio for the wavelength disturbance amplitude which leads to break-up. The time required for break-up, t_b, can now be computed using the above equations as

$$t_b = \frac{1}{q} \ln\left(\alpha_b / \alpha_0 \right) \tag{6.15}$$

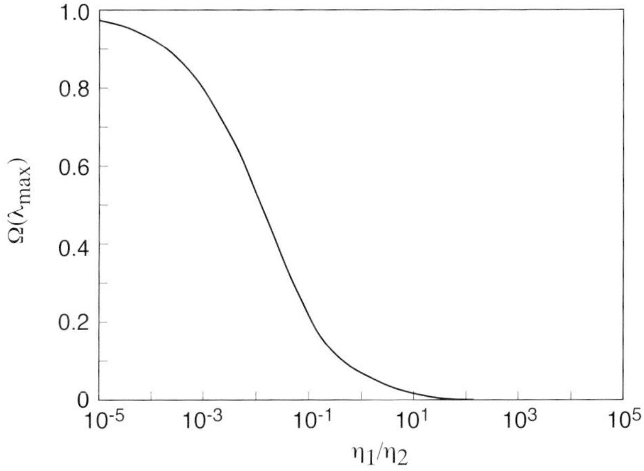

Figure 6.33 Dominant growth rate of interfacial disturbances as a function of viscosity ratio.

where α_b is the amplitude at break-up which for a sinusoidal disturbance is $\alpha_b = \sqrt{2/3} R_0$. The break-up time decreases as the critical capillary number is exceeded. The reduced break-up time t_b^* can be approximated using [14]

$$t_b^* = t_b \left(\frac{Ca}{Ca_{crit}} \right)^{-0.63} \tag{6.16}$$

As mentioned before, surface tension plays a large role in the mixing process, especially when dealing with dispersive mixing, when the capillary number approaches its critical value. Because of the stretching of the interfacial area, due to distributive mixing, the local radii of the suspended components decrease, causing surface tension to play a role in the process. It should also be noted that once the capillary number assumes a value below the critical Ca, only slight deformations occur and internal circulation maintains an equilibrium elliptical droplet shape in the flow field as schematically represented in Fig. 6.34. At that point, the mixing process reduces to the distribution of the dispersed droplets. Analytical and numerical investigations of stable droplet shapes, for $Ca < Ca_{crit}$ in simple shear flow have been made by several investigators [15–17]. Figure 6.32 also shows that at viscosity ratios above 4 simple shear flows are not able to break-up fluid droplets.

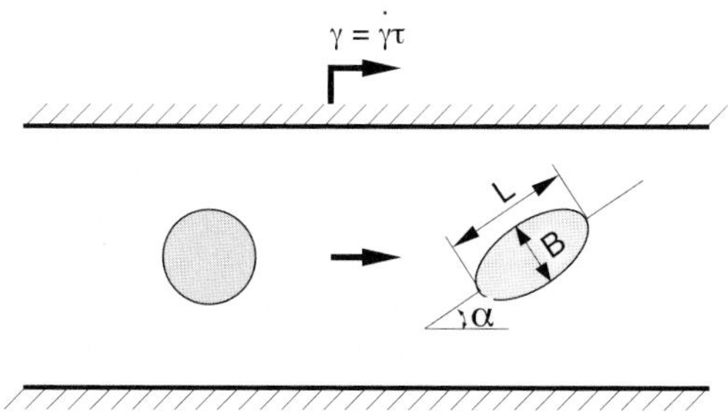

Figure 6.34 Schematic of droplet deformation in simple shear flow.

6.2.3 Mixing Devices

The final properties of a polymer component are heavily influenced by the blending or mixing process that takes place during processing or as a separate step in the manufacturing process. As mentioned earlier, when measuring the quality of mixing it is also necessary to evaluate the efficiency of mixing. For example, the amount of power required to achieve the highest mixing quality for a blend may be unrealistic or unachievable. This section presents some of the most commonly used mixing devices encountered in polymer processing.

In general, mixers can be classified in two categories: internal batch mixers and continuous mixers. Internal batch mixers, such as the Banbury type mixer, are the oldest type of mixing devices in polymer processing and are still widely used in the rubber compounding industry. Industry often also uses continous mixers because they combine mixing in addition to their normal processing tasks. Typical examples are single and twin screw extruders that often have mixing heads or kneading blocks incorporated into their system.

6.2.3.1 Static Mixers

Static mixers or motionless mixers are pressure-driven continuous mixing devices through which the melt is pumped, rotated, and divided, leading to effective mixing without the need for movable parts and mixing heads. One of the most commonly used static mixers is the twisted tape static mixer schematically shown in Fig. 6.35. Figure 6.36 [19] shows computed streamlines relative to the twist in the wall. As the fluid is rotated by the dividing wall, the interfaces between the fluids increase. The interfaces are then re-oriented by 90° once the material enters a new section. Figure 6.36 shows a typical trajectory of a particle as it travels on a streamline in section N of the static mixer and ends on a different streamline after entering the next section, N+1. The stretching-re-orientation sequence is repeated until the number of striations is so high that a seemingly homogeneous mixture is achieved. Figure 6.37 shows a sequence of cuts down a Kenics static mixer[5]. From the figure it can be seen that the number of striations increases from section to section by 2, 4, 8, 16, 32, etc., which can be computed using

$$N = 2^n \tag{6.17}$$

where N is the number of striations and n is the number of sections in the mixer.

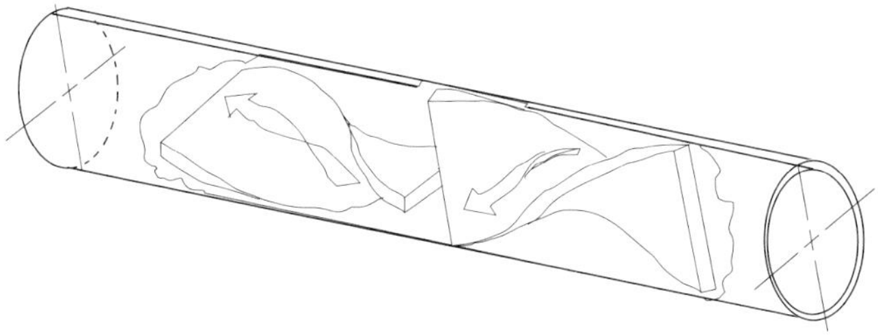

Figure 6.35 Schematic diagram of a Kenics static mixer.

5 Courtesy Chemineer, Inc., North Andover, Massachusetts.

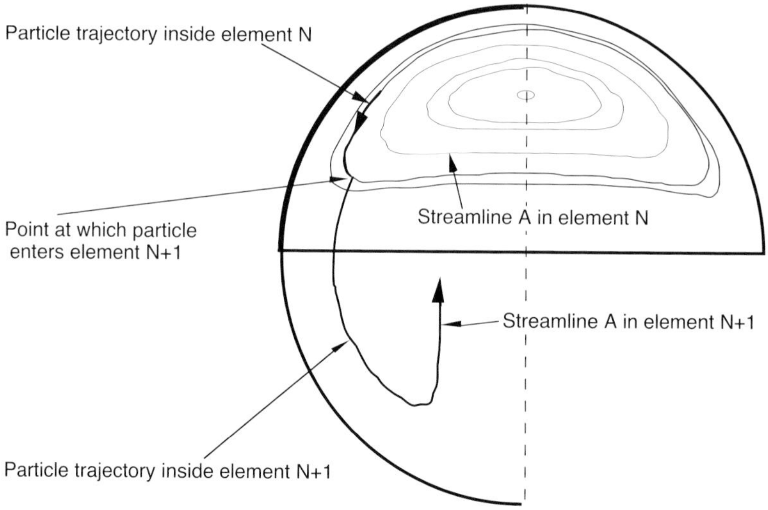

Figure 6.36 Simulated streamlines inside a Kenics static mixer section.

Figure 6.37 Experimental progression of the layering of colored resins in a Kenics static mixer.

6.2.3.2 Banbury Mixer

The Banbury type mixer, schematically shown in Fig. 6.38, is perhaps the most commonly used internal batch mixer. Internal batch mixers are high intensity mixers that generate complex shearing and elongational flows which work especially well in the dispersion of solid particle agglomerates within polymer matrices. One of the most common applications for high intensity internal batch mixing is the break-up of carbon black agglomerates into rubber compounds. The dispersion of agglomerates is strongly dependent on mixing time, rotor speed, temperature, and rotor blade geometry [18]. Figure 6.39 [15, 21] shows the fraction of undispersed carbon black as a function of time in a Banbury mixer at 77 rpm and 100 °C. The broken line in the figure represents the fraction of particles smaller than 500 nm.

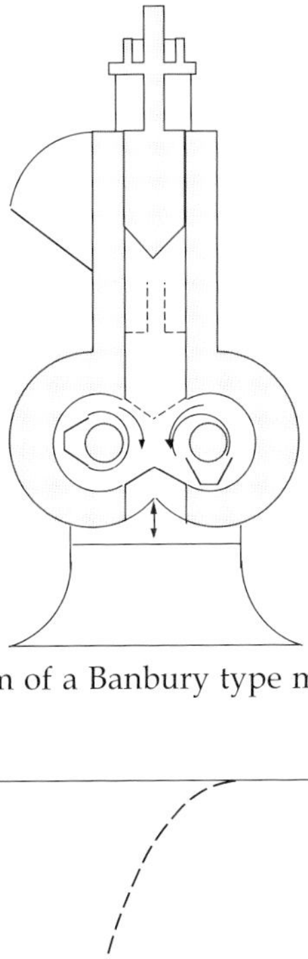

Figure 6.38 Schematic diagram of a Banbury type mixer.

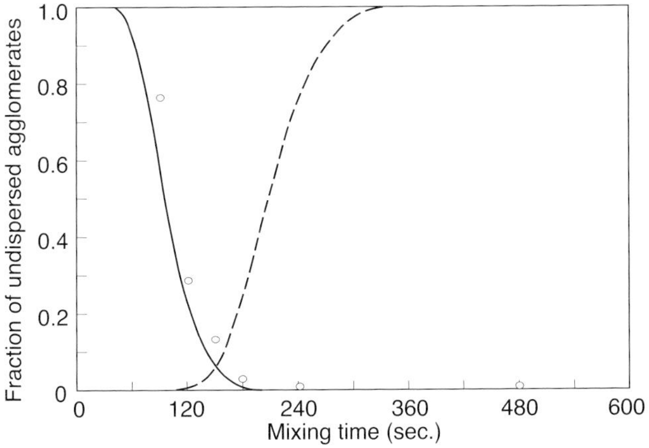

Figure 6.39 Fraction of undispersed carbon black, of size above 9 μm, as a function of mixing time inside a Banbury mixer. (O) denotes experimental results and solid line theoretical predictions. Broken line denotes the fraction of aggregates of size below 500 nm.

6.2.3.3 Mixing in Single Screw Extruders

Mixing caused by the cross-channel flow component can be further enhanced by introducing pins in the flow channel. These pins can either sit on the screw as shown in Fig. 6.40 [22] or on the barrel as shown in Fig. 6.41 [23]. The extruder with the adjustable pins on the barrel is generally referred to as QSM-extruder[6]. In both cases the pins disturb the flow by re-orienting the surfaces between fluids and by creating new surfaces by splitting the flow. Figure 6.42 presents a photograph of the channel contents of a QSM-extruder[7]. The photograph clearly demonstrates the re-orientation of the layers as the material flows past the pins. The pin type extruder is especially necessary for the mixing of high viscosity materials such as rubber compounds; thus, it is often called a cold feed rubber extruder. This machine is widely used in the production of rubber profiles of any shape and size.

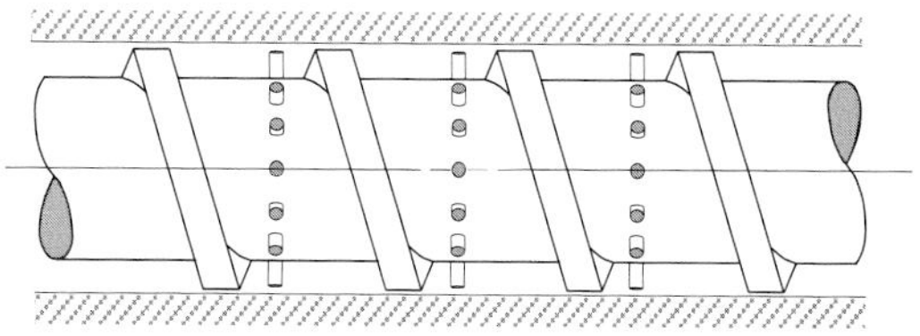

Figure 6.40 Pin mixing section on the screw of a single screw extruder.

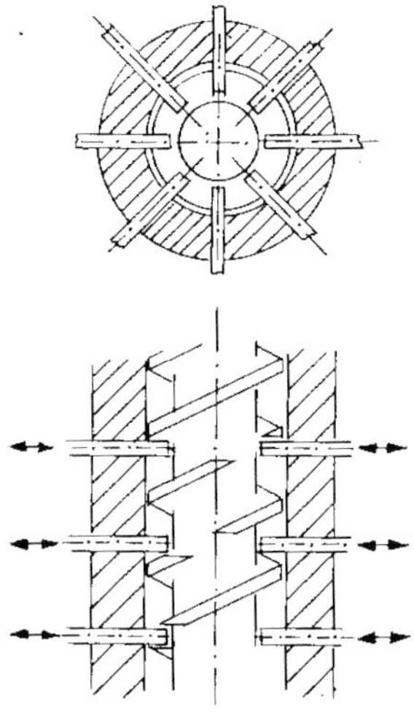

Figure 6.41 Pin barrel extruder (Quer Strom Misch Extruder).

6 QSM comes from the German *Quer Strom Misch* which translates into cross-flow mixing.

7 Courtesy of Paul Troester Maschinenfabrik, Hannover, Germany.

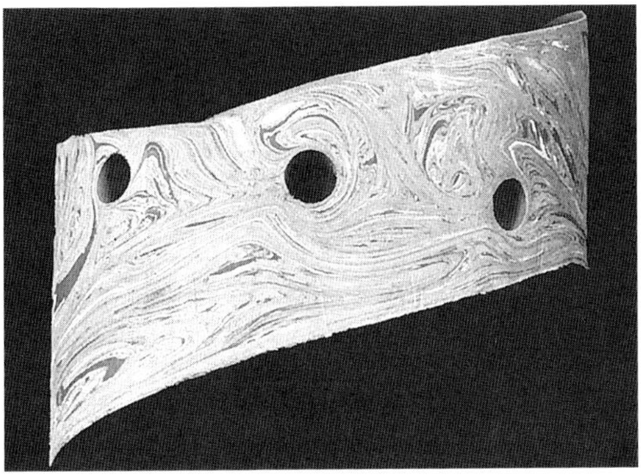

Figure 6.42 Photograph of the unwrapped channel contents of a pin barrel extruder.

For lower viscosity fluids, such as thermoplastic polymer melts, the mixing action caused by the cross-flow is often not sufficient to re-orient, distribute, and disperse the mixture, making it necessary to use special mixing sections. Re-orientation of the interfaces between primary and secondary fluids and distributive mixing can be induced by any disruption in the flow channel. Figure 6.43 [22] presents commonly used distributive mixing heads for single screw extruders. These mixing heads introduce several disruptions in the flow field which have proven to perform well in mixing.

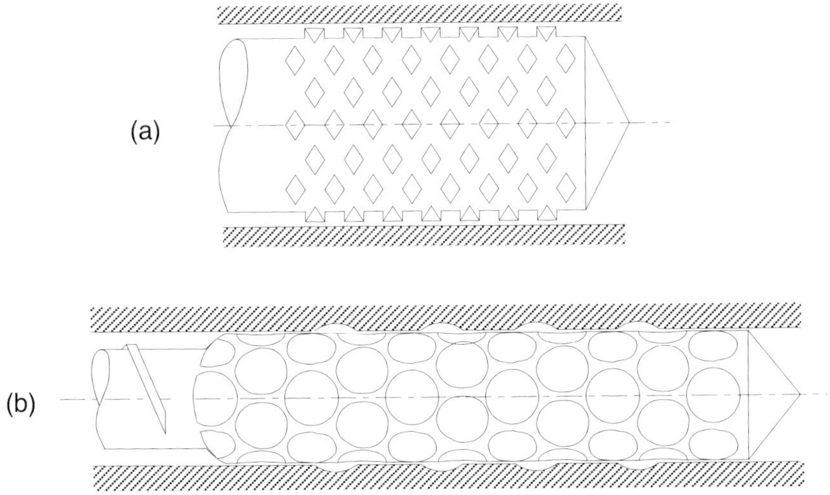

Figure 6.43 Distributive mixing sections: (a) Pineapple mixing section, (b) cavity transfer mixing section.

As mentioned earlier, dispersive mixing is required when breaking down particle agglomerates or when surface tension effects exist between primary and secondary fluids in the mixture. To disperse such systems, the mixture must be subjected to large stresses. Barrier-type screws are often sufficient to apply high stresses to the polymer melt. However, more intensive mixing can be applied by using a mixing head. When using barrier-type screws or a mixing head as shown in Fig. 6.44 [22] the mixture is forced through narrow gaps, causing high stresses in the melt. It should be noted that dispersive as well as distributive mixing heads result in a resistance to the flow, which results in viscous heating and pressure losses during extrusion.

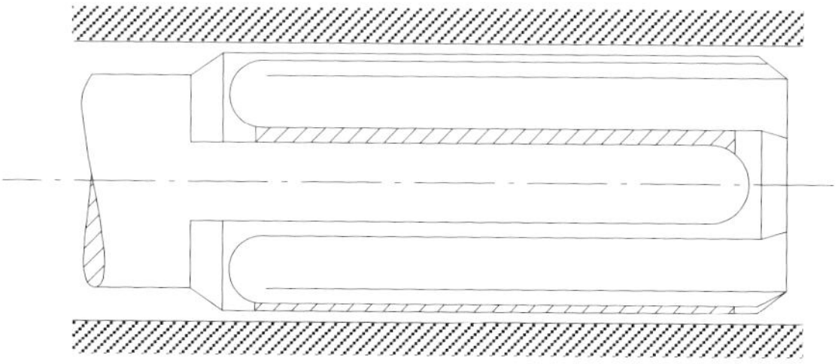

Figure 6.44 Maddock or Union Carbide mixing section.

6.2.3.4 Cokneader

The cokneader is a single screw extruder with pins on the barrel and a screw that oscillates in the axial direction. Figure 6.45 shows a schematic diagram of a cokneader. The pins on the barrel practically wipe the entire surface of the screw, making it the only self-cleaning single-screw extruder. This results in a reduced residence time, which makes it appropriate for processing thermally sensitive materials. The pins on the barrel also disrupt the solid bed creating a dispersed melting [24] which improves the overall melting rate while reducing the overall temperature in the material.

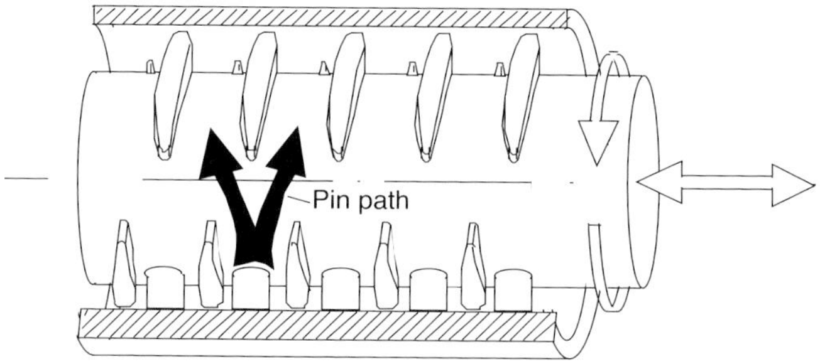

Figure 6.45 Schematic diagram of a cokneader.

A simplified analysis of a cokneader gives a number of striations per *L/D* of [25]

$$N_s = 2^{12} \tag{6.18}$$

which means that over a section of 4D the number of striations is $2^{12}(4) = 2.8E14$. A detailed discussion on the cokneader is given by Rauwendaal [25] and Elemans [26].

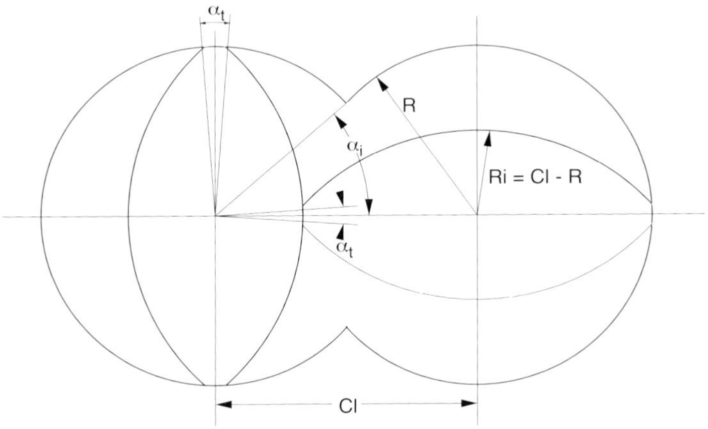

Figure 6.46 Geometry description of a double-flighted, co-rotating, self-cleaning, twin screw extruder.

6.2.3.5 Twin Screw Extruders

In the past two decades, twin screw extruders have developed into the best available continuous mixing devices. In general, they can be classified into intermeshing or non-intermeshing, and co-rotating or counter-rotating twin screw extruders[8]. The intermeshing twin screw extruders render a self-cleaning effect which evens-out the residence time of the polymer in the extruder. The self-cleaning geometry for a co-rotating double flighted twin screw extruder is shown in Fig. 6.46. The main characteristic of this type of configuration is that the surfaces of the screws are sliding past each other, constantly removing the polymer that is stuck to the screw.

In the last two decades, the co-rotating twin screw extruder systems have established themselves as efficient continuos mixers, including reactive extrusion. In essence, the co-rotating systems have a high pumping efficiency caused by the double transport action of the two screws. Counter- rotating

8 A complete overview of twin screw extruders is given by White, J.L., *Twin Screw Extrusion-Technology and Principles*, Hanser Publishers, Munich, (1990).

systems generate high stresses because of the calendering action between the screws, making them efficient machines to disperse pigments and lubricants[9].

Several studies have been performed to evaluate the mixing capabilities of twin screw extruders. Noteworthy are two recent studies performed by Lim and White [27, 28] that evaluated the morphology development in a 30.7 mm diameter screw co-rotating [17] and a 34 mm diameter screw counter-rotating [18] intermeshing twin screw extruder. In both studies they dry-mixed 75/25 blend of polyethylene and polyamide 6 pellets that were fed into the hopper at 15 kg/h. Small samples were taken along the axis of the extruder and evaluated using optical and electron microscopy.

Figure 6.47 shows the morphology development along the screws at positions marked A, B, C, and D for a counter-rotating twin screw extruder configuration without special mixing elements. The dispersion of the blend becomes visible by the reduction of the characteristic size of the polyamide 6 phase. Figure 6.48 is a plot of the weight average and number average domain size of the polyamide 6 phase along the screw axis. The weight average phase size at the end of the extruder was measured to be 10 μm and the number average 6 μm. By replacing sections of the screw with one kneading-pump element and three special mixing elements, the final weight average phase size was reduced to 2.2 μm and the number average to 1.8 μm, as shown in Fig. 6.49.

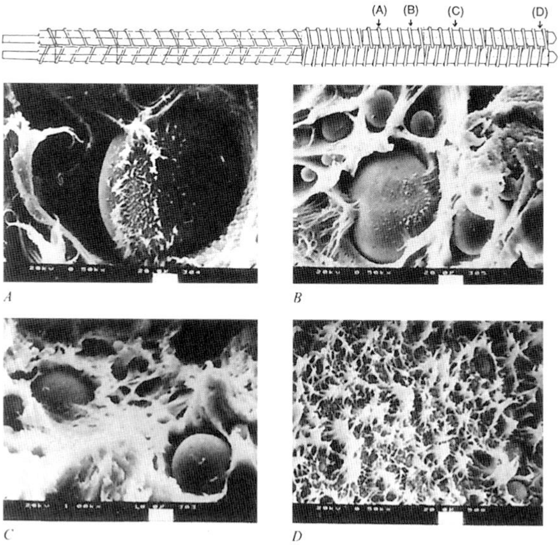

Figure 6.47 Morphology development inside a counter-rotating twin screw extruder.

9 There seems to be considerable disagreement about co- versus counter-rotating twin screw extruders between different groups in the polymer processing industry and academic community.

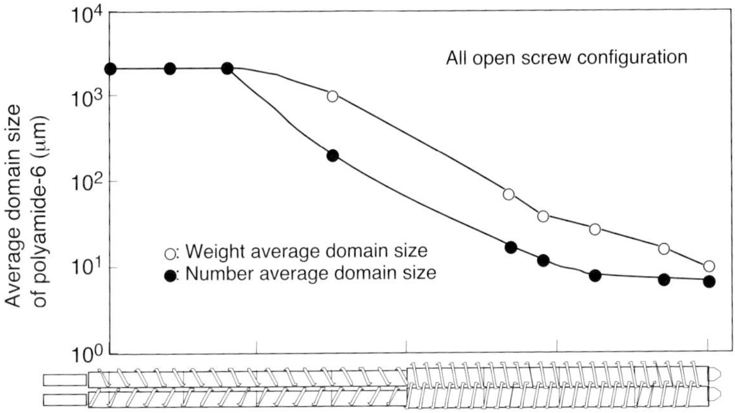

Figure 6.48 Number and weight average of polyamide 6 domain sizes along the screws for a counter-rotating twin screw extruder.

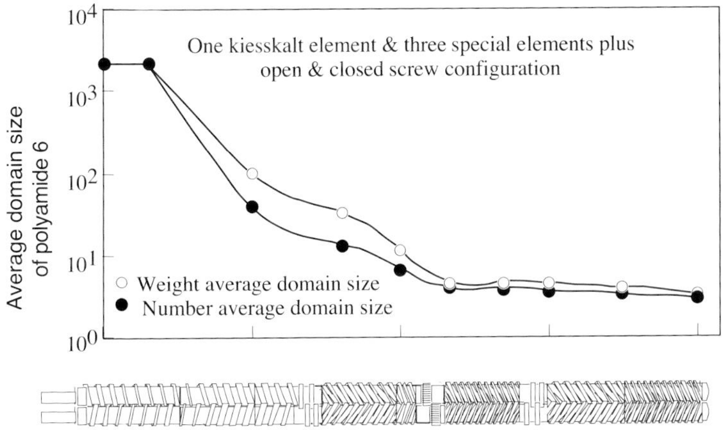

Figure 6.49 Number and weight average of polyamide 6 domain sizes along the screws for a counter-rotating twin screw extruder with special mixing elements.

Using a co-rotating twin screw extruder with three kneading disk blocks, a final morphology with polyamide 6 weight average phase sizes of 2.6 μm was achieved. Figure 6.50 shows the morphology development along the axis of the screws. When comparing the outcome of both counter-rotating (Fig. 6.49) and co-rotating (Fig. 6.50), it is clear that both extruders achieve a similar final mixing quality. However, the counter-rotating extruder achieved the final morphology much earlier in the screw than the co-rotating twin screw extruder. A possible explanation for this is that the blend traveling through the counter-rotating configuration melted earlier than in the co-rotating geometry. In addition the phase size was slightly smaller, possibly due to the calendering effect between the screws in the counter-rotating system.

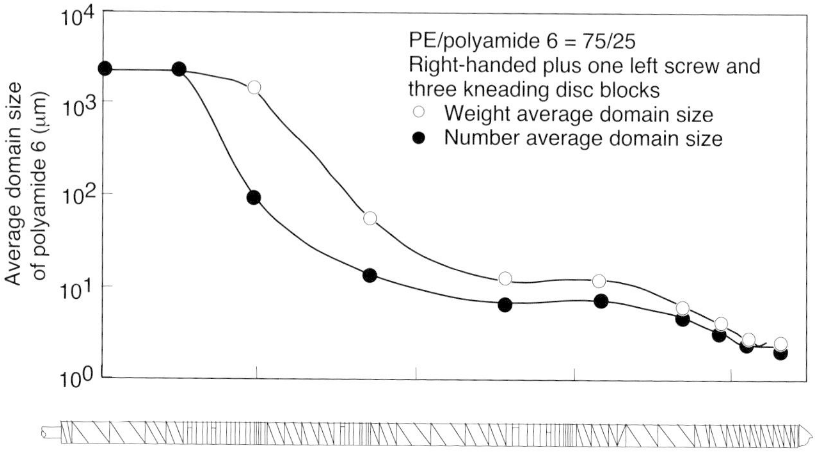

Figure 6.50 Number and weight average of polyamide 6 domain sizes along the screws for a co-rotating twin screw extruder with special mixing elements.

6.2.4 Energy Consumption During Mixing

Energy consumption is of extreme importance when assessing and comparing various mixing devices. High energy requirements for optimal mixing mean high costs and expensive equipment. The power consumption per unit volume of a deforming Newtonian fluid is given by [29]

$$p = 2\mu\left[\left(\frac{\partial v_x}{\partial x}\right)^2 + \left(\frac{\partial v_y}{\partial y}\right)^2 + \left(\frac{\partial v_z}{\partial z}\right)^2\right] + \mu\left[\left(\frac{\partial v_x}{\partial y} + \frac{\partial v_y}{\partial x}\right)^2 + \left(\frac{\partial v_y}{\partial z} + \frac{\partial v_z}{\partial y}\right)^2 + \left(\frac{\partial v_x}{\partial z} + \frac{\partial v_z}{\partial x}\right)^2\right] \quad (6.19)$$

Erwin [30] used the above equation to assess the energy input requirements for different types of mixing flows: simple shear, pure shear, and extensional flows. Table 6.4 presents flow fields and energy requirements for various flows described by Erwin [30]. For example, to produce a mixture such that $A/A_0 = 10^4$ in time $t_0 = 100$ s, for a fluid with viscosity $\mu = 10^4\,\text{Pa}-\text{s}$, in a mixer which deforms the fluid with an elongational flow, one needs 96 kJ/m^3 of energy input. Since the flow is steady, this requires a power input of 0.96 kW/m^3 for 100 s. In a mixer that deforms the fluid in a biaxial extensional flow the energy required is 24 kJ/m^3 with 0.24 kW/m^3 of power input. For the same amount of mixing, a mixer which deforms the fluid in pure shear requires an energy input of 40 kJ/m^3 or a steady power input of 0.4 kW/m^3 for 100 s. A device that deforms the fluid in simple shear requires a total energy input of 4x107 kJ/m^3 or a steady power input of 40,000 kW/m^3 for a 100 second period to achieve the same amount of mixing.

Table 6.4 Energy Input Requirements for Various Flow Mixers

Flow type	Flow field	Power	Energy input
Extensional flow (elongational)	$v_x = Gx$ $v_y = -Gy/2$ $v_z = -Gz/2$	$3\mu G^2$	$\frac{12\mu}{t_0}\left(\ln\left(\frac{5A}{4A_0}\right)\right)^2$
Extensional flow (biaxial)	$v_x = -Gx$ $v_y = Gy/2$ $v_z = Gz/2$	$3\mu G^2$	$\frac{3\mu}{t_0}\left(\ln\left(\frac{5A}{4A_0}\right)\right)^2$
Pure shear	$v_x = -hx$ $v_y = -Hy$ $v_z = 0$	$2\mu(h^2 + H^2)$	$\frac{4\mu}{t_0}\left(\ln\left(2\frac{A}{A_0}\right)\right)^2$
Simple shear	$v_x = -Gy$ $v_y = 0$ $v_z = 0$	μG^2	$\frac{4\mu}{t_0}\left(\frac{A}{A_0}\right)^2$

From this it is clear that, in terms of energy and power consumption, simple shear flows are significantly inferior to extensional flows.

6.2.5 Mixing Quality and Efficiency

In addition to the flow number, strain, and capillary number, several parameters have been developed by various researchers in the polymer industry to quantify the efficiency of the mixing processes. Some have used experimentally measured parameters while others have used mixing parameters which are easily calculated from computer simulation.

A parameter used in visual experiments is the batch homogenization time (BHT). This parameter is defined as the time it takes for a material to become homogeneously colored inside the mixing chamber after a small sample of colored pigment is placed near the center of the mixer. A downfall to this technique is that the observed homogenized time can be quite subjective.

To describe the state of the dispersion of fillers in a composite material, Suetsugu [31] used a dispersion index defined as:

$$\text{Dispersion index} = 1 - \phi_a \tag{6.20}$$

where ϕ_a is a dimensionless area that the agglomerates occupy and is defined by:

$$\phi_a = \frac{\pi}{4A\phi}\sum d_i^2 n_i \tag{6.21}$$

where A is the area under observation, ϕ the volume fraction of the filler, d_i the diameter of the agglomerate and n_i the number agglomerates. The dispersion index ranges between 0 for the worst case of dispersion and 1 where no agglomerates remain in the system.

A commonly used method to analyze the mixing capabilities of the extruder is the residence time distribution (RTD). It is calculated by monitoring the output of the extruder with the input of a secondary component. Two common response techniques are the step input response and the pulse input response shown in Fig. 6.51 [22]. The response of the input gives information on the mixing and conveying performance of the extruder. The RTD response to a pulse input for an ideal mixing situation is shown in Fig. 6.52. The figure shows a quick response to the input with a constant volume fraction of the secondary component until there is no material left.

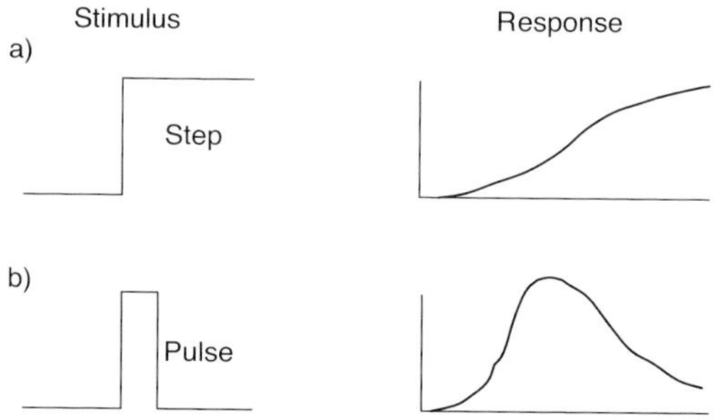

Figure 6.51 Step input and pulse input residence time distribution responses.

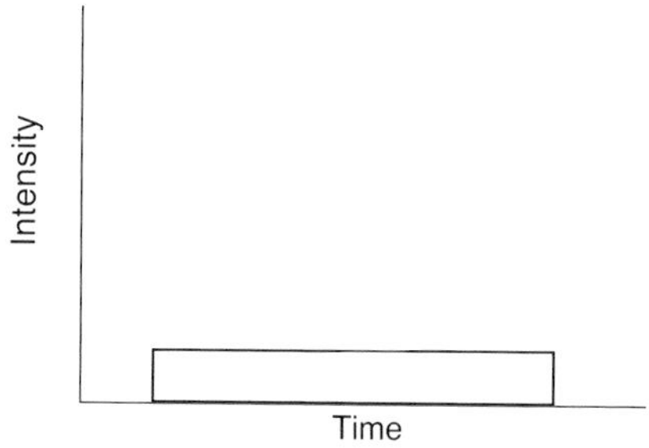

Figure 6.52 Ideal residence time distribution response to a pulse input.

By the use of a computer simulation, velocities, velocity gradients, and particle tracking can be computed with some degree of accuracy—depending on the computational method and assumptions made. Using information from a computer simulation, several methods to quantify mixing have been developed. Poincaré sections are often used to describe the particle paths during the mixing process. The Poincaré section shows the trajectory of several particles during the mixing process. They can be very useful in locating stagnation points, recirculation regions, and detecting symmetric flow patterns where no exchange exists across the planes of symmetry—all issues that hinder mixing.

6.2.6 Plasticization

Solvents, commonly referred to as plasticizers, are sometimes mixed into a polymer to dramatically alter its rheological and/or mechanical properties. Plasticizers are used as a processing aid since they have the same impact as raising the temperature of the material. Hence, the lowered viscosities at lower temperatures reduce the risk of thermal degradation during processing. For example, cellulose nitrite would thermally degrade during processing without the use of a plasticizer.

Plasticizers are more commonly used to alter a polymer's mechanical properties such as stiffness, toughness, and strength. For example, adding a plasticizer such as dioctylphthalate (DOP) to PVC can reduce its stiffness by three orders of magnitude and lower its glass transition temperature to -35 ^{o}C. In fact, a highly plasticized PVC is rubbery at room temperature. Table 6.5 [45] presents some common plasticizers with the polymers they plasticize, and their applications.

Since moisture is easily absorbed by polyamides, slightly modifying their mechanical behavior, it can be said that water acts as a plasticizing agent with these materials. Figure 6.53 shows the equilibrium water content for polyamide 6 and 66 as a function of the ambient relative humidity[10]. This moisture absorption causes the polyamide to expand or swell as shown in Fig. 6.54[11].

10 Courtesy of Bayer AG, Germany.

11 *Ibid.*

Table 6.5 Commercial Plasticizers and Their Applications

Plasticizer	Polymers	Plasticizer type
Di-octyl phthalate (DOP)	Polyvinyl chloride and copolymers	General purpose, primary plasticizer
Tricresyl phosphate (TCP)	Polyvinyl chloride and copolymers, cellulose acetate, cellulose nitrate	Flame retardant, primary plasticizer
Di-octyl adipate (DOA)	Polyvinyl chloride, cellulose acetate, butyrate	Low temperature plasticizer
Di-octyl sebacate (DOS)	Polyvinyl chloride , cellulose acetate, butyrate	Secondary plasticizer
Adipic acid polyesters (MW = 1500-3000)	Polyvinyl chloride	Non-migratory secondary plasticizer
Sebacic acid polyesters (MW = 1500-3000)	Polyvinyl chloride	Non-migratory secondary plasticizer
Chlorinated paraffins (%CI = 40-70) (MW = 600-1000)	Most polymers	Flame retardant, plasticizer extenders
Bi- and terphenyls (also hydrogenated)	Aromatic polyesters	Various
N-ethyl-toluene sulphonamide	Polyamides	General purpose, primary plasticizer
Sulphonamide-formaldehyde resins	Polyamides	Non-migratory secondary plasticizers

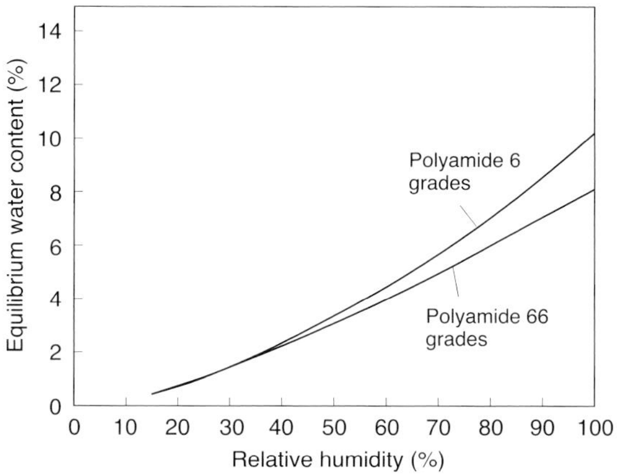

Figure 6.53 Equilibrium water content as a function of relative humidity for polyamide 6 and polyamide 66.

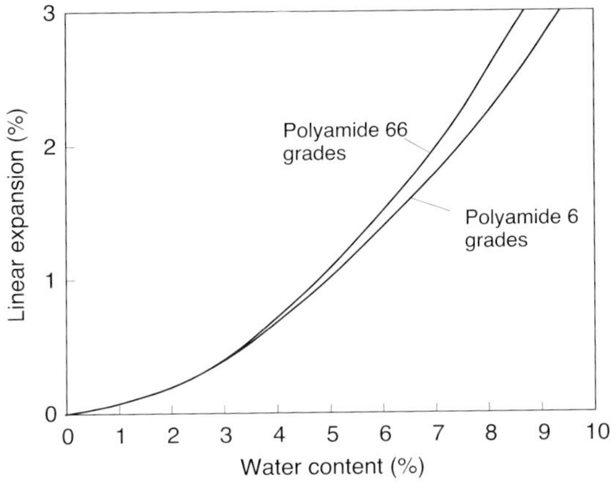

Figure 6.54 Linear expansion as a function of water content for polyamide 6 and polyamide 66.

The behavior of polymers toward solvents depends in great part on the nature of the solvent and on the structure of the polymer molecules. If the basic building block of the macromolecule and the solvent molecule are the same or of similar nature, then the absorption of a solution will lead to swelling. If a sufficient amount is added, the polymer will dissolve in the solvent. Crystalline regions of a semi-crystalline thermoplastic are usually not affected by solvents, whereas amorphous regions are easily penetrated. In addition, the degree of cross-linking in thermosets and elastomers has a great influence on whether a material can be permeated by solvents. The shorter the distances between the linked molecules, the less solvent molecules can permeate and give mobility to chain segments. While elastomers can swell in the presence of a solvent, highly cross-linked thermosets do not swell or dissolve.

The amount of solvent that is absorbed depends not only on the chemical structure of the two materials but also on the temperature. Since an increase in temperature reduces the covalent forces of the polymer, solubility becomes higher. Although it is difficult to determine the solubility of polymers, there are some rules to estimate it. The simplest rule is: same dissolves same (i.e., when both—polymer and solvent—have the same valence forces, solubility exists).

The solubility of a polymer and a solvent can be addressed from a thermodynamic point of view using the familiar Gibbs free energy equation

$$\Delta G = \Delta H - T\Delta S \tag{6.22}$$

where ΔG is the change in free energy, ΔH is the change in enthalpy, ΔS the change in entropy and T the temperature. If ΔG in Eq. 6.22 is negative solubility, is possible. A positive ΔH suggests that polymer and the solvent do not "want" to mix, which means solubility can only occur if $\Delta H < T\,\Delta S$. On the other hand, $\Delta H \approx 0$ implies that solubility is the natural lower energy state. Since the entropy change when dissolving a polymer is very small, the determining factor if a solution will occur or not is the change in enthalpy, ΔH. Hildebrand and Scott [46] proposed a useful equation that estimates the change in enthalpy during the formation of a solution. The Hildebrand equation is stated by

$$\Delta H = V\left(\left(\frac{\Delta E_1}{V_1}\right)^{1/2} - \left(\frac{\Delta E_2}{V_2}\right)^{1/2}\right)^2 \phi_1\phi_2 \tag{6.23}$$

where V is the total volume of the mixture, V_1 and V_2 the volumes of the solvent and polymer, ΔE_1 and ΔE_2 their energy of evaporation and, ϕ_1 and ϕ_2 their volume fractions. Equation 6.23 can be simplified to

$$\Delta H = V(\delta_1 - \delta_2)^2 \phi_1\phi_2 \tag{6.24}$$

where δ is called the solubility parameter and is defined by

$$\delta = \left(\frac{\Delta E}{V}\right)^{1/2} \tag{6.25}$$

If the solubility parameter of the substances are nearly equal they will dissolve. A rule-of-thumb can be used that if $|\delta_1 - \delta_2| < 1\ (\text{cal/cm}^3)^{1/2}$ solubility will occur [47]. The units $(\text{cal/cm}^3)^{1/2}$ are usually referred to as Hildebrands. Solubility parameters for various polymers are presented in Table 6.6 [48], and for various solvents in Table 6.7 [48].

Figure 6.55 [49] shows a schematic diagram of swelling and dissolving behavior of cross-linked and uncross-linked polymers as a function of the solubility or *solubility parameter*, δ_s, of the solvent. When the solubility parameter of the polymer and the solvent approach each other, the uncross-linked polymer becomes unconditionally soluble. However, if the same polymer is cross-linked then it is only capable of swelling. The amount of swelling depends on the degree of cross-linking.

Table 6.6 Solubility Parameter for Various Polymers

Polymer	$\delta\left(\mathrm{cal/cm^3}\right)^{1/2}$
Polytetrafluoroethylene	6.2
Polyethylene	7.9
Polypropylene	8.0
Polyisobutylene	8.1
Polyisoprene	8.3
Polybutadiene	8.6
Polystyrene	9.1
Poly(vinyl acetate)	9.4
Poly(methyl methacrylate)	9.5
Polycarbonate	9.9
Polysulphone	9.9
Poly(vinyl chloride)	10.1
Polyethylene terephthalate	10.2
Polyamide 6	11.0
Cellulose nitrate	11.5
Poly(vinylidene chloride)	12.2
Polyamide 66	13.6
Polyacrylonitrile	15.4

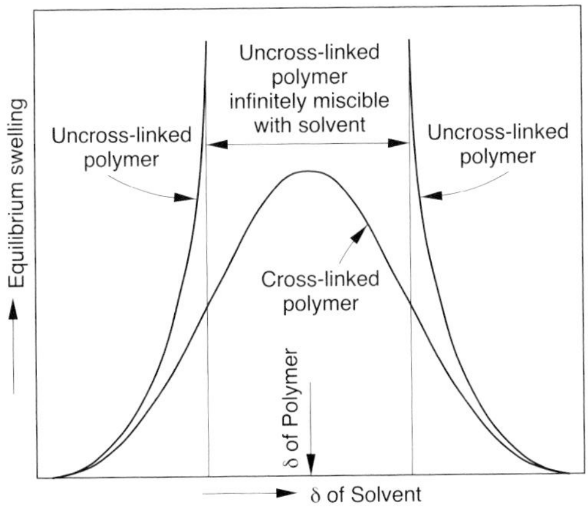

Figure 6.55 Equilibrium swelling as a function of solubility parameter for cross-linked and uncross-linked polymers.

Table 6.7 Solubility Parameter of Various Plasticizers and Solvents

Solvent	$\delta\left(\text{cal}/\text{cm}^3\right)^{1/2}$
Acetone	10.0
Benzene	9.1
Di-butoxyethyl phthalate (Dronisol)	8.0
n-butyl alcohol	11.4
Sec-butyl alcohol	10.8
Butyl stearate	7.5
Chlorobenzene	9.6
Cyclohexanone	9.9
Dibutyl phenyl phosphate	8.7
Dibutyl phthalate	9.3
Dibutyl sebacate	9.2
Diethyl phthalate	10.0
Di-n-hexyl phthalate	8.9
Diisodecyl phthalate	7.2
Dimethyl phthalate	10.7
Dioctyl adipate	8.7
Dioctyl phthalate (DOP)	7.9
Dioctyl sebacate	8.6
Dipropyl phthalate	9.7
Ethyl acetate	9.1
Ethyl alcohol	12.7
Ethylene glycol	14.2
2-ethylhexyl diphenyl phosphate (Santicizer 141)	8.4
N-ethyl-toluenesulfonamide (Santicizer 8)	11.9
Hydrogenated terphenyl (HB-40)	9.0
Kronisol	8.0
Methanol	14.5
Methyl ethyl ketone	9.3
Nitromethane	12.7
n-propyl alcohol	11.9
Toluene	8.9
Tributyl phophate	8.2
1,1,2-trichloro-1,2,2-trifluoroethane (freon 113)	7.2
Trichloromethane (chloroform)	9.2
Tricresyl phosphate	9.0
Triphenyl phosphate	9.2
Water	23.4
Xylene	8.8

6.3 Injection Molding

Injection molding is the most important process used to manufacture plastic products. Today, more than one-third of all thermoplastic materials are injection molded and more than half of all polymer processing equipment is for injection molding. The injection molding process is ideally suited to manufacture mass-produced parts of complex shapes requiring precise dimensions. The process goes back to 1872 when the Hyatt brothers patented their stuffing machine to inject cellulose into molds. However, today's injection molding machines are mainly related to the reciprocating screw injection molding machine patented in 1956. A modern injection molding machine with its most important elements is shown in Fig. 6.56. The components of the injection molding machine are the plasticating unit, clamping unit, and the mold.

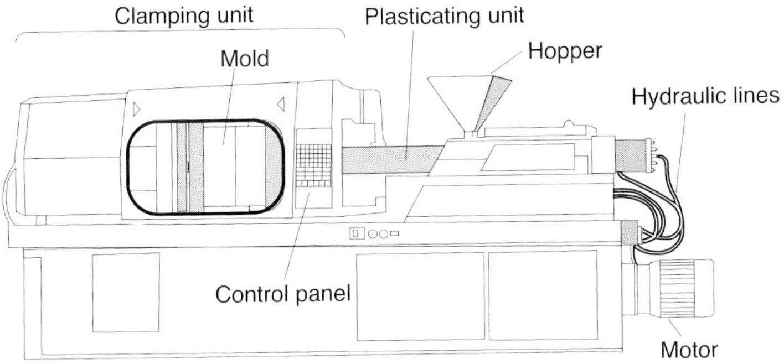

Figure 6.56 Schematic of an injection molding machine.

Today, injection molding machines are classified by the following international convention[12]

Manufacturer $T\,/\,P$

where T is the clamping force in metric tons and P is defined as

$$P = \frac{V_{max}\,p_{max}}{1000} \tag{6.26}$$

where V_{max} is the maximum shot size in cm3 and p_{max} is the maximum injection pressure in bar. The clamping forced T can be as low as 1 metric ton for small machines, and as high as 11,000 tons.

12 The old US convention uses MANUFACTURER T-v where T is the clamping force in British tons and v the shot size in ounces of polystyrene.

6.3.1 The Injection Molding Cycle

The sequence of events during the injection molding of a plastic part, as shown in Fig. 6.57, is called the injection molding cycle. The cycle begins when the mold closes, followed by the injection of the polymer into the mold cavity. Once the cavity is filled, a holding pressure is maintained to compensate for material shrinkage. In the next step, the screw turns, feeding the next shot to the front of the screw. This causes the screw to retract as the next shot is prepared. Once the part is sufficiently cool, the mold opens and the part is ejected.

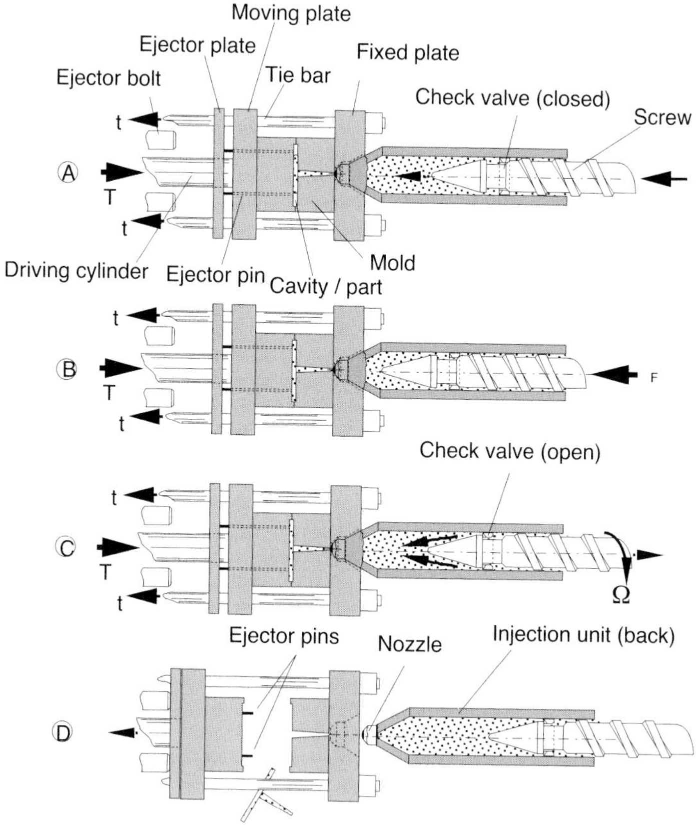

Figure 6.57 Sequence of events during an injection molding cycle.

Figure 6.58 presents the sequence of events during the injection molding cycle. The figure shows that the cycle time is dominated by the cooling of the part inside the mold cavity. The total cycle time can be calculated using

$$t_{cycle} = t_{closing} + t_{cooling} + t_{ejection} \tag{6.27}$$

where the closing and ejection times, $t_{closing}$ and $t_{ejection}$, can last from a fraction of second to a few seconds, depending on the size of the mold and machine. The cooling times, which dominate the process, depend on the maximum thickness of the part. The cooling time for a plate-like part of thickness h can be estimated using

$$t_{cooling} = \frac{h^2}{\pi\alpha}\ln\left(\frac{8}{\pi^2}\frac{T_m - T_w}{T_D - T_w}\right) \tag{6.28}$$

and for a cylindrical geometry of diameter D using

$$t_{cooling} = \frac{D^2}{23.14\alpha}\ln\left(0.692\frac{T_m - T_w}{T_D - T_w}\right) \tag{6.29}$$

where T_m represents the temperature of the injected melt, T_w the temperature of the mold wall, T_D the average temperature at ejection, and α the thermal diffusivity.

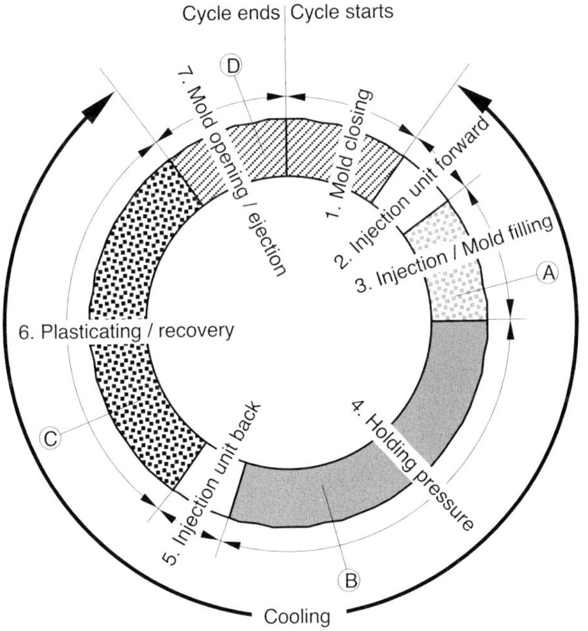

Figure 6.58 Injection molding cycle.

Using the average part temperature history and the cavity pressure history, the process can be followed and assessed using the PvT diagram as depicted in Fig. 6.59 [32-33]. To follow the process on the PvT diagram, we must transfer both the temperature and the pressure at matching times. The diagram reveals four basic processes: an isothermal injection (0-1) with pressure rising to the holding pressure (1-2), an isobaric cooling process during the holding cycle (2-3), an isochoric cooling after the gate freezes with a pressure drop to atmospheric (3-4), and then isobaric cooling to room temperature (4-5).

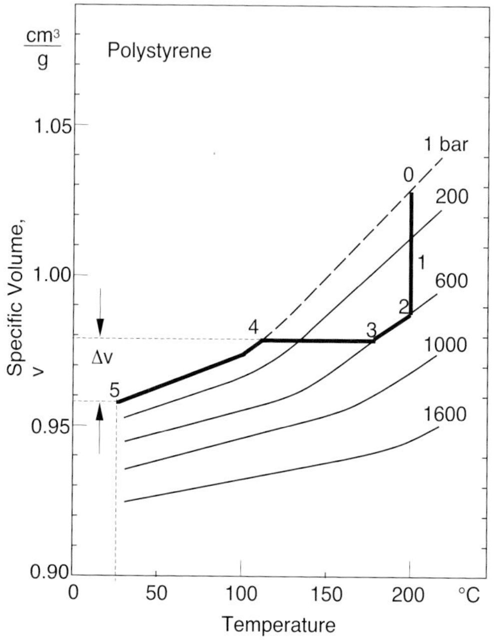

Figure 6.59 Trace of an injection molding cycle in a PvT diagram.

The point on the PvT diagram where the final isobaric cooling begins (4), controls the total part shrinkage, Δv. This point is influenced by the two main processing conditions - the melt temperature, T_m, and the holding pressure, P_H, as depicted in Fig. 6.60. Here the process in Fig. 6.59 is compared to one with a higher holding pressure. Of course, there is an infinite combination of conditions that render acceptable parts, bound by minimum and maximum temperatures and pressures. Figure 6.61 presents the molding diagram with all limiting conditions. The melt temperature is bound by a low temperature that results in a short shot or unfilled cavity and a high temperature that leads to material degradation. The hold pressure is bound by a low pressure that leads to excessive shrinkage or low part weight, and a high pressure that results in

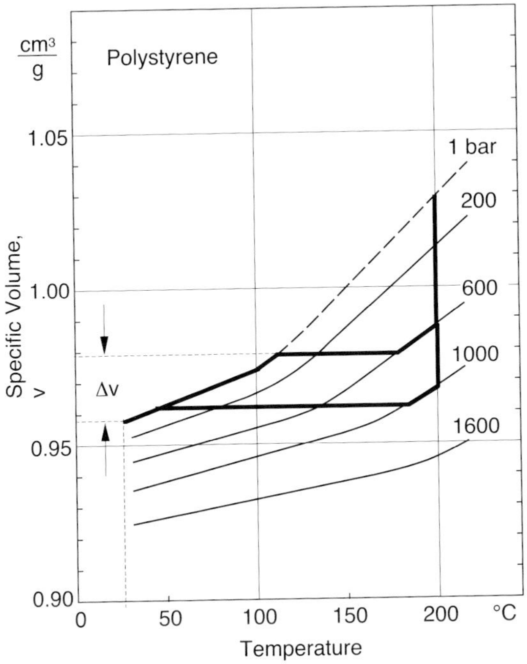

Figure 6.60 Trace of two different injection molding cycles in a PvT diagram.

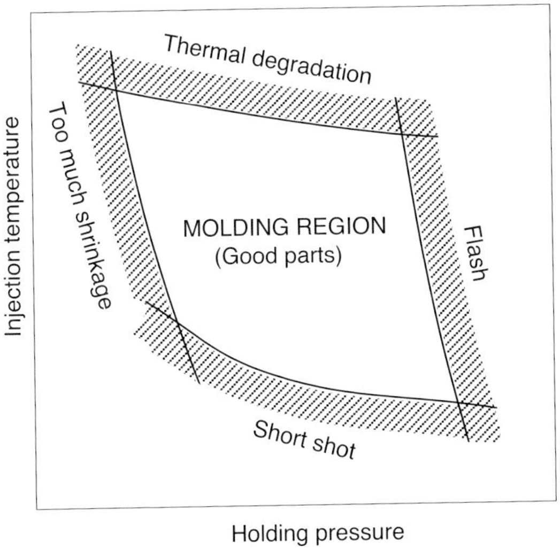

Figure 6.61 The molding diagram.

flash. Flash results when the cavity pressure force exceeds the machine clamping force, leading to melt flow across the mold parting line. The holding pressure determines the corresponding clamping force required to size the injection molding machine. An experienced polymer processing engineer can usually determine which injection molding machine is appropriate for a specific application. For the untrained polymer processing engineer, finding this appropriate holding pressure and its corresponding mold clamping force can be difficult.

With difficulty one can control and predict the component's shape and residual stresses at room temperature. For example, sink marks in the final product are caused by material shrinkage during cooling, and residual stresses can lead to environmental stress cracking under certain conditions [35].

Warpage in the final product is often caused by processing conditions that lead to asymmetric residual stress distributions through the part thickness. The formation of residual stresses in injection molded parts is attributed to two major coupled factors: cooling and flow stresses. The first and most important is the residual stress formed as a result of rapid cooling which leads to large temperature variations.

6.3.2 The Injection Molding Machine

6.3.2.1 The Plasticating and Injection Unit

A plasticating and injection unit is shown in Fig. 6.62. The major tasks of the plasticating unit are to melt the polymer, to accumulate the melt in the screw chamber, to inject the melt into the cavity, and to maintain the holding pressure during cooling.

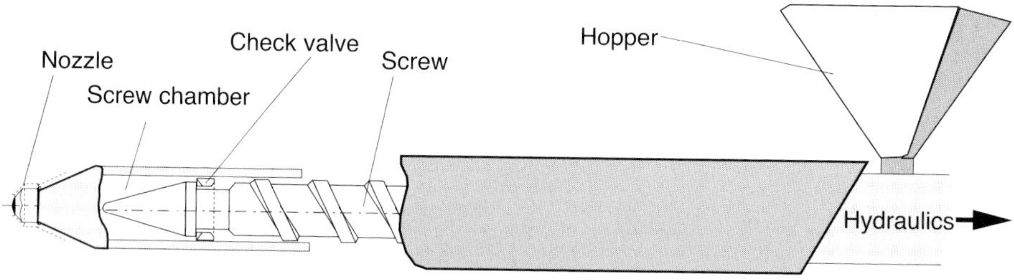

Figure 6.62 Schematic of the plasticating unit.

The main elements of the plasticating unit follow:

- Hopper
- Screw
- Heater bands
- Check valve
- Nozzle

The hopper, heating bands, and the screw are similar to a plasticating single screw extruder, except that the screw in an injection molding machine can slide back and forth to allow for melt accumulation and injection. This characteristic gives it the name reciprocating screw. For qquality purposes, the maximum stroke in a reciprocating screw should be set smaller than 3D.

Although the most common screw used in injection molding machines is the three-zone plasticating screw, two stage vented screws are often used to extract moisture and monomer gases just after the melting stage.

The check valve, or non-return valve, is at the end of the screw and enables it to work as a plunger during injection and packing without allowing polymer melt back flow into the screw channel. A check valve and its function during operation is depicted in Fig. 6.57, and in Fig. 6.62. A high quality check valve allows less then 5% of the melt back into the screw channel during injection and packing.

The nozzle is at the end of the plasticating unit and fits tightly against the sprue bushing during injection. The nozzle type is either open or shut-off. The open nozzle is the simplest, rendering the lowest pressure consumption.

6.3.2.2 The Clamping Unit

The job of a clamping unit in an injection molding machine is to open and close the mold, and to close the mold tightly to avoid flash during the filling and holding. Modern injection molding machines have two predominant clamping types: mechanical and hydraulic.

Figure 6.63 presents a toggle mechanism in the open and closed mold positions. Although the toggle is essentially a mechanical device, it is actuated by a hydraulic cylinder. The advantage of using a toggle mechanism is that, as the mold approaches closure, the available closing force increases and the closing decelerates significantly. However, the toggle mechanism only transmits its maximum closing force when the system is fully extended.

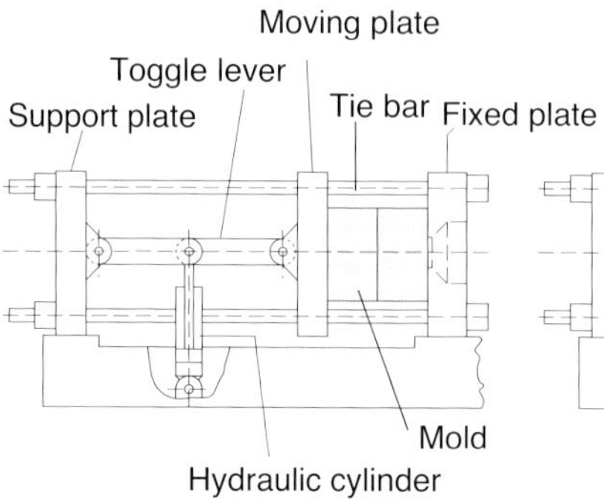

Figure 6.63 Clamping unit with a toggle mechanism.

Figure 6.64 presents a schematic of a hydraulic clamping unit in the open and closed positions. The advantages of the hydraulic system is that a maximum clamping force is attained at any mold closing position and that the system can take different mold sizes without major system adjustments.

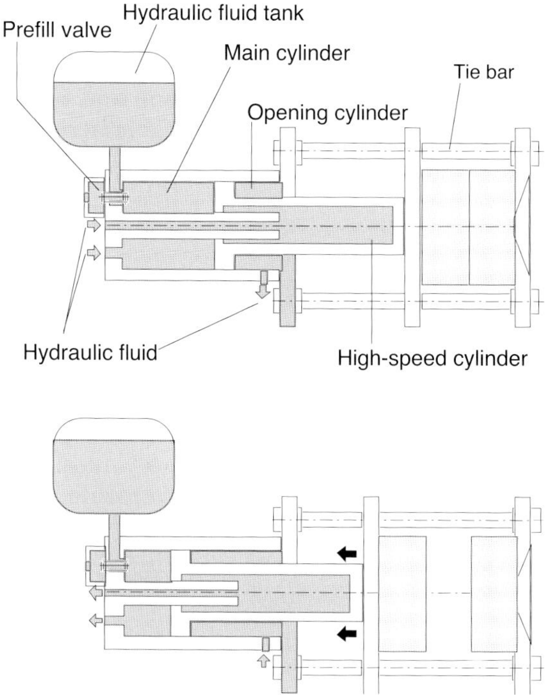

Figure 6.64 Hydraulic clamping unit.

6.3.2.3 The Mold Cavity

The central point in an injection molding machine is the mold. The mold distributes polymer melt into and throughout the cavities, shapes the part, cools the melt and, ejects the finished product. As depicted in Fig. 6.65, the mold is custom-made and consists of the following elements:

- Sprue and runner system
- Gate
- Mold cavity
- Cooling system (thermoplastics)
- Ejector system

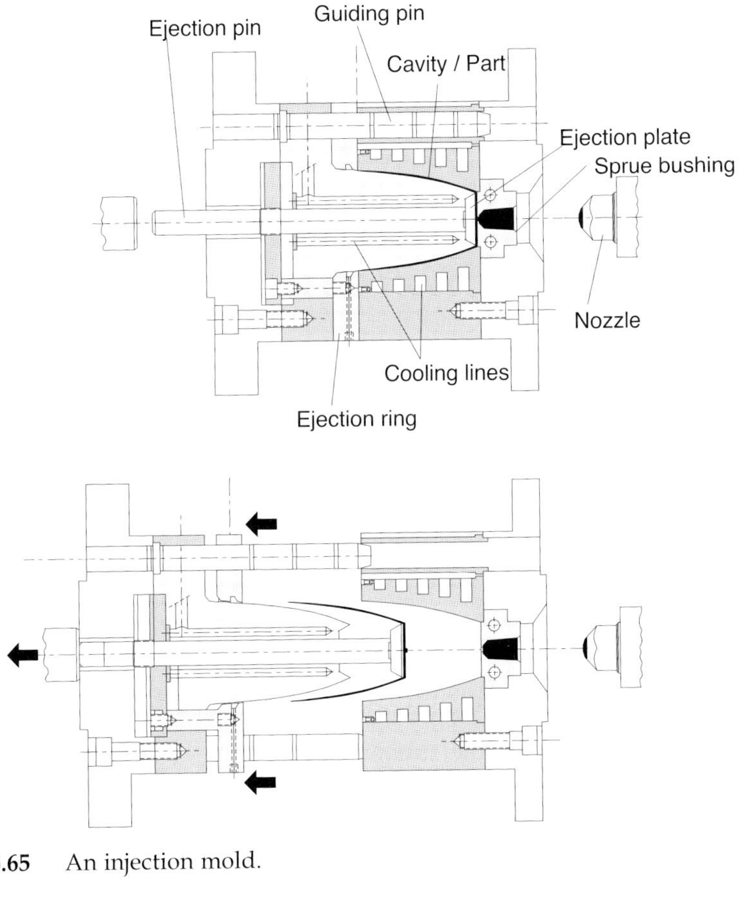

Figure 6.65 An injection mold.

During mold filling, the melt flows through the sprue and is distributed into the cavities by the runners, as seen in Fig. 6.66.

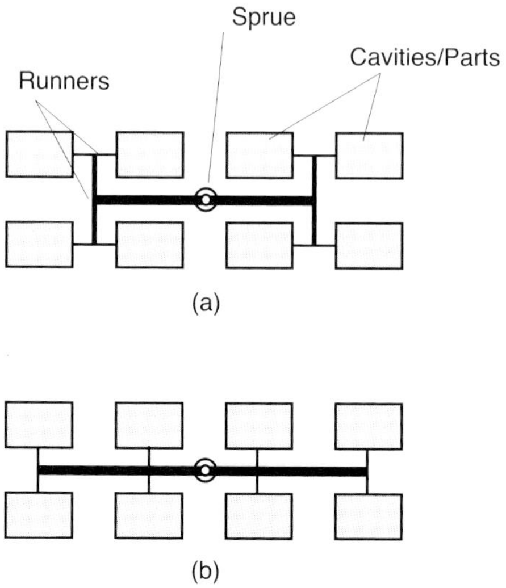

Figure 6.66 Schematic of different runner system arrangements.

The runner system in Fig. 6.66 (a) is symmetric where all cavities fill at the same time causing the polymer to fill all cavities in the same way. The disadvantage of this balanced runner system is that the flow paths are long, leading to high material and pressure consumption. On the other hand, the asymmetric runner system shown in Fig. 6.66 (b) leads to parts of different quality. Equal filling of the mold cavities can also be achieved by varying runner diameters. There are two types of runner systems - cold and hot. Cold runners are ejected with the part, and are trimmed after mold removal. The advantage of the cold runner is lower mold cost. The hot runner keeps the polymer at its melt temperature. The material stays in the runner system after ejection, and is injected into the cavity in the following cycle. There are two types of hot runner system: externally and internally heated. The externally heated runners have a heating element surrounding the runner that keeps the polymer isothermal. The internally heated runners have a heating element running along the center of the runner, maintaining a polymer melt that is warmer at its center and possibly solidified along the outer runner surface. Although a hot runner system considerably increases mold cost, its advantages include elimination of trim and lower pressures for injection.

When large items are injection molded, the sprue sometimes serves as the gate, as shown in Fig. 6.67. The sprue must be subsequently trimmed, often requiring further surface finishing. On the other hand, a pin-type gate (Fig. 6.67) is a small orifice that connects the sprue or the runners to the mold cavity. The part is easily broken off from such a gate, leaving only a small mark that usually does not require finishing. Other types of gates, also shown in Fig. 6.67, are film gates, used to eliminate orientation, and disk or diaphragm gates for symmetric parts such as compact discs.

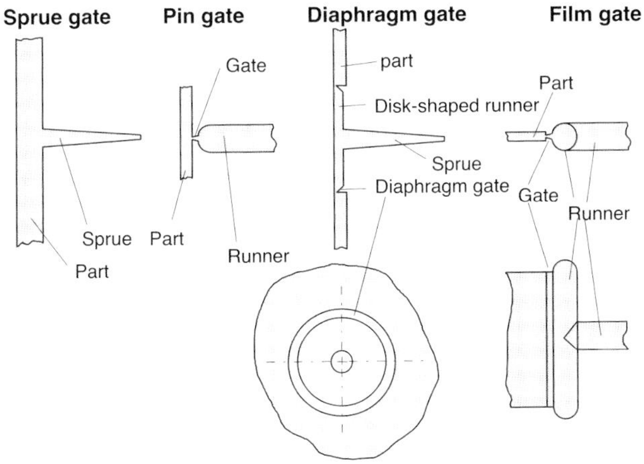

Figure 6.67 Schematic of different gating systems.

6.3.3 Related Injection Molding Processes

Although most injection molding processes are covered by the conventional process description discussed earlier in this chapter, there are several important molding variations including:

- Multi-color
- Multi-component
- Co-injection
- Gas-assisted
- Injection-compression

Multi-component injection molding occurs when two or more polymers, or equal polymers of different color, are injected through different runner and gate systems at different stages during the molding process. Each component is injected using its own plasticating unit. The molds are often located on a turntable. Multi-color automotive stop lights are molded this way.

In principle, the multi-component injection molding process is the same as the multi-color process. Here, either two incompatible materials are molded or one component is cooled sufficiently so that the two components do not adhere to each other. For example, to mold a ball and socket system, the socket of the linkage is molded first. The socket component is allowed to cool somewhat and the ball part is injected inside it. This results in a perfectly movable system. This type of injection molding process is used to replace tedious assembling tasks and is becoming popular in countries where labor costs are high. In addition, today, a widely used application is the multi-component injection of a hard and a soft polymer such as polypropylene with a thermoplastic elastomer.

In contrast to multi-color and multi-component injection molding, co-injection molding uses the same gate and runner system. Here, the component that ends as the outer skin of the part is injected first, followed by the core component. The core component displaces the first and a combination of the no-slip condition between polymer and mold and the freezing of the melt creates a sandwiched structure as depicted in Fig. 6.68.

In principle, the gas-assisted injection molding process is similar to co-injection molding. Here, the second or core component is nitrogen, which is injected through a needle into the polymer melt, blowing the melt out of the way and depositing it against the mold surfaces.

Injection-compression molding first injects the material into a partially opened mold, and then squeezes the material by closing the mold. Injection-compression molding is used for polymer products that require a high quality surface finish, such as compact discs and other optically demanding components because it practically eliminates tangential molecular orientation.

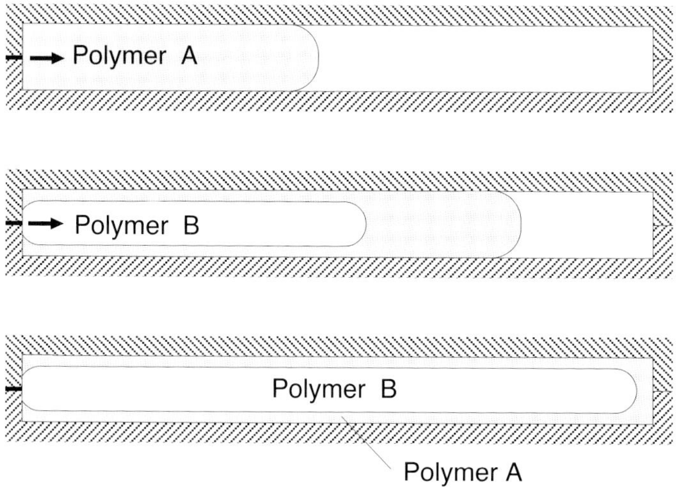

Figure 6.68 Schematic of the co-injection molding process.

6.4 Secondary Shaping

Secondary shaping operations such as extrusion blow molding, film blowing, and fiber spinning occur immediately after the extrusion profile emerges from the die. The thermoforming process is performed on sheets or plates previously extruded and solidified. In general, secondary shaping operations consist of mechanical stretching or forming of a preformed cylinder, sheet, or membrane.

6.4.1 Fiber Spinning

Fiber spinning is used to manufacture synthetic fibers. During fiber spinning, a filament is continuously extruded through an orifice and stretched to diameters of 100 μm and smaller. The process is schematically depicted in Fig. 6.69. The molten polymer is first extruded through a filter or *screen pack*, to eliminate small contaminants. The melt is then extruded through a spinneret, a die composed of multiple orifices. A spinneret can have between one and 10,000 holes. The fibers are then drawn to their final diameter, solidified, and wound onto a spool. The solidification takes place either in a water bath or by forced convection. When the fiber solidifies in a water bath, the extrudate undergoes an adiabatic stretch before cooling begins in the bath. The forced convection cooling, which is more commonly used, leads to a non-isothermal spinning process.

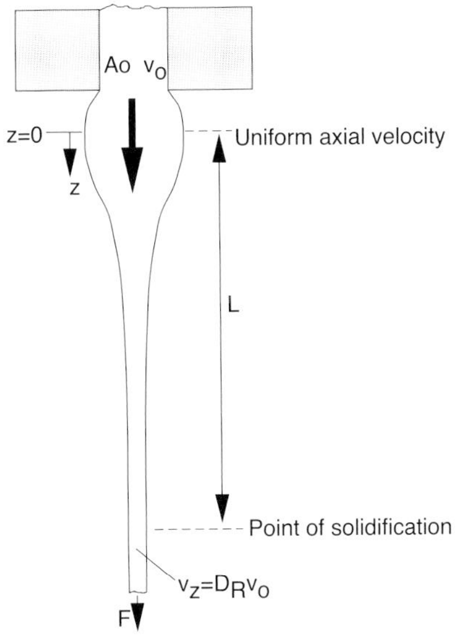

Figure 6.69 The fiber spinning process.

The drawing and cooling processes determine the morphology and mechanical properties of the final fiber. For example, ultra high molecular weight HDPE fibers with high degrees of orientation in the axial direction can have the stiffness of steel with today's fiber spinning technology.

Of major concern during fiber spinning are the instabilities that arise during drawing, such as brittle fracture, Rayleigh disturbances, and draw resonance. Brittle fracture occurs when the elongational stress exceeds the melt strength of the drawn polymer melt. The instabilities caused by Rayleigh disturbances are like those causing filament break-up during dispersive mixing as discussed in Chapter 5. Draw resonance appears under certain conditions and manifests itself as periodic fluctuations that result in diameter oscillation.

6.4.2 Film Production

6.4.2.1 Cast Film Extrusion

In a cast film extrusion process, a thin film is extruded through a slit onto a chilled, highly polished, turning roll where it is quenched from one side. The speed of the roller controls the draw ratio and final film thickness. The film is then sent to a second roller for cooling of the other side. Finally, the film passes through a system of rollers and is wound onto a roll. A typical film casting process is depicted in Fig. 6.70 and 6.71. The cast film extrusion process exhibits stability problems similar to those encountered in fiber spinning [36].

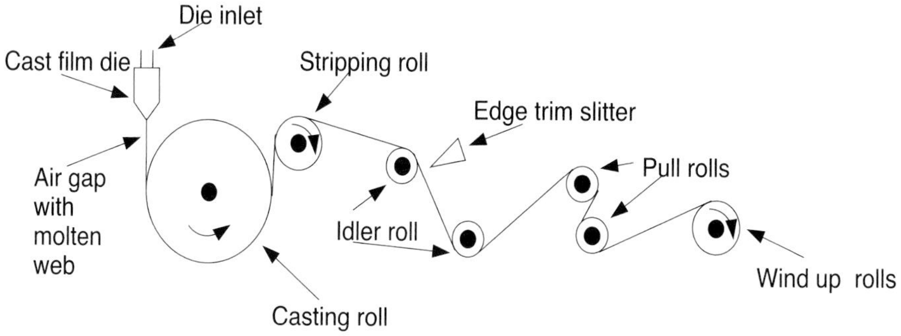

Figure 6.70 Schematic of a film casting operation.

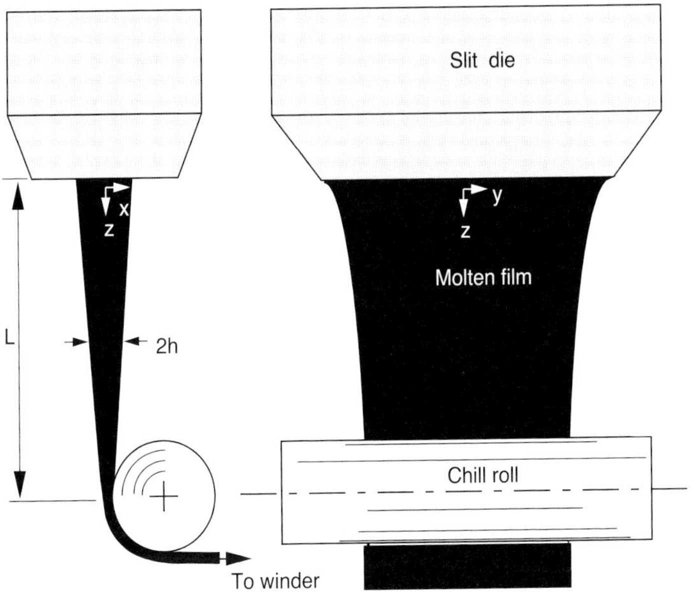

Figure 6.71 Film casting.

6.4.2.2 Film Blowing

In film blowing, a tubular cross-section is extruded through an annular die, normally a spiral die, and is drawn and inflated until the freezing line is reached. Beyond this point, the stretching is practically negligible. The process is schematically depicted in Fig. 6.72 [37] and Fig. 6.73. The advantage of film blowing over casting is that the induced biaxial stretching renders a stronger and less permeable film. Film blowing is mainly used with less expensive materials such as polyolefins. Polymers with lower viscosity such as PA and PET are better manufactured using the cast film process.

The extruded tubular profile passes through one or two air rings to cool the material. The tube interior is maintained at a certain pressure by blowing air into the tube through a small orifice in the die mandrel. The air is retained in the tubular film, or bubble, by collapsing the film well above its freeze-off point and tightly pinching it between rollers. The size of the tubular film is calibrated between the air ring and the collapsing rolls.

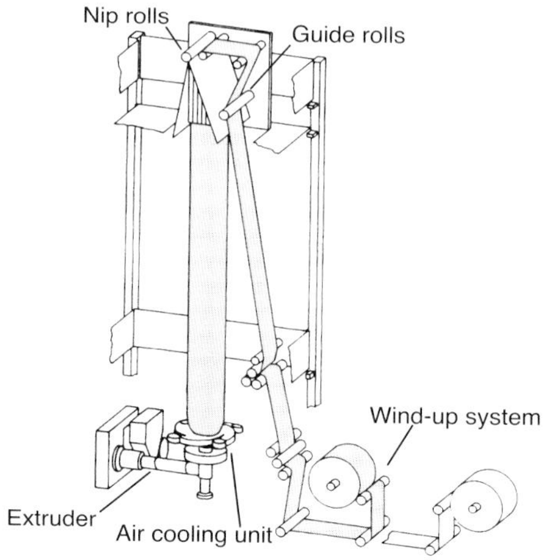

Figure 6.72 Film blowing.

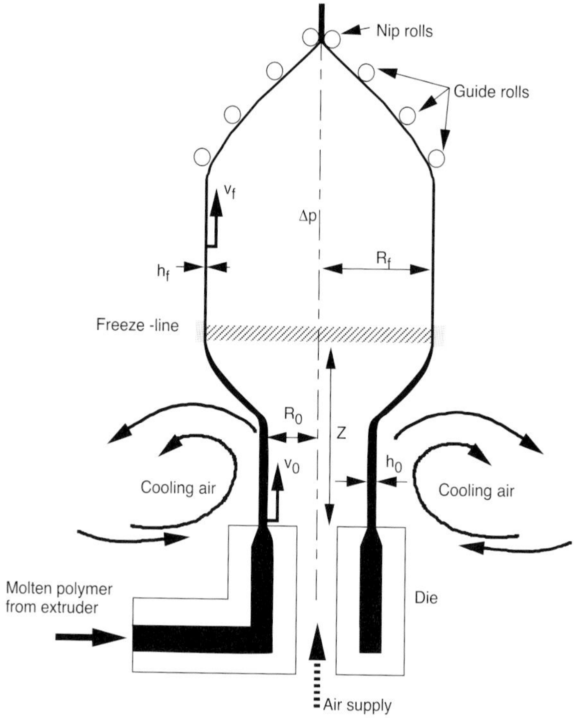

Figure 6.73 Film blowing.

6.4.3 Blow Molding

The predecessor of the blow molding process was the blowing press developed by Hyatt and Burroughs in the 1860s to manufacture hollow celluloid articles. Polystyrene was the first synthetic polymer used for blow molding during World War II and polyethylene was the first material to be implemented in commercial applications. Until the late 1950s, the main application for blow molding was the manufacture of LDPE articles such as squeeze bottles.

Blow molding produces hollow articles that do not require a homogeneous thickness distribution. Today, HDPE, LDPE, PP, PET, and PVC are the most common materials used for blow molding.

6.4.3.1 Extrusion Blow Molding

In extrusion blow molding, a *parison* or tubular profile is extruded and inflated into a cavity with the specified geometry. The blown article is held inside the cavity until it is sufficiently cool. Figure 6.74 [38] presents a schematic of the steps in blow molding.

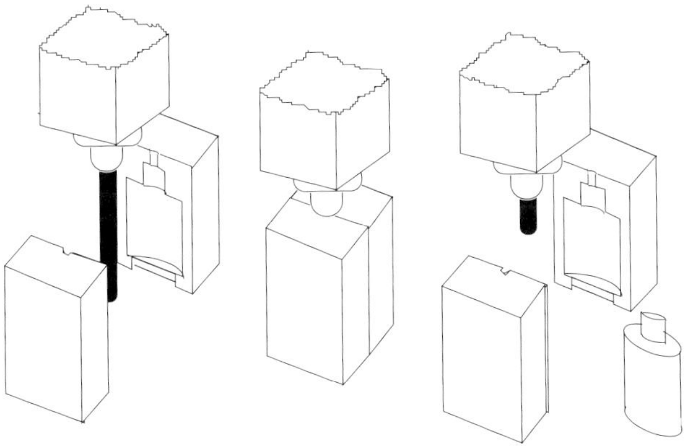

Figure 6.74 Schematic of the extrusion blow molding process.

During blow molding, one must generate the appropriate parison length such that the trim material is minimized. Another means of saving material is generating a parison of variable thickness, usually referred to as *parison programming*, such that an article with an evenly distributed wall thickness is achieved after stretching the material. An example of a programmed parison and finished bottle thickness distribution is presented in Fig. 6.75 [39].

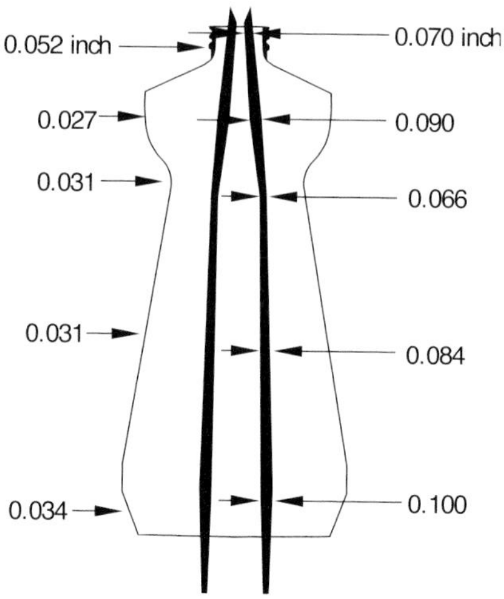

Figure 6.75 Wall thickness distribution in the parison and the part.

A parison of variable thickness can be generated by moving the mandrel vertically during extrusion as shown in Fig. 6.76. A thinner wall not only results in material savings but also reduces the cycle time due to the shorter required cooling times.

As expected, the largest portion of the cycle time is the cooling of the blow molded container in the mold cavity. Most machines wor with multiple molds in order to increase production. Rotary molds are often used in conjunction with vertical or horizontal rotating tables (Fig. 6.77 [37]).

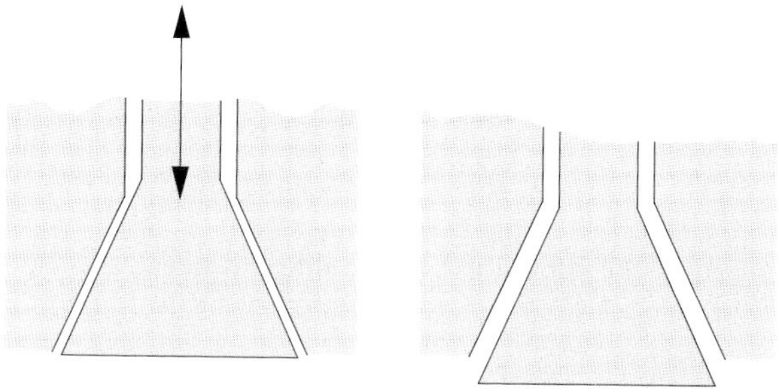

Figure 6.76 Moving mandrel used to generate a programmed parison.

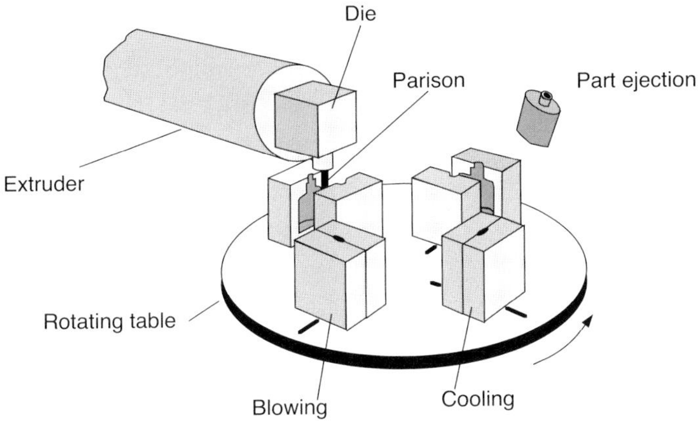

Figure 6.77 Schematic of an extrusion blow molder with a rotating table.

6.4.3.2 Injection Blow Molding

Injection blow molding depicted in Fig. 6.78 [38] begins by injection molding the parison onto a core and into a mold with finished bottle threads. The formed parison has a thickness distribution that leads to reduced thickness variations throughout the container. Before blowing the parison into the cavity, it can be mechanically stretched to orient molecules axially, Fig. 6.79 [38]. The subsequent blowing operation introduces tangential orientation. A container with biaxial molecular orientation exhibits higher optical (clarity) and mechanical properties and lower permeability. In the injection blow molding process one can go directly from injection to blowing or one can have a re-heating stage in-between.

The advantages of injection blow molding over extrusion blow molding are:

- Pinch-off and therefore post-mold trimming are eliminated
- Controlled container wall thickness
- Dimensional control of the neck and screw-top of bottles and containers

Disadvantages include higher initial mold cost, the need for both injection and blow molding units and lower volume production.

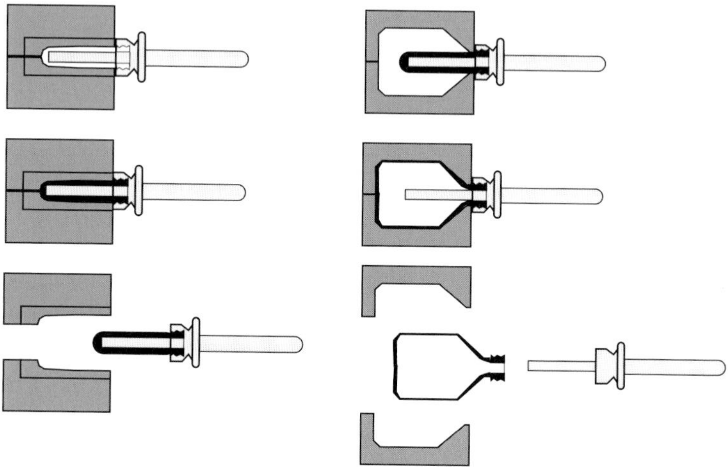

Figure 6.78 Injection blow molding.

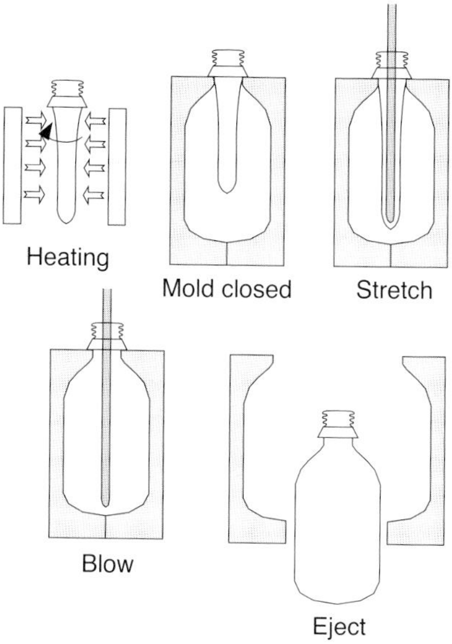

Figure 6.79 Stretch blow molding.

6.4.3.3 Thermoforming

Thermoforming is an important secondary shaping method of plastic film and sheet. Thermoforming consists of warming the plastic sheet and forming it into a cavity or over a tool using vacuum, air pressure, and mechanical means. During the 18th century, tortoiseshells and hooves were thermoformed into combs and other shapes. The process was refined during the mid-19th century to thermoform various cellulose nitrate articles. During World War II, thermoforming was used to manufacture acrylic aircraft cockpit enclosures, canopies, and windshields, as well as translucent covers for outdoor neon signs. During the 1950s, the process made an impact in the mass production of cups, blister packs, and other packaging commodities. Today, in addition to packaging, thermoforming is used to manufacture refrigerator liners, pick-up truck cargo box liners, shower stalls, bathtubs, as well as automotive trunk liners, glove compartments, and door panels.

A typical thermoforming process is presented in Fig. 6.80 [37]. The process begins by heating the plastic sheet slightly above the glass transition temperature, for amorphous polymers, or slightly below the melting point, for semi-crystalline materials. Although, both amorphous and semi-crystalline polymers are used for thermoforming, the process is easiest with amorphous polymers because they have a wide rubbery temperature range above the glass transition temperature. At these temperatures the polymer is easily shaped, but still has enough rigidity to hold the heated sheet without much sagging. Most semi-crystalline polymers lose their strength rapidly once the crystalline structure breaks up above the melting temperature.

The heating is achieved using radiative heaters and the temperature reached during heating must be high enough for sheet shaping, but low enough so the sheets do not droop into the heaters. One key requirement for successful thermoforming is to bring the sheet to a uniform forming temperature. The sheet is then shaped into the cavity over the tool. This can be accomplished in several ways. Most commonly a vacuum sucks the sheet onto the tool, stretching the sheet until it contacts the tool surface. The main problem here is the irregular thickness distribution that arises throughout the part. Hence, the main concern of the process engineer is to optimize the system such that the differences in thickness throughout the part are minimized. This can be accomplished in many ways but most commonly by plug-assist. Here, as the plug pushes the sheet into the cavity, only the parts of the sheet not touching the plug-assist stretch. Since the unstretched portions of the sheet must remain hot for subsequent stretching, the plug-assist is made of a low thermal conductivity material such as wood or hard rubber. The initial stretch is followed by a vacuum for final shaping. Once cooled, the product is removed.

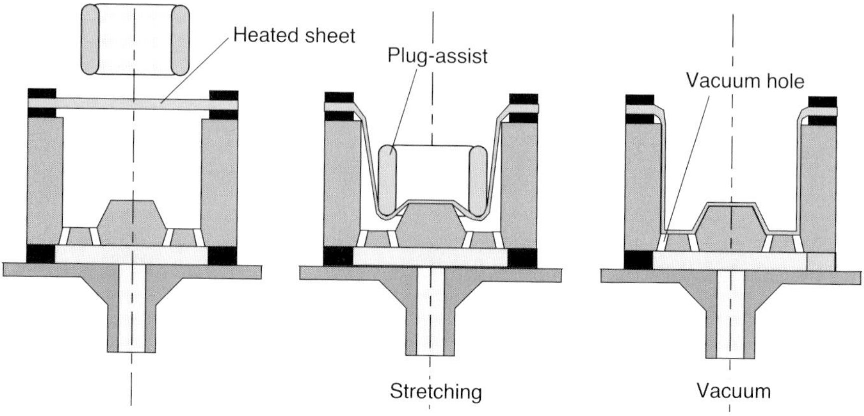

Figure 6.80 Plug-assist thermoforming using vacuum.

To reduce thickness variations in the product, the sheet can be pre stretched by forming a bubble at the beginning of the process. This is schematically depicted in Fig. 6.81 [37]. The mold is raised into the bubble, or a plug-assist pushes the bubble into the cavity, and a vacuum finishes the process.

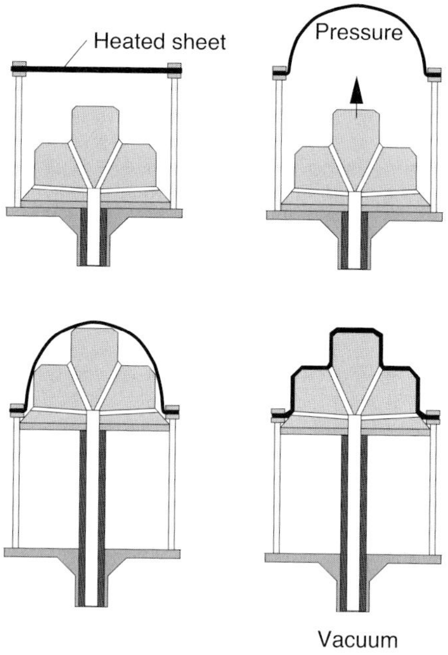

Figure 6.81 Reverse draw thermoforming with plug-assist and vacuum.

One of the main reasons for the rapid growth and high volume of thermoformed products is that the tooling costs for a thermoforming mold are much lower than for injection molding.

6.5 Calendering

In a calender line, the polymer melt is transformed into films and sheets by squeezing it between pairs of co-rotating high precision rollers. Calenders are also used to produce certain surface textures which may be required for different applications. Today, calendering lines are used to manufacture PVC sheet, floor covering, rubber sheet, and rubber tires. They are also used to texture or emboss surfaces. When producing PVC sheet and film, calender lines have a great advantage over extrusion processes because of the shorter residence times, resulting in a lower requirement for stabilizer. This can be cost effective since stabilizers are a major part of the overall expense of processing these polymers.

Figure 6.82 [37] presents a typical calender line for manufacturing PVC sheet. A typical system is composed of:

- Plasticating unit
- Calender
- Cooling unit
- Accumulator
- Wind-up station

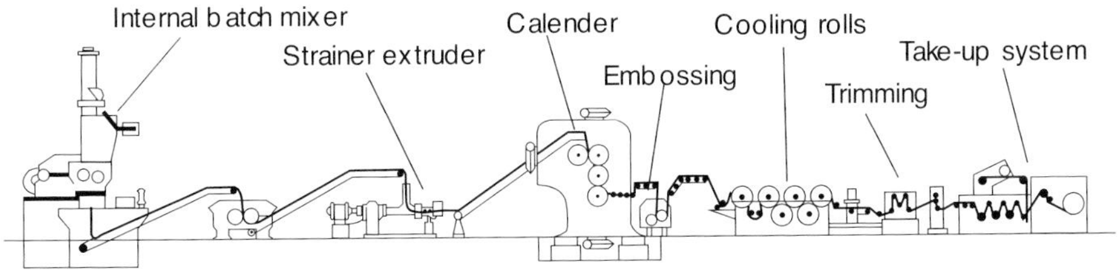

Figure 6.82 Schematic of a typical calendering process (Berstorff GmbH).

In the plasticating unit, which is represented by the internal batch mixer and the strainer extruder, the material is melted and mixed and is fed in a continuous stream between the nip of the first two rolls. In another variation of the process, the mixing may take place elsewhere, and the material is simply reheated on the roll mill. Once the material is fed to the mill, the first pair of rolls control the feeding rate, while subsequent rolls in the calender calibrate the sheet thickness. Most calender systems have four rolls as does the one in Fig. 6.82, which is an inverted L- or F-type system. Other typical roll arrangements are shown in Fig. 6.83 and 6.84. After passing through the main calender, the sheet can be passed through a secondary calendering operation for embossing. The sheet is then passed through a series of chilling rolls where it is cooled from both sides in an alternating fashion. After cooling, the film or sheet is wound.

One of the major concerns in a calendering system is generating a film or sheet with a uniform thickness distribution with tolerances as low as ±0.005 mm. To achieve this, the dimensions of the rolls must be precise. It is also necessary to compensate for roll bowing resulting from high pressures in the nip region. Roll bowing is a structural problem that can be mitigated by placing the rolls in a slightly crossed pattern, rather than completely parallel, or by applying moments to the roll ends to counteract the separating forces in the nip region.

Calendering can be modeled by assuming steady state, laminar flow and isothermal conditions.

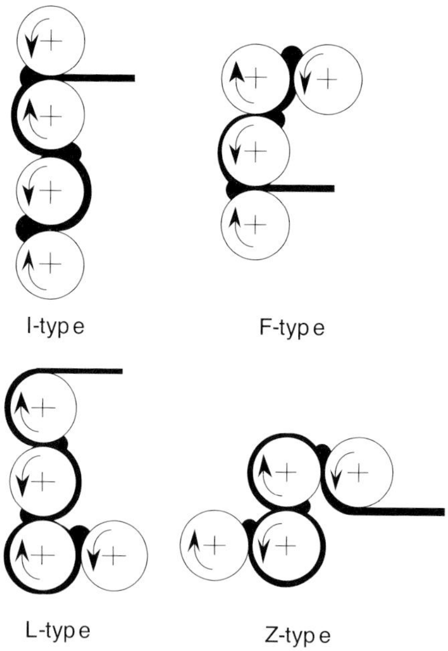

Figure 6.83 Calender arrangements.

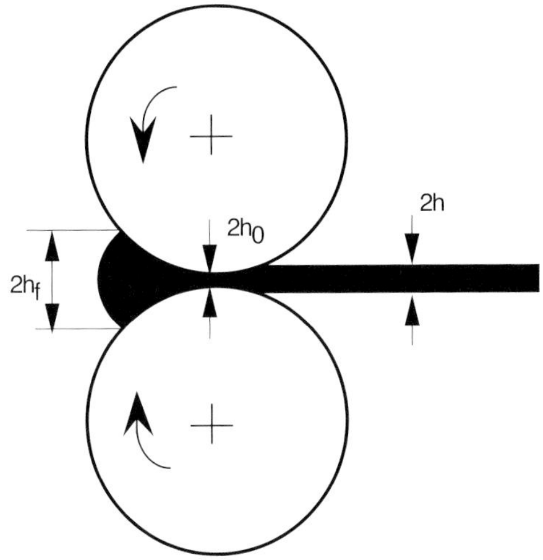

Figure 6.84 Schematic of the calendering process.

6.6 Coating

In coating a liquid film is continuously deposited on a moving, flexible or rigid substrate. Coating is done on metal, paper, photographic films, audio and video tapes, and adhesive tapes. Typical coating processes include *wire coating, dip coating, knife coating, roll coating, slide coating*, and *curtain coating*.

In wire coating, a wire is continuously coated with a polymer melt by pulling the wire through an extrusion die. The polymer resin is deposited onto the wire using the drag flow generated by the moving wire and sometimes a pressure flow generated by the back pressure of the extruder. The process is schematically depicted in Fig. 6.85[13]. The second normal stress differences, generated by the high shear deformation in the die, help keep the wire centered in the annulus [40].

Dip coating is the simplest and oldest coating operation. Here, a substrate is continuously dipped into a fluid and withdrawn with one or both sides coated with the fluid. Dip coating can also be used to coat individual objects that are dipped and withdrawn from the fluid. The fluid viscosity and density and the speed and angle of the surface determine the coating thickness.

13 Other wire coating processes extrude a tubular sleeve which adheres to the wire via stretching and vacuum. This is called tube coating.

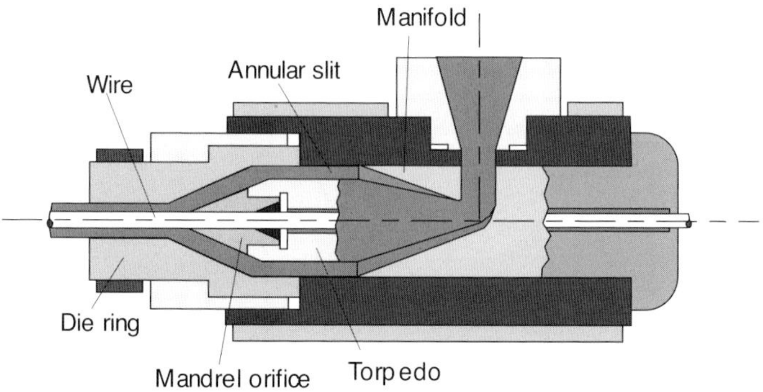

Figure 6.85 Wire coating process.

Knife coating, depicted in Fig. 6.86, consists of metering the coating material onto the substrate from a pool of material, using a fixed rigid or flexible knife. The knife can be normal to the substrate or angled and the bottom edge can be flat or tapered. The thickness of the coating is nearly half the gap between the knife edge and the moving substrate or web. A major advantage of a knife edge coating system is its simplicity and relatively low maintenance.

Roll coating consists of passing a substrate and the coating simultaneously through the nip region between two rollers. The physics governing this process is similar to calendering, except that the fluid adheres to both the substrate and the opposing roll. The coating material is a low viscosity fluid, such as a polymer solution or paint and is picked up from a bath by the lower roll and applied to one side of the substrate. The thickness of the coating can be as low as a few μm and is controlled by the viscosity of the coating liquid and the nip dimension. This process can be configured as either forward roll coating for co-rotating rolls or reverse roll coating for counter-rotating rolls (Fig. 6.87). The reverse roll coating process delivers the most accurate coating thicknesses.

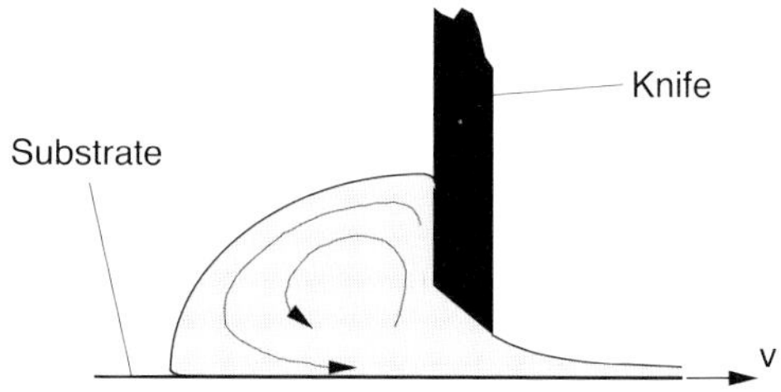

Figure 6.86 Schematic of a knife coating process.

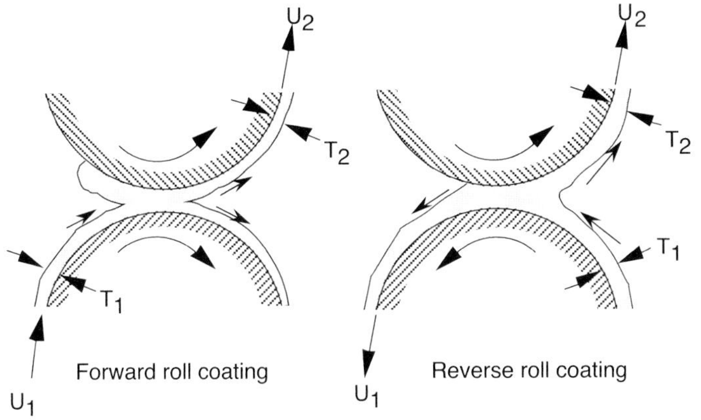

Figure 6.87 Schematic of forward and reverse roll coating processes.

Slide coating and curtain coating, schematically depicted in Fig. 6.88, are commonly used to apply multi-layered coatings. However, curtain coating has also been widely used to apply single layers of coatings to cardboard sheet. In both methods, the coating fluid is pre-metered.

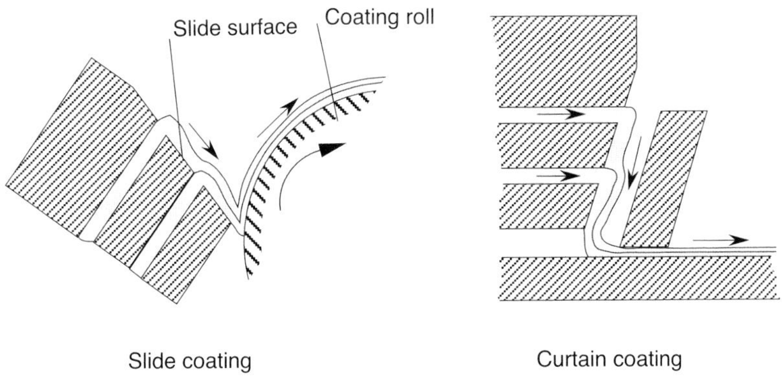

Figure 6.88 Slide and curtain coating.

6.7 Compression Molding

Compression molding is widely used in the automotive industry to produce parts that are large, thin, lightweight, strong, and stiff. It is also used in the household goods and electrical industries. Compression molded parts are formed by squeezing a charge, often glass fiber reinforced, inside a mold cavity, as depicted in Fig. 6.89. The matrix can be either a thermoset or thermoplastic. The oldest and still widest used material for compression molded products is phenolic. The thermoset materials used to manufacture fiber reinforced compression molded articles is unsaturated polyester sheet or bulk, reinforced with glass fibers, known as sheet molding compound (SMC) or bulk molding compound (BMC). In SMC, the 25 mm long reinforcing fibers are randomly oriented in the plane of the sheet and make up for 20-30% of the molding compound's volume fraction. A schematic diagram of an SMC production line is depicted in Fig. 8.90 [41]. When producing SMC, the chopped glass fibers are sandwiched between two carrier films previously coated with unsaturated polyester-filler matrix. A fiber reinforced thermoplastic charge is often called a glass mat reinforced thermoplastic (GMT) charge. The most common GMT matrix is polypropylene. More recently, long fiber reinforced themoplastics (LFT) have become common. Here, one squeezes sausage shaped charges deposited on the mold by an extruder.

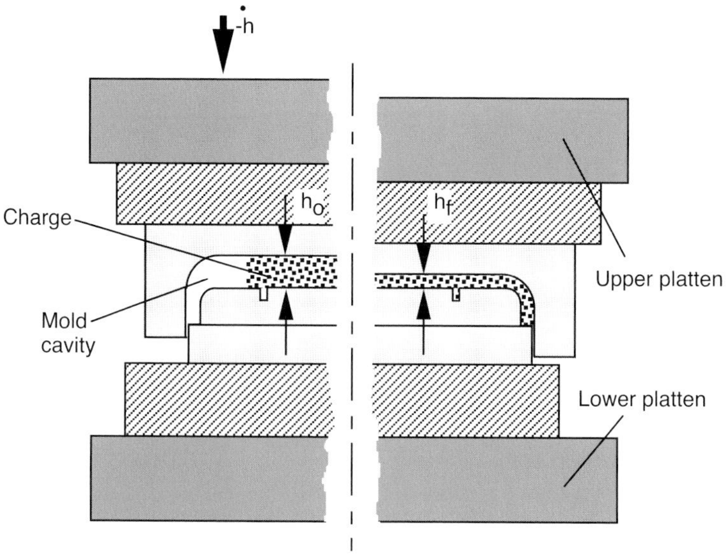

Figure 6.89 Compression molding process (h_0 = charge thickness, h_f = part thickness, and $\dot{h}$= closing speed).

During processing of thermoset charges, the SMC blank is cut from a preformed roll and is placed between heated cavity surfaces. Generally, the mold is charged with 1 to 4 layers of SMC, each layer about 3 mm thick, which initially cover about half the mold cavity's surface. During molding, the initially randomly oriented glass fibers orient, leading to anisotropic properties in the finished product. When processing GMT charges, the preforms are cut and heated between radiative heaters. Once heated, they are placed inside a cooled mold that rapidly closes and squeezes the charges before they cool and solidify.

One of the main advantages of the compression molding process is the low fiber attrition during processing. Here, relatively long fibers can flow in the melt without the fiber damage common during plastication and cavity filling in injection molding.

An alternate process is injection-compression molding. Here, a charge is injected through a large gate followed by a compression cycle. The material used in the injection compression molding process is called bulk molding compound (BMC), which is reinforced with shorter fibers, generally 1 cm long, with an unsaturated polyester matrix. The main benefit of injection compression molding over compression molding is automation. The combination of injection and compression molding leads to lower degrees of fiber orientation and fiber attrition compared to injection molding.

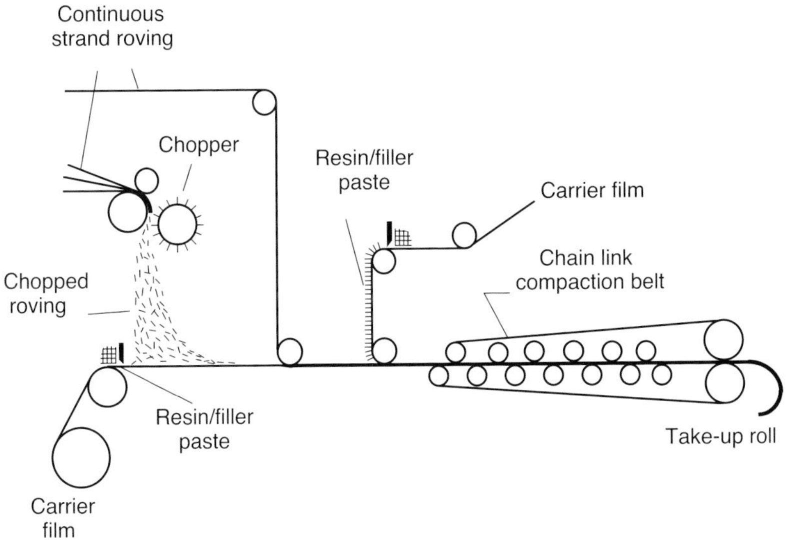

Figure 6.90 SMC production line.

6.8 Foaming

In foam or a foamed polymer, a cellular or porous structure has been generated through the addition and reaction of *physical* or *chemical blowing agents*. The basic steps of foaming are cell nucleation, expansion or cell growth, and cell stabilization. Nucleation occurs when, at a given temperature and pressure, the solubility of a gas is reduced, leading to saturation, expelling the excess gas to form a bubble. Nucleating agents, such as powdered metal oxides, are used for initial bubble formation. The bubbles reach an equilibrium shape when their inside pressure balances their surface tension and surrounding pressures. The cells formed can be completely enclosed (closed cell) or can be interconnected (open cell).

A physical foaming process is one where a gas such as nitrogen or carbon dioxide is introduced into the polymer melt. Physical foaming also occurs after heating a melt that contains a low boiling point fluid, causing it to vaporize. For example, the heat-induced volatilization of low-boiling-point liquids, such as pentane and heptane, is used to produce polystyrene foams. Also, foaming occurs during volatilization from the exothermic reaction of gases produced during polymerization such as the production of carbon dioxide during the reaction of isocyanate with water. Physical blowing agents are added to the plasticating zone of the extruder or molding machine. The most widely used physical blowing agent is nitrogen. Liquid blowing agents are often added to the polymer in the plasticating unit or the die.

Chemical blowing agents are usually powders introduced in the hopper of the molding machine or extruder. Chemical foaming occurs when the blowing agent thermally decomposes, releasing large amounts of gas. The most widely used chemical blowing agent for polyolefin is azodicarbonamide.

In mechanical foaming, a gas dissolved in a polymer expands upon reduction of the processing pressure.

The foamed structures commonly generated are either homogeneous foams or integral foams. Figure 6.91 [42] presents the various types of foams and their corresponding characteristic density distributions. In integral foam, the unfoamed skin surrounds the foamed inner core. This type of foam can be achieved during injection molding and extrusion and it replaces the sandwiched structure also shown in Fig. 6.91.

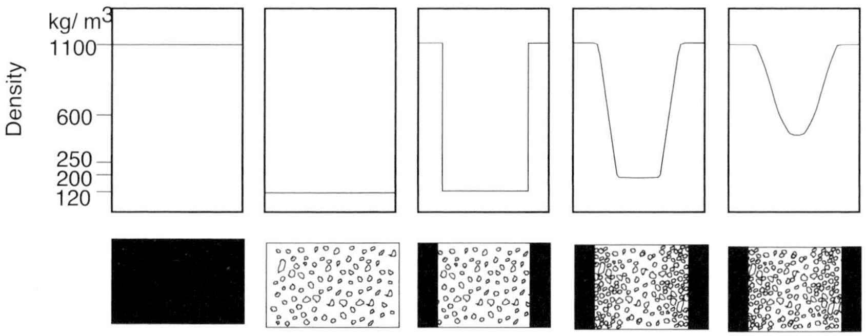

Figure 6.91 Schematic of various foam structures.

Today, foams are of great commercial importance and are primarily used in packaging and as heat and noise insulating materials. Examples of foamed materials are polyurethane foams, expanded polystyrene (EPS) and expanded polypropylene particle foam (EPP).

Polyurethane foam is perhaps the most common foaming material and is a typical example of a chemical foaming technique. Here, two low viscosity components, a polyol and an isocyanate, are mixed with a blowing agent such as pentane. When manufacturing semi-finished products the mixture is deposited on a moving conveyor belt where it is allowed to rise, like a loaf of bread contained whithin shaped paper guides. The result is a continuous polyurethane block that can be used, among others, in the upholstery and matress industries.

The basic material to produce expanded polystyrene products are small pearls produced by suspension styrene polymerization with 6-7% of pentane as a blowing agent. To process the pearls they are placed in pre-expanding machines heated with steam until their temperature reaches 80 to 100 °C. To enhance their expansion, the pearls are allowed to cool in a vacuum and allowed to age and dry in ventilated storage silos before the shaping operation. Polystyrene foam is is used extensively in packaging, but its uses also extend to the construction industry as a thermal insulating material, as well as for shock absorption in children's safety seats and bicycle helmets.

Expanded polypropylene particle foam is similar in to EPS but is characterized by its excellent impact absorption and chemical resistance. Its applications are primarely in the automotive industry as bumper cores, sun visors and knee cushions, to name a few.

6.9 Rotational Molding

Rotational molding is used to make hollow objects. In rotational molding, a carefully measured amount of powdered polymer, typically polyethylene, is placed in a mold. The mold is then closed and placed in an oven where the mold turns about two axes as the polymer melts, as depicted in Fig. 6.92. During heating and melting, which occur at oven temperatures between 250 and 450 °C, the polymer is deposited evenly on the mold's surface. To ensure uniform thickness, the axes of rotation should not coincide with the centroid of the molded product. The mold is then cooled and the solid part is removed from the mold cavity. The parts can be as thick as 1 cm, and still be manufactured with relatively low residual stresses. The reduced residual stress and the controlled dimensional stability of the rotational molded product depend in great part on the cooling rate after the mold is removed from the oven. A mold that is cooled too fast yields warped parts. Usually, a mold is first cooled with air to start the cooling slowly, followed by a water spray for faster cooling.

The main advantages of rotational molding over blow molding are the uniform part thickness and the low cost involved in manufacturing the mold. In addition, large parts such as play structures or kayaks can be manufactured more economically than with injection molding or blow molding. The main disadvantage of the process is the long cycle time for heating and cooling of the mold and polymer.

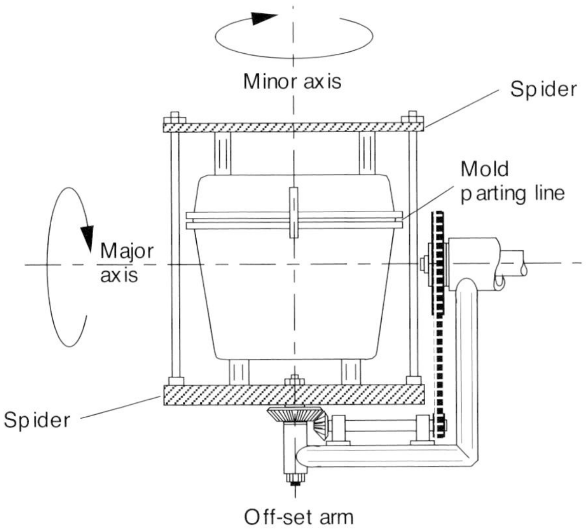

Figure 6.92 Schematic of the rotational molding process.

Figure 6.93 presents the air temperature inside the mold in a typical rotational molding cycle for polyethylene powders [43]. The process can be divided into six distinct phases:

1. Induction or initial air temperature rise
2. Melting and sintering
3. Bubble removal and densification
4. Pre-cooling
5. Crystallization of the polymer melt
6. Final cooling

The induction time can be significantly reduced by pre-heating the powder, and the bubble removal and cooling stage can be shortened by pressurizing the material inside the mold. The melting and sintering of the powder during rotational molding depends on the rheology and geometry of the particles. This phenomenon was studied in depth by Bellehumeur and Vlachopoulos [44].

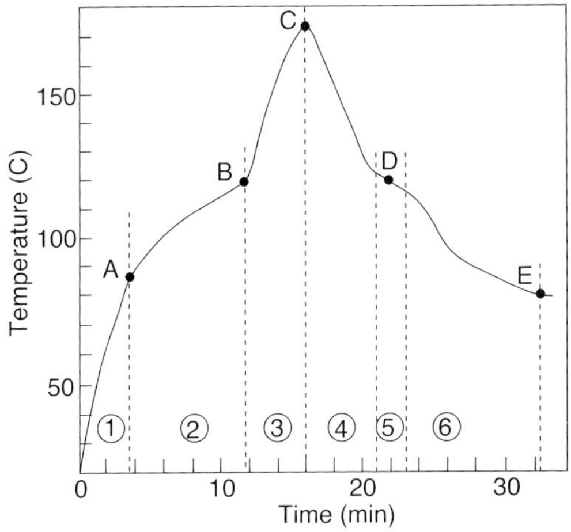

Figure 6.93 Typical air temperature in the mold while rotomolding polyethylene parts.

Examples

6.1 You are to use a 45 mm diameter single screw extruder to create a polymer polycarbonate/polypropylene polymer blend. The maximum screw rotation is 160 rpm and the screw channel depth is 4mm. Assuming a barrel temperature of 300 ^oC, a surface tension, between the two polymers of $8x10^{-3}$ N/m, and using the viscosity curves given in in the appedix of this book, determine:

- If one can disperse 20% PC into 80% PP
- If one can disperse 20% PP into 80% PC
- The minimum size of the dispersed phase

We start this problem by first calculating the average speed in the extruder using

$$v_0 = \pi Dn = \pi(45)(160)(1/60) = 377\,\text{mm/s}$$

which results in an average rate of deformation of

$$\dot{\gamma} = \frac{v_0}{h} = \frac{377\,\text{mm/s}}{4\,\text{mm}} = 94\text{s}^{-1}$$

From the viscosity curves we get $\eta_{PC} \approx 600\,\text{Pa}-\text{s}$ and $\eta_{PP} \approx 150\,\text{Pa}-\text{s}$. Using Fig. 6.32 we can deduce that one cannot disperse polycarbonte into polypropylene using a single screw extruder that only induces shear deformation, since $\eta_{PC}/\eta_{PP} > 4$. On the other hand, one can disperse polypropylene into polycarbonate using the same single screw extruder.

Using Fig.6.32 we can see that dispersive mixing for a $\eta_{PC}/\eta_{PP} > 0.25$ will occur at $Ca_{crit} \approx 0.7$. Hence, neglecting the effects of coalescence we can calculate the minimun size of the dispersed phase using

$$Ca_{crit} = 0.7 = \frac{\tau R}{\sigma_s} = \frac{600(94)R}{8x10^{-3}} \rightarrow D = 2R = 0.2\mu m$$

To achieve this dispersion we must maintain the stresses for an extended period.

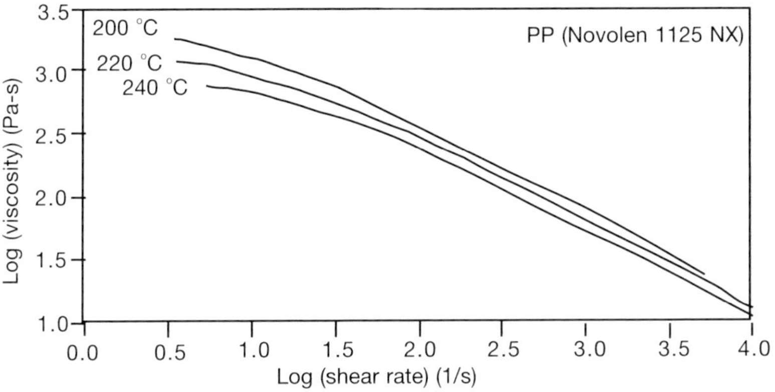

Figure 6.94 Viscosity curves for a polypropylene.

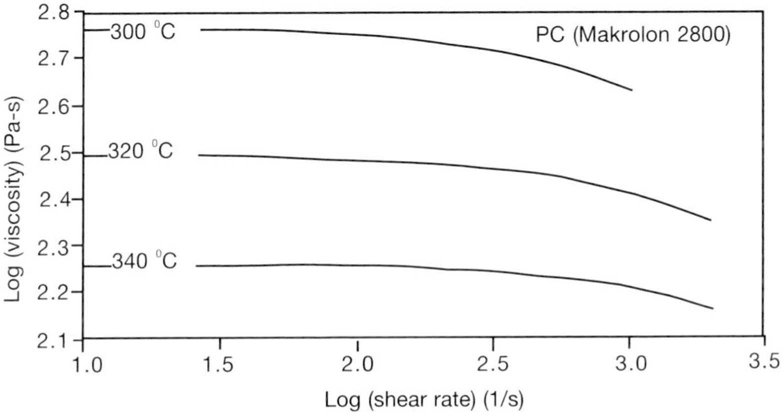

Figure 6.95 Viscosity curves for a polycarbonate.

6.2 You are to determine the maximum clamping force and injection pressure required to mold an ABS suitcase with a filling time, t_{fill}, of 2.5 seconds. Use the dimensions shown in Fig. 6.96, an injection temperature, T_i, of 227 °C (500 K), and a mold temperature, T_m, of 27 °C (300 K). The properties necessary for the calculations are also given below.

Table 6.8 Properties for ABS

$n = 0.29$	$\rho = 1020 \text{kg}/\text{m}^3$
$m_0 = 29x10^6 \text{Pa}-\text{s}^{\text{n}}$	$C_p = 2343 \text{J}/\text{Kg}/\text{K}$
$a = 0.01369/\text{K}$	$k = 0.184 \text{W}/\text{m}/\text{K}$

To aid the polymer processing engineer in finding required injection pressures and corresponding mold clamping forces, Stevenson [4] generated a set of dimensionless groups and corresponding graphs for non-isothermal mold filling of non-Newtonian polymer melts. We start this problem by first laying the suitcase flat and determining the required geometric factors (Fig. 6.97). From the suitcase geometry, the longest flow path, , is 0.6 m and the radius of the projected area, R_p, is 0.32 m.

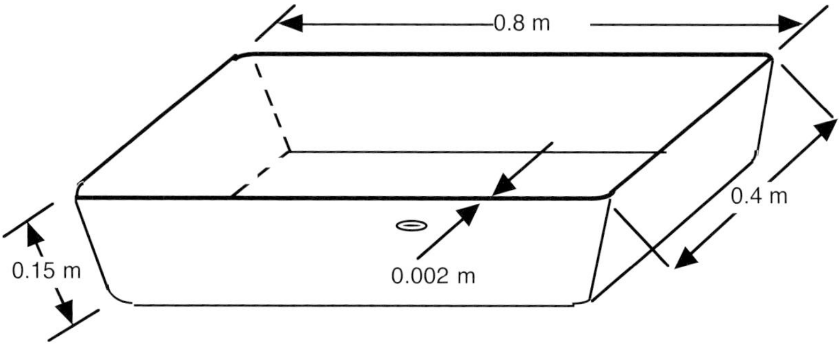

Figure 6.96 Suitcase geometry.

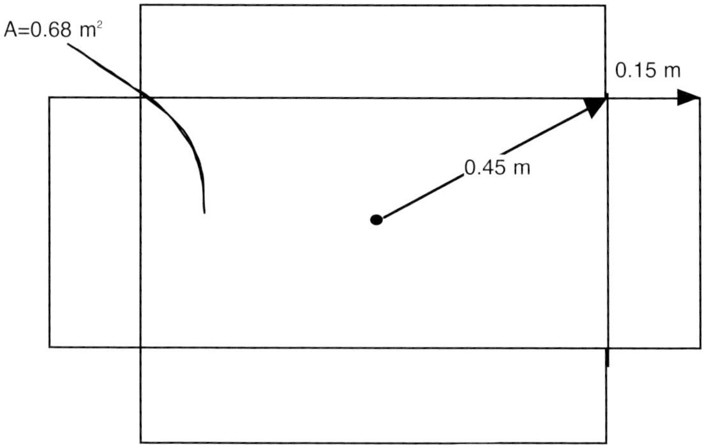

Figure 6.97 Layed-flat suitcase.

Using the notation in Fig. 6.25 and a viscosity defined by $\eta = m_0 e^{-a(T-T_{ref})} |\dot{\gamma}|^{n-1}$, four dimensionless groups are defined.

- The dimensionless temperature β determines the intensity of the coupling between the energy equation and the momentum balance. It is defined by

$$\beta = a(T_i - T_m) \tag{6.30}$$

  where T_i and T_m are the injection and mold temperatures respectively.

- The dimensionless time is the ratio of the filling time, t_{fill}, and the time for thermal equilibrium via conduction, defined by

$$\tau = \frac{t_{fill} K}{h^2 \rho C_p} \tag{6.31}$$

- The Brinkman number Br is the ratio of the energy generated by viscous dissipation and the energy transported by conduction. For a non-isothermal, non-Newtonian model it is

$$Br = \frac{m_0 e^{-aT_i} h^2}{k(T_i - T_m)} \left(\frac{R}{t_{fill} h} \right)^{n+1} \tag{6.32}$$

- The power-law index n of the Ostwald and deWaale model reflects the shear thinnng behavior of the polymer melt. Once the dimensionless parameter are calculated, the dimensionless injection pressures $(\Delta p / \Delta p_I)$ and dimensionless clamping forces $(F/F_I)$are read from Figs. 6.98 to 6.101. The isothermal pressure and force are computed using

$$\Delta p_I = \frac{m_0 e^{-aT_i}}{1-n} \left[\frac{1+2n}{2n} \frac{R}{t_{fill} h} \right]^n \left(\frac{R}{h} \right) \tag{6.33}$$

and

$$F_I = \pi R^2 \left(\frac{1-n}{3-n} \right) \Delta p_I \tag{6.34}$$

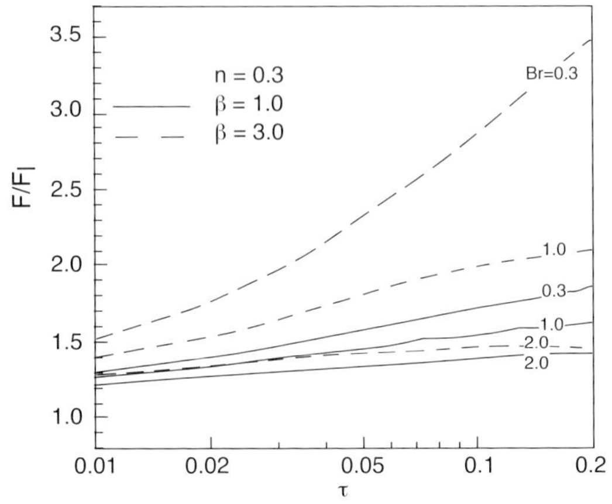

Figure 6.98 Dimensionless clamping force versus dimensionless groups.

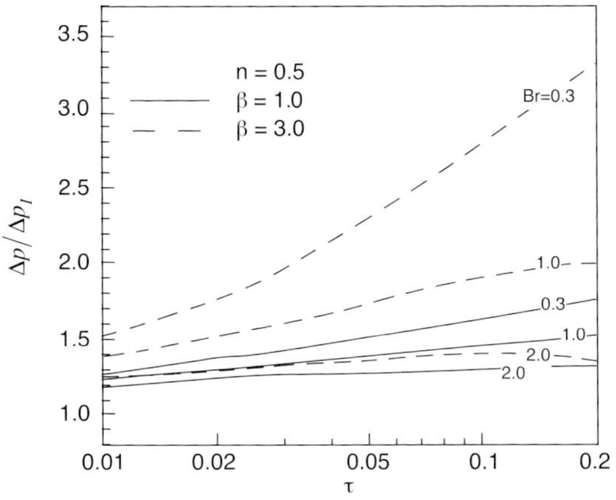

Figure 6.99 Dimensionless clamping force versus dimensionless groups.

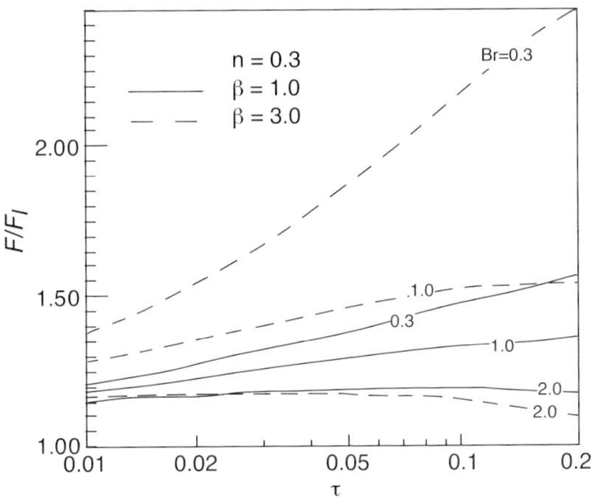

Figure 6.100 Dimensionless injection pressure versus dimensionless groups.

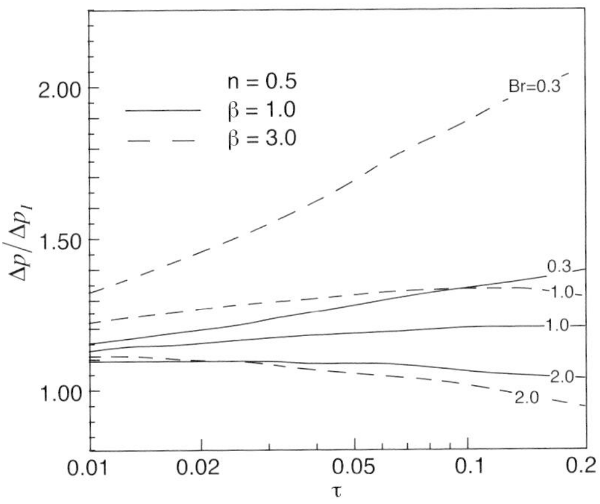

Figure 6.101 Dimensionless injection pressure versus dimensionless groups.

Using the data given for ABS and the dimensions for the laid-flat suitcase we can compute the dimensionless groups as

$$\beta = 0.01369(500 - 300) = 2.74$$

$$\tau = \frac{2.5(0.184)}{(0.001)^2(1020)(2343)} = 0.192$$

$$Br = \frac{(29x10^6)e^{-0.01369(500)}(0.001)^2}{0.184(500 - 300)}\left(\frac{0.6}{2.5(0.001)}\right)^{0.29+1} = 0.987$$

The isothermal injection pressure and clampng force are computed using eqs. 6.30 to 6.34.

$$\Delta p_I = \frac{29x10^6 e^{-0.01369(500)}}{1-0.29}\left(\frac{1+2(0.29)}{2(0.29)}\frac{0.6}{2.5(0.001)}\right)^{0.29}\left(\frac{0.6}{0.001}\right) = 171\text{MPa}$$

$$F_I = \pi(0.6)^2\left(\frac{1-0.29}{3-0.29}\right)(17.1x10^7) = 50.7x10^6\text{N}$$

We now look up $\Delta p / \Delta p_I$ and F / F_I in Figs. 6.98 to 6.101. Since little change occurs between n = 0.3 and n = 0.5, we choose n = 0.3. However, for other values of n we can interpolate or extrapolate. For $\beta = 2.74$, we interpolate between 1 and 3 as

$$\beta = 1 \rightarrow \Delta p / \Delta p_I = 1.36 \text{ and } F / F_I = 1.65$$

$$\beta = 3 \rightarrow \Delta p / \Delta p_I = 1.55 \text{ and } F / F_I = 2.1$$

$$\beta = 2.74 \rightarrow \Delta p / \Delta p_I = 1.53 \text{ and } F / F_I = 2.04$$

$$\Delta p = \left({}^{\Delta p}\!/_{\Delta p_I}\right)\Delta p_I = 262\text{MPa} = 2{,}620\text{bar}$$

$$F = \left({}^{F}\!/_{F_I}\right)F_I = 10.3x10^7\text{N} = 10{,}300\,\text{metric tons}$$

Since the part are exceeds the projected area, Fig. 6.102 can be used to correct the computed clamping force. The clamping force can be corrected forn an Rp= 0.32 m using Fig. 6.102 and $R_p / R = 0.53$.

$$F_{projected} = (0.52)10,300 = 5,356 \text{metric tons}$$

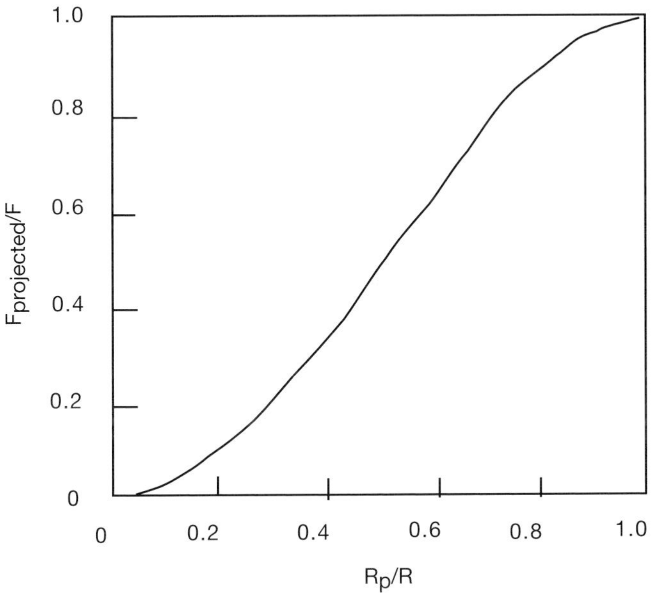

Figure 6.102 Clamping force correction for the projected area.

For our suitcase cover, where the total volume is 1,360 cc and total part area is 0.68 m^2, the above numbers are too high. A usefull rulo-of-thumb is a maximun allowable clamping force of 2 tons/in^2. Here, we have gratly exceeded that number. Normally, around 3,000 metric tons/m^2 are allowed in commercial injection molding machines. For example, a typical injection molding machine[14] with a shot size of 2,000 cc has a maximun clamping force of 630 metric tons with a maximun injection pressure of 1,400 bar. A machine with much larger clamping forces and injection pressures is suitable for much larger parts. For example, a mchaine with a shot size of 19,000 cc allows a maximum clamping force of 6,000 metric tons with a maximum injection pressure of 1,700 bar. For this example we must reduce the pressure and

14 MINIFLOW, Injection Molding Simulation, The Madison Groups PPRC, Madison, WI.

clamping force requirements. This can be acomplished by increasing the injection and mold temperatures or by reducing the filling time. Recommended injection temperatures for ABS are between 210 and 240 °C and recommended mold temperatures are between 40 and 90 °C.[15] As can be seen, there is room for adjustment in the processing condition, so one must repeat the above procedure using new conditions.

6.3 Let us consider the multi-cavity injection molding process shown in Fig. 6.103. To achieve equal part quality, the filling time for all cavities must be balanced. For the case in question, we need to balance the cavities by solving for the runner radius. For a balanced runner system the flow rates into all cavities must match. For a given flow rate Q, length L, and radius R_1, we can also solve for the pressure at the runner system junctures. Assuming an isothermal flow of a non-Newtonian shear thinning polymer with viscosity η, we can compute the radius for a part molded of polystyrene with a consistency index of $2.8x10^4$ $\text{Pa}-\text{s}^n$ and a power law index (n) of 0.28.

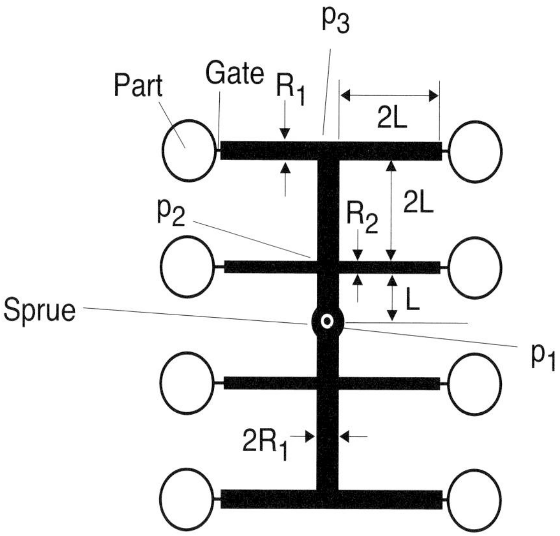

Figure 6.103 Runner system lay out.

The flow through each runner section is governed by equation 5.46, and the various sections can be representied using:

15 A good reference for such values is the CAMPUS® material data bank. Books such as H. Domininghaus, *Plastics for Engineers*, Hanser Publishers (1992), Munich are also recommended.

Section 1: $4Q = \left(\frac{\pi(2R_1)^3}{s+3}\right)\left(\frac{2R_1(P_1 - P_2)}{2\text{mL}}\right)^s$

Section 2: $2Q = \left(\frac{\pi(2R_1)^3}{s+3}\right)\left(\frac{2R_1(P_2 - P_3)}{2\text{m}(2\text{L})}\right)^s$

Section 3: $Q = \left(\frac{\pi R_2^3}{s+3}\right)\left(\frac{R_2(P_2 - 0)}{2m(2L)}\right)^s$

Section 4: $Q = \left(\frac{\pi R_1^3}{s+3}\right)\left(\frac{R_1(P_3 - 0)}{2\text{m}(2\text{L})}\right)^s$

Using values of $L = 10\,\text{cm}$, $R_1 = 4\,\text{mm}$ mm, and $Q = 20\,cm^3/s$, the unknown parameters, P_1, P_2, P_3, and R_2 can be obtained using the preciding equations. The equation are non-linear and must be solved in an iterative manner. For the given values, a radius, R_2, of 3.4 mm would result in a balanced runner system, with pressures P_1 = 265.7 bar, P_2 = 230.3 bar, and P_3 = 171.9 bar. For comparison, if one had assumed a Newtonian model with the same consistency index and a power law index of 1.0 a radius, R_2, of 3.9 mm would have resulted, with much higher required pressures of P_1 = 13,926 bar, P_2 = 12,533 bar, and P_3 = 11,140 bar. The difference is due to shear thinning.

Problems

6.1 You are to extrude a 100 mm wide high density polyethylene sheet using a 40 mm diameter single screw extruder with distributive as well as dispersive mixing heads. The screw characteristic curve is shown in Fig. 6.104[16].

The die can be approximated with a 100 mm wide and 100 mm long slit. On the graph below draw the die characteristic curves for dies with 1mm and 1.5mm thick slits. Will it be feasible to extrude a sheet through a 1.5mm thick slit? If yes, what screw speed would you choose? What about a 1 mm thick slit.

16 Courtesy of ICIPC, Medellín, Colombia.

Why do your die characteristic curves cross over the ones shown in the graph? Note that the data in the graph was measured experimentally with a variable restriction (valve) die.

Typical power law constants for HDPE at 180 °C are m=20,000 Pa-s^n and n=0.41. Use a specific gravity for HDPE of 0.95.

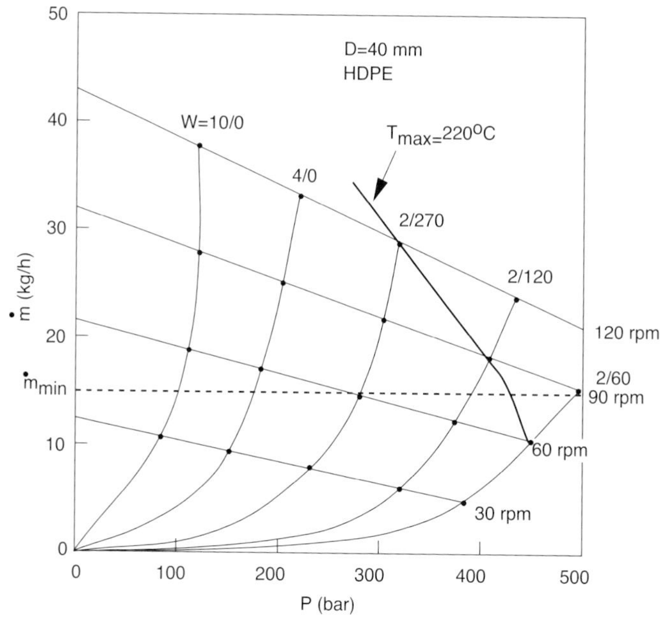

Figure 6.104 Process characteristic curves for a 45 mm diameter conventional extruder.

6.2 Estimate the striation thickness of a 3mm diameter pigmented polystyrene pellet in a polystyrene matrix after traveling through a 20 turn, 45 mm diameter, 5 mm constant channel depth, single screw extruder. Assume open discharge conditions. Use 100 rpm rotational speed.

6.3 Someone in your company proposes to use an existing square pitch 150 mm diameter plasticating single screw extruder as a mixing device for a 40/60 PS/PP polymer blend. The metering section is 5 turns long and has a channel depth of 10 mm. Will dispersion of the polystyrene occur for a screw rotation of 60 rpm? Assume open discharge and a temperature of 220 °C. Use viscosity data given in example 6.1 and Fig. 6.106.

6.4 Someone in your company proposes to use an existing square pitch 150 mm diameter plasticating single screw extruder as a mixing device for a 40/60 PP/PS polymer blend. The metering section is 5 turns long and has a channel depth of 10 mm. Will dispersion of the polypropylene occur for a screw rotation of 60 rpm? Will there be enough time for dispersion? Assume open discharge and a temperature of 220 ^{o}C. Use viscosity data given in problems 6.3 and below.

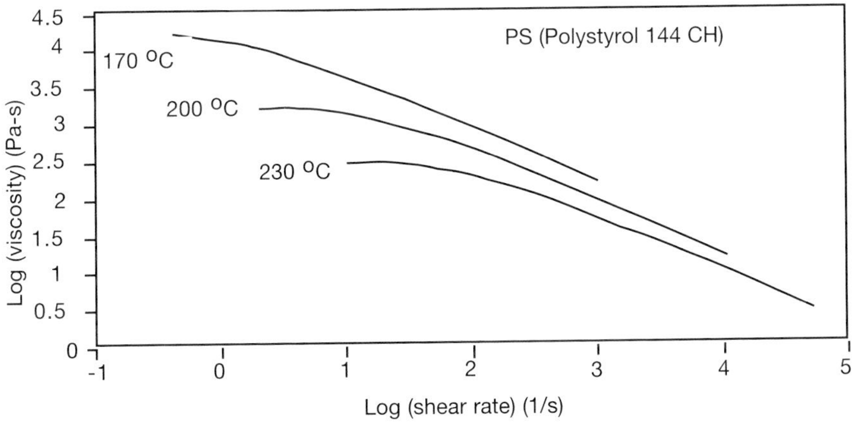

Figure 6.105 Viscosity curves for a polystyrene.

6.5 A thin polyamide 66 component is injection molded under the following conditions:

- The melt is injected at 275 ^{o}C to a maximum pack/hold pressure of 800 bar.
- The 800 bar pack/hold pressure is maintained until the gate freezes off, at which point the part is at an average temperature of 175 ^{o}C.
- The pressure drops to 1 bar as the part cools inside the cavity.
- The part is removed from the mold and cooled to 25 ^{o}C.
- Draw the whole process on the PvT diagram.
- Estimate the final part thickness if the mold thickness is 1 mm. For thin injection molded parts, most of the shrinkage leads to part thickness reduction.

6.6 Does the screw and die characteristic curves, correspond to a conventional or a grooved single screw extruder?

A die can be approximated with a 1 mm diameter and 30 mm long capillary. On the graph below draw the die characteristic curve for the given die?

Typical power law constants for HDPE at 180 °C are m=20,000 Pa-s^n and n=0.41. Use a specific gravity for HDPE of 0.95.

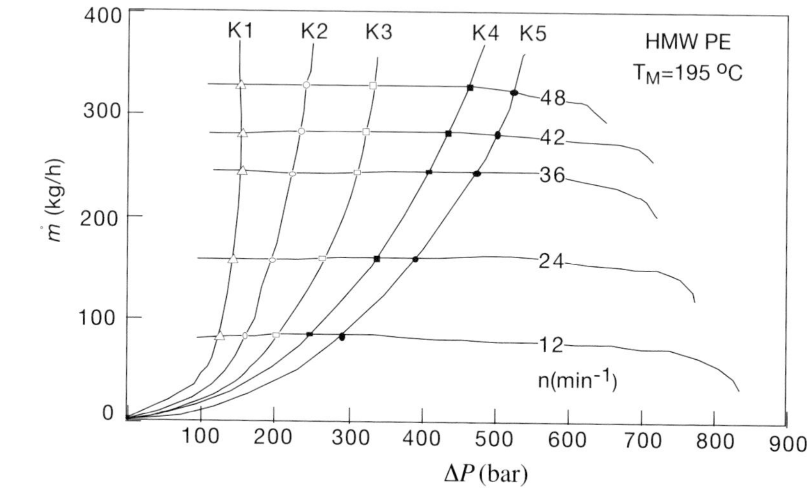

Figure 6.106 Process characteristic curves for a 45 mm diameter extruder.

6.7 An internal batch mixer maintains shear rates, $\dot{\gamma}$, of 100 s^{-1} for extended periods. In the mixer you want to disperse LDPE in a PS matrix at 170 °C. What is the size of the dispersed phase? Will the PS still be transparent? Use viscosity data given in Problem 6.4 and below.

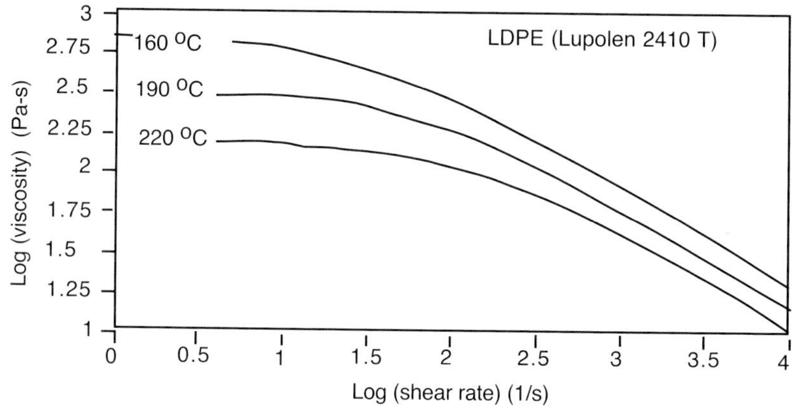

Figure 6.107 Viscosity curves for a low density polyethylene.

6.8 Design a balanced runner system for the mold in Example 6.3 if you are to injection mold a polystyrene product. Assume a power-law model with a consistency index, m, of 2.8×10^4 Pa-s^n, and a power-law index, n, of 0.28.

6.9 Estimate the cooling time for the ABS suitcase presented in Example 6.2 if demolding occurs when the average part temperature is below 60 ^oC.

6.10 What are the required clamping force and injection pressure if the filling time in Example 6.2 is increased from 2.5 s to 3 s?

6.11 What are the required clamping force and injection pressure if the mold temperature in Example 6.2 is increased from 27 ^oC to 90 ^oC ?

6.12 What are the required clamping force and injection pressure if the injection temperature in Example 6.2 is increased from 227 to 240 ^oC?

6.13 Measure and plot the wall thickness distribution on a PE-HD one gallon milk container.

6.14 Measure and plot the wall thickness distribution on a small thermoformed individual coffee cream container.

References

1. Tadmor, Z., and Gogos, C.G., Principles of Polymer Processing, John Wiley & Sons, New York, (1979).
2. Rauwendaal, C., *Mixing in Polymer Processing*, Marcel Dekker, Inc., New York, (1991).
3. Osswald, T.A., Turng, L.S., and Gramann, P.J., Injection Molding Handbook, Hanser Publishers, Munich, (2001).
4. Menges, G., and Predöhl, W., *Plastverarbeiter, 20*, 79, (1969).
5. Menges, G., and Predöhl, W., *Plastverarbeiter, 20*, 188, (1969).
6. Scott, C.E., and Macosko, C.W., *Polymer Bulletin, 26*, 341, (1991).
7. Gramann, P.J., Stradins, L., and Osswald, T.A., *Intern. Polymer Processing*, 8, 287, (1993).
8. Osswald, T.A., *Polymer Processing Fundamentals*, Hanser Publishers, Munich, (1998).
9. Erwin, L., *Polym. Eng. & Sci., 18*, 572, (1978).
10. Erwin, L., *Polym. Eng. & Sci., 18*, 738, (1978).
11. Ng, K.Y., Master of Science Thesis, University of Wisconsin-Madison, (1979).
12. Tadmor, Z., *Ind. Eng. Fundam., 15*, 346, (1976).

13. Cheng, J., and Manas-Zloczower, I., *Internat. Polym. Proc.*, 5, 178, (1990).
14. Grace, H.P., *Chem. Eng. Commun., 14*, 225, (1982).
15. Cox, R.G., *J. Fluid Mech.*, 37, 3, 601–623, (1969).
16. Bentley, B.J. and Leal, L.G., *J. Fluid Mech.*, 167, 241–283, (1986).
17. Stone, H.A. and Leal, L.G., *J. Fluid Mech.*, 198, 399–427, (1989).
18. Biswas, A., and Osswald, T.A., unpublished research, (1994).
19. Gramann, P.J., M.S. Thesis, University of Wisconsin-Madison, (1991).
20. Manas-Zloczower, I., Nir, A., and Tadmor, Z., *Rubber Chemistry and Technology*, 55, 1250, (1983).
21. Boonstra, B.B., and Medalia, A.I., *Rubber Age*, March and April, (1963).
22. Rauwendaal, C., *Polymer Extrusion*, Hanser Publishers, Munich, (1990).
23. Menges, G., and Harms, E., *Kautschuk und Gummi, Kunststoffe, 25,* 469, (1972).
24. Rauwendaal, C., *SPE ANTEC Tech. Pap., 39*, 2232, (1993).
25. Rauwendaal, C., *Mixing in Reciprocating Extruders*, A chapter in *Mixing and Compounding of Polymers*, Eds. I. Manas-Zloczower and Z. Tadmor, Hanser Publishers, Munich, (1994).
26. Elemans, P.H.M., Modeling of the cokneater, A chapter in *Mixing and Compounding of Polymers*, Eds. I. Manas-Zloczower and Z. Tadmor, Hanser Publishers, Munich, (1994).
27. Lim, S. and White, J.L., *Intern. Polymer Processing, 8*, 119, (1993).
28. Lim, S. and White, J.L., *Intern. Polymer Processing, 9,* 33, (1994).
29. Bird, R.B., Steward, W.E., and Lightfoot, E.N., *Transport Phenomena*, John Wiley & Sons, New York, (1960).
30. Erwin, L., *Polym. Eng. & Sci., 18*, 1044, (1978).
31. Suetsugu, *Intern. Polymer Processing, 5*, 184, (1990).
32. Greener, J., *Polym. Eng. Sci., 26*, 886 (1986).
33. Michaeli, W., and Lauterbach, M., *Kunststoffe , 79*, 852 (1989).
34. Stevenson, J.F., *Polym. Eng. Sci., 18*, 577 (1978).
35. Osswald, T.A., and Menges, G., *Materials Science of Polymers for Engineers,* Hanser Publishers, Munich (1996).
36. Anturkar, N.R., and Co, A., *J. Non-Newtonian Fluid Mech., 28*, 287 (1988).
37. Menges, G., *Einführung in die Kunststoffverarbeitung,* Hanser Publishers, Munich (1986).
38. Rosato, D.V., *Blow Molding Handbook*, Hanser Publishers, Munich (1989).
39. Modern Plastics Encyclopedia, 53, McGraw-Hill , New York (1976).
40. Tadmor, Z., and Bird, R.B., *Polym.Eng.Sci, 14*, 124 (1973).
41. Denton, D.L, *The Mechanical Properties of an SMC-R50 Composite*, Owens-Corning Fiberglas Corporation (1979).
42. Shutov, F.A., *Integral/Structural Polymer Foams*, Springer-Verlag, Berlin (1986).
43. Crawford, R.J., *Rotational Molding of Plastics*, Research Studies Press, Somerset (1992).
44. Bellehumeur, C.T., and Vlachopoulos, J., *SPE 56th Antec,* (1998).
45. Mascia, L., *The Role of Additives in Plastics*, John Wiley & Sons, New York, (1974).
46. Hildebrand, J. and Scott, R.L., *The Solubility of Non-Electrolytes*, 3rd Ed., Reinhold Publishing Co., New York, (1949).
47. Rosen, S.L., *Fundamental Principles of Polymeric Materials*, 2nd. Ed., John Wiley & Sons, Inc., New York, (1993).

48. Janssen, J.M.H., Ph.D. Thesis, Eindhoven University of Technology, The Netherlands, (1993).
49. van Krevelen, D. W., and Hoftyzer, P.J., *Properties of Polymers*, 2nd ed., Elsevier, Amsterdam, (1976).

7

Anisotropy Development During Processing

The mechanical properties and dimensional stability of a molded polymer part are strongly dependent upon the anisotropy of the finished part[1]. The structure of the final part, in turn, is influenced by the design of the mold cavity, e.g., the type and position of the gate, and by the various processing conditions such as injection speed, melt or compound temperatures, mold cooling or heating rates, and others. The amount and type of filler or reinforcing material also has a great influence on the quality of the final part.

This chapter discusses the development of anisotropy during processing of thermoset and thermoplastic polymer parts and presents basic analyses that can be used to estimate anisotropy in the final product.

7.1 Orientation in the Final Part

During processing the molecules, fillers, and fibers are oriented in the flow and greatly affect the properties of the final part. Since there are large differences in the processing of thermoplastic and thermoset polymers, the two will be discussed individually in the next two sections.

7.1.1 Processing Thermoplastic Polymers

When thermoplastic components are manufactured, the polymer molecules become oriented. The molecular orientation is induced by the deformation of the polymer melt during processing. The flexible molecular chains get stretched, and because of their entanglement they cannot relax fast enough before the part cools and solidifies. At lower processing temperatures this phenomenon is multiplied, leading to even higher degrees of molecular

1 This chapter primarily deals with the anisotropy development of injection and compression molded parts.

orientation. This orientation is felt in the stiffness and strength properties of the polymer component. Orientation also gives rise to *birefringence*, or *double refraction*, a phenomenon discussed in Chapter 12. The various degrees of molecular orientation and the different main directions of orientation in the material introduce a variable refractive index field, n(x,y,z), throughout the part. The value of the refractive index, n, depends on the relative orientation of the molecules, or the molecular axes, to the direction of the light shining through the part.

As polarized light travels through a part, a series of colored lines called *isochromatics* become visible or appear as shown in Fig. 7.1 [1]. The isochromatics are lines of equal molecular orientation and numbered from zero, at the region of no orientation, up with increasing degrees of orientation. A zero degree of orientation is usually the place in the mold that fills last and the degree of orientation increases towards the gate. Figure 7.2 shows schematically how molecular orientation is related to birefringence. The layers of highest orientation are near the outer surfaces of the part with orientation increasing towards the gate.

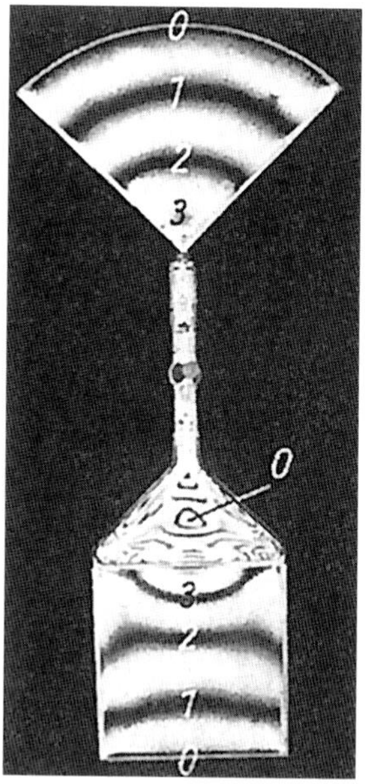

Figure 7.1 Isochromatics in a polycarbonate specimen of 1.7 mm wall thickness.

The degree of orientation increases and decreases depending on the various processing conditions and materials. For example, Fig. 7.3 [1] shows quarter disks of various wall thicknesses, molded out of four different materials:

polycarbonate, cellulose acetate, polystyrene, and polymethyl methacrylate. We see that for all materials the degree of orientation increases with decreasing wall thickness. An explanation for this is that the velocity gradients increase when wall thickness decreases. In subsequent sections of this chapter we discuss how orientation is directly related to velocity gradients.

Orientation is also related to the process used to manufacture the part. For example, Fig. 7.4 [1] shows two injection molded polycarbonate parts molded with different injection molders: a piston-type and a screw-type machine. It is obvious that the cover made with the piston-type injection molder has much higher degrees of molecular orientation than the one manufactured using the screw-type injection molder. Destructive tests revealed that it was impossible to produce a cover that is sufficiently crack-proof when molded with a piston-type molding machine.

The articles in Figs. 7.1, 7.3, and 7.4 were injection molded – a common processing method for thermoplastic polymers. Early studies have already shown that a molecular orientation distribution exists across the thickness of thin injection molded parts [2]. Figure 7.5 [2] shows the shrinkage distribution in longitudinal and transverse flow directions of two different plates. The curves demonstrate the degree of anisotropy that develops during injection molding, and the influence of the geometry of the part on this anisotropy.

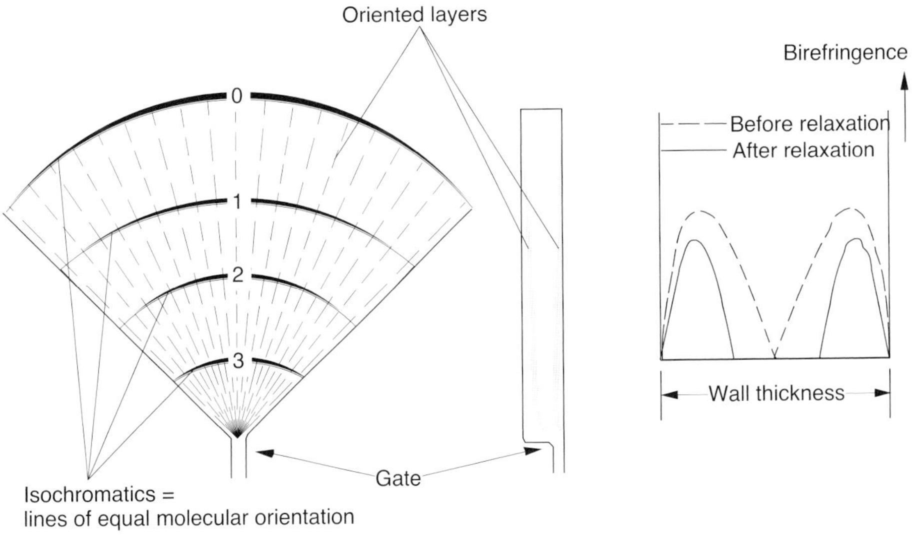

Figure 7.2 Orientation birefringence in a quarter disk.

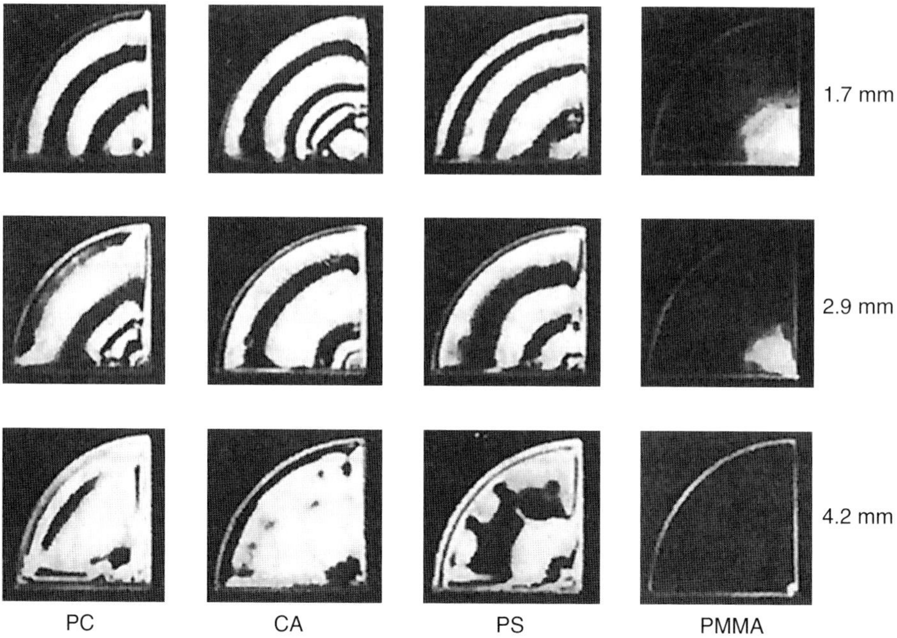

Figure 7.3 Isochromatics in polycarbonate, cellulose acetate, polystyrene. and polymethyl methacrylate quarter disks of various thicknesses.

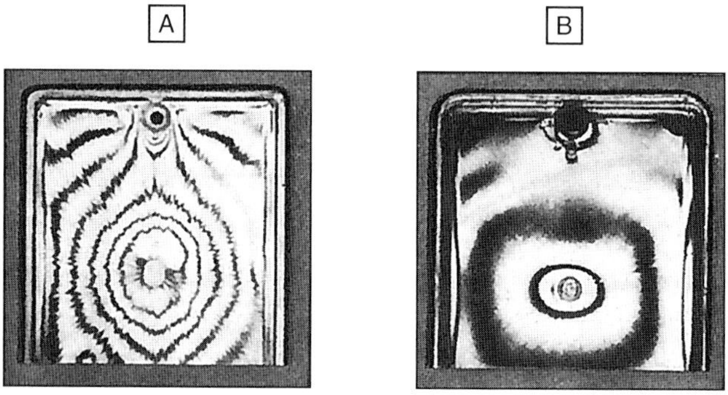

Figure 7.4 Isochromatics in polycarbonate parts molded with (left) piston-type and (right) screw-type injection molding machines.

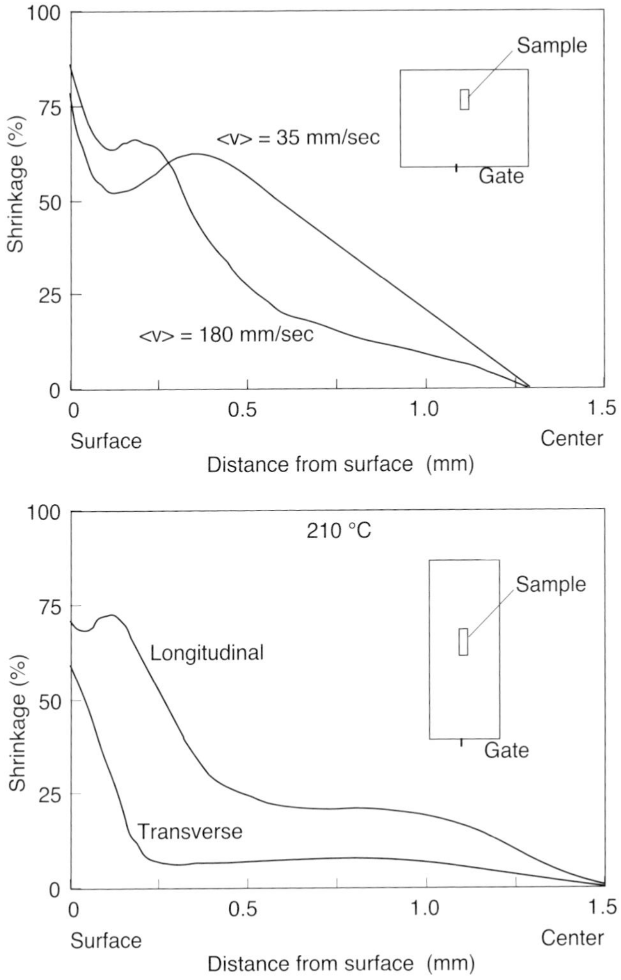

Figure 7.5 Shrinkage distribution of injection molded polystyrene plates.

An example of how to use the birefringence pattern of polymer parts to detect severe problems is in the manufacture of polycarbonate compact disks [3, 4]. Figure 7.6 shows the birefringence distribution in the *rz*-plane of a 1.2 mm thick disk molded with polycarbonate. The figure shows how the birefringence is highest at the surface of the disk and lowest just below the surface. Towards the inside of the disk the birefringence rises again and drops somewhat toward the central core of the disk. A similar phenomenon was observed in glass fiber reinforced [5–7] and liquid crystalline polymer [8] injection molded parts which show large variations in fiber and molecular orientation through the thickness.

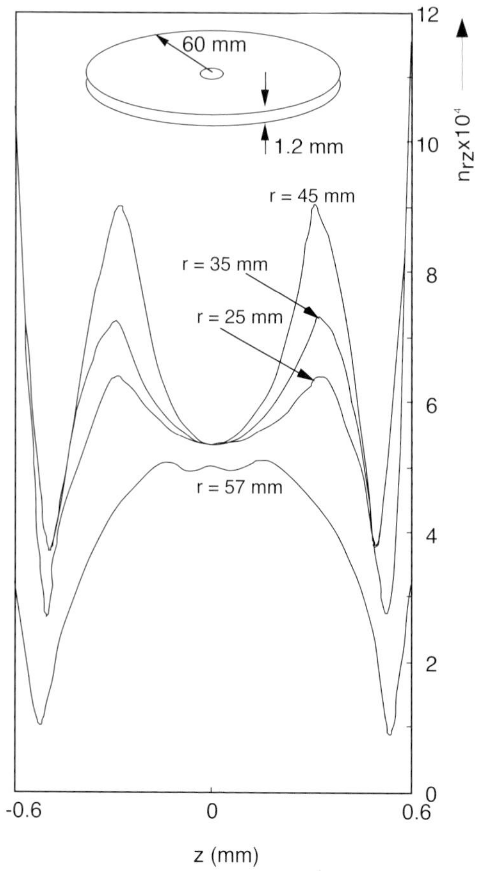

Figure 7.6 Birefringence distribution in the *rz*-plane at various radius positions. Numbers indicate radial position.

All these recent findings support earlier claims that molecular or filler orientation in injection molded parts can be divided into seven layers schematically represented in Fig. 7.7 [1]. The seven layers may be described as follows:

- Two thin outer layers with a biaxial orientation, random in the plane of the disk;
- Two thick layers next to the outer layers with a main orientation in the flow direction;
- Two thin randomly oriented transition layers next to the center core;
- One thick center layer with a main orientation in the circumferential direction.

There are three mechanisms that lead to high degrees of orientation in injection molded parts: fountain flow effect, radial flow, and holding pressure induced flow.

The *fountain flow effect* [9] is caused by the no-slip condition on the mold walls, which forces material from the center of the part to flow outward to the mold surfaces as shown in Fig. 7.8 [10]. As the figure schematically represents, the melt that flows inside the cavity freezes upon contact with the cooler mold walls. The melt that subsequently enters the cavity flows between the frozen layers, forcing the melt skin at the front to stretch and unroll onto the cool wall where it freezes instantly. The molecules which move past the free flow front are oriented in the flow direction and laid on the cooled mold surface which freezes them into place, though allowing some relaxation of the molecules after solidification. Using computer simulation, the fountain flow effect has been extensively studied in past few years [11]. Figures 7.9a and b [12] show simulated instantaneous velocity vectors and streamlines during the isothermal mold filling of a Newtonian fluid.[2] and Figs. 7.9c and d show the velocity vectors relative to the moving flow front. Figure 7.10 [13] presents the predicted shape and position of the tracer relative to the flow front along with the streamlines for a non-Newtonian non-isothermal fluid model. The square tracer mark is stretched as it flows past the free flow front, and is deposited against the mold wall, pulled upward again and eventually deformed into a V-shaped geometry. Eventually, the movement of the outer layer is stopped as it cools and solidifies.

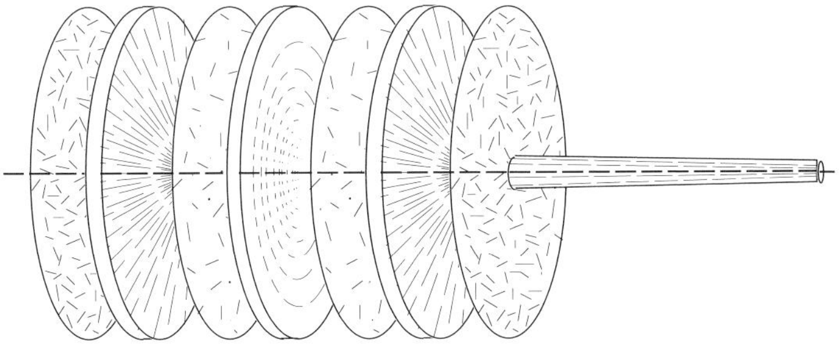

Figure 7.7 Filler orientation in seven layers of a centrally injected disk.

2 The isothermal and Newtonian analysis should only serve to explain the mechanisms of fountain flow. The non-isothermal nature of the injection molding process plays a significant role in the orientation of the final part and should not be left out in the analysis of the real process.

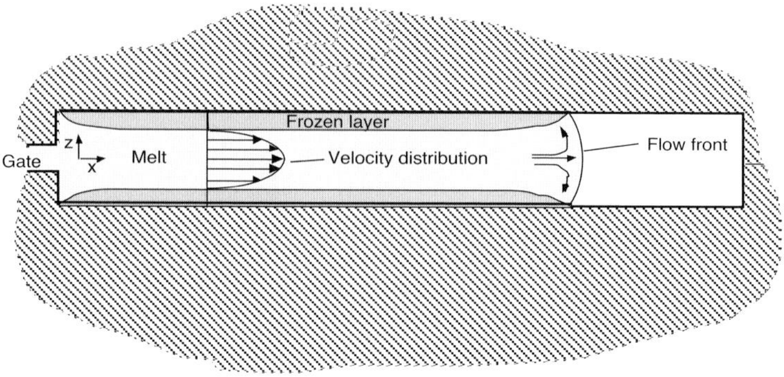

Figure 7.8 Flow and solidification mechanisms through the thickness during injection molding.

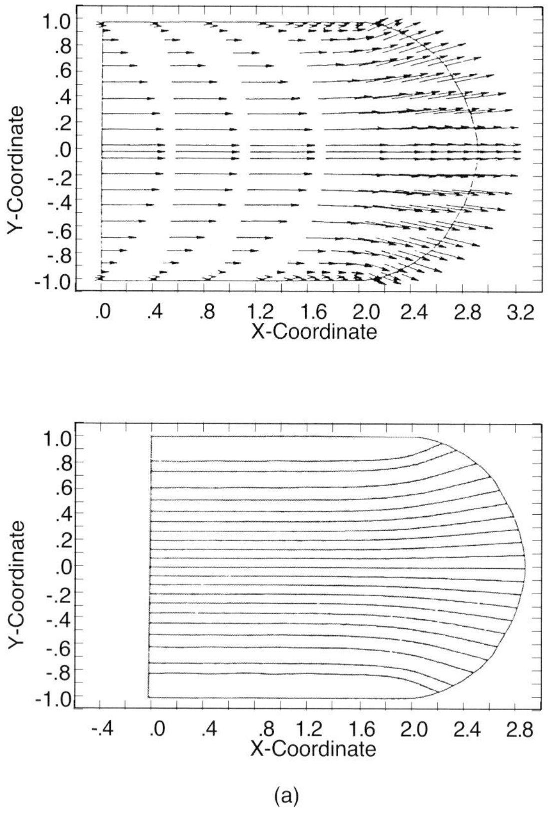

Figure 7.9 Fountain flow effect: (a) Actual velocity vectors and streamlines. *(continued)*

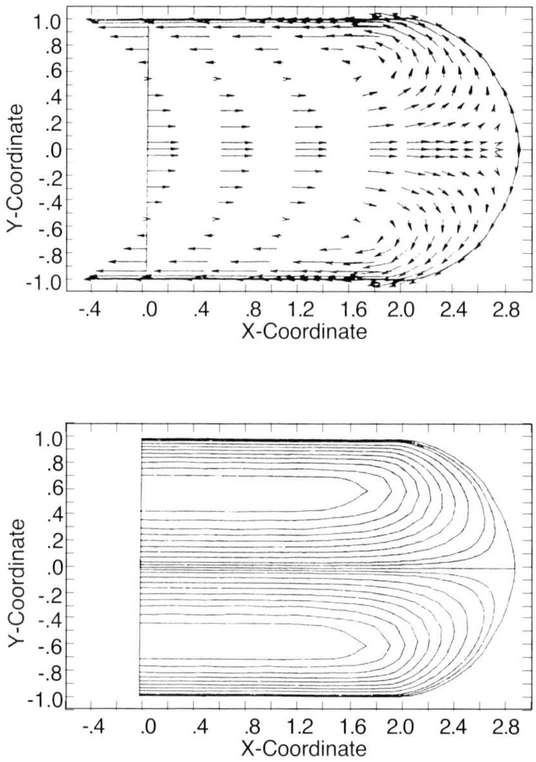

(b)

Figure 7.9 *(continued)* Fountain flow effect: (b) relative to the moving front velocity vectors and streamlines.

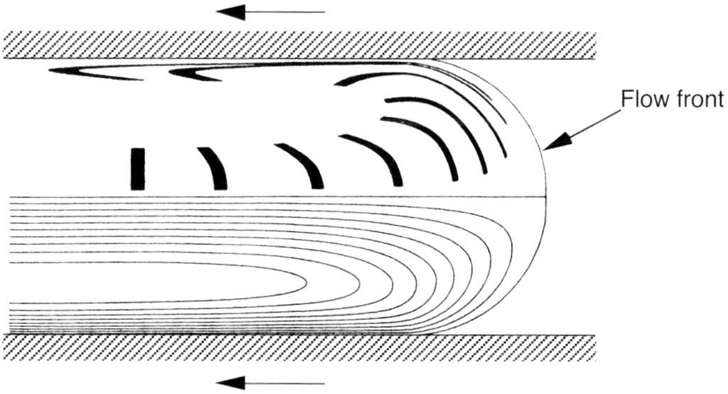

Figure 7.10 Deformation history of a fluid element and streamlines for frame of reference that moves with the flow front.

Radial flow is the second mechanism that often leads to orientation perpendicular to the flow direction in the central layer of an injection molded part. This mechanism is schematically represented in Fig. 7.11. As the figure suggests, the material that enters through the gate is transversely stretched while it radially expands as it flows away from the gate. This flow is well represented in today's commonly used commercial injection mold filling software.

Finally, the flow induced by the holding pressure as the part cools leads to additional orientation in the final part. This flow is responsible for the spikes in the curves shown in Figs. 7.5 and 7.6.

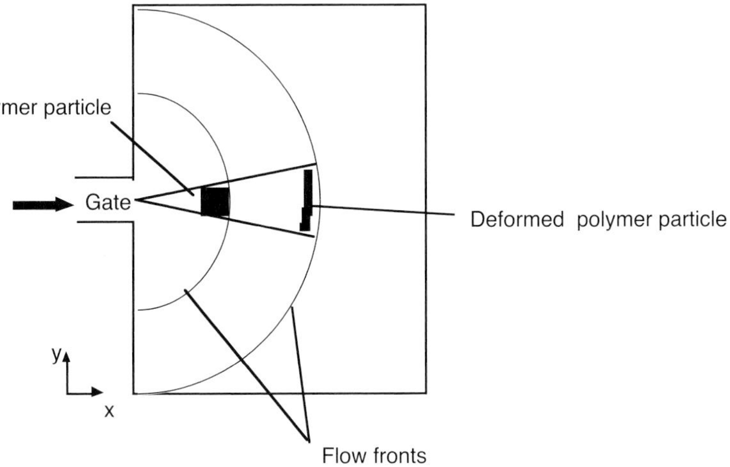

Figure 7.11 Deformation of the polymer melt during injection molding.

7.1.2 Processing Thermoset Polymers

During the manufacture of thermoset parts there is no molecular orientation because of the cross-linking that occurs during the solidification or curing reaction. A thermoset polymer solidifies as it undergoes an exothermic reaction and forms a tight network of inter-connected molecules.

However, many thermoset polymers are reinforced with filler materials such as glass fiber, wood flour, etc. These composites are molded via transfer molding, compression molding, or injection-compression molding. The properties of the final part are dependent on the filler orientation. In addition, the thermal expansion coefficients and the shrinkage of these polymers are highly dependent on the type and volume fraction of filler being used. Different forms of orientation may lead to varying strain fields, which may cause warpage in the final part. This topic will be discussed in the next chapter.

During the processing of filled thermoset polymers the material deforms uniformly through the thickness with slip occurring at the mold surface as shown schematically in Fig. 7.12 [14]. Several researchers have studied the development of fiber orientation during transfer molding and compression molding of sheet molding compound (SMC) parts [15]. During compression molding, a thin SMC charge is placed in a heated mold cavity and squeezed until the charge covers the entire mold surface. An SMC charge is composed of a polyester resin with around 10% by volume of calcium carbonate filler and 20–50% by volume glass fiber content. The fibers are usually 25 mm long and the final part thickness is 1–5 mm. Hence, the fiber orientation can be described with a planar orientation distribution function.

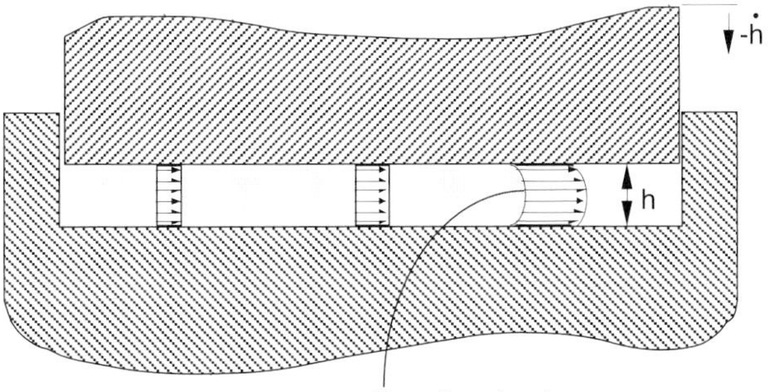

Figure 7.12 Velocity distribution during compression molding with slip between material and mold surface.

To determine the relationship between deformation and final orientation in compression molded parts, it is common to mold rectangular plates with various degrees of extensional flow, as shown in Fig. 7.13. These plates are molded with a small fraction of their glass fibers impregnated with lead so that they become visible in a radiograph. Figure 7.14 shows a computer generated picture from a radiograph, taken from a plate where the initial charge coverage was 33% [15, 16]. In Fig. 7.14, about 2000 fibers are visible which through digitizing techniques resulted in the histogram presented in Fig. 7.15 and depict the fiber orientation distribution in the plate. Such distribution functions are very common in compression or transfer molded parts and lead to high degrees of anisotropy throughout a part.

Furthermore, under certain circumstances, filler orientation may lead to crack formation as shown in Fig. 7.16 [1]. Here, the part was transfer molded through two gates which led to a knitline and filler orientation shown in the figure. Knitlines are cracklike regions where there are few or no fibers bridging across, lowering the strength across the region to that of the matrix material. A better way to mold the part of Fig. 7.16 would be to inject the material through a ring-type gate which would result in an orientation distribution mainly in the circumferential direction.

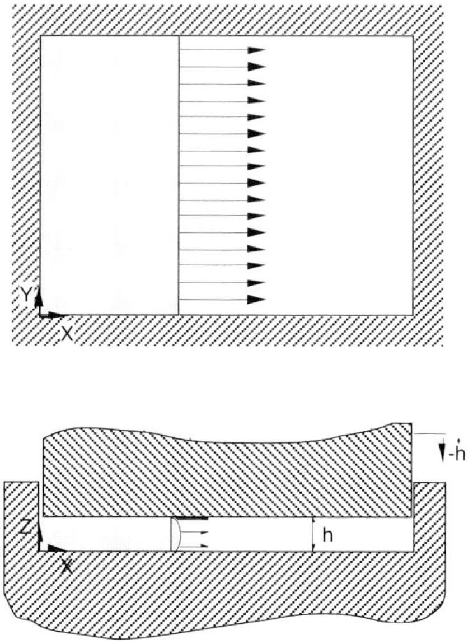

Figure 7.13 Schematic of extensional flow during compression molding.

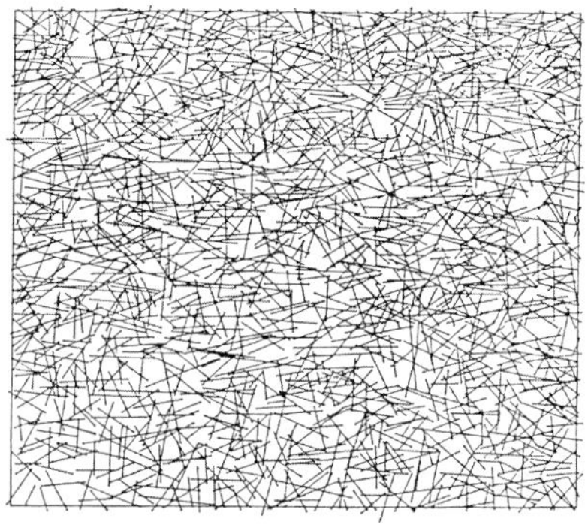

Figure 7.14 Computer redrawn plot of fibers in a radiograph of a rectangular SMC plate.

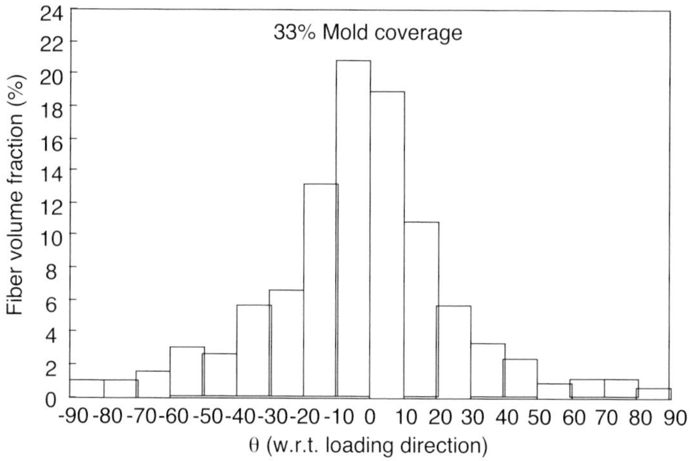

Figure 7.15 Measured fiber orientation distribution histogram in a plate with 33% initial mold coverage and extensional flow during mold filling.

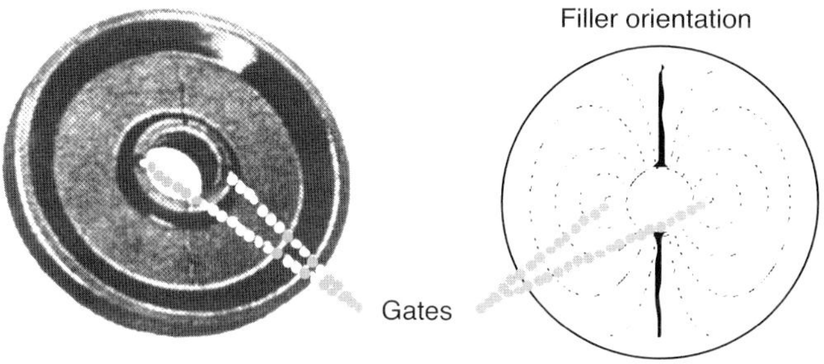

Figure 7.16 Formation of knitlines in a fiber filled thermoset pulley.

In compression molding, knitlines are common when multiple charges are placed inside the mold cavity or when charges with re-entrant corners are used, as shown in Fig. 7.17 [18]. However, a re-entrant corner does not always imply the formation of a knitline. For example, when squeezing a very thick charge, an equibiaxial deformation results and knitline formation is avoided. On the other hand, a very thin charge will have a friction dominated flow leading to knitline formation at the beginning of flow. Knitlines may also form when there are large differences in part thickness and when the material flows around thin regions as demonstrated in Fig. 7.18. Here, a crack forms as the material flows past the thinner section of the body panel. It is interesting to point out that usually the thin region will eventually be punched out to give room to headlights, door handles, etc.

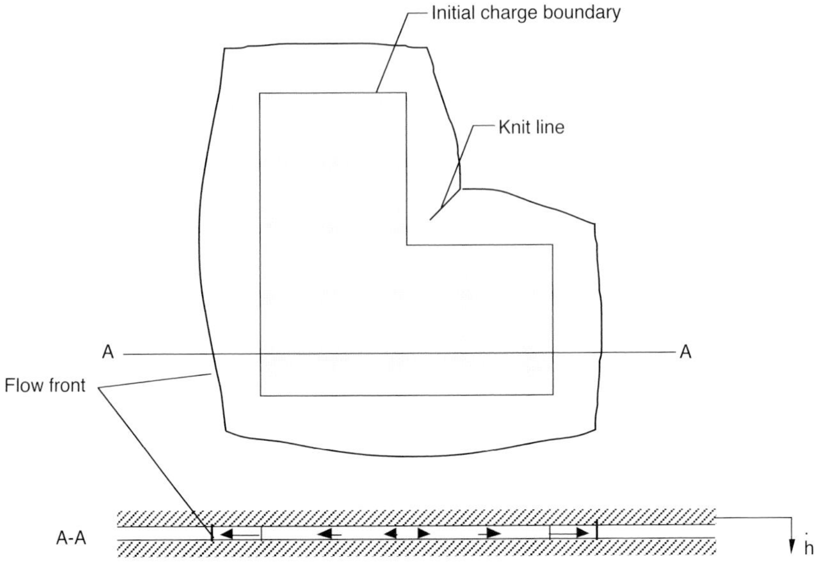

Figure 7.17 Knitline formation in an L-shaped charge for a squeeze ratio of 2.

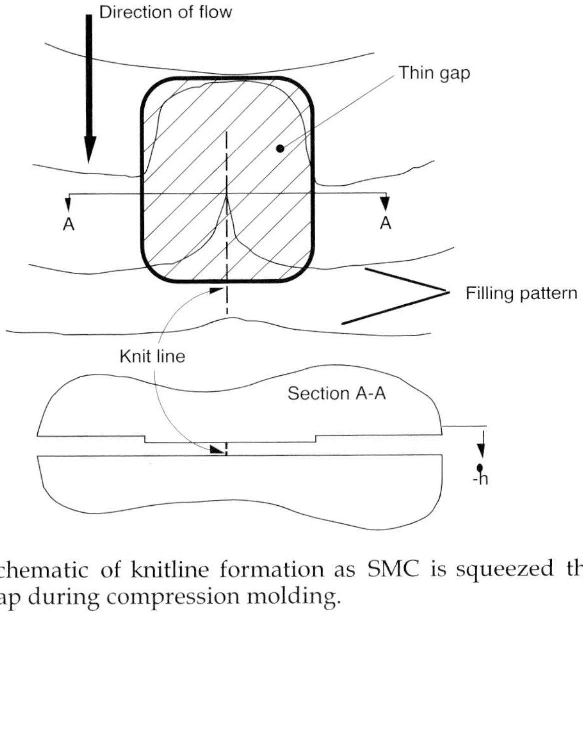

Figure 7.18 Schematic of knitline formation as SMC is squeezed through a narrow gap during compression molding.

7.2 Predicting Orientation in the Final Part

In general, the orientation of a particle, such as a fiber, is described by two angles, ϕ and θ, as shown in Fig. 7.19. These angles change in time as the polymer flows through a die or is stretched or sheared during mold filling. In many cases the angular orientation of a particle can be described by a single angle, ϕ. This is true since many complex three-dimensional flows can be reduced to planar flows such as the squeezing flow shown in Figs. 7.12 and 7.13 or the channel flow shown in Fig. 7.20. In squeezing flow, the z-dimension is very small compared the other dimensions, whereas in channel flow, the z-dimension is much larger than the other dimensions. Channel flow is often encountered inside extrusion dies, and squeezing flow is common in compression molding where the fiber length is much larger than the thickness of the part. In both cases the fiber is allowed to rotate about the z-axis, with the channel flow having a three-dimensional orientation and the squeezing flow a planar orientation distribution. Due to the simplicity of planar orientation distributions and their applicability to a wide range of applications, we will limit our discussion to two-dimensional systems which can be handled with planar models. However, it should be pointed out that for many polymer articles, such as injection molded parts, a planar orientation is not sufficient to describe the angular position of the fillers or molecules. Since the topic of three-dimensional orientation distribution function is still in the research stages and is therefore beyond the scope of this book, the reader is encouraged to read the literature [6, 7].

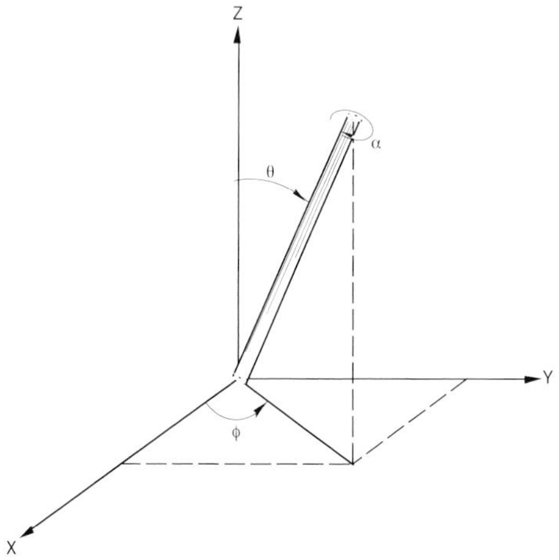

Figure 7.19 Fiber in a three-dimensional coordinate system.

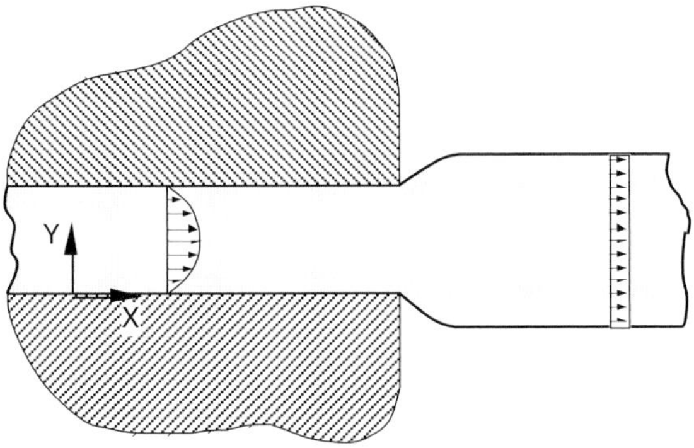

Figure 7.20 Schematic of two-dimensional channel flow.

7.2.1 Planar Orientation Distribution Function

The state of particle orientation at a point can be fully described by an orientation distribution function, $\psi(\phi,x,y,t)$. The distribution is defined such that the probability of a particle, located at x,y at time t, being oriented between angles ϕ_1 and ϕ_2, is given by.

$$P(\phi_1 < \phi < \phi_2) = \int_{\phi_1}^{\phi_2} \psi(\phi,x,y,t)\,d\phi \tag{7.1}$$

This is graphically depicted in Fig. 7.21. For simplicity, the x,y,t from the orientation distribution function can be dropped.

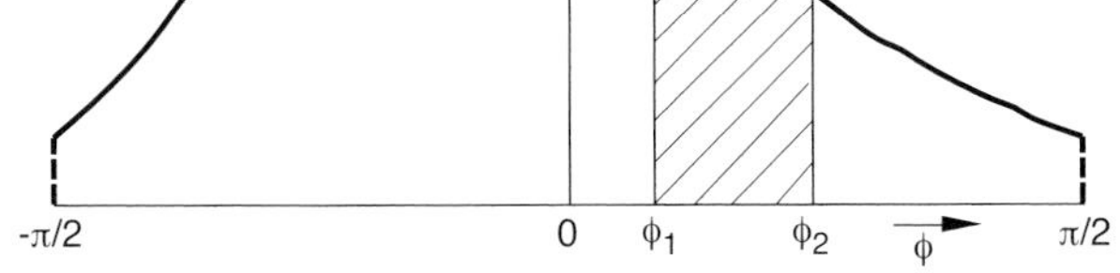

Figure 7.21 Orientation distribution function.

Since one end of a particle is indistinguishable from the other, the orientation distribution function must be periodic:

$$\psi(\phi) = \psi(\phi + \pi) \tag{7.2}$$

Since all particles are located between $-\pi/2$ and $\pi/2$, the orientation distribution function must be normalized such that

$$\int_{-\pi/2}^{\pi/2} \psi(\phi)\, d\phi = 1 \tag{7.3}$$

The orientation distribution function changes constantly as the particles travel within a deforming fluid element. Assuming the fiber density is homogeneous throughout the fluid and remains that way during processing[3], a balance around a differential element in the distribution function can be performed. This is graphically represented in Fig. 7.22. Here, the rate of change of the fiber density of the differential element, shown in the figure, should be the difference between the number of particles that move into and out of the control volume in a short time period Δt. This can be written as

$$\frac{\psi(\phi)\Delta\phi}{\Delta t} = \psi(\phi)\dot{\phi}(\phi) - \psi(\phi + \Delta\phi)\dot{\phi}(\phi + \Delta\phi) \tag{7.4}$$

Letting Δt and $\Delta\phi \rightarrow 0$ reduces Eq. 7.4 to

$$\frac{\partial \psi}{\partial t} = -\frac{\partial}{\partial \phi}\left(\psi \dot{\phi}\right) \tag{7.5}$$

This expression is known as the *fiber density continuity equation*. It states that a fiber which moves out of one angular position must move into a neighboring one, conserving the total number of fibers. If the initial distribution function, ψ_0, is known, an expression for the angular velocity of the particle, $\dot{\phi}$, must be found to solve for Eq. 7.5 and determine how the distribution function varies in time. The next sections present various models that can be used to determine the angular rotation of a slender, fiberlike particle.

3 It is common knowledge that the fiber density is not constant throughout the part. However, this assumption is reasonable for predicting fiber orientation distribution functions.

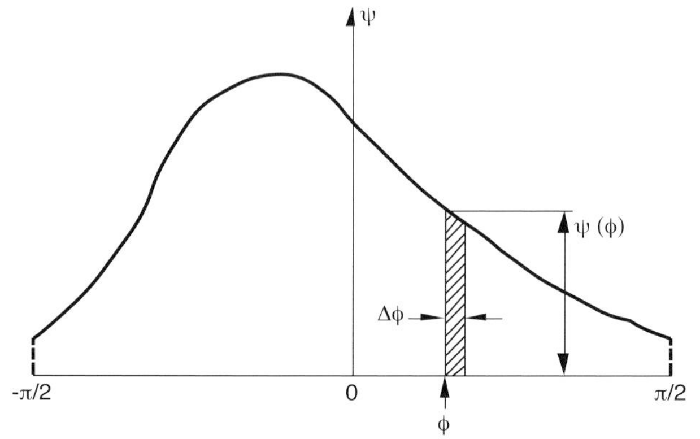

Figure 7.22 Differential element in an orientation distribution function.

7.2.2 Single Particle Motion

The motion of molecules, particles, or fibers can often be described by the motion of a rigid single rod in a planar flow. The analysis is furthermore simplified assuming that the rod is of infinite aspect ratio, that is, the ratio of length to diameter, L/D, is infinite. Using the notation in Fig. 7.23a, the fiber-end velocities can be broken into x and y components and rotational speed of the rod can be computed as a function of velocity gradients and its angular position:

$$\dot{\phi} = -\cos\phi \sin\phi \frac{\partial v_x}{\partial x} - \sin^2\phi \frac{\partial v_x}{\partial y} + \cos^2\phi \frac{\partial v_y}{\partial x} + \sin\phi \cos\phi \frac{\partial v_y}{\partial y} \tag{7.6}$$

Applying this equation to a simple shear flow as shown in Fig. 7.24, the rotational speed reduces to

$$\dot{\phi} = -\frac{\partial v_x}{\partial y} \sin^2\phi \tag{7.7}$$

Figure 7.25 shows the rotational speed, $\dot{\phi}$, as a function of angular position, ϕ, and Figure 7.26 shows the angular position of a fiber as a function of time for a fiber with an initial angular position of 90°. It should be clear that for this model ($L/D = \infty$) all fibers will eventually reach their 0° position and stay there.

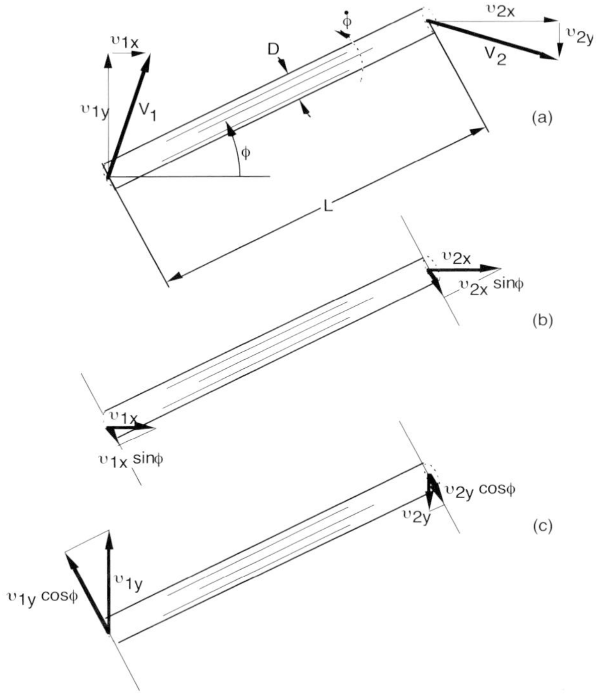

Figure 7.23 Fiber motion in planar flows.

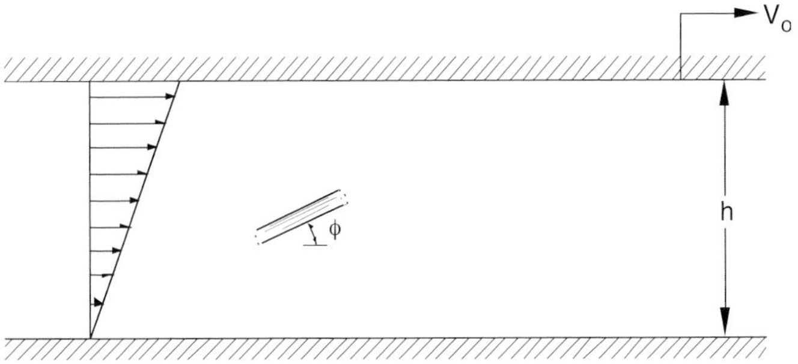

Figure 7.24 Schematic of fiber motion in simple shear flow.

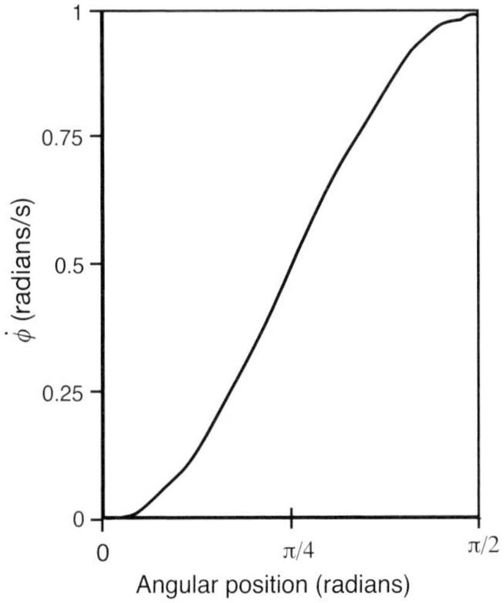

Figure 7.25 Rotational speed of a fiber with $L/D = \infty$ in a simple shear flow.

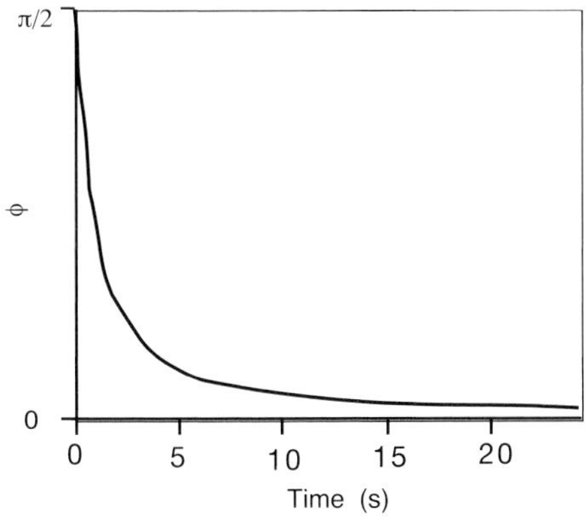

Figure 7.26 Angular position of a fiber with $L/D = \infty$ in a simple shear flow.

7.2.3 Jeffery's Model

The rotational speed of a single particle, as described in Eq. 7.6, is only valid for an infinite L/D ratio, since the model does not include the rotational speed contribution caused by the thickness dimension of the particle. The thickness term was included in the classical equation derived by Jeffery [19] and describes the rotational motion of a single ellipsoidal particle. His equation was later modified to account for the motion of cylindrical particles or rods [20, 21] and is written as

$$\dot{\phi} = \frac{r_e^2}{r_e^2+1}\left(-\sin\phi\cos\phi\frac{\partial v_x}{\partial x} - \sin^2\phi\frac{\partial v_x}{\partial y} + \cos^2\phi\frac{\partial v_y}{\partial x} + \sin\phi\cos\phi\frac{\partial v_y}{\partial y}\right)$$
$$-\frac{1}{r_e^2+1}\left(-\sin\phi\cos\phi\frac{\partial v_x}{\partial x} - \cos^2\phi\frac{\partial v_x}{\partial y} + \sin^2\phi\frac{\partial v_y}{\partial x} + \sin\phi\cos\phi\frac{\partial v_y}{\partial y}\right) \quad (7.8)$$

Here, r_e is the ratio of the major dimensions of the particle, *L/D*. Note that for the infinite *L/D* case, Eq. 7.8 reduces to Eq. 7.6.

7.2.4 Folgar–Tucker Model

A simplification of the Jeffery model is that a dilute suspension is assumed (i.e., very few fibers or fillers are present and do not interact with each other during flow). In polymer processing, this is usually not a valid assumption. In compression molding, for example, in a charge with 20–50% fiber content by volume the fibers are so closely packed that one cannot see through a resinless bed of fibers, even for very thin parts. This means that as a fiber rotates during flow, it bumps into its neighbors making the fiber–fiber interaction a major inhibitor of fiber rotation.

Folgar and Tucker [21, 22] derived a model for the orientation behavior of fibers in concentrated suspensions. For the case of planar flow Folgar and Tucker's model is as follows:

$$\dot{\phi} = \frac{-C_I\dot{\gamma}}{\psi}\frac{\partial\psi}{\partial\phi} - \cos\phi\sin\phi\frac{\partial v_x}{\partial x} - \sin^2\phi\frac{\partial v_x}{\partial y} + \cos^2\phi\frac{\partial v_y}{\partial x} + \sin\phi\cos\phi\frac{\partial v_y}{\partial y} \quad (7.9)$$

Here, $\dot{\gamma}$ is the *magnitude of the strain rate tensor* and C_I is a phenomenological coefficient which models the interactions between the fibers. Folgar and Tucker's *interaction coefficient*, C_I, varies between 0, for a fiber without interaction with its neighbors, and 1, for a closely packed bed of fibers. For a fiber reinforced polyester resin mat with 20–50% volume fiber content, C_I is usually between 0.03 and 0.06. When Eq. 7.9 is substituted into Eq. 7.5, the transient governing equation for fiber orientation distribution with fiber interaction built-in, becomes

$$\frac{\partial\psi}{\partial t} = -C_I\dot{\gamma}\frac{\partial^2\psi}{\partial\phi^2} - \frac{\partial\psi}{\partial\phi}\left(sc\frac{\partial v_x}{\partial x} - s^2\frac{\partial v_x}{\partial y} + c^2\frac{\partial v_y}{\partial x} + sc\frac{\partial v_y}{\partial y}\right)$$
$$-\psi\frac{\partial}{\partial\phi}\left(sc\frac{\partial v_x}{\partial x} - s^2\frac{\partial v_x}{\partial y} + c^2\frac{\partial v_y}{\partial x} + sc\frac{\partial v_y}{\partial y}\right) \quad (7.10)$$

where s and c represent sin ϕ and cos ϕ, respectively. The Folgar–Tucker model can easily be solved numerically. Using fiber reinforced thermoset composites as an example, the numerical solution of fiber orientation is discussed in the next section.

7.2.5 Tensor Representation of Fiber Orientation

Advani and Tucker [17, 23] developed a more efficient method to represent fiber orientation using orientation tensors. Their technique dramatically reduced the computational requirements when solving orientation problems using the Folgar–Tucker model.

Instead of representing the orientation of a fiber by its angle , ϕ, Advani and Tucker used the components of a unit vector p directed along the axis of the fiber. The components of p are related to ϕ,

$$p_1 = \cos\phi \quad \text{and} \tag{7.11a}$$

$$p_2 = \sin\phi \tag{7.11b}$$

where, p_i (i=1,2) are the two-dimensional Cartesian components of p. A suitably compact and general description of fiber orientation state is provided by the tensor of the form

$$a_{ij} = \langle p_i p_j \rangle \quad \text{and} \tag{7.12}$$

$$a_{ijkl} = \langle p_i p_j p_k p_l \rangle \tag{7.13}$$

Here the angle brackets < > represent an average overall possible direction of p, weighted by the probability distribution function, and a_{ij} is called the second-order orientation tensor and a_{ijkl} the fourth-order tensor. The properties of these tensors are discussed extensively by Advani and Tucker [24]. For the present, note that a_{ij} is symmetric and its trace equals unity. The advantage of using the tensor representation is that only a few numbers are required to describe the orientation state at any point in space. For planar orientations there are four components of a_{ij}, but only two are independent. Advani and Tucker were concerned with planar orientation in SMC only, and used a_{11} and a_{22} to describe the direction and distribution of orientation at a point. Once the orientation tensor a_{ij} is known, the mechanical properties of the composite can be predicted.

The Folgar–Tucker's model for single fiber motion in a concentrated suspension can be combined with the equation of continuity to produce an *equation of change* for the probability function and/or the orientation tensor [17, 25 ,26]. The result of second-order orientation tensors is

$$\frac{Da_{ij}}{Dt} = -\frac{1}{2}\left(\omega_{ik}a_{kj} - a_{ik}\omega_{kj}\right) + \frac{1}{2}\lambda\left(\dot{\gamma}_{ik}a_{kj} + a_{ik}\dot{\gamma}_{k} - 2\dot{\gamma}_{kl}a_{ijkl}\right) + 2C_1\dot{\gamma}\left(\delta_{ij} - \alpha a_{ij}\right) \qquad (7.14)$$

where δ_{ij} is the unit tensor and α equals 3 for three-dimensional orientation and 2 for planar orientation. Here, ω_{ij} and γ_{ij} are the velocity and the rate of deformation tensors, defined in terms of velocity gradients as

$$\omega_{ij} = \frac{\partial v_j}{\partial x_i} - \frac{\partial v_i}{\partial v_j} \quad \text{and} \qquad (7.15)$$

$$\dot{\gamma}_{ij} = \frac{\partial v_j}{\partial x_i} + \frac{\partial v_i}{\partial v_j} \qquad (7.16)$$

The material derivative in Eq. 7.14 appears on the left-hand side because the fibers are convected with the fluid. This casts the model of Folgar and Tucker into a useful form for computer simulation.

To calculate components of a_{ij} from Eq. 7.14, a_{ijkl} must be replaced by a suitable closure approximation. Combinations of the unit tensor and the components of a_{ij} can be used to form the approximation. Various closure approximations in planar and three-dimensional flow fields have been extensively tested by Advani and Tucker [17], Verleye and Dupret [27], and Cintra and Tucker [28]. It has been shown that Cintra and Tucker's *orthotropic closure approximation* performs best.

To obtain the orientation state of the fibers during mold filling simulation, Eq. 7.14 can be solved in the context of a finite element/control volume approach filling simulation. Only two equations for a_{11} and a_{12} need be solved. The other components depend on these, and can be replaced on the right-hand side of Eq. 7.14 using $a_{21} = a_{12}$ and $a_{22} = 1 - a_{11}$.

Nodes that lie within the charge at any given time are treated with a conventional Galerkin finite element method. The spatial orientation field is discretized using nodal values of the independent tensor components a_{11} and a_{12} together with element shape function.

The same mesh and linear shape functions that were used in the filling simulation are used for fiber orientation. The resulting finite element equations for fiber orientation may be compactly expressed in a matrix form as

$$\begin{bmatrix} C_{ij} & 0 \\ 0 & C_{ij} \end{bmatrix} \begin{Bmatrix} \dot{a}_{11j} \\ \dot{a}_{12j} \end{Bmatrix} + \begin{bmatrix} K_{Iij} & K_{IIIij} \\ K_{IIij} & K_{IVij} \end{bmatrix} \begin{Bmatrix} a_{11j} \\ a_{12j} \end{Bmatrix} = \begin{Bmatrix} R_i \\ Q_i \end{Bmatrix} \qquad (7.17)$$

Here the dot denotes ordinary differentiation with respect to time. This nonlinear system of ordinary differential equation was solved using fully implicit time stepping and a Newton-Raphson technique. The initial condition is provided by the orientation state of the fibers in the initial charge.

Advani [17] compared their model to the experiments and found that, overall, there is a good agreement between experimental and simulation results.

7.2.5.1 Predicting Orientation in Complex Parts Using Computer Simulation

Today, computer simulation is commonly used to predict mold filling, fiber orientation, thermal history, residual stresses and warpage in complex parts.

In injection molding, researchers are making progress on solving three-dimensional orientation for complex realistic applications [27, 29]. Crochet and co-workers have solved for the non-isothermal, non-Newtonian filling and fiber orientation in non-planar injection molded parts. They used the Hele–Shaw model [30] to simulate the mold filling and Advani and Tucker's tensor representation for the fiber orientation distribution in the final part. They divided the injection molded part into layers and included the fountain flow effect in the heat transfer and fiber orientation calculations. Figure 7.27 presents the fixed finite element mesh used to represent a 100 x 40 x 1 mm plate and the filling pattern during molding. Figure 7.28 presents the isotherms, the instant of fill, in three layers of the plate shown in Fig. 7.27, and Fig. 7.29 shows the fiber orientation distribution for the same layers.

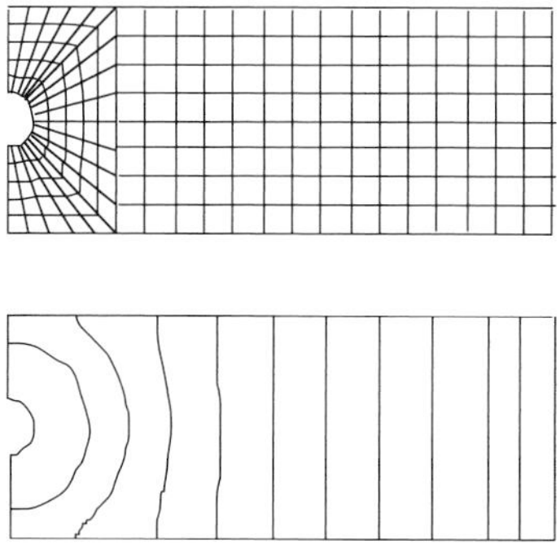

Figure 7.27 Fixed finite element mesh used to represent a 100 x 40 x 1 mm plate and temporary mesh adapted to represent the polymer melt at an arbitrary time during filling.

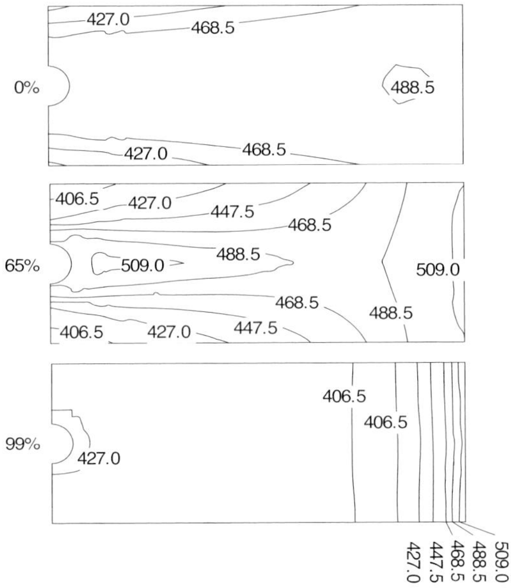

Figure 7.28 Isotherms in three layers at, 0 (center-line), 0.65 and 0.99 mm, the instant of fill.

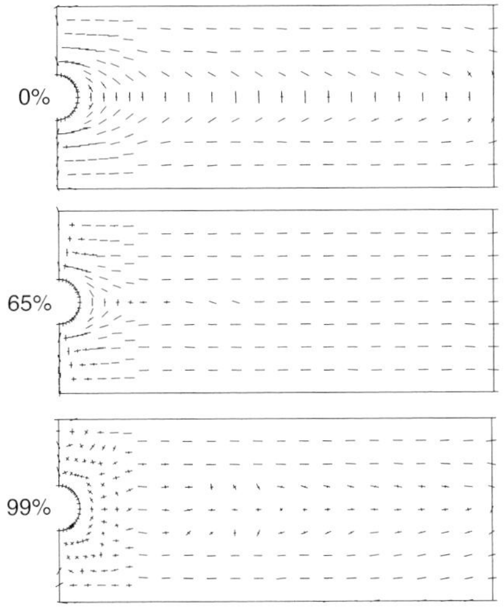

Figure 7.29 Fiber orientation in the same three layers shown in Fig. 7.28.

Because planar flow governs the compression molding process, the models described earlier work very well to represent the orientation of the fibers during processing and of the final part. The Folgar–Tucker model, Eq. 7.10, is usually solved using the finite difference technique and the velocity gradients in the equation are obtained from mold filling simulation. The initial condition is supplied by fitting ψ_i $(t = 0)$ to the measured initial orientation state. For sheet molding compound charges the starting fiber orientation distribution is usually random, or $\psi_i = 1/\pi$.

The model has proven to work well compared to experiments done with extensional flows described in Section 7.1.2. Figure 7.30 compares the measured fiber orientation distributions to the calculated distributions using the Folgar–Tucker model for cases with 67%, 50%, and 33% initial charge mold coverage. To illustrate the effect of fiber orientation on material properties of the final part, Fig. 7.31 [31] shows how the fiber orientation presented in Fig. 7.30 affects the stiffness of the plates.

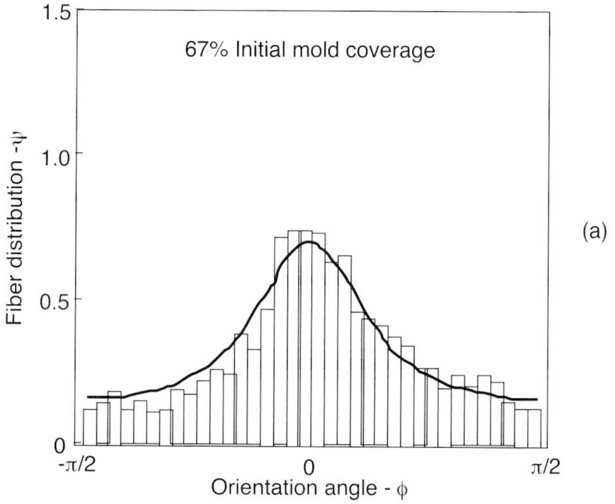

Figure 7.30 Comparison of predicted and experimental fiber orientation distributions for SMC experiments with (a) 67% initial mold coverage and $C_I = 0.04$, and (b) 50% initial mold coverage and $C_I = 0.04$ and (c) 33% initial mold coverage and $C_I = 0.04$. *(continued)*

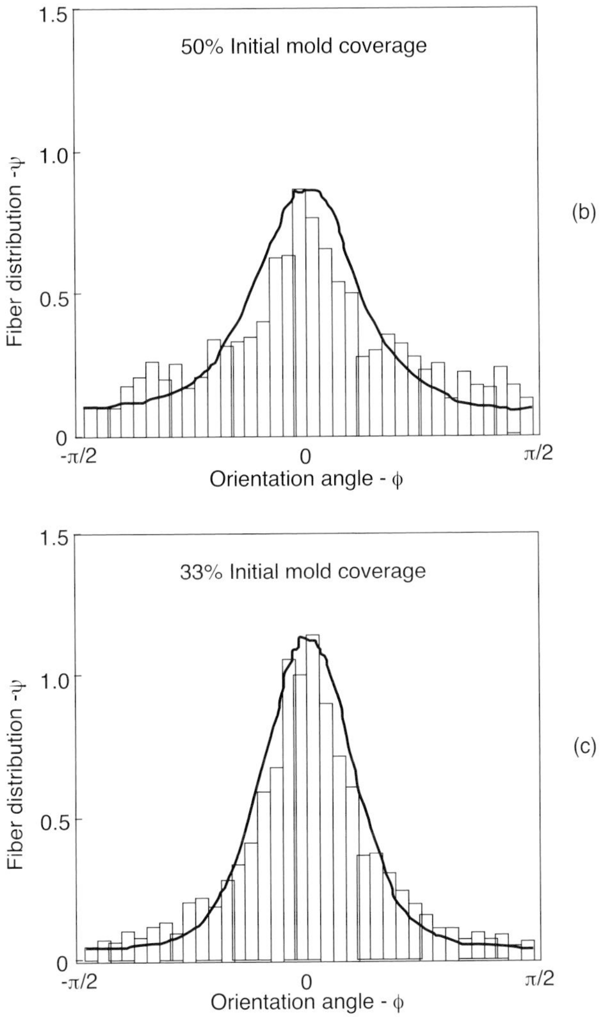

Figure 7.30 *(continued)* Comparison of predicted and experimental fiber orientation distributions for SMC experiments with (a) 67% initial mold coverage and $C_I = 0.04$, and (b) 50% initial mold coverage and $C_I = 0.04$, and (c) 33% initial mold coverage and $C_I = 0.04$.

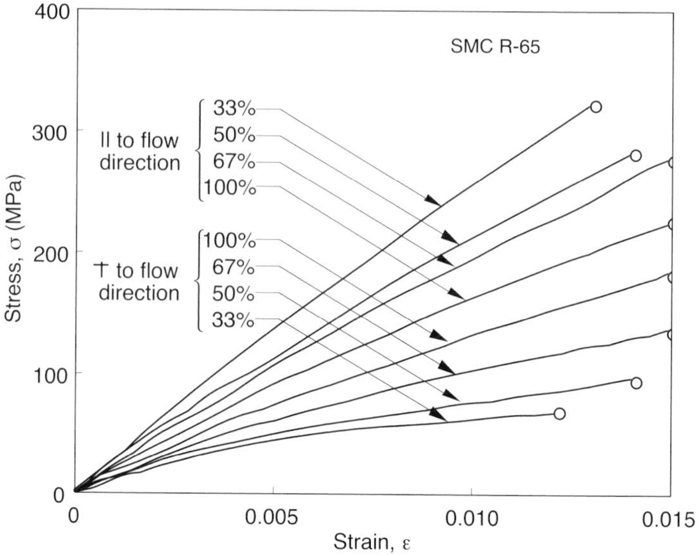

Figure 7.31 Stress-strain curves of 65% glass by volume SMC for various degrees of deformation.

The Folgar–Tucker model has been implemented into various, commercially available compression mold filling simulation programs. To illustrate the prediction of fiber orientation distribution in realistic polymer products, the compression molding process of a truck fender will be used as an example. To compute the fiber orientation, the filling pattern must first be computed. This is usually done by using the control volume approach [32]. The initial charge location and filling pattern during compression molding of the fender is shown in Fig. 7.32, and the finite element discretization with which the process was simulated is shown in Fig. 7.33. The fiber orientation distribution field, computed with the Folgar–Tucker model for the compression molded automotive fender under the above conditions is also shown in Fig. 7.33 [33]. For clarity, the orientation distribution function was plotted in polar coordinates from 0 to 2π and in the center of each finite element that is used for mold filling computation. For more detail about fiber orientation simulation the reader is encouraged to read the literature [34].

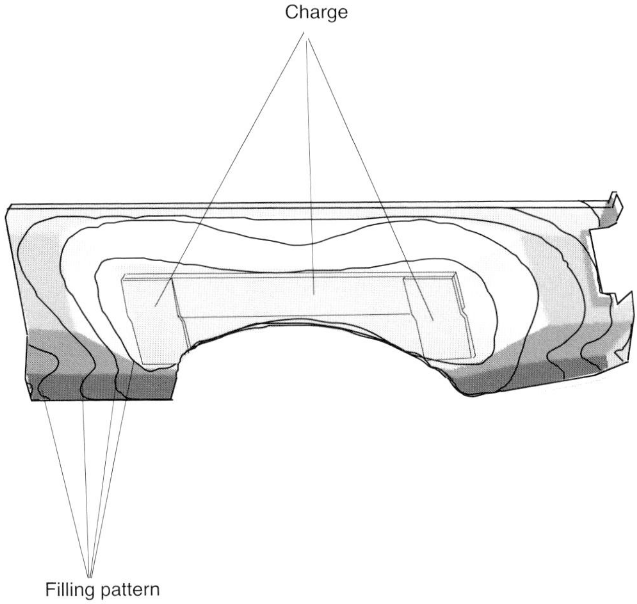

Figure 7.32 Initial charge and filling pattern during compression molding of an automotive fender.

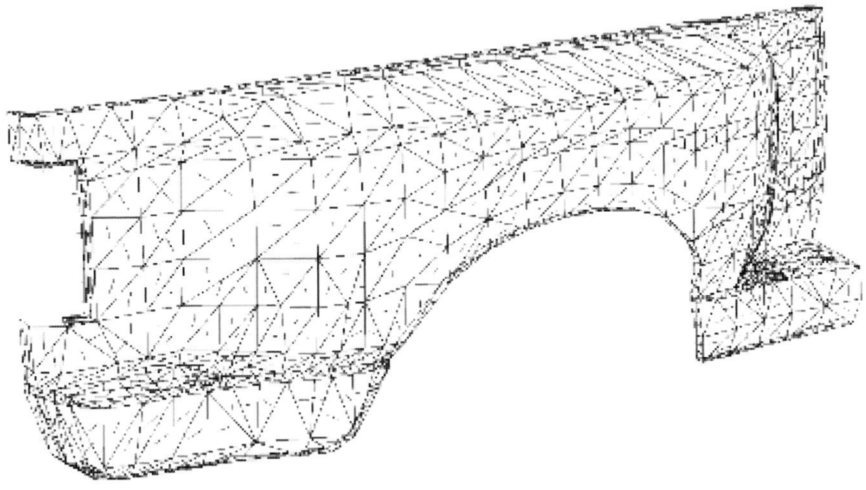

Figure 7.33 Fiber orientation distribution in a compression molded automotive fender.

7.3 Fiber Damage

One important aspect when processing fiber reinforced polymers is fiber damage or *fiber attrition*. This is especially true during injection molding where high shear stresses are present. As the polymer is melted and pumped inside the screw section of the injection molding machine and as it is forced through the narrow gate, most fibers shorten in length, reducing the properties of the final part (e.g., stiffness and strength).

Figure 7.34 helps explain the mechanism responsible for fiber breakage. The figure shows two fibers rotating in a simple shear flow: fiber "a", which is moving out of its 0° position, has a compressive loading and fiber "b", which is moving into its 0° position, has a tensile loading. It is clear that the tensile loading is not large enough to cause any fiber damage, but the compressive loading is potentially large enough to buckle and break the fiber. A common equation exists that relates a critical shear stress, τ_{crit}, to elastic modulus, E_f, and to the *L/D* ratio of the fibers

$$\tau_{\text{crit}} = \frac{\ln(2L/D) - 1.75}{2(L/D)^4} E_f \tag{7.18}$$

where τ_{crit} is the stress required to buckle the fiber. When the stresses are above τ_{crit} the fiber *L/D* ratio is reduced. Figure 7.35 shows a dimensionless plot of critical stress versus *L/D* ratio of a fiber as computed using Eq. 7.18. It is worthwhile to point out that although Eq. 7.18 predicts *L/D* ratios for certain stress levels, it does not include the uncertainty which leads to fiber *L/D* ratio distributions– very common in fiber filled systems.

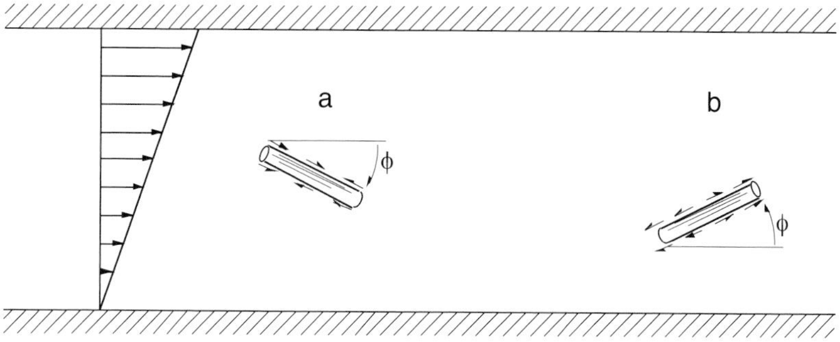

Figure 7.34 Fiber in compression and tension as it rotates during simple shear flow.

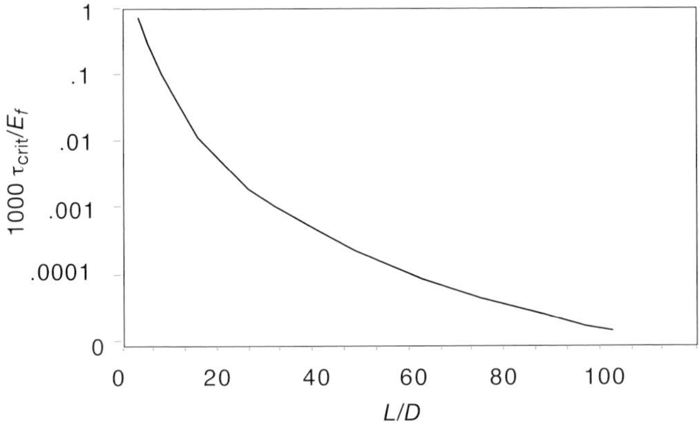

Figure 7.35 Critical stress, τ_{crit}, versus fiber *L/D* ratio.

Figure 7.36 presents findings by Thieltges [35] where he demonstrates that during injection molding most of the fiber damage occurs in the transition section of the plasticating screw. Lesser effects of fiber damage were measured in the metering section of the screw and in the throttle valve of the plasticating machine. The damage observed inside the mold cavity was marginal. However, the small damage observed inside the mold cavity is of great importance since the fibers flowing inside the cavity underwent the highest stresses, further reducing their *L/D* ratios. Bailey and Kraft [36] also found that fiber length distribution in the injection molded part is not uniform. For example, the skin region of the molding contained much shorter fibers than the core region.

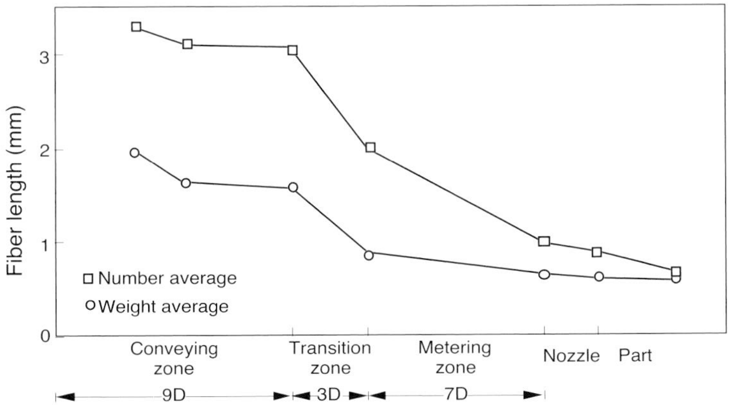

Figure 7.36 Fiber damage measured in the plasticating screw, throttle valve and mold during injection molding of a polypropylene plate with 40% fiber content by weight.

Another mechanism responsible for fiber damage is explained in Fig. 7.37 [37], where the fibers that stick out of partially molten pellets are bent, buckled, and sheared-off during plastication.

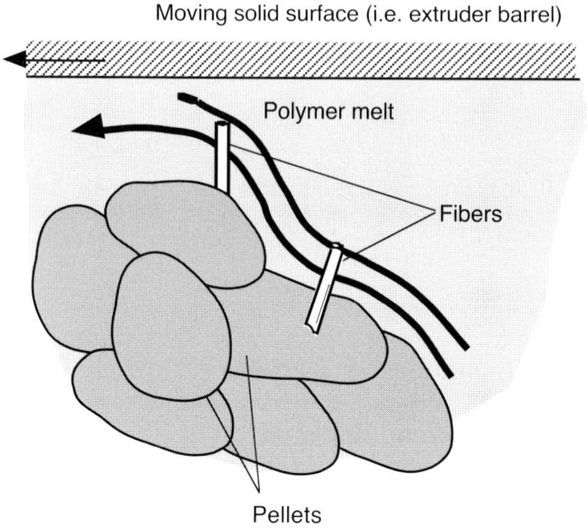

Figure 7.37 Fiber damage mechanism is the interface between solid and melt.

Examples

7.1 Estimate the fiber length after extruding polyamide 66 reinforced with 33% glass fiber. The initial fiber *L/D*=300, and the fiber diameter is 10 μm. The extruder diameter is 45 mm and the channel depth in the metering section is 5 mm. The extruder rotational speed is 200 rpm. The stiffness of the glass fiber is 90 GPa.

When solving this problem we can assume a simple shear flow in the metering section of the extruder. The maximum speed is computed using

$$v_0 = R\Omega = 22.5\,\text{mm}(200\,\text{rpm})\left(\frac{2\pi}{rev}\right)\left(\frac{1\,\text{min}}{60\,\text{s}}\right) = 471\,\text{mm/s}$$

and the shear with

$$\dot{\gamma} = v_0/h = 471\text{mm/s}/5\text{mm} = 94\text{s}^{-1}$$

At a rate of deformation of 100 s^{-1} the viscosity of PA66 is approximately 66 Pa-s leading to a deviatoric stress of 6204 Pa. We can now either use Eq. 7.18 or Fig. 7.35 with a critical stress equal to 6204 Pa, to find a corresponding *L/D* of approximately 70.

7.2 Consider a fiber of aspect ratio, *L/D*, of 10 moving in a simple shear flow. Plot the rotational speed of the fiber as a function of angular position, and the angular position as function of time.

If we apply Eq. 7.8 to simple shear flow, it reduces to

$$\dot{\phi} = \frac{1}{r_e^2+1}\cos^2\phi\frac{\partial v_x}{\partial y} - \frac{r_e^2}{r_e^2+1}\sin^2\phi\frac{\partial v_x}{\partial y} \tag{7.19}$$

When r_e = 10 – typical for a very short, damaged fiber in fiber reinforced composite parts – the equation for rotational speed is

$$\dot{\phi} = 0.01\cos^2\phi\frac{\partial v_x}{\partial y} - 0.99\sin^2\phi\frac{\partial v_x}{\partial y} \tag{7.20}$$

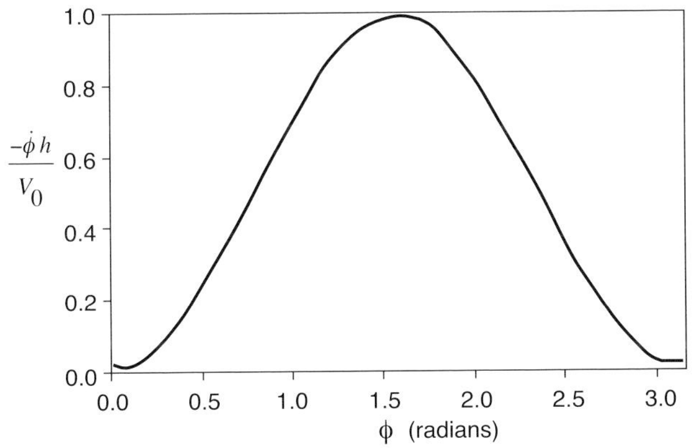

Figure 7.38 Rotational speed of a fiber with *L/D* = 10 in a simple shear flow, computed using Jeffery's model.

Figure 7.39 shows the rotation speed, $\dot{\phi}$, as a function of angular position. From Eq. 7.20 it is obvious that, at angular position of 0°, there is a very small rotational speed of $0.01\partial v_x/\partial y$. As the angle moves out of its 0° position it rapidly increases in speed to a maximum of

$0.990\partial v_x/\partial y$ at 90°. Hence, most of the time a fiber is oriented in its 0° position. Figure 7.38 shows this effect by plotting the angular position of the fiber with respect to time. The effect is further increased for higher L/D ratios. Therefore, in shear dominated flows, most of the fibers will be oriented in the direction of the shear plane on which they are traveling. In polymer processes such as extrusion and injection molding, the main mode of deformation is shear. The relationship in Eq. 7.19 and the behavior seen in Fig. 7.39 apply for each plane with its individual gradient $\partial v_x/\partial y$. In processes such as fiber spinning and compression molding, the main modes of deformation are elongational and a similar analysis as done above may be performed.

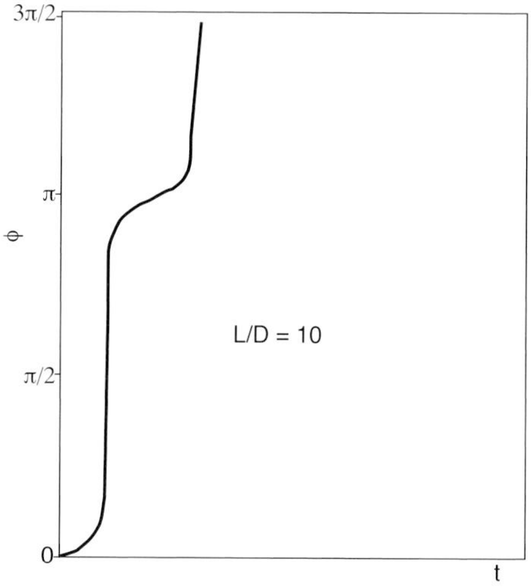

Figure 7.39 Angular position of a fiber with L/D = 10 in a simple shear flow, computed using Jefferey's model.

Problems

7.1 Consider the compression molding process of a circular sheet molding compound (SMC) charge. The charge is composed of 30% by volume 1 in. long glass fibers in an unsaturated polyester resin. The flow field when compression molding a circular charge is described with the following equations: $v_x = x/2$ and $v_y = y/2$. How will the fibers orient if the initial planar orientation is random? Explain.

7.2 Does a lower melt temperature during injection molding lead to higher or lower degrees of molecular orientation? Why?

7.3 Would you use polystyrene to injection mold CD's? Why?

7.4 When manufacturing a component that includes a small lens, would you place the injection gate near the lens edge, or as far as possible from the lens? Why?

7.5 Find an injection molded transparent polystyrene component and observe it through cross-polarizers. Determine the location of the gate. Point out the places of highest and lowest orientation within the part.

7.6 Is there a direct relation between the fringe patterns that you see when observing a transparent component through cross-polarizers and the residual stresses that built up during manufacturing? Why?

7.7 A 45 mm single screw extruder is used to extrude plates manufactured of polyamide 66 filled with 33% glass fiber by volume. Given that the extruder is rotating at 120 rpm and that the channel depth in the metering section is 5 mm, estimate the average final L/D of the fibers.

7.8 Work out example 7.1 for an aspect ratio, L/D, of 100. Plot the rotational speed as a function angular position, and angular position as a function of time.

7.9 The charge location and filling pattern during compression mold filling of a truck wind deflector is presented in Fig. 7.40 [38]. The part was molded using two sheet molding compound (SMC) charges, one 8 layer charge and another 9 layer charge.

Is there a knitline in the finished product? If yes, sketch a diagram with its approximate location in the finished part.

Point out the areas of maximum fiber orientation.

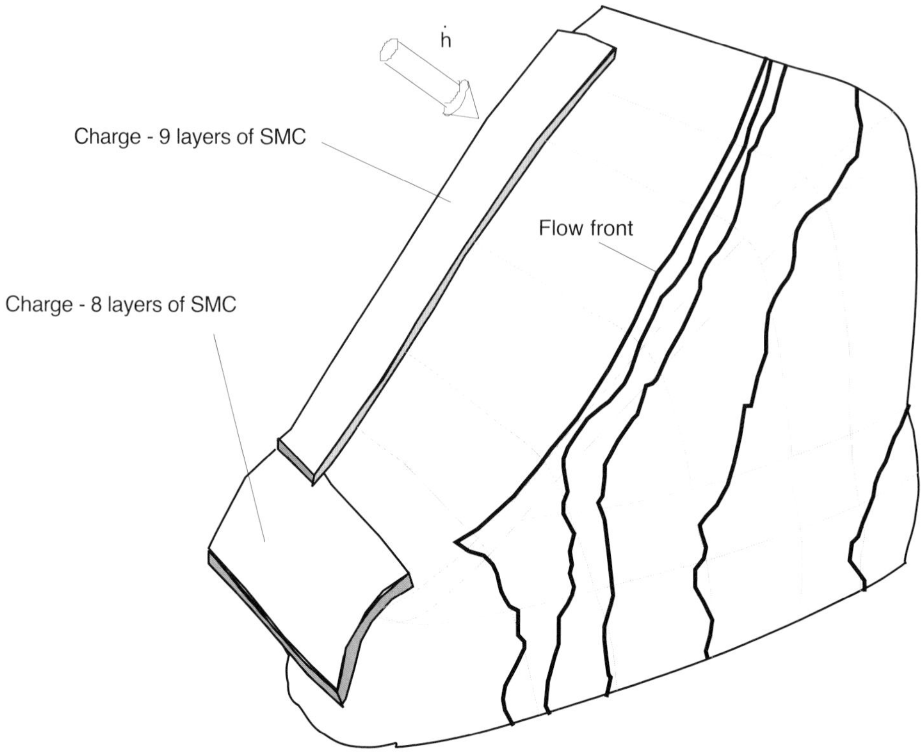

Figure 7.40 Actual mold filling pattern during compression molding of an SMC truck wind deflector

References

1. Woebken, W., *Kunststoffe, 51*, 547, (1961).
2. Menges, G., and Wübken, W., *SPE, 31st ANTEC,* (1973).
3. Wimberger-Friedl, R., *Polym. Eng. Sci.*, 30, 813, (1990).
4. Wimberger-Friedl, R., Ph.D. Thesis, Eindhoven University of Technology, The Netherlands, (1991).
5. Menges, G., and Geisbüsch, P., *Colloid & Polymer Science, 260*, 73, (1982).
6. Bay, R.S., and Tucker III, C.L., *Polym. Comp., 13*, 317, (1992a).
7. Bay, R.S., and Tucker III, C.L., *Polym. Comp., 13*, 322, (1992b).
8. Menges, G., Schacht, T., Becker, H., and Ott, S., *Intern. Polymer Processing, 2*, 77, (1987).
9. Leibfried, D., Ph.D. Thesis, IKV, RWTH-Aachen, Germany, (1970).
10. Wübken, G., Ph.D. Thesis, IKV, RWTH-Aachen, (1974).
11. Mavrides, H., Hrymak, A.N., and Vlachopoulos, J., *Polym. Eng. Sci.*, 26, 449, (1986).

12. Ibid.
13. Mavrides, H., Hrymak, A.N., and Vlachopoulos, J., *J. Rheol.*, 32, 639, (1988).
14. Barone, M.R., and Caulk, D.A., *J. Appl. Mech.*, 361, (1986).
15. Lee, C.-C., Folgar, F., and Tucker III, C.L., *J. Eng. Ind.*, 186, (1984).
16. Jackson, W.C., Advani, S.G., and Tucker III, C.L., *J. Comp. Mat.*, *20*, 539, (1986).
17. Advani, S.G., Ph.D. Thesis, University of Illinois at Urbana-Champaign, (1987).
18. Barone, M.R., and Osswald, T.A., *Polym. Comp.*, *9*, 158, (1988).
19. Jeffery, G.B., *Proc. Roy. Soc.*, *A102*, 161, (1922).
20. Burgers, J.M., *Verh. K. Akad. Wet.*, *16*, 8, (1938).
21. Folgar, F.P., Ph.D. Thesis, University of Illinois at Urbana-Champaign, (1983).
22. Folgar, F.P., and Tucker III, C.L., *J. Reinf. Plast. Comp.*, 3, 98, (1984).
23. Advani, S.G. and Tucker III, C.L., *Polym. Comp.*, *11*, 164, (1990).
24. Advani, S.G. and Tucker III, C.L., *J. Rheol.*, 31 (1987).
25. Folgar, F. and Tucker III, C.L., *J. Reinf. Plast. Comp.*, *3*,98 (1984).
26. Jackson, W.C., Advani, S.G., and Tucker III, C.L., *J. Comp. Mat.*, *20*, :539, (1986).
27. Verleye, V., and Dupret, F., *Proc. ASME WAM*, New Orleans, (1993).
28. Cintra, J. S., and Tucker III, C.L., *J. Rheol.*, forthcoming(1995).
29. Crochet, M.J., Dupret, F., and Verleye, V., *Flow and Rheology in Polymer Composites Manufacturing*, Ed. S.G. Advani, Elsevier, Amsterdam, (1994).
30. Hele-Shaw, H.S., *Proc. Roy. Inst.*, *16*, 49, (1899).
31. Chen, C.Y., and Tucker III, C.L., *J. Reinf, Compos.*, *3*, 120, (1984).
32. Osswald, T.A., and Tucker III, C.L., *Int. Polym. Process.*, 5 79, (1989).
33. Gramann, P.J., Sun, E.M., and Osswald, T.A., *SPE 52nd Antec*, (1994).
34. Tucker III, C.L., and Advani, S.G., *Flow and Rheology in Polymer Composites Manufacturing*, Ed. Advani, Elsevier, Amsterdam, (1994).
35. Thieltges, H.-P., Ph.D. Thesis, RWTH-Aachen, Germany, (1992).
36. Bailey, R., and Kraft, H., *Intern. Polym. Proc.*, 2, 94, (1987).
37. Mittal, R.K., Gupta, V.B., and Sharma, P.K., *Composites Sciences and Tech.*, 31, 295, (1988).
38. Osswald, T.A., Ph.D. Thesis, University of Illinois at Urbana-Champaign, (1986).

8

Solidification of Polymers

Solidification is the process in which a material undergoes a phase change and hardens. The phase change occurs as a result of either a reduction in material temperature or a chemical curing reaction. As discussed in previous chapters, a thermoplastic polymer hardens as the temperature of the material is lowered below either the melting temperature for a semi-crystalline polymer or the glass transition temperature for an amorphous thermoplastic. A thermoplastic has the ability to soften again as the temperature of the material is raised above the solidification temperature. On the other hand, the solidification of a leads to cross-linking of molecules. The effects of cross-linkage are irreversible and lead to a network that hinders the free movement of the polymer chains independent of the material temperature.

8.1 Solidification of Thermoplastics

The term "solidification" is often misused to describe the hardening of amorphous thermoplastics. The solidification of most materials is defined at a discrete temperature, whereas amorphous polymers do not exhibit a sharp transition between the liquid and the solid states. Instead, an amorphous thermoplastic polymer vitrifies as the material temperature drops below the glass transition temperature, T_g. A semi-crystalline polymer does have a distinct transition temperature between the melt and the solid state, the melting temperature, T_m.

8.1.1 Thermodynamics During Cooling

As heat is removed from a polymer melt, the molecules loose their ability to move freely, thus making the melt highly viscous. As amorphous polymers cool, the molecules slowly become closer packed, thus changing the viscous material into a leathery or rubberlike substance. Once the material has cooled below the glass transition temperature, T_g, the polymer becomes stiff and brittle. At the glass transition temperature, the specific volume and enthalpy curves experience a significant change in slope. This can be seen for polystyrene in the enthalpy-temperature curve shown in Fig. 8.1. With semi-crystalline thermoplastics, at a crystallization temperature near the melting temperature, the molecules start arranging themselves in small crystalline and amorphous regions, creating a very complicated morphology. During the process of crystalline structure formation, a quantum of energy, often called *heat of crystallization* or *heat of fusion*, is released and must be conducted out of the material before the cooling process can continue. The heat of fusion is reflected in the shape of the enthalpy-temperature curve as shown for polyamide 6.6, polyethylene and polypropylene in Fig. 8.1. At the onset of crystalline growth, the material becomes rubbery yet not brittle, since the amorphous regions are still above the glass transition temperature. As seen earlier, the glass transition temperature for some semi-crystalline polymers is far below room temperature, making them tougher than amorphous polymers. For common semi-crystalline polymers, the degree of crystallization can be between 30 and 70%. This means that 30–70% of the molecules form crystals and the rest remain in an amorphous state. The degree of crystallization is highest for those materials with short molecules since they can crystallize faster and more easily.

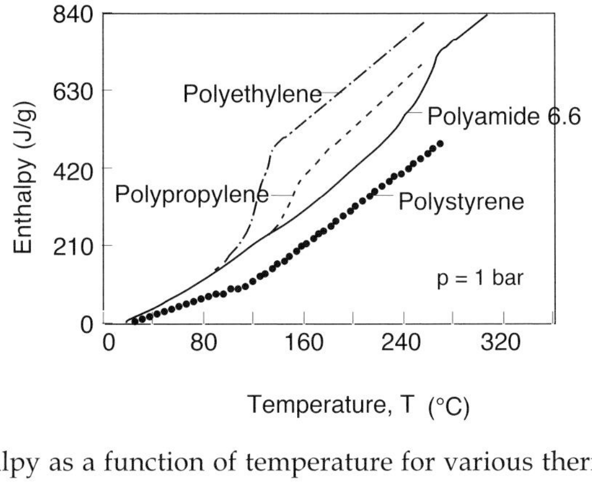

Figure 8.1 Enthalpy as a function of temperature for various thermoplastics.

Figure 8.2 [1] shows the volumetric temperature dependence of a polymer. In the melt state, the chains have "empty spaces" in which molecules can move freely. Hence, undercooled polymer molecules can still move as long as space is available. The point at which this free movement ends for a molecule or segment of chains is called the glass transition temperature or solidification point. As pointed out in Fig. 8.2, the free volume is frozen-in as well. In the case of crystallization, ideally, the volume should jump to a lower specific volume. However even here, small amorphous regions remain which permit a slow flow or material creep. This free volume reduces to nothing at absolute zero temperature at which heat transport can no longer occur.

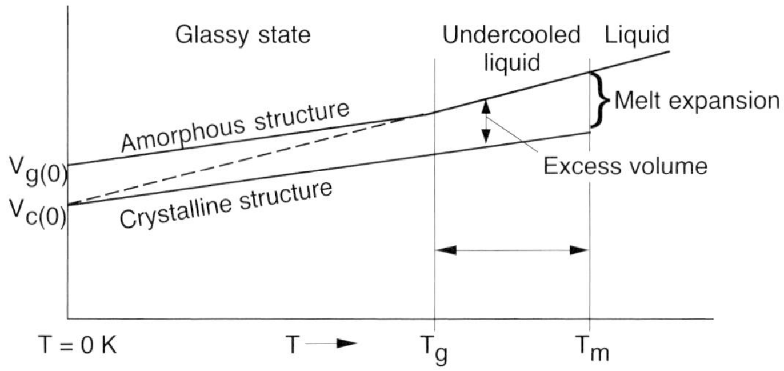

Figure 8.2 Thermal expansion model for thermoplastic polymers.

The specific volume of a polymer changes with pressure even at the glass transition temperature. This is demonstrated for an amorphous thermoplastic in Fig. 8.3 and for a semi-crystalline thermoplastic in Fig. 8.4.

It should be noted here that the size of the frozen-in free volume depends on the rate at which a material is cooled; high cooling rates result in a large free volume. In practice this is very important. When the frozen-in free volume is large, the part is less brittle. On the other hand, high cooling rates lead to parts that are highly permeable, which may allow the diffusion of gases or liquids through container walls. The cooling rate is also directly related to the dimensional stability of the final part. The effect of high cooling rates can often be mitigated by heating the part to a temperature that enables the molecules to move freely; this will allow further crystallization by additional chain folding. This process has a great effect on the structure and properties of the crystals and is referred to as *annealing*. In general, this only signifies a qualitative improvement of polymer parts. It also affects shrinkage and warpage during service life of a polymer component, especially when thermally loaded.

All these aspects have a great impact on processing. For example, when extruding amorphous thermoplastic profiles, the material can be sufficiently cooled inside the die so that the extrudate has enough rigidity to carry its own weight as it is pulled away from the die. Semi-crystalline polymers with low molecular weights have a viscosity above the melting temperature that is too low to be able to withstand their own weight as the extrudate exits the die. Temperatures below the melting temperature, T_m, however cannot be used due to solidification inside the die. Similar problems are encountered in the thermoforming process in which the material must be heated to a point so that it can be formed into its final shape, yet has to be able to withstand its own weight.

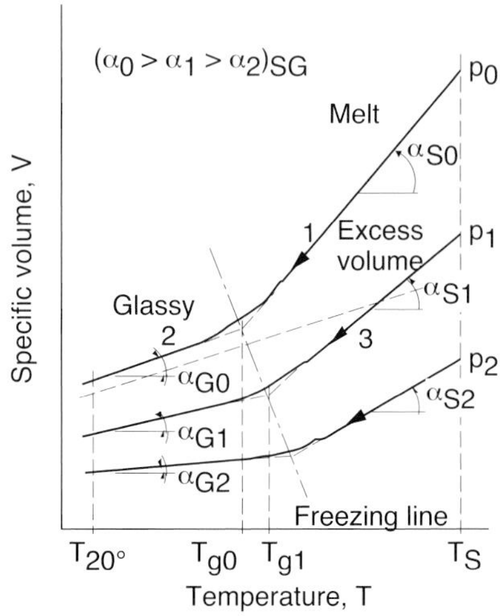

Figure 8.3 Schematic of a p-v-T diagram for amorphous thermoplastics.

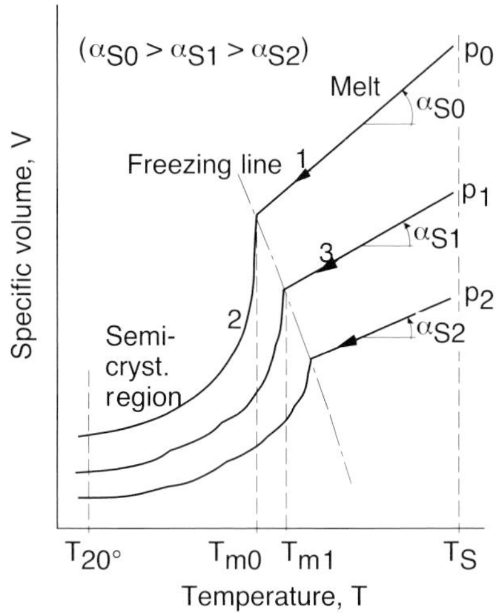

Figure 8.4 Schematic of a p-v-T diagram for semi-crystalline thermoplastics.

Semi-crystalline polymers are also at a disadvantage in the injection molding process. Because of the heat needed for crystallization, more heat must be removed to solidify the part; and since there is more shrinkage, longer packing times and larger pressures must be employed. All this implies longer cycle times and more shrinkage. High cooling rates during injection molding of semi-crystalline polymers will reduce the degree of crystallization. However, the amorphous state of the polymer molecules may lead to some crystallization after the process, which will result in further shrinkage and warpage of the final part. It is quite common to follow the whole injection molding process in the p-v-T diagrams presented in Figs. 8.3 and 8.4, and thus predict how much the molded component has shrunk.

8.1.2 Morphological Structure

Morphology is the order or arrangement of the polymer structure. The possible "order" between a molecule or molecule segment and its neighbors can vary from a very ordered highly crystalline polymeric structure to an amorphous structure (i.e., a structure in greatest disorder or random). The possible range of order and disorder is clearly depicted on the left side of Fig. 8.5. For example, a purely amorphous polymer is formed only by the non-crystalline or amorphous chain structure, whereas the semi-crystalline polymer is formed by a combination of all the structures represented in Fig. 8.5.

The image of a semi-crystalline structure as shown in the middle of Fig. 8.5, can be captured with an electron microscope. A macroscopic structure, shown in the right hand side of the figure, can be captured with an optical microscope. An optical microscope can capture the coarser macro-morphological structure such as the spherulites in semi-crystalline polymers.

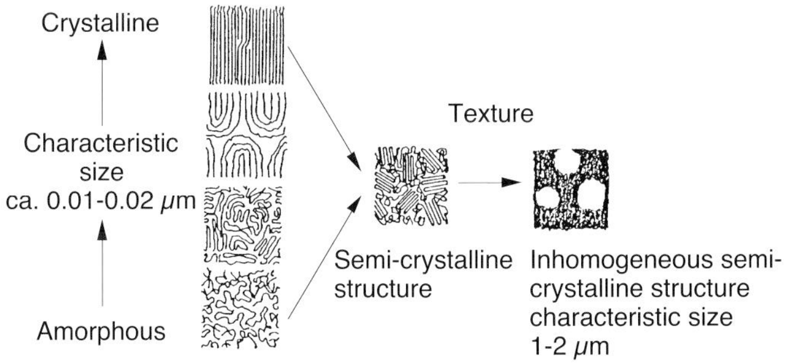

Figure 8.5 Schematic diagram of possible molecular structure which occur in thermoplastic polymers.

An amorphous polymer is defined as having a purely random structure. However it is not quite clear if a "purely amorphous" polymer as such exists. Electron microscopic observations have shown amorphous polymers that are composed of relatively stiff chains, show a certain degree of macromolecular structure and order, for example, globular regions or fibrilitic structures. Nevertheless, these types of amorphous polymers are still found to be optically isotropic. Even polymers with soft and flexible macromolecules, such as polyisoprene which was first considered to be random, sometimes show band-like and globular regions. These bundle-like structures are relatively weak and short-lived when the material experiences stresses. The shear thinning viscosity effect of polymers sometimes is attributed to the breaking of such macromolecular structures.

8.1.3 Crystallization

Early on, before the existence of macromolecules had been recognized, the presence of highly crystalline structures had been suspected. Such structures were discovered when undercooling or when stretching cellulose and natural rubber. Later, it was found that a crystalline order also existed in synthetic macromolecular materials such as polyamides, polyethylenes, and polyvinyls. Because of the polymolecularity of macromolecular materials, a 100% degree of crystallization cannot be achieved. Hence, these polymers are referred to as semi-crystalline. It is common to assume that the semi-crystalline structures are formed by small regions of alignment or crystallites connected by random or amorphous polymer molecules.

With the use of electron microscopes and sophisticated optical microscopes the various existing crystalline structures are now well recognized. They can be listed as follows:

- *Single crystals.* These can form in solutions and help in the study of crystal formation. Here, plate-like crystals and sometimes whiskers are generated.
- *Spherulites.* As a polymer melt solidifies, several folded chain lamellae spherulites form which are up to 0.1 mm in diameter. A typical example of a spherulitic structure is shown in Fig. 8.6 [2]. The spherulitic growth in a polypropylene melt is shown in Fig. 8.7 [3].
- *Deformed crystals.* If a semi-crystalline polymer is deformed while undergoing crystallization, oriented lamellae form instead of spherulites.
- *Shish-kebab.* In addition to spherulitic crystals, which are formed by plate- and ribbonlike structures, there are also shish-kebab, crystals which are formed by circular plates and whiskers. Shish-kebab structures are generated when the melt undergoes a shear deformation during solidification. A typical example of a shish-kebab crystal is shown in Fig. 8.8 [4].

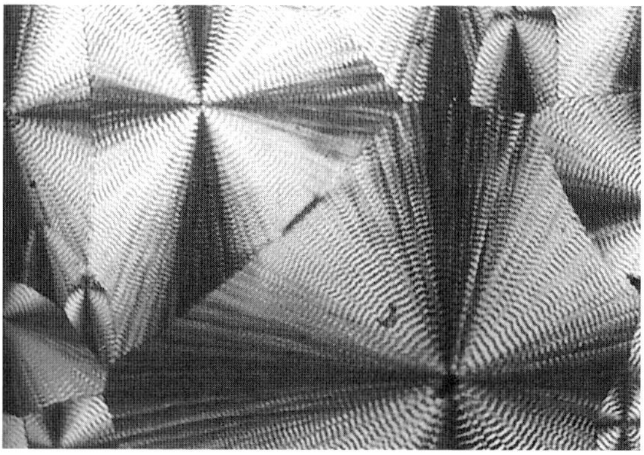

Figure 8.6 Polarized microscopic image of the spherulitic structure in polypropylene.

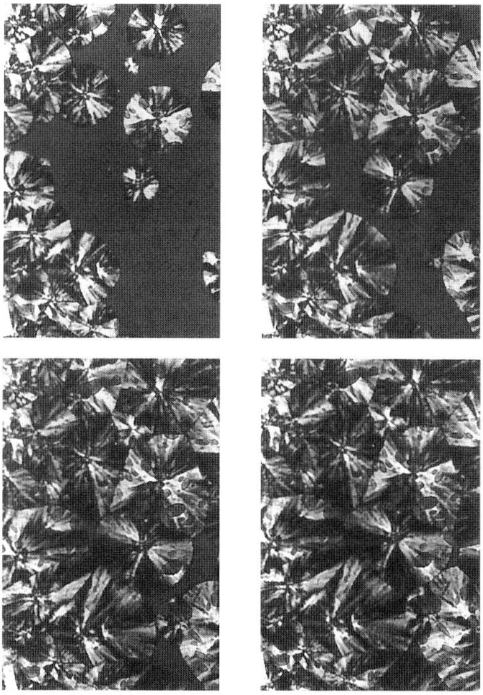

Figure 8.7 Development of the spherulitic structure in polypropylene. Images were taken at 30 s intervals.

The crystallization fraction can be described by the *Avrami equation* [5], written as follows:

$$x(t) = 1 - e^{-Zt^n} \tag{8.1}$$

where Z is a molecular weight and temperature dependent crystallization rate and n the Avrami exponent. However, since a polymer cannot reach 100% crystallization the above equation should be multiplied by the maximum possible degree of crystallization, x_∞.

$$x(t) = x_\infty\left(1 - e^{-Zt^n}\right) \tag{8.2}$$

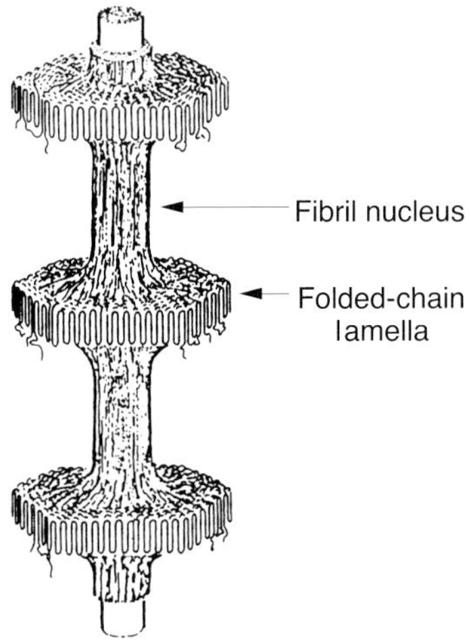

Figure 8.8 Model of the shish-kebab morphology.

The Avrami exponent, n, ranges between 1 and 4 depending on the type of nucleation and growth. For example, the Avrami exponent for spherulitic growth from sporadic nuclei is around 4, disclike growth 3, and rodlike growth 2. If the growth is activated from instantaneous nuclei, the Avrami exponent is lowered by 1.0 for all cases. The crystalline growth rate of various polymers differ significantly from one to another. This is demonstrated in Table 8.1 which shows the maximum growth rate for various thermoplastics. The crystalline mass fraction can be measured experimentally with a differential scanning calorimeter (DSC).

A more in-depth coverage of crystallization and structure development during processing is given by Eder and Janeschitz-Kriegl [7].

Table 8.1 Maximum Crystalline Growth Rate and Maximum Degree of Crystallinity for Various Thermoplastics

Polymer	Growth rate (μ/min)	Maximum crystallinity (%)
Polyethylene	>1000	80
Polyamide 66	1000	70
Polyamide 6	200	35
Isotactic polypropylene	20	63
Polyethylene terephthalate	7	50
Isotactic polystyrene	0.30	32
Polycarbonate	0.01	25

8.1.4 Heat Transfer During Solidification

Since polymer parts are generally thin, the energy equation[1] can be simplified to a one-dimensional problem. Thus, using the coordinate description shown in Fig. 8.9 the energy equation can be reduced to

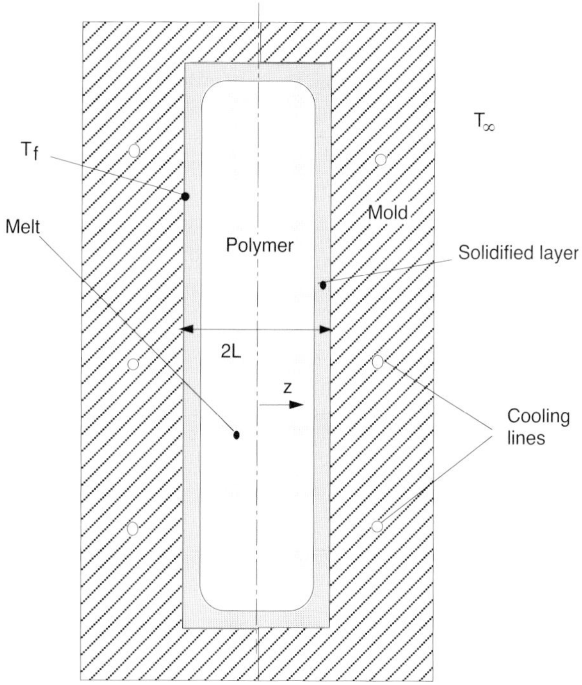

Figure 8.9 Schematic diagram of polymer melt inside an injection mold.

1 The energy equation is discussed in Chapter 3 and can be found in its complete form in Appendix A.

$$\rho C_p \frac{\partial T}{\partial t} = k \frac{\partial^2 T}{\partial z^2} \tag{8.3}$$

Another assumption— and to reduce warpage, usually a requirement— is a symmetry boundary condition:

$$\frac{\partial T}{\partial z} = 0 \quad \text{at} \quad z = 0 \tag{8.4}$$

If the sheet is cooled via forced convection or the part is inside a perfectly cooled mold, the final temperature of the part can be assumed to be the second boundary condition:

$$T = T_f \tag{8.5}$$

A typical temperature history for a polystyrene plate, its properties presented in Table 8.2 [8], is shown in Fig. 8.10. Once the material's temperature drops below the glass transition temperature, T_g, it can be considered solidified. This is shown schematically in Fig. 8.11. Of importance here is the position of the solidification front, $X(t)$. Once the solidification front equals the plate's dimension L, the solidification process is complete. From Fig. 8.10 it can be shown that the rate of solidification decreases as the solidified front moves further away from the cooled surface. For amorphous thermoplastics, the well-known *Neumann solution* can be used to estimate the growth of the glassy or solidified layer. The Neumann solution is written as

$$X(t) \propto \sqrt{\alpha t} \tag{8.6}$$

Table 8.2 Material Properties for Polystyrene

K	=	0.117 W/mK
C_p	=	1185 J/kgK
ρ	=	1040 kg/m^3
T_g	=	80 °C
E	=	3.2E9 Pa
ν	=	0.33

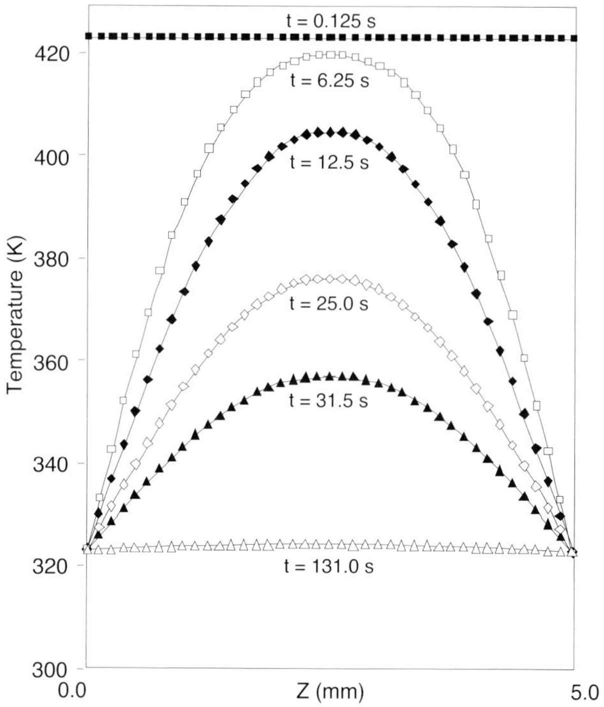

Figure 8.10 Temperature history of polystyrene cooled inside a 5 mm thick mold.

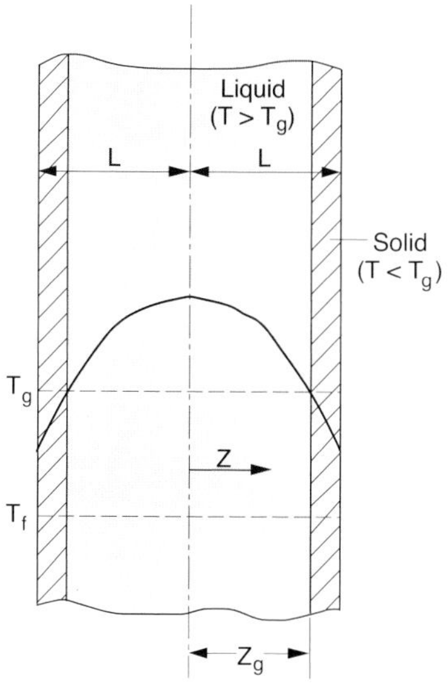

Figure 8.11 Schematic diagram of the cooling process of a polymer plate.

where α is the thermal diffusivity of the polymer. It must be pointed out here that for the Neumann solution, the growth rate of the solidified layer is infinite as time goes to zero.

The solidification process in a semi-crystalline materials is a bit more complicated due to the heat of fusion or heat of crystallization, nucleation rate, etc. When measuring the specific heat as the material crystallizes, a peak which represents the heat of fusion is detected (see Fig. 4.12). Figure 8.12 shows the calculated temperature distribution in a semi-crystalline polypropylene plate during cooling. The material properties used for the calculations are shown in Table 8.3 [8]. Here, the material that is below the melting temperature, T_m, is considered solid[2]. Experimental evidence [9] has demonstrated that the growth rate of the crystallized layer in semi-crystalline polymers is finite. This is mainly due to the fact that at the beginning the nucleation occurs at a finite rate. Hence, the Neumann solution presented in Eq. 8.6 as well as the widely used *Stefan condition* [10], do not hold for semi-crystalline polymers. This is clearly demonstrated in Fig. 8.13 [10] which presents measured thickness of crystallized layers as a function of time for polypropylene plates quenched at three different temperatures. For further reading on this important topic the reader is encouraged to consult the literature [11, 12].

Table 8.3 Material Properties for Polypropylene

K	=	0.117 W/mK
$C_{p, solid}$	=	1800 J/kgK
$C_{p, melt}$	=	2300 J/kgK
ρ	=	930 kg/m^3
T_g	=	-18 °C
T_m	=	186 °C
λ	=	209 kJ/kg

2 It is well-known that the growth of the crystalline layer in semi-crystalline polymers is maximal somewhat below the melting temperature, at a temperature T_c. The growth speed of nuclei is zero at the melting temperature and at the glass transition temperature.

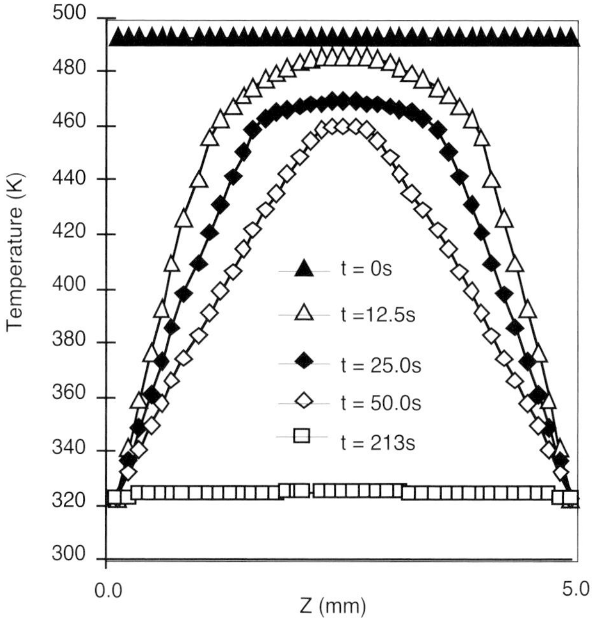

Figure 8.12 Temperature history of polypropylene cooled inside a 5 mm thick mold.

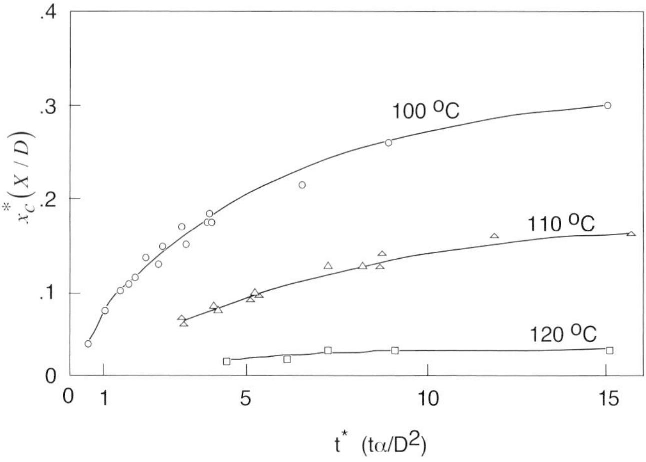

Figure 8.13 Dimensionless thickness of the crystallized layers as a function of dimensionless time for various temperatures of the quenching surface.

8.2 Solidification of Thermosets

The solidification process of thermosets, such as phenolics, unsaturated polyesters, epoxy resins, and polyurethanes is dominated by an exothermic chemical reaction called curing reaction. A curing reaction is an irreversible process that results in a structure of molecules that are more or less cross-linked. Some thermosets cure under heat and others cure at room temperature. Thermosets that cure at room temperature are those for which the reaction starts immediately after mixing two components, where the mixing is usually part of the process. However, even with these thermosets, the reaction is accelerated by the heat released during the chemical reaction, or the *exotherm*. In addition, it is also possible to activate cross-linking by absorption of moisture or radiation, such as ultra-violet, electron beam, and laser energy sources [13].

In processing, thermosets are often grouped into three distinct categories, namely those that undergo a *heat activated cure,* those that are dominated by a *mixing activated cure,* and those which are activated by the absorption of humidity or radiation. Examples of heat activated thermosets are phenolics; examples of mixing activated cure are epoxy resins and polyurethane.

8.2.1 Curing Reaction

In a cured thermoset, the molecules are rigid, formed by short groups that are connected by randomly distributed links. The fully reacted or solidified thermosetting polymer does not react to heat as observed with thermoplastic polymers. A thermoset may soften somewhat upon heating and but then degrades at high temperatures. Due to the high cross-link density, a thermoset component behaves as an elastic material over a large range of temperatures. However, it is brittle with breaking strains of usually 1 to 3%. The most common example is phenolic, one of the most rigid thermosets, which consists of carbon atoms with large aromatic rings that impede motion, making it stiff and brittle. Its general structure after cross-linking is given in Figs. 3.22 and 3.23.

Thermosets can be broken down into three categories: thermosets which cure via *condensation polymerization,* those that undergo *addition polymerization* and those that cure via *free radical polymerization.*

Condensation polymerization is defined as the growth process that results from combining two or more monomers with reactive end-groups, and leads to by-products such as an alcohol, water, an acid, etc. A common thermoset that polymerizes or solidifies via condensation polymerization is phenol formaldehyde, discussed in Chapter 3. The by-product of the reaction when making phenolics is water. Examples of addition polymerization are polyurethanes and epoxies.

An example of a cross-linking reaction of a thermoset by *free radical reaction* is the co-polymerization of unsaturated polyester with styrene molecules, shown in Fig. 8.14. The molecules contain several carbon-carbon double bonds which act as cross-linking sites during curing. An example of the resulting network after the chemical reaction is shown in Fig. 8.15.

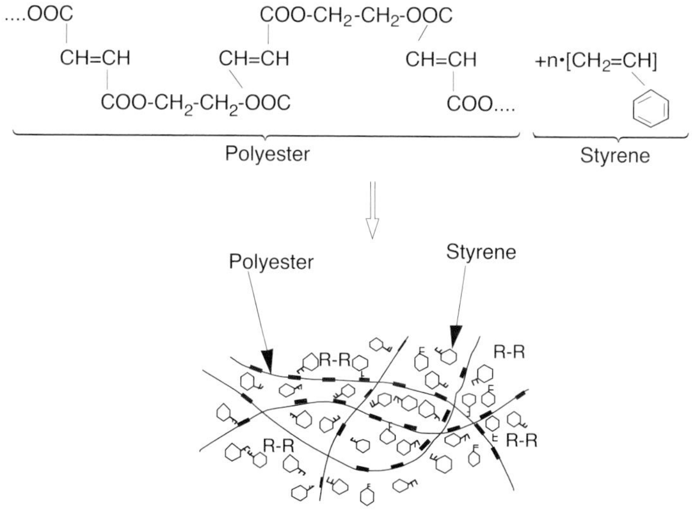

Figure 8.14 Symbolic and schematic representations of uncured unsaturated polyester.

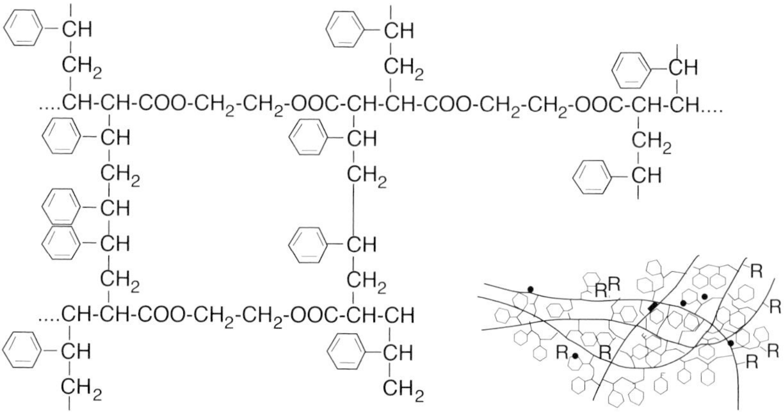

Figure 8.15 Symbolic and schematic representations of cured unsaturated polyester.

8.2.2 Cure Kinetics

As discussed earlier, in processing thermosets can be grouped into two general categories: *heat activated cure* and *mixing activated cure* thermosets. However, no matter which category a thermoset belongs to, its curing reaction can be described by the reaction between two chemical groups denoted by A and B which link two segments of a polymer chain. The reaction can be followed by tracing the concentration of unreacted *As* or *Bs*, C_A or C_B. If the initial concentration of *As* and *Bs* is defined as C_{A_0} and C_{B_0}, the degree of cure can be described by

$$C^* = \frac{C_{A_0} - C_A}{C_{A_0}} \tag{8.7}$$

The degree of cure or conversion, C^*, equals zero when there has been no reaction and equals one when all *As* have reacted and the reaction is complete. However, it is impossible to monitor reacted and unreacted *As* and *Bs* during the curing reaction of a thermoset polymer. It is known though that the exothermic heat released during curing can be used to monitor the conversion, C^*. When small samples of an unreacted thermoset polymer are placed in a differential scanning calorimeter (DSC), each at a different temperature, every sample will release the same amount of heat, Q_T. This occurs because every cross-linking that occurs during a reaction releases a small amount of energy in the form of heat. For example, Fig. 8.16 [14] shows the heat rate released during isothermal cure of a vinyl ester at various temperatures.

The degree of cure can be defined by the following relation

$$C^* = \frac{Q}{Q_T} \tag{8.8}$$

where Q is the heat released up to an arbitrary time τ, and is defined by

$$Q = \int_0^{\tau} \dot{Q} dt \tag{8.9}$$

DSC data is commonly fitted to empirical models that accurately describe the curing reaction. Hence, the rate of cure can be described by the exotherm, $\dot{Q}$, and the total heat released during the curing reaction, Q_T, as

$$\frac{dC^*}{dt} = \frac{\dot{Q}}{Q_T} \tag{8.10}$$

With the use of Eq. 8.10, it is now easy to take the DSC data and find the models that describe the curing reaction.

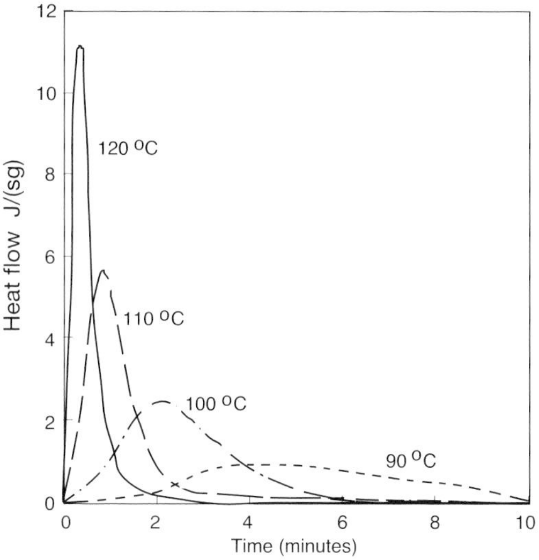

Figure 8.16 DSC scans of the isothermal curing reaction of vinyl ester at various temperatures.

During cure, thermoset resins exhibit three distinct phases; viscous liquid, gel, and solid. Each of these three stages is marked by dramatic changes in the thermomechanical properties of the resin. The transformation of a reactive thermosetting liquid to a glassy solid generally involves two distinct macroscopic transitions: molecular gelation and vitrification. Molecular gelation is defined as the time or temperature at which covalent bonds connect across the resin to form a three-dimensional network which gives rise to long range elastic behavior in the macroscopic fluid. This point is also referred to as the gel point, where $C^* = C_g$. As a thermosetting resin cures, the cross-linking begins to hinder molecular movement, leading to a rise in the glass transition temperature. Eventually, when T_g nears the processing temperature, the rate of curing reduces significantly, eventually dominated by diffusion. At this point the resin has reached its vitrification point. Figure 8.17, which presents degree of cure as a function of time, illustrates how an epoxy resin reaches a maximum degree of cure at various processing temperatures. The resin processed at 200 °C reaches 100% cure since the glass transition temperature of fully cure epoxy is 190 °C, less than the processing temperature. On the other hand, the sample processed at 180 °C reaches 97% cure and the one processed at 160 °C only reaches 87% cure.

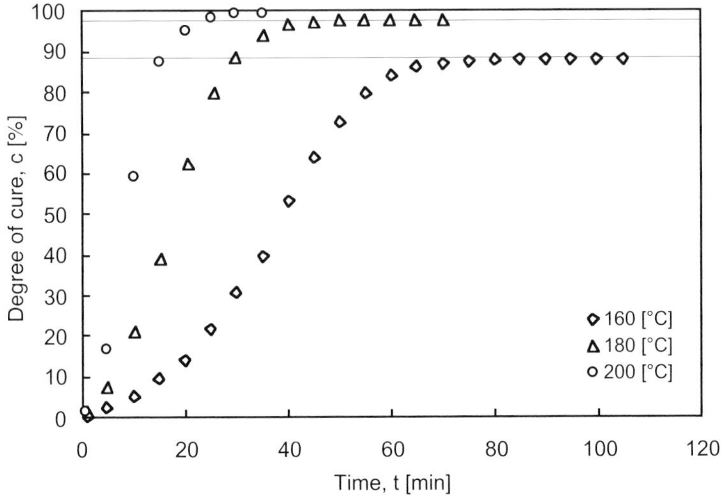

Figure 8.17 Degree of cure as a function time for an epoxy resin measured using isothermal DSC.

Figures 8.16 and 8.17 also illustrate how the curing reaction is accelerated as the processing temperature is increased. The curing reaction of thermally cured thermoset resins is not immediate, thus the blend can be stored in a refrigerator for a short period of time without having any significant curing reaction.

The behavior of curing thermosetting resins can be represented with the generalized time-temperature-transformation (TTT) cure diagram developed by Enns and Gillham [16]; it can be used to relate the material properties of thermosets as a function of time and the processing temperature as shown in Fig. 8.18.

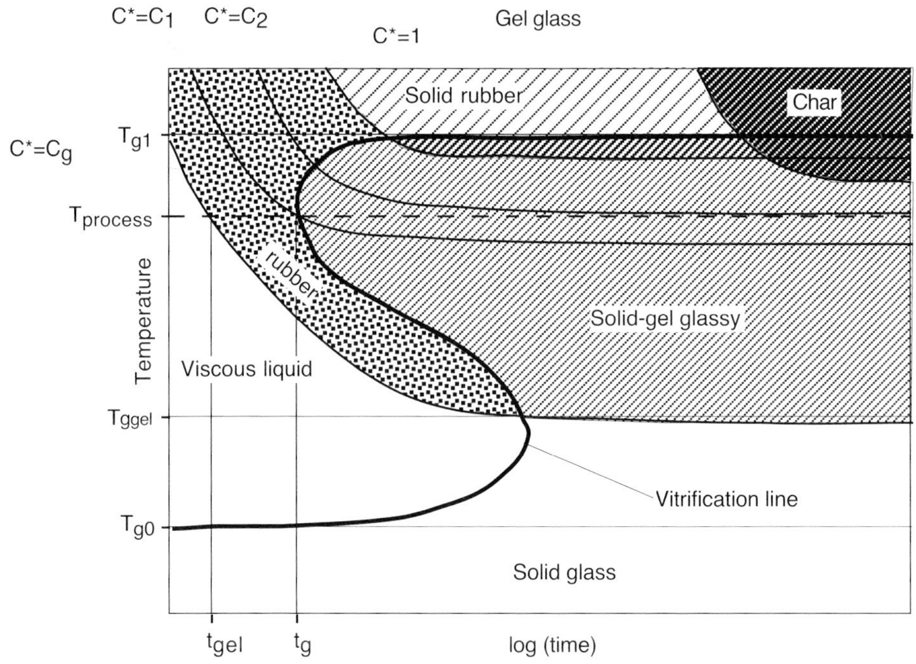

Figure 8.18 Time-temperature-transformation (TTT) diagram for a thermoset

The diagram presents various lines that represent constant degrees of cure. The curve labeled $C^* = C_g$ represents the gel point and $C^* = 1$ the fully cured resin. Both curves have their corresponding glass transition temperatures, T_{g1} and $T_{g\,gel}$, for the glass transition temperature of the fully cured resin and at its gel point, respectively. The glass transition temperature of the uncured resin, T_{g0}, and an S-shaped curve labeled "vitrification line," are also depicted. The *vitrification line* represents the boundary where the glass transition temperature becomes the processing temperature. Hence, to the left of the vitrification curve the curing process is controlled by a very slow diffusion process. The TTT-diagram shows an arbitrary process temperature. The material being processed reaches the gel point at $t = t_{gel}$ and the vitrification line at $t = t_g$. At this point the material has reached a degree of cure of C_1 and glass transition temperature of the resin is equal to the processing temperature. The material continues to cure very slowly (diffusion controlled) until it reaches a degree of cure just below C_2. There are also various regions labeled in the diagram. The one labeled "viscous liquid" is the one where the resin is found from the beginning of processing until the gel point has been reached. The flow and deformation that occurs during processing or shaping must occur within this region. The region labeled "char" must be avoided during processing, since at high processing temperatures the polymer will eventually undergo thermal degradation.

The model that best represents the curing kinetics of thermosetting resins as reflected in a TTT-diagram is a diffusion modified Kamal-Sourour reaction model [17, 18, 19]. To model autocatalytic cure kinetics, the model can be applied as

$$\frac{dC^*}{dt} = \left(k_1 + k_2 C^{*m}\right)\left(1 - C^*\right)^n \tag{8.11}$$

where m and n are reaction orders, and k_1 and k_2 are constants defined by

$$\frac{1}{k_i} = \frac{1}{k_1^c} + \frac{1}{k_D} \tag{8.12}$$

Here, k_i^c are Arrhenius overall rate constants defined by

$$k_1^c = a_1 e^{-E_1/RT} \tag{8.13}$$

and

$$k_2^c = a_2 e^{-E_2/RT} \tag{8.14}$$

where a_1 and a_2 are fitting parameters, E_1 and E_2, activation energies and R the ideal gas constant. The constant k_D in Eq. 8.12 is the diffusion rate constant defined as

$$k_D = a_D e^{-E_D/RT} e^{-b/f} \tag{8.15}$$

where a_D and b are adjustable parameters, E_D is the activation energy of the diffusion process, and f is the equilibrium fractional free volume given by

$$f = 0.00048\left(T - T_g\right) + 0.025 \tag{8.16}$$

where T_g is the instantaneous glass transition temperature during cure. Equation (8.12) shows that the overall rate constant is governed at one extreme by the Arrhenius rate constant when $k_D >> k_i^c$, which is the case prior to vitrification, and at the other extreme by the diffusion rate constant when $k_D << k_i^c$, which is the case well after vitrification. For a system exhibiting a unique one-to-one relationship between the glass transition temperature and conversion, DiBenedetto's equation [20] is one of the easiest approaches for stoichiometric ratios to express this relationship using only a single parameter as

$$T_g = T_{g0} + \frac{\left(T_{g1} - T_{g0}\right)\lambda C^*}{1-(1-\lambda)C^*} \tag{8.17}$$

where T_{g0} is the glass transition temperature of the uncured resin, T_{g1} is the glass transition temperature of the fully reacted network, and λ is a structure dependent parameter theoretically equated to

$$\lambda = \frac{\Delta C_{p0}}{\Delta C_{p1}} \tag{8.18}$$

The values of ΔC_{p0} and ΔC_{p1} are the differences in the heat capacity between the glassy and rubbery state for the uncured resin and the fully cured network, respectively. However the parameter λ can also be used as fitting parameter.

Mixing activated cure materials such as polyurethanes will instantly start releasing exothermic heat after the mixture of its two components has occurred. The proposed *Castro-Macosko curing model* accurately fits this behavior and is written as [15]

$$\frac{dC^*}{dt} = k_0 e^{-E/RT}\left(1-C^*\right)^2 \tag{8.19}$$

8.2.3 Heat Transfer During Cure

A well-known problem in thick section components is that the thermal and curing gradients become more complicated and difficult to analyze since the temperature and curing behavior of the part is highly dependent on both the mold temperature and part geometry [21, 22]. A thicker part will result in higher temperatures and a more complex cure distribution during processing. This phenomenon becomes a major concern during the manufacture of thick components since high temperatures may lead to thermal degradation. A relatively easy way to check temperatures that arise during molding and curing or demolding times is desired. For example, a one-dimensional form of the energy equation that includes the exothermic energy generated during curing can be solved:

$$\rho C_p \frac{\partial T}{\partial t} = k\frac{\partial^2 T}{\partial z^2} + \rho\dot{Q} \tag{8.20}$$

Assuming the material is confined between two mold halves at equal temperatures, the use of a symmetric boundary condition at the center of the part is valid:

$$\frac{\partial T}{\partial z} = 0 \quad \text{at} \quad z = 0 \tag{8.21}$$

and

$$T = T_m \tag{8.22}$$

at the mold wall.

With the use of the finite difference technique and a six constant model that represents dC^*/dt, Barone and Caulk [23] solved Eqs. 8.20–8.22 for the curing of sheet molding compound (SMC). The SMC was composed of an unsaturated polyester resin with 40.7% calcium carbonate and 30% glass fiber by weight. Figures 8.19 and 8.20 show typical temperature and degree of cure distributions, respectively, during the solidification of a 10 mm thick part as computed by Barone and Caulk. In Fig. 8.19, the temperature rise resulting from exothermic reaction is obvious. This temperature rise increases in thicker parts and with increasing mold temperatures. Figure 8.21 is a plot of the time to reach 80% cure versus thickness of the part for various mold temperatures. The shaded area represents the conditions at which the internal temperature within the part exceeds 200 °C because of the exothermic reaction. Temperatures above 200 °C can lead to material degradation and high residual stresses in the final part.

Improper processing conditions can result in a non-uniform curing distribution which may lead to voids, cracks, or imperfections inside the part. It is of great importance to know the appropriate processing conditions which will both avoid the over-heating problem and speed up the manufacturing process.

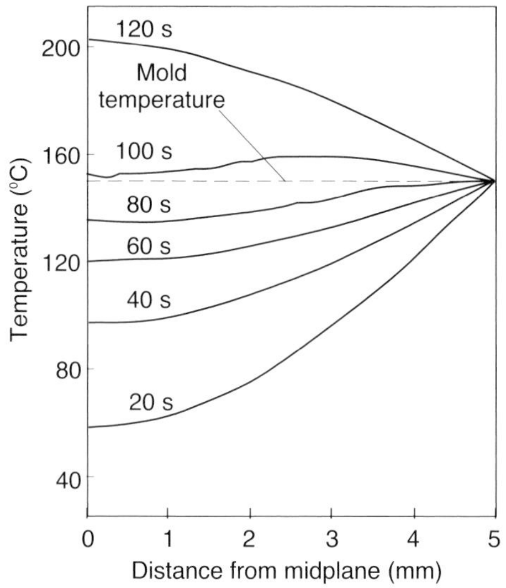

Figure 8.19 Temperature profile history of a 10 mm thick SMC plate.

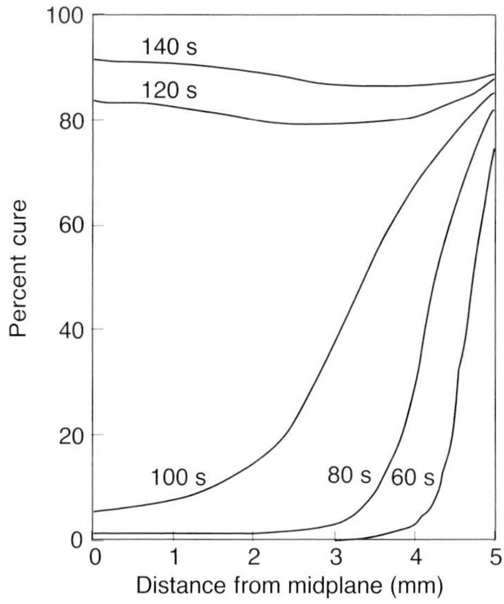

Figure 8.20 Curing profile history of a 10 mm thick SMC plate.

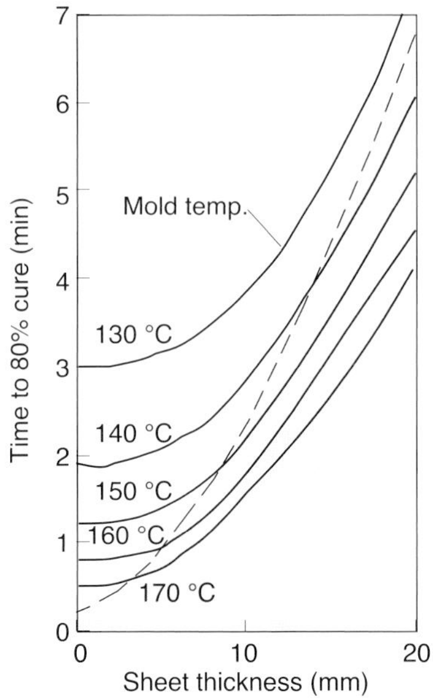

Figure 8.21 Cure times versus plate thickness for various mold temperatures. Shaded region represents the conditions at which thermal degradation may occur.

8.3 Residual Stresses and Warpage of Polymeric Parts

Some major problems encountered when molding polymeric parts are the control and prediction of the component's shape at room temperature. For example, the resulting *sink marks* in the final product are caused by the shrinkage of the material during cooling[3] or curing. A common geometry that usually leads to a sink mark is a ribbed structure. The size of the sink mark, which is often only a cosmetic problem, is not only related to the material and processing conditions but also to the geometry of the part. A rib that is thick in relation to the flange thickness will result in significant sinking on the flat side of the part.

Warpage in the final product is often caused by processing conditions that cause unsymmetric residual stress distributions through the thickness of the part. Thermoplastic parts most affected by residual stresses are those that are manufactured with the injection molding process. The formation of residual stresses in injection molded parts is attributed to two major coupled factors:

3 In injection molding one can mitigate this problem by continuously pumping polymer melt into the mold cavity as the part cools until the gate freezes shut.

cooling and flow stresses. The first and most important factor is the residual stress that is formed because of the rapid cooling or quenching of the part inside the mold cavity. As will be discussed and explained later in this chapter, this dominant factor is the reason why most thermoplastic parts have residual stresses that are tensile in the central core of the part and compressive on the surface. Typical residual stress distributions are shown in Fig. 8.22 [24], which presents experimental[4] results for PMMA and PS plates cooled at different conditions.

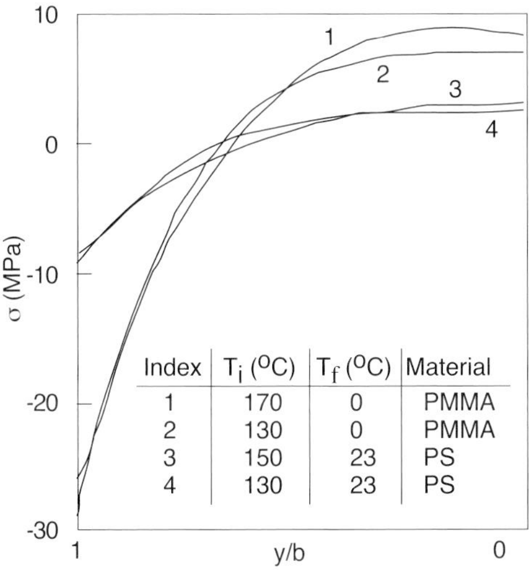

Figure 8.22 Residual stress distribution for 3 mm thick PMMA plates cooled from 170 °C and 130 °C to 0 °C, and for 2.6 mm thick PS plates cooled from 150 °C and 130 °C to 23 °C.

Residual stresses in injection molded parts are also formed by the shear and normal stresses that exist during flow of the polymer melt inside the mold cavity during the filling and packing stage. These tensile flow induced stresses are often very small compared to the stresses that build up during cooling. However, at low injection temperatures, these stresses can be significant in size, possibly leading to parts with tensile residual stresses on the surface. Figure 8.23 [25] demonstrates this concept with PS plates molded at different injection temperatures. The figure presents residual stress distributions through the thickness of the plate perpendicular and parallel to the flow direction. Isayev [24, 25] has also demonstrated that flow stresses are maximum near the gate. The resulting tensile residual stresses are of particular concern since they may lead to *stress cracking* of the polymer component.

4 The experimental residual stress distributions where directly computed from curvature measurements obtained by the layer removal method.

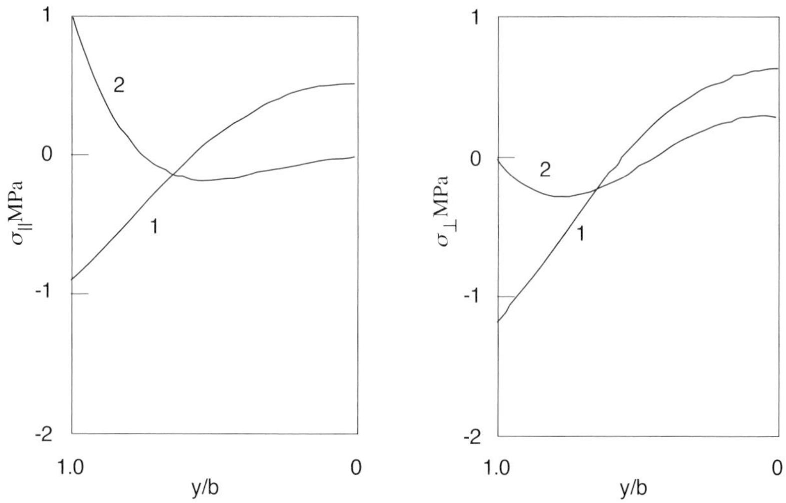

Figure 8.23 Residual stress distribution parallel and perpendicular to the flow direction for a 2.54 mm thick PS plate cooled from 244 °C (1) and 210 °C (2) to 60 °C

The development of models and simulations to predict shrinkage and warpage in the manufacturing of plastic parts is necessary to understand and control the complex thermomechanical behavior the material undergoes during processing. Shrinkage and warpage result from material inhomogeneities and anisotropy caused by mold filling, molecular or fiber orientation, curing or solidification behavior, poor thermal mold lay-out, and improper processing conditions. Shrinkage and warpage are directly related to residual stresses. Transient thermal or solidification behavior as well as material anisotropies can lead to the build-up of residual stresses during manufacturing. Such process-induced residual stresses can significantly affect the mechanical performance of a component by inducing warpage or initiating cracks and delamination in composite parts. It is hoped that an accurate prediction of the molding process and the generation of residual stresses will allow for the design of better molds with appropriate processing conditions.

This section presents basic concepts of the thermomechanical behavior during the manufacturing process of polymeric parts. The formation of residual stresses during the fabrication of plastic parts is introduced first, followed by a review of simple models used to compute residual stresses and associated warpage of plates and beams under different thermal loadings. Several models, which characterize the transient mechanical and thermomechanical behavior of thermoplastic polymers will be reviewed and discussed next. Using these existing models, residual stresses, shrinkage, and warpage of injection molded thermoplastic parts can be predicted. Furthermore, results from the literature are presented. Since thermoset polymers behave quite differently from thermoplastic polymers during molding, other models need to be introduced to compute the thermomechanical behavior of thermoset polymers. Based on these models, results for predicting residual stresses and the resulting shrinkage and warpage for both thin and thick thermoset parts are also discussed.

8.3.1 Residual Stress Models

The formation of residual stresses is most critical during the solidification of polymer components inside injection molds. To illustrate residual stress build-up during injection molding, the plate shaped injection molding cavity shown in Fig. 8.9 is considered. As a first order approximation, it can be assumed that a hard polymer shell forms around the melt pool as the material is quenched by the cool mold surfaces. Neglecting the packing stage during the injection molding cycle, this rigid frame contains the polymer as it cools and shrinks during solidification. The shrinkage of the polymer is, in part, compensated by the deflection of the rigid surfaces, a deformation that occurs with little effort. In fact, if the packing stage is left out, it is a common experimental observation that between 85 to 90% of the polymer's volumetric changes are compensated by shrinkage through the thickness of the part [26]. To understand which material properties, boundary conditions, and processing conditions affect the residual stresses in a solidified polymer component, the cooling process of an injection molded amorphous polymer plate, as it cools inside the mold cavity, will be considered. For simplicity, in the following analysis we include only the thermal stresses which result from the solidification of an injection molded article as it is quenched from an initial temperature, T_i, to a final temperature, T_f (Fig. 8.9). However, it is important to point out again that, in injection molded parts, the solidification process starts during mold filling, and that flow continues during the post-filling or packing stage. This results in frozen-in flow stresses that are of the same order as the thermal stresses. Baaijens [27] calculated the residual stresses in injection molded parts, including the viscoelastic behavior of the polymer and the flow and thermal stresses. With his calculations, he demonstrated that the flow induced stresses are significant and that a major portion of them stems back to the post-filling stage during injection molding. This is in agreement with experimental evidence from Isayev [24] and Wimberger-Friedl [28].

In Fig. 8.9 the plate thickness, 2L, denotes the characteristic dimension across the z-direction and is considered to be much smaller than its other dimensions. This is a common assumption for most polymer parts. It is assumed that the polymer behaves as a viscous liquid above T_g and as an elastic solid below T_g. The resulting residual stresses form because the cooling of the plate, from the outside to the inside, cause the outer layers to solidify first without any resistance from the hot liquid core. As the inner layers solidify and cool, their shrinkage is resisted by the solidified outer surface, thus, leading to a residual stress which is tensile in the center and compressive at the surface. Hence, the residual stress build-up must depend on material and process dependent temperatures, space, thermal properties, elastic properties, and on time. This can be expressed as

$$\sigma = \sigma\left\{T_i - T_f, T_g - T_f, L, z, \beta, k, h, \alpha, E, v, t\right\} \tag{8.23}$$

where β is the thermal expansion coefficient, k the thermal conductivity, α the thermal diffusivity, E the elastic modulus, v Poisson's ratio, and t time. Using the dimensional analysis and assuming that stress relaxation effects are negligible, the final residual stress can be written as

$$\frac{\sigma(1-v)}{E} = f\left(Bi, \hat{z}, \varepsilon_T\right) \tag{8.24}$$

where $\hat{z}$ is a dimensionless coordinate defined by z/L and B_i is the Biot number defined by the ratio of convective heat removal to heat conduction and is calculated with

$$Bi = \frac{hL}{k} \tag{8.25}$$

A large Biot number signifies a process where the heat is removed from the surface of the part at a high rate. This is typical of fast quench processes, which result in both high temperature gradients and residual stresses. Predicted temperature distributions in a process with a large Biot number is shown in Fig. 8.24a. On the other hand, a low Biot number describes a process where the heat is removed from the part's surface at a very low rate, resulting in parts with fairly constant temperatures. Predicted temperature fields, for low Biot number processes, as shown in Fig. 8.24b, will lead to low residual stresses in the final part.

The third quantity, ε_T, found in Eq. 8.24 is the thermal strain that will lead to residual stress. It is a quantity that measures the influence of processing conditions on residual stress formation and is defined by

$$\varepsilon_T = \beta\left(T_g - T_f\right) \tag{8.26}$$

The limits of the thermal strain are described by

$$\varepsilon_T = 0 \tag{8.27}$$

if $T_f = T_g$ and

$$\varepsilon_T = \text{Maximun} \tag{8.28}$$

if $T_i = T_g$

(a)

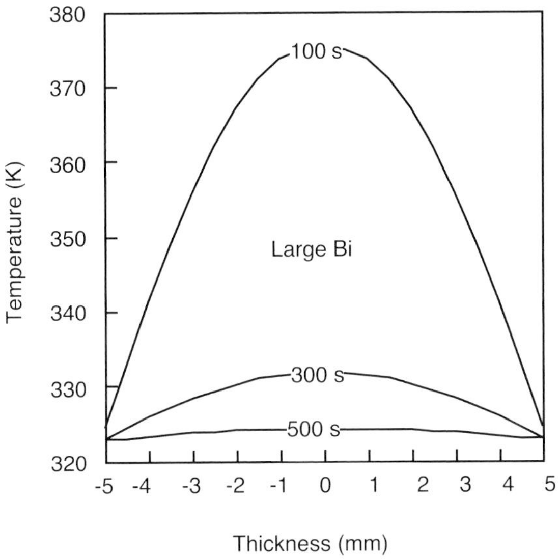

(b)

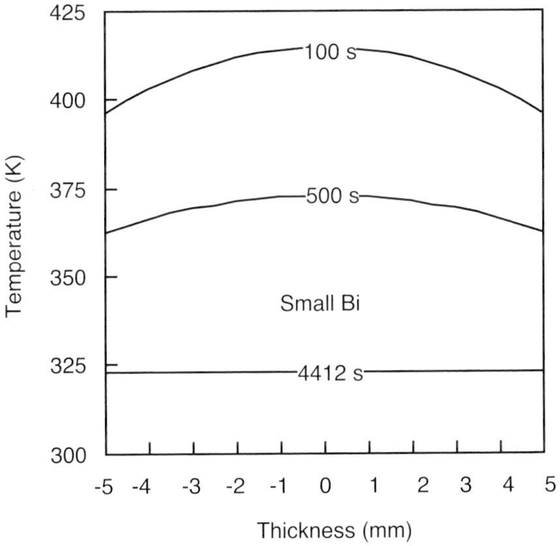

Figure 8.24 Effect of Biot number on the temperature distribution history in a cooling plate: (a) large Bi (b) small Bi.

These limits can be explained. A polymer that is only allowed to cool to T_g, where $T_f = T_m$, does not have a chance to build up any residual stresses, since these can only exist below the glass transition temperature and not in the liquid state. On the other hand, a polymer that is initially at the glass transition temperature and cools to another temperature perceives all its strain in the solid state, hence, conceivably translating them completely into stresses.

8.3.1.1 Residual Stress Model Without Phase Change Effects

The parabolic temperature distribution which is present once the part has solidified will lead to a parabolic residual stress distribution that is compressive in the outer surfaces of the component and tensile in the inner core. Assuming no residual stress build-up during phase change, a simple function based on the parabolic temperature distribution, can be used to approximate the residual stress distribution in thin sections [28]:

$$\sigma = \frac{2}{3}\frac{E\beta}{1-\nu}\left(T_s - T_f\right)\left(\frac{6z^2}{4L^2} - \frac{1}{2}\right) \tag{8.29}$$

Here, T_S denotes the solidification temperature: the glass transition temperature for amorphous thermoplastics, or the melting temperature for semi-crystalline polymers. Equation 8.29 was derived by assuming static equilibrium (e.g., the integral of the stresses through the thickness must be zero). The full derivation is left out here, since a more general approach is presented in the next section. Figure 8.23 [29] compares the compressive stresses measured on the surface of PMMA samples to Eq. 8.29.

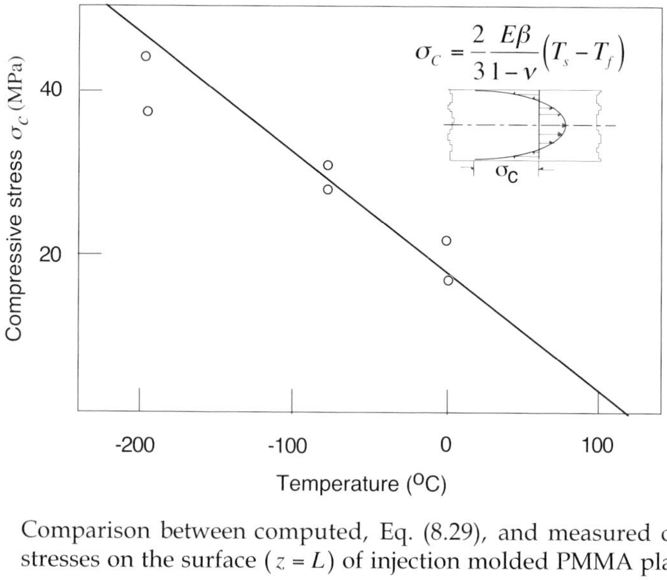

Figure 8.25 Comparison between computed, Eq. (8.29), and measured compressive stresses on the surface ($z = L$) of injection molded PMMA plates.

8.3.1.2 Model to Predict Residual Stresses with Phase Change Effects

As the plate shown in Fig. 8.9 cools, it develops a solidified layer that continues to grow until the whole plate hardens. Figure 8.11 shows a cross-section of the plate at an arbitrary point in time. At any instance of time, t, the location that has just reached T_g is defined as $Z_g(t)$. To solve for the residual stress distribution, the energy equation, Eq. 8.3, must be solved while satisfying the force balance equation within the solidified material. At the centerline, the symmetry boundary condition can be used, Eq. 8.4, and a convective boundary condition on the outer surface of the plate is needed.

$$h\left(T_s - T_f\right) = -k\frac{\partial T}{\partial z} \quad \text{at} \quad z = L \tag{8.30}$$

The strain at any time and position is usually defined as the sum of its elastic, thermal, and viscous strain components:

$$\varepsilon(t) = \varepsilon_E + \varepsilon_{th} + \varepsilon_v \tag{8.31}$$

where ε_E is the elastic strain, ε_v the viscous strain and ε_{th} the thermal strain, which occurs only after the material is below the glass transition temperature. The thermal strain can be written as

$$\varepsilon_{th} = \beta\left(T(z,t) - T_g\right) \tag{8.32}$$

The viscous strain is the strain the layer undergoes just before solidifying, caused by thermal contraction and viscous flow. The viscous strain occurs under a negligible stress and is not felt by the layer that has just solidified. Each layer has a different viscous strain equal to the overall strain of the plate, $\varepsilon(t)$, the instant that layer has solidified, which makes the viscous strain a function of space. Solving for the elastic strain results in

$$\sigma(z,t) = \frac{E}{1-\nu}\left(\varepsilon(t) - \beta\left(T(z,t) - T_g\right) - \varepsilon_v(z)\right) \tag{8.33}$$

To solve for the total strain of the plate, the stresses must approach equilibrium and add up to zero, as

$$\int_{-L}^{L} \sigma(z,t)\,dz = 0 \tag{8.34}$$

Since the plate can be considered symmetric and the stresses are zero above the glass transition temperature, we can write

$$\int_{z_g(t)}^{L} \sigma(z,t)dz = 0 \tag{8.35}$$

Substituting Eq. 8.33 into Eq. 8.35 gives

$$\int_{z_g(t)}^{L} \left(\varepsilon(t) - \beta\left(T(z,t) - T_g\right) - \varepsilon_v(z)\right)dz = 0 \tag{8.36}$$

However, the plate's total strain is constant through the thickness and can be integrated out and solved for as

$$\varepsilon(t) = \frac{1}{L - z_g(t)} \int_{z_g(t)}^{L} \left(\beta\left(T(z,t) - T_g\right) + \varepsilon_v(z)\right)dz \tag{8.37}$$

The total strain $\varepsilon(t)$ and its viscous component ε_v are unknown but equal to each other and can be found by solving both Eq. 8.37 and the energy equation, Eq. 8.3, with a convective boundary condition. The energy equation can be solved numerically by using the finite difference method. The same grid points used for the energy equation can be used for the integration of Eq. 8.37. The solution is achieved in successive time steps from the beginning of cooling until the whole plate has reached the glass transition temperature, at which point the whole viscous strain distribution is known. Now, the part needs to be cooled until its final temperature distribution of $T = T_f$ has been reached. The final residual stress distribution can be computed as

$$\sigma(z) = \frac{E}{1-\nu}\left(\varepsilon_{tot} - \beta\left(T_f - T_g\right) - \varepsilon_v(z)\right) \tag{8.38}$$

The total strain of the plate is unknown and can be found by solving the equilibrium equation, Eq. 8.34, using the final residual stress distribution as

$$\varepsilon_{tot} = \frac{1}{L}\int_{0}^{L} \left(\beta\left(T_f - T_g\right) - \varepsilon_v(z)\right)dz \tag{8.39}$$

Figures 8.24 and 8.25 show residual stress distributions for several Biot numbers and values of Θ_g, respectively. The value Θ_g is the dimensionless temperature that leads to a residual stress build-up and is defined by

$$\Theta_g = \frac{T_g - T_f}{T_i - T_f} \tag{8.40}$$

In Figure 8.25 the model described in this section is also compared to the residual stress distribution of Eq. 8.29. In this comparison, the influence of phase change effects becomes evident.

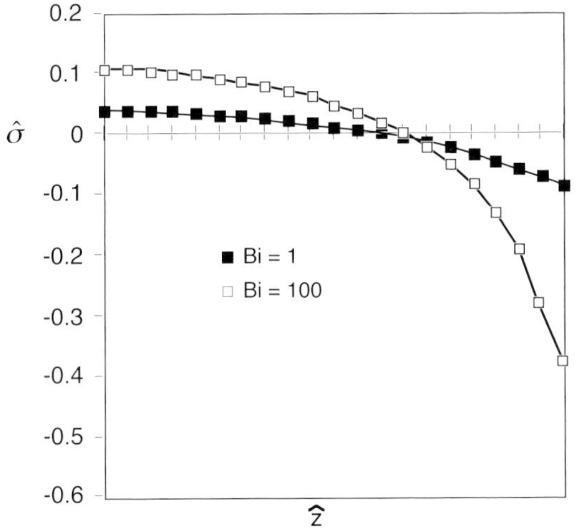

Figure 8.26 Residual stress distributions as a function of Biot number in a polystyrene plate after cooling.

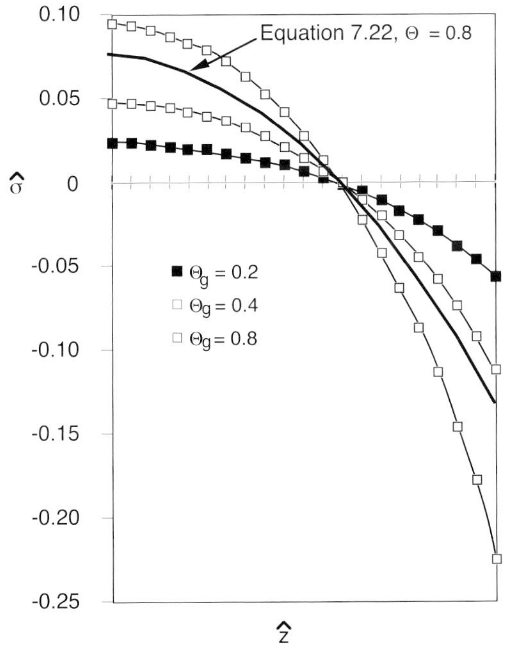

Figure 8.27 Residual stress distributions as a function of thermal boundary conditions in a polystyrene plate after cooling.

Equations 8.37–8.39 can be modified and solved together with the energy equation for thermosets, Eq. 8.20, to compute the residual distributions in thermosetting parts. Here, T_g must be replaced with the temperature of the material at the time it solidified (e.g., when its conversion was 80%).

8.3.2 Other Simple Models to Predict Residual Stresses and Warpage

In practice, the complexity of part geometry and cooling channel design in the manufacture of plastic parts can result in unsymmetric mold wall temperature variations which in turn lead to warpage of the part after it is ejected from the mold. Here, a simple model is presented to evaluate residual stress and warpage of a flat plate caused by uneven mold temperatures (Fig. 8.26a), temperature rise due to exothermic curing reaction in thermoset parts (Fig. 8.26b), and by unsymmetric stress distributions in laminated composites (Fig. 8.26c). In general, the stress distribution has to satisfy the equilibrium equation as defined in Eq. 8.33 where the stress–strain relation is defined as

$$\sigma(z) = \frac{E}{(1-\nu)}\left(\varepsilon_{tot} - \beta\Delta T(z) - \varepsilon_v(z)\right) \tag{8.41}$$

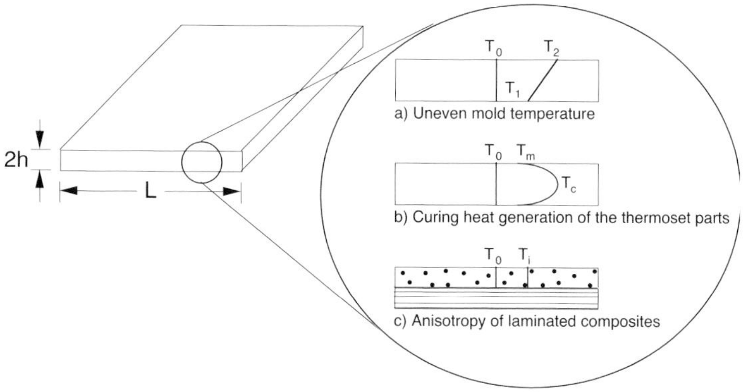

Figure 8.28 Possible causes of residual stress build-up across the thickness of a part.

Here, ε_{tot} is the total or actual shrinkage of the plate, ΔT the change in temperature, E Young's modulus, ν Poisson's ratio, and β the thermal expansion coefficient. For simplicity, the viscous strain is often neglected, assuming the part is thin enough that it solidifies at once. Based on classical shell theory and using the stress distribution, one can additionally compute a thermal moment as follows:

$$M = w\int_{-L}^{L} \sigma(z) z dz \tag{8.42}$$

where w is the width of the plate. In the following analyses the above Eqs. 8.39–8.42 will be used to compute residual stress and warpage for various cases. If the part is fixed as it is cooled to its final temperature, the total strain, ε_{tot}, is zero. In such a case, residual stress is dominated by thermal strain.

8.3.2.1 Uneven Mold Temperature

During molding, the mold wall surface temperatures may vary due to improper thermal mold layout (Fig. 8.26a) with variations typically in the order of 10 °C. Furthermore, the temperatures on the mold surface may vary depending on where the heating or cooling lines are positioned. However, in the current example, we assume that this effect is negligible. The amount of warpage caused by temperature variations between the two mold halves can easily be computed using the equations of the last section. The temperature field across the thickness of a part can be described by

$$T = \left[\frac{1}{2}(T_1 + T_2) + \frac{1}{2}(T_1 - T_2)\frac{z}{L}\right] \tag{8.43}$$

After substituting Eqs. 8.34 and 8.43 into Eq. 8.41, the stress distribution throughout the thickness can be obtained:

$$\sigma = \frac{\beta E}{2(1-\nu)}\left[(T_2 - T_1)\frac{z}{L}\right] \tag{8.44}$$

Substituting Eq. 8.44 into Eq. 8.42, the thermal moment M becomes

$$M = \frac{w\beta E L^2}{3(1-\nu)}(T_2 - T_1) \tag{8.45}$$

For a part whose width, w, is much smaller than its length, ℓ, we can assume a cantilevered geometry as shown in Fig. 8.27. For this geometry the deflection, δ, that results from a moment, M, can be written as

$$\delta = \frac{M\ell^2}{2EI} \tag{8.46}$$

where the area moment of inertia, I, is written as

$$I = \frac{1}{12} w(2L)^3 \tag{8.47}$$

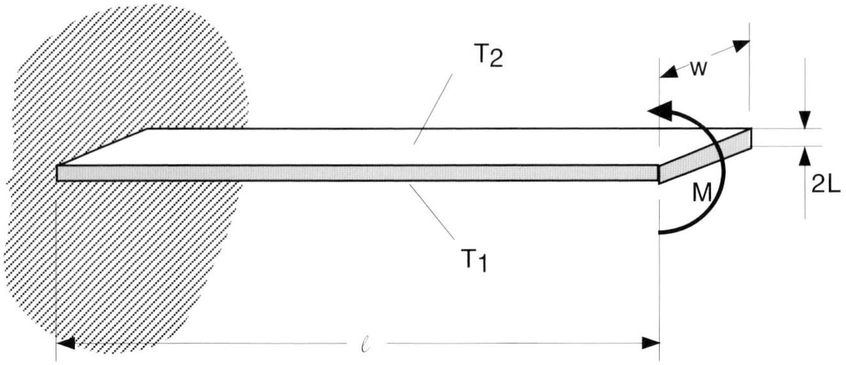

Figure 8.29 Simplified geometry of a thin slender part

Substituting Eqs. 8.45 and 8.47 into Eq. 8.46 we get

$$\delta = \frac{\beta \ell^2}{4L} (T_2 - T_1) \tag{8.48}$$

8.3.2.2 Residual Stress in a Thin Thermoset Part

In addition to uneven mold temperatures, the exothermic curing reaction is a known problem when manufacturing thermoset parts. Heat release during such reactions can cause the transient temperature inside the part to be higher than the mold wall temperatures. Typical temperature and curing history plots are shown in Figs. 8.17 and 8.18. Note that such a temperature distribution dominates the final residual stress distribution. Here, a simple elastic model is presented to approximate the residual stress distribution. The temperature field in the calculation can be described by a parabolic curve

$$T(z) = T_c + (T_m - T_c)\frac{z^2}{L^2} \tag{8.49}$$

where T_c and T_m are the temperature at the center of the plate and the mold surface, respectively.

We assume the part to be cooled elastically to room temperature, T_0. Using Eq. 8.41, the stress distribution can be expressed by

$$\sigma(z) = \frac{E}{(1-v)}\left[\varepsilon_{tot} - \beta\left(T_0 - T(z)\right)\right] \tag{8.50}$$

Since the above equation has to satisfy the equilibrium equation, Eq. 8.34, the total strain can be obtained by integrating the stress field across the thickness

$$\varepsilon_{tot} = \frac{\beta}{L}\left[T_0 - \frac{2}{3}T_c - \frac{1}{3}T_m\right] \tag{8.51}$$

Substituting Eq. 8.51 into Eq. 8.39, the residual stress distribution becomes

$$\sigma(z) = \frac{\beta E}{(1-v)}\left[\frac{1}{3}(T_c - T_m) - (T_c - T_m)\frac{z^2}{L^2}\right] \tag{8.52}$$

After defining the non-dimensional parameters, $\hat{\sigma}$ and $\hat{z}$, as

$$\hat{\sigma} \equiv \frac{(1-v)\sigma}{E\beta(T_c - T_m)} \tag{8.53}$$

and

$$\hat{z} \equiv \frac{z}{L} \tag{8.54}$$

Eq. 8.52 can be normalized as

$$\hat{\sigma}(\hat{z}) = \frac{1}{3} - \hat{z}^2 \tag{8.55}$$

The non-dimensional residual stress distribution across the thickness of the part is depicted in Fig. 8.27. The stresses in the outer layer are compressive while tensile stresses are found in the inner layers. A maximum compressive non-dimensional stress of 2/3 is located on the surface and a maximum tensile stress of 1/3 occurs at the center of the part.

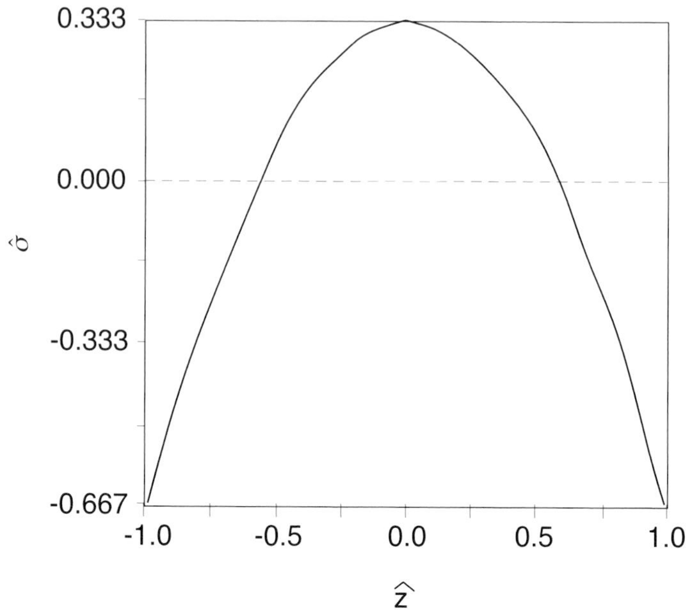

Figure 8.30 Normalized residual stress distribution in a thin thermoset part.

8.3.2.3 Anisotropy Induced Curvature Change

In the manufacturing of large and thin laminate structures or fiber reinforced composite parts with a large fiber–length/part–thickness ratio, the final part exhibits a higher thermal expansion coefficient in the thickness direction than in the surface direction. If the part is curved, it will undergo an angular distortion, as shown in Fig. 8.28, which is a consequence of the anisotropy of the composites. This phenomenon is usually called the *spring-forward* effect or *anisotropy induced curvature change* [32]. Through–thickness thermal strains, which are caused by different thermal expansion coefficients, can lead to an angle distortion of a cylindrical shell experiencing a temperature change. As demonstrated in Fig. 8.28, when a curved part undergoes a temperature change of ΔT, the curved angle, θ, will change by $\Delta\theta$. The resulting $\Delta\theta$, therefore, is dependent on the angle θ, the temperature change ΔT, and the difference of the thermal expansion coefficients in the r and θ directions [33]

$$\Delta\theta = (\beta_r - \beta_\theta)\theta\Delta T = \Delta\beta\,\theta\Delta T \tag{8.56}$$

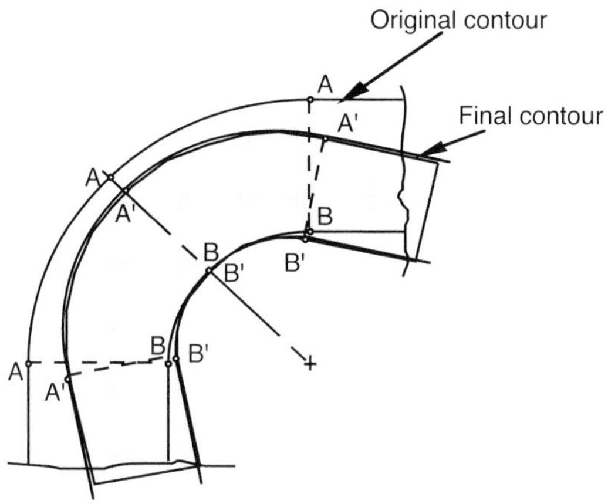

Figure 8.31 Schematic diagram of the spring-forward effect.

In plate analysis, the inclusion of anisotropies that lead to curvature changes is very involved. Hence, it is easier to introduce the curvature change by an equivalent thermal moment [34]:

$$M = \frac{E}{(1-\nu)} \frac{\Delta\beta \Delta T}{R} \frac{L^3}{12} \tag{8.57}$$

where R represents the local radius of curvature.

8.3.3 Predicting Warpage in Actual Parts

Shrinkage and warpage are directly related to residual stresses which result from locally varying strain fields that occur during the curing or solidification stage of a manufacturing process. Such strain gradients are caused by nonuniform thermomechanical properties and temperature variations inside the mold cavity. Shrinkage due to cure can also play a dominant role in the residual stress development in thermosetting polymers and becomes important for fiber reinforced thermosets when are a concern sink marks appearing in thick sections or ribbed parts.

When processing thermoplastic materials, shrinkage and warpage in a final product depend on the molecular orientation and residual stresses that form during processing. The molecular or fiber orientation and the residual stresses inside the part in turn depend on the flow and heat transfer during the mold filling, packing, and cooling stage of the injection molding process. Kabanemi et al. [35] used a three-dimensional finite element approach to solve the thermal history and residual stress build-up in injection molded parts. To predict the residual stress in the finished part, they characterized the thermomechanical response of the polymer from melt to room temperature, or from the p-v-T behavior to stress–strain behavior. Bushko and Stokes [36, 37] used a thermorheologically simple thermo-viscoelastic material model to predict residual stresses and warpage in flat plates. With their model, they found that packing pressure had a significant effect on the shrinkage of the final part but little effect on the residual stress build-up. Wang and co-workers have developed unified simulation programs to model the filling and post-filling stages in injection molding [38–41]. In their models they perform a simultaneous analysis of heat transfer, compressible fluid flow, fiber orientation, and residual stress build-up in the material during flow and cooling using finite element/control volume approach for flow, finite difference techniques for heat transfer, and finite element methods for fiber orientation and thermomechanical analysis.

The shrinkage and warpage in thin compression molded fiber reinforced thermoset plates were predicted by various researchers [42] using fully three-dimensional finite element models and simplified finite element plate models. More recently [4, 44], the through–thickness properties, temperature, and curing variations that led to warpage have been represented with equivalent moments. By eliminating the thickness dimensions from their analysis, they significantly reduced computation costs and maintained agreement with experimental results. At the same time, they were able to use the same finite element meshes used in common commercial codes to predict the mold filling and the fiber orientation in the final part.

The governing equations used for the stress analysis of polymer components are derived using the principle of virtual work. Here, the stresses are represented as a function of local strain and residual stress $\{\sigma_0\}$.

$$\{\sigma\} = [E]\{\varepsilon\} - [E]\{\varepsilon_{tot}\} + \{\sigma_0\} \tag{8.58}$$

In Eq. 8.58 the material tensor [E] is anisotropic and temperature or degree of cure dependent and $\{\varepsilon_{tot}\}$ is the total internal strain that occurs due to curing, cooling or heating during a time step. Two kinds of internal strains should be included when simulating the thermomechanical behavior of polymer parts. One is a thermal strain caused by temperature change and the other is a curing strain resulting from cross-linking polymerization of thermoset resins. Thus, the total internal strain can be expressed by

$$\{\varepsilon_{tot}\} = \{\varepsilon_0^{th}\} + \{\varepsilon_0^{c}\} \tag{8.59}$$

Here, superscript *th* denotes the thermal strain and *c* the curing strain. The thermal strains can be represented in terms of temperature change and thermal expansion coefficients

$$\left\{\varepsilon_0^{th}\right\}T = \Delta T\left\{\alpha_{xx}\ \alpha_{yy}\ \alpha_{zz}\ 0\ 0\ 0\right\} \tag{8.60}$$

The anisotropic thermal expansion coefficient, caused by fiber orientation, is perhaps the largest cause of warpage in fiber reinforced parts. Figure 8.29 demonstrates how, for typical thermoset composite parts, the thermal shrinkage parallel to the main orientation direction is about half of that normal to the main orientation direction[5].

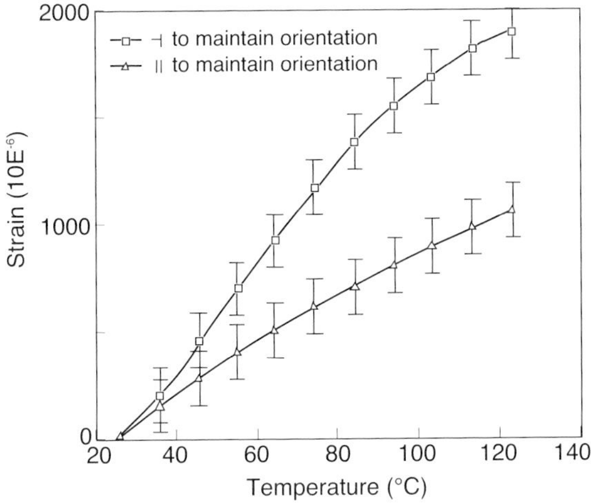

Figure 8.32 Experimentally measured thermal strains in an SMC plate with a fiber orientation distribution that resulted from a 25% initial mold coverage charge.

To calculate the residual stress development during the manufacturing process, the heat transfer equation is coupled to the stress–strain analysis through constitutive equations. Figure 8.30 compares the mold geometry with part geometry for the truck fender shown in Chapter 7, after mold removal and cooling, computed using the above model. The fiber content by volume in the part was 21% (ϕ=0.21) and the material properties for the glass fiber and the unsaturated polyester resin are listed in Table 8.4.

5 The thermal shrinkage was measured from a rectangular plate molded with a charge that covered 25% of the mold surface and that was allowed to flow only in one direction.

Table 8.4 Mechanical and Thermomechanical Properties for Various Materials

	Fiberglass	Polyester	Epoxy
E (MPa)	7.3×10^4	2.75×10^3	4.1×10^3
v	0.25	0.34	0.37
β (mm/mm/K)	5.0×10^{-6}	3.7×10^{-5}	5.76×10^{-5}

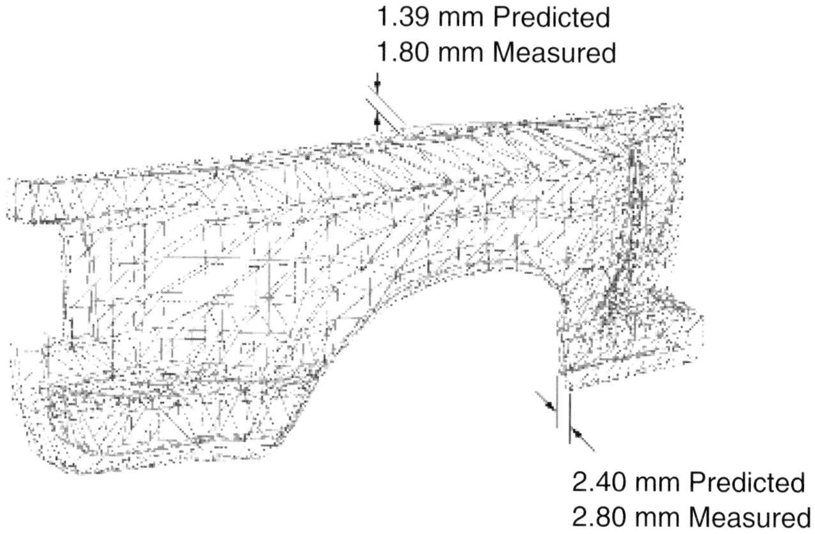

Figure 8.33 Simulated displacements of an automotive body panel. Displacements were magnified by a factor of 20.

Minimizing warpage is one of the biggest concerns for the design engineer. One way to reduce warpage is by introducing a temperature gradient between the upper and lower mold halves. Again, this through–thickness temperature gradient will introduce a thermal moment which increases or decreases the warpage. Also, by changing the formulation of the polyester resin, the coefficient of thermal expansion of the matrix can be reduced, making it similar to the coefficient of the glass fiber. Theoretically, reduction of the coefficients for the matrix would decrease the in–plane differential shrinkage, which in turn could help reduce the final warpage. Furthermore, the fiber content also has a great effect on the deformation of a body panel. Here, although the warpage is actually caused by the existence of fibers inside the resin, the increase of fiber content adds to the stiffness of the part which in turn reduces warpage. Further reduction in warpage can also be achieved by changing the size and location of the initial charge, a trial-and-error solution, which is still the most feasible with today's technology.

Examples

8.1 Three small samples of an epoxy resin were subjected to a temperature rise from 50 °C to 300 °C in a differential scanning calorimeter using three different heating rates: 2.78 K/min (5F/min), 5.66 K/min (10 F/min) and 11.11 K/min (20 F/min). Table 8.5 presents the exothermal heat released during the tests. Determine the total heat of reaction using the data measured with the 2.78 K/min heating rate. Plot the degree of cure as a function of time.

In order to compute the degree of cure as a function of time we use Eq. 8.8

$$C_i^* = \frac{Q_i}{Q_T}$$

where

$$Q_i = \int_0^{\tau_i} \dot{Q} dt = \sum_{j=1}^{i} \dot{Q}_j \Delta t$$

and

$$Q_T = \int_0^{\infty} \dot{Q} dt = \sum_{j=1}^{n} \dot{Q}_j \Delta t$$

where n is the total number of data points and Δt is the time step. Since the data are given in terms of temperature, we must transform the temperature step to time using the heating rate 2.78 K/min using

$$\Delta t = \frac{10\text{K}}{2.78\text{K}/\text{min}}$$

The above leads to a total heat of reaction, Q_T, of 785.4 J/g or 785.4 kJ/kg. The degree of cure as a function of temperature is plotted in Fig. 8.31.

Table 8.5 DSC Data for an Epoxy Resin

T (°C)	$\dot{Q}\left(\frac{\text{J}}{\text{g}\cdot\text{min}}\right)$		
	2.78 K/min	5.66 K/min	11.11 K/min
50	0.	0.	0.
60	0.	0.	0.
70	0.01	0.	0.01
80	0.02	0.01	0.02
90	0.03	0.02	0.03
100	0.06	0.04	0.06
110	0.12	0.09	0.12
120	0.22	0.16	0.21
130	0.38	0.28	0.37
140	0.65	0.49	0.62
150	1.11	0.82	1.03
160	1.91	1.35	1.66
170	3.38	2.23	2.668
180	6.23	3.69	4.24
190	12.21	6.23	6.76
200	24.97	10.86	10.94
210	47.10	19.76	18.08
220	60.09	37.23	30.79
230	39.40	68.36	53.89
240	15.0	103.36	94.37
250	4.29	100.09	152.83
260	0.98	55.65	196.48
270	0.17	19.99	168.68
280	0.018	5.17	90.54
290	0.	0.92	31.62
300	0.	0.09	7.06

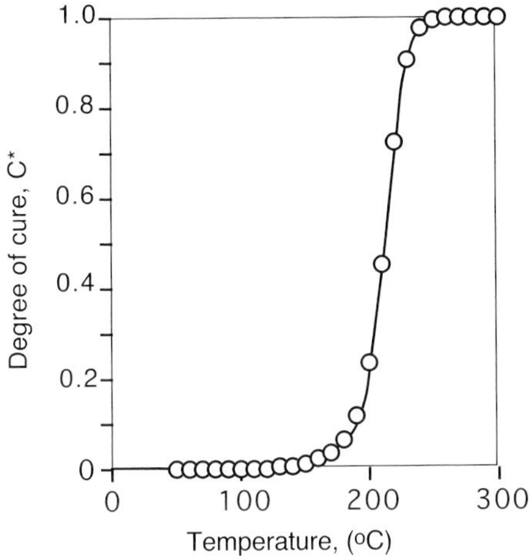

Figure 8.34 Degree of cure as function of temperature for a heating rate of 2.78 K/min.

Problems

8.1 Determine the constants for the curing model presented in Eq. 8.12 for the vinyl ester presented in Fig. 8.15.

8.2 Write a finite difference program to solve for the transient temperature distribution in a cooling amorphous polymer plate. Test the program with the data given for polystyrene in Table 8.2 for a 2 mm thick plate.

8.3 Using the momentum balance presented in Eq. 8.43, derive the residual stress distribution, Eq. 8.38, in a cooling polymer plate neglecting the effects of phase change.

8.4 A 1 mm thick polystyrene ruler is injection molded. For a resin with a glass transition temperature around 100 ^{o}C and a final ambient temperature of 20 ^{o}C, estimate the residual stress at the surface of the part shortly after demolding. Assume a Young's modulus of 3200 MPa and a thermal expansion coefficient of 7×10^{-5} mm/mm/^{o}C. Neglect phase change effects.

8.5 For a standard 30 cm long, 3 cm wide, and 1 mm thick ruler, how much warpage would occur with a 5 ^{o}C temperature difference between the mold halves? Assume a Young's modulus of 3200 MPa and a thermal expansion coefficient of 7x10^{-5} mm/mm/^{o}C.

8.6 Write a finite difference program to solve for the transient temperature distribution in a cooling amorphous polymer plate. Solve Eqs. 8.31-8.39 along with the predicted temperatures to estimate the residual stress in the final plate. Compare the residual stress distribution for a 2 and a 5 mm thick polystyrene plate.

8.7 A long and thin unsaturated polyester - unidirectional glass fiber composite plate (1000x20x2 mm) is molded in a thermally unbalanced mold. One mold side is a 140 ^{o}C and the other at 150 ^{o}C. Assuming the temperature distribution is at steady state the instant the part solidifies, what is the residual stress distribution in the plate before it is allowed to warp. $E_{composite} = 20\text{GPa}$. Towards which direction would the plate warp, the hot side or cooler side?

8.8 Figure 8.31 presents a schematic of the cross-section of an SMC pick-up truck box. In the finished product, the average through the thickness thermal expansion coefficient, β_3, is 2x10^{-5} mm/mm/K, and the average planar thermal expansion coefficient, β_1 or β_2 is 8x10^{-6} mm/mm/K. Estimate the amount and direction that the side walls will warp, if any.

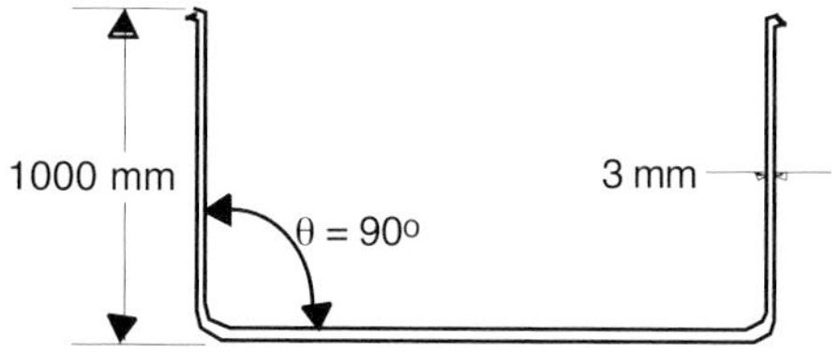

Figure 8.35 Schematic of an SMC pick-up truck cross-section.

8.9 Using the data given in Example 8.1, plot the degree of cure for heating rates of 5.66 K/min and 1.11 K/min. What are the characteristic curing times for all three heating rates.

References

1. van Krevelen, D.W., and Hoftyzer, P.J., *Properties of Polymers*, Elsevier Scientific Publishing Company, Amsterdam, (1976).
2. Wagner, H., Internal report, AEG, Kassel, Germany, (1974).
3. Menges, G., and Winkel, E., *Kunststoffe, 72*, 2, 91, (1982).
4. Tadmor, Z., and Gogos, C.G., *Principles of Polymer Processing*, John Wiley & Sons, New York, (1979).
5. Avrami, M., *J. Chem. Phys., 7*, 1103, (1939).
6. Sharples, A., Polymer Science, *A Materials Science Handbook*, Chapter 4, A.D. Jenkins, Ed., North Holland Publishing Company, Amsterdam, (1972).
7. Eder., G., and Janeschitz-Kriegl, H., *Material Science and Technology, Vol. 18.*, Ed. H.E.H. Meijer, Verlag Chemie, Weinheim, (1995).
8. Brandrup, J., and Immergut, E.H., *Polymer Handbook*, John Wiley & Sons, New York, (1975).
9. Krobath, G., Liedauer, S., and Janeschitz-Kriegl, H., *Polymer Bulletin, 14*, 1, (1985).
10. Stefan, J., *Ann. Phys. and Chem., N.F., 42*, 269, (1891).
11. Eder, G., and Janeschitz-Kriegl, H., *Polymer Bulletin, 11*, 93, (1984).
12. Janeschitz-Kriegl, H., and Krobath, G., *Intern. Polym. Process., 3*, 175, (1988).
13. Randell, D.R., *Radiation Curing Polymers*, Burlington House, London, (1987).
14. Palmese, G.R., Andersen, O., and Karbhari, V.M., *Advanced Composites X: Proceedings of the 10th Annual ASM/ESD Advance Composites Conference*, Dearborn, MI, ASM International, Material Park, (1994).
15. Macosko, C.W., *RIM Fundamentals of Reaction Injection Molding*, Hanser Publishers, Munich, (1989).
16. Enns, J.B. and Gillham, J.K., "Time-Temperature-Transformation (TTT) Cure Diagram: Modeling the Cure Behavior of Thermosets", *J. Appl. Polym. Sci.*, 28, 2567 (1983).
17. Kamal, M.R., "Thermoset Characterization for Moldability Analysis", *Polym. Eng. Sci., 14* (3), 231 (1974).
18. Rabinowitch, E., *Trans. Faraday Soc., 33*, 1225 (1937).
19. Havlicek, I and Dusek, K., in *Crosslinked Epoxies*, 417, B. Sedlacek and J. Kahovec, Eds., Walter de Gruyter, New York (1987).
20. DiBenedetto, A.T., *J. Polym. Sci. Polym. Phys.*, ed., *25*, 1949 (1987).
21. Bogetti, T.A. and Gillespie, J.W., *45th SPI Conf. Proc.,* (1990).
22. Bogetti, T.A. and Gillespie, J.W., *21st Int. SAMPE Tech. Conf.,* (1989).
23. Barone, M.R. and Caulk, D.A., *Int. J. Heat Mass Transfer, 22* , 1021, (1979).
24. Isayev, A.I., and Crouthamel, D.L., *Polym. Plast. Technol., 22*, 177, (1984).
25. Isayev, A.I., *Polym. Eng. Sci., 23*, 271, (1983).
26. Wübken, G., Ph.D. Thesis, IKV, RWTH-Aachen, Germany, (1974).
27. Baaijens, F.P.T., *Rheol. Acta, 30*, 284, (1991).
28. Wimberger-Friedl, R., *Polym. Eng. Sci.*, 30 , 813, (1990).
29. Ehrenstein, G.W., *Polymer-Werkstoffe*, Hanser Publishers, Munich, (1978).
30. S. P. Timoshenko and Krieger, S. W., *Mechanics of Composite Materials*, McGraw-Hill, New York, (1959).

31. J. H. Faupel and Fisher, F. E., *Engineering Design*, McGraw-Hill, New York, (1981).
32. C.-C. Lee, GenCorp Research, Akron OH, Personal communication (1994).
33. O'Neill, J.M., Rogers, T.G., and Spencer, A.J.M., *Math. Eng. Ind., 2,* 65, (1988).
34. Tseng, S.C. and Osswald, T.A., *Polymer Composites*, *15*, 270, (1994).
35. Kabanemi, K.K., and Crochet, M.J., *Intern. Polym. Proc, 7*, 60, (1992).
36. Bushko, W.C., and Stokes, V.K., *Polym. Eng. Sci., 35,* 351, (1995).
37. Bushko, W.C., and Stokes, V.K., *Polym. Eng. Sci., 35,* 365, (1995).
38. Chiang, H.H., Hieber, C.A., and Wang, K.K., *Polym. Eng. Sci., 31*, 116, (1991).
39. Chiang, H.H., Hieber, C.A., and Wang, K.K., *Polym. Eng. Sci., 31*, 125, (1991).
40. Chiang, H.H., Himasekhar, K., Santhanam, N., and Wang, K.K., *J. Eng. Mater.Tech., 115*, 37, (1993).
41. Chiang, H.H., Santhanam, N., Himasekhar, K., and Wang, K.K., *Advances in Computer Aided Engineering (CAE) of Polymer Processing, MD-Vol. 49,* ASME, New York, (1994).
42. Osswald, T.A., *J. Thermoplast. Comp. Mater., 4*, 173, (1991).
43. Tseng, S.C., Ph.D. Thesis, Dept. of Mech. Eng., University of Wisconsin-Madison, (1993).
44. Tseng, S.C. and Osswald, T.A., *Polymer Composites, 15*, 270, (1994).

Part III

Engineering Design Properties

9

Mechanical Behavior of Polymers

Polymeric materials are implemented into various designs because of their low cost, processability, and desirable material properties. Of interest to the design engineer are the short and long-term responses of a loaded component. Properties for short-term responses are usually acquired through short-term tensile tests and impact tests, whereas long-term responses depend on properties measured using techniques such as the creep and the dynamic test.

9.1 Basic Concepts of Stress and Strain

Strictly speaking, polymers cannot be modeled using linear theory of elasticity. However, the stress-strain response of a linear elastic model for the polymer component can suffice in the evaluation of a design and the prediction of the behavior of the component during loading.

For a full three-dimensional model, as shown for a small material element in Fig. 9.1, there are six components of stress and strain. The stress-strain relation for a linear elastic material is defined by the following equations:

$$\sigma_{xx} = EI_{\varepsilon} + 2G\varepsilon_{xx} \quad (9.1)$$

$$\sigma_{yy} = EI_{\varepsilon} + 2G\varepsilon_{yy} \quad (9.2)$$

$$\sigma_{zz} = EI_{\varepsilon} + 2G\varepsilon_{zz} \quad (9.3)$$

$$\tau_{xy} = G\gamma_{xy} \quad (9.4)$$

$$\tau_{yz} = G\gamma_{yz} \quad (9.5)$$

$$\tau_{zx} = G\gamma_{zx} \quad (9.6)$$

where

$$E = \frac{\nu E}{(1+\nu)(1-2\nu)} \tag{9.7}$$

and I_ε is the first invariant of the strain tensor and represents the volumetric expansion of the material which is defined by

$$I_\varepsilon = \varepsilon_{xx} + \varepsilon_{yy} + \varepsilon_{zz} \tag{9.8}$$

The elastic constants E, ν and G represent the modulus of elasticity, Poisson's ratio and shear modulus, respectively. The shear modulus, or modulus of rigidity, can be written in terms of E and ν as

$$G = \frac{E}{2(1+\nu)} \tag{9.9}$$

The above equations can be simplified for different geometries and load cases. Two of the most important simplified models, the plane stress and plane strain models, are discussed below.

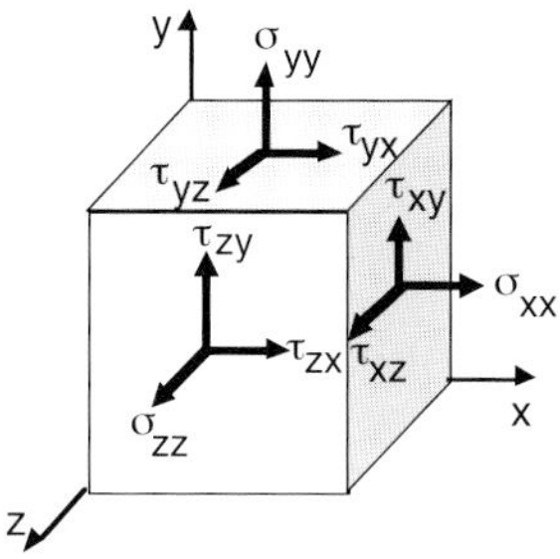

Figure 9.1 Differential material element with coordinate and stress definition.

9.1.1 Plane Stress

A common model describing the geometry and loading of many components is the plane stress model. The model reduces the problem to two dimensions by assuming that the geometry of the part can be described on the x-y plane with a relatively small thickness in the z direction. In such a case $\sigma_{zz} = \tau_{zx} = \tau_{yz} = 0$ and Eqs. 9.1-9.6 reduce to

$$\sigma_{xx} = \frac{E}{1-\nu^2}\left(\varepsilon_{xx} + \nu\varepsilon_{yy}\right) \tag{9.10}$$

$$\sigma_{yy} = \frac{E}{1-\nu^2}\left(\nu\varepsilon_{xx} + \varepsilon_{yy}\right) \quad \text{and} \tag{9.11}$$

$$\tau_{xy} = G\gamma_{xy} \tag{9.12}$$

9.1.2 Plane Strain

Another common model used to describe components is the plane strain model. Similar to the plane stress model, the geometry can be described on an x-y plane with an infinite thickness in the z direction. This problem is also two-dimensional, with negligible strain in the z direction but with a resultant σ_{zz}. For this case, Eqs. 9.1–9.8 reduce to

$$\sigma_{xx} = \frac{E(1-v)}{(1+v)(1-2v)}\left(\varepsilon_{xx} + \frac{v}{1-v}\varepsilon_{yy}\right) \tag{9.13}$$

$$\sigma_{yy} = \frac{E(1-v)}{(1+v)(1-2v)}\left(\frac{v}{1-v}\varepsilon_{xx} + \varepsilon_{yy}\right) \quad \text{and} \tag{9.14}$$

$$\tau_{xy} = G\gamma_{xy} \tag{9.15}$$

9.2 The Short-Term Tensile Test

The most commonly used mechanical test is the short-term stress-strain tensile test. Stress-strain curves for selected polymers are displayed in Fig. 9.2 [1]. For comparison, the figure also presents stress-strain curves for copper and steel. It becomes evident from Fig. 9.2 that although they have much lower tensile strengths, many engineering polymers exhibit much higher strains at break.

The next two sections discuss the short-term tensile test for elastomers and thermoplastic polymers separately. The main reason for identifying two separate topics is that the deformation of a cross-linked elastomer and an uncross-linked thermoplastic vary greatly. The deformation in a cross-linked polymer is in general reversible, whereas the deformation in typical uncross-linked polymers is associated with molecular chain relaxation, which makes the process time-dependent, and sometimes irreversible.

9.2.1 Rubber Elasticity

The main feature of elastomeric materials is that they can undergo very large and reversible deformations. This is because the curled-up polymer chains stretch during deformation but are hindered in sliding past each other by the cross-links between the molecules. Once a load is released, most of the molecules return to their coiled shape. As an elastomeric polymer component is deformed, the slope of the stress-strain curve drops significantly as the uncurled molecules provide less resistance and entanglement, allowing them to move more freely. Eventually, at deformations of about 400%, the slope starts to increase since the polymer chains are fully stretched. This is followed by polymer chain breakage or crystallization which ends with fracture of the component. Stress-deformation curves for natural rubber (NR) [2] and a rubber compound [3] composed of 70 parts of styrene-butadiene-rubber (SBR) and 30 parts of natural rubber are presented in Fig. 9.3. Because of the large deformations, typically several hundred percent, the stress-strain data are usually expressed in terms of extension ratio, λ, defined by

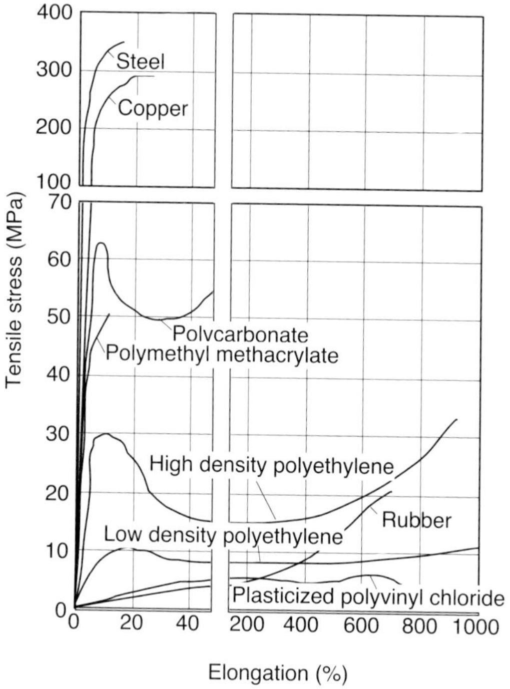

Figure 9.2 Tensile stress-strain curves for several materials.

$$\lambda = \frac{L}{L_0} \tag{9.16}$$

where L represents the instantaneous length and L_0 the initial length of the specimen.

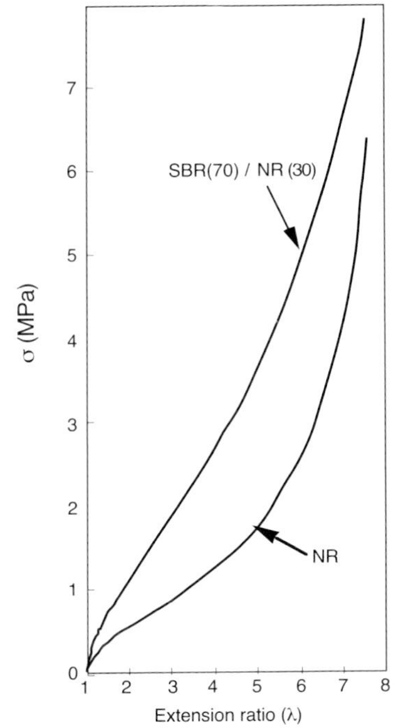

Figure 9.3 Experimental stress-extension curves for NR and a SBR/NR compound.

Based on kinetic theory of rubber elasticity [2, 4] simple expressions can be derived to predict the stress as a function of extension. For a component in uniaxial extension, or compression, the stress can be computed as[1]

$$\sigma = G_0\left(\lambda - \frac{1}{\lambda^2}\right) \tag{9.17}$$

where G_0 is the shear modulus at zero extension, which for rubbers can be approximated by

$$G_0 = \frac{E_0}{3} \tag{9.18}$$

with E_0 as the elastic tensile modulus at zero extension.

Figure 9.4 [2] compares the kinetic theory model with the experimental data for natural rubber, presented in Fig. 9.3. The agreement is good up to about 50% extension (λ = 1.5). However, Eq. 9.17 can be used to approximate the

1 A similar equation exists for equibiaxial extension (inflation) of thin sheets. This equation is written as follows $\sigma = G_0\left(\lambda^2 - 1/\lambda^4\right)$.

stress-strain behavior up to 600% extension (λ = 7.0). For compression, the model agrees much better with experiments, as shown for natural rubber in Fig. 9.5 [2]. Fortunately, rubber products are rarely deformed more than 25% in compression or tension, a fact which often justifies the use of Eq. 9.17.

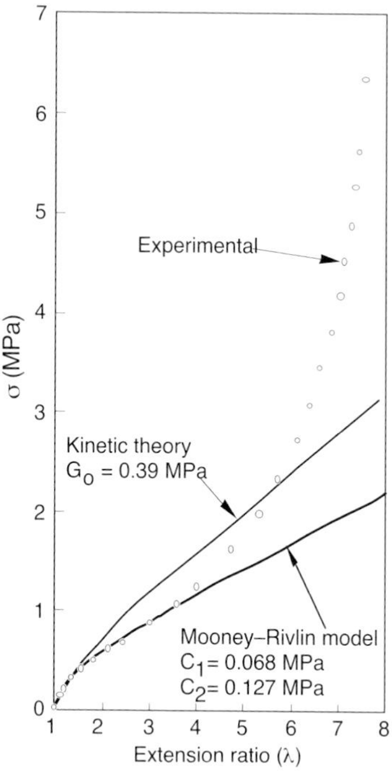

Figure 9.4 Comparison of theoretical and experimental stress-extension curves for natural rubber.

A more complex model representing the deformation behavior of elastomers in the region in which the stress-strain curve is reversible is the *Mooney-Rivlin* equation [5, 6] written as

$$\sigma = 2\left(\lambda - \frac{1}{\lambda^2}\right)\left(C_1 + \frac{C_2}{\lambda}\right) \tag{9.19}$$

which can be rearranged to give

$$\frac{\sigma}{2\left(\lambda - \frac{1}{\lambda^2}\right)} = C_1 + \frac{C_2}{\lambda} \tag{9.20}$$

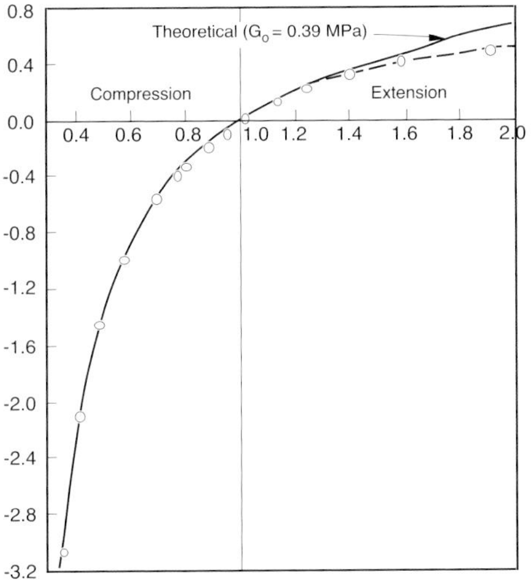

Figure 9.5 Experimental and theoretical stress-extension and compression curves for natural rubber.

A plot of of the *reduced stress*, $\sigma/2(\lambda - 1/\lambda^2)$, versus $1/\lambda$ is usually referred to as a *Mooney plot* and should be linear with a slope of $C_2$and an ordinate of $(C_1 + C_2)$at $1/\lambda = 1$. A typical Mooney plot is presented in Fig. 9.6 [7] for a natural rubber with different formulations and times of vulcanization.

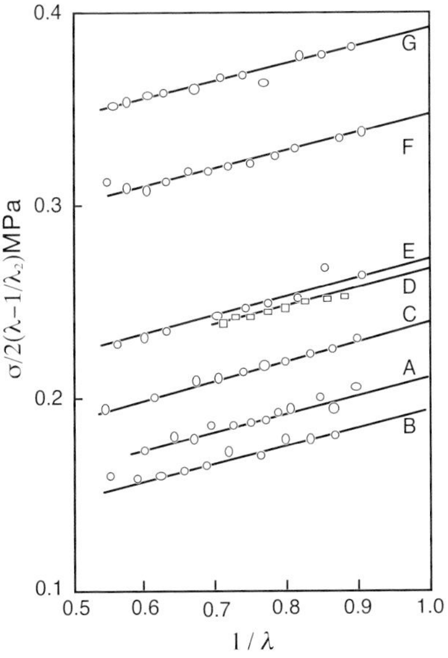

Figure 9.6 Mooney plots for rubber with various degrees of vulcanization.

The description of the various rubber formulations tested are presented in Table 9.1 [7]. It can be seen that C_2 shows little change, even with different rubber composition, and is approximately 0.1 Mpa. On the other hand, C_1 changes with degree of vulcanization and composition. A comparison between the Mooney-Rivlin model, the kinetic theory model, and experimental data for natural rubber is found in Fig. 9.4. For this material, the Mooney-Rivlin model represents the experimental data quite well up to extension ratios of 3.5.

Table 9.1 Compounding Details of the Vulcanizates Used by Gumbrell et al.[a]

Mix	A	B	C	D	E	F	G
Rubber	100	100	100	100	100	100	100
Sulfur	3.0	3.0	3.0	3.0	3.25	4.0	4.0
Zinc oxide	5.0	5.0	5.0	5.0	5.0	5.0	5.0
Stearic acid	1.0	1.0	1.0	1.0	1.0	1.0	1.0
Benzthiazyl disulphide	0.5	0.5	0.5	0.75		1.0	0.5
Mercaptobenz-thiazyl-disulphide		0.5	0.5	0.25		0.3	
Zinc dimethyl dithiocarbamate				0.1		0.15	
Diphinyl guanidine					1.25		1.0
Antioxidant	1.0	1.0	1.0	1.0	1.0	1.0	1.0
Time of vulcanization at 141.5 °C (min)	45	10	30	20	60	10	12

[a] Reference [7.]

Finally, it should be noted that the stiffness and strength of rubber is increased by filling with carbon black. The most common expression for describing the effect of carbon black content on the modulus of rubber was originally derived by Guth and Simha [8] for the viscosity of particle suspensions, and later used by Guth [9] to predict the modulus of filled polymers. The Guth equation can be written as

$$\frac{G_f}{G_0} = 1 + 2.5\phi + 14.1\phi^2 \qquad (9.21)$$

where G_f is the shear modulus of the filled material and ϕ the volume fraction of particulate filler. The above expression is compared to experiments [10, 11] in Fig. 9.7.

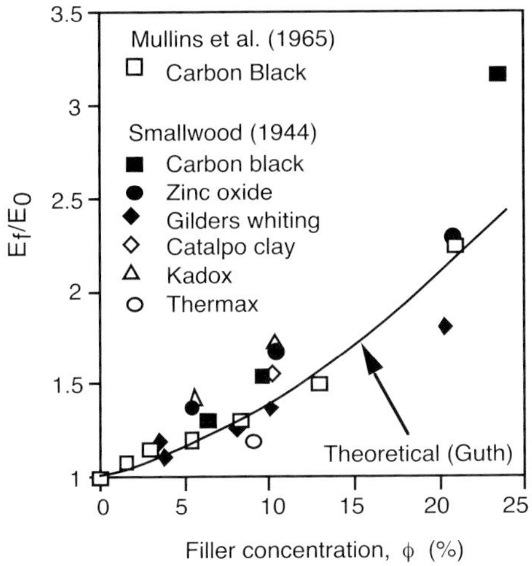

Figure 9.7 Effect of filler on modulus of natural rubber.

9.2.2 The Tensile Test and Thermoplastic Polymers

Of all the mechanical tests done on thermoplastic polymers, the tensile test is the least understood, and the results are often misinterpreted and misused. Since the test was inherited from other materials that have linear elastic stress-strain responses, it is often inappropriate for testing polymers. However, standardized tests such as DIN 53457 and ASTM D638 are available to evaluate the stress-strain behavior of polymeric materials.

The DIN 53457, for example, is performed at a constant elongational strain rate of 1% per minute, and the resulting data are used to determine the *short-term modulus*. The ASTM D638 test also uses one rate of deformation per material to measure the modulus; a slow speed for brittle materials and a fast speed for ductile ones. However, these tests do not reflect the actual rate of deformation experienced by the narrow portion of the test specimen, making it difficult to maintain a constant speed within the region of interest.

Extensive work was done by Knausenberger and Menges [12] where the rate of deformation of the test specimen is maintained constant. This is achieved by optically measuring the deformation on the specimen itself, as schematically demonstrated in Fig. 9.8 [12], and using that information as a feedback to control the elongational speed of the testing machine. The Knausenberger-Menges technique allows the testing engineer to measure the stress-strain response at various strain rates, where in each test the rate of deformation in the narrow section of the test specimen is accurately controlled. The resulting data can be used to determine the viscoelastic properties of polymers ranging from impact to long-term responses. A typical test performed on PMMA at various strain rates at room temperature is shown in Fig. 9.9. The increased curvature in the results with slow elongational speeds suggests that stress relaxation plays a significant role during the test.

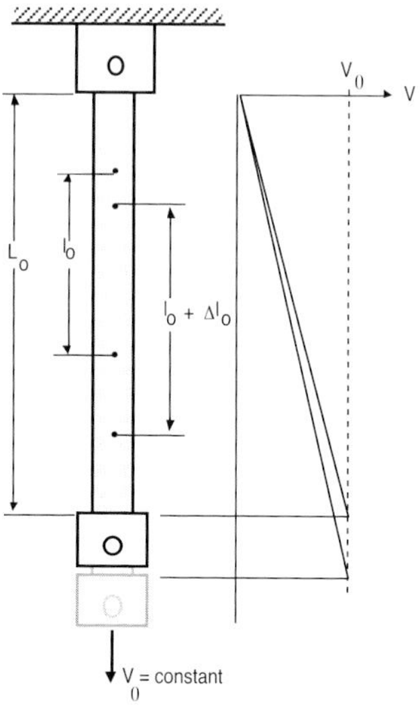

Figure 9.8 Flat tensile bar with velocity distribution.

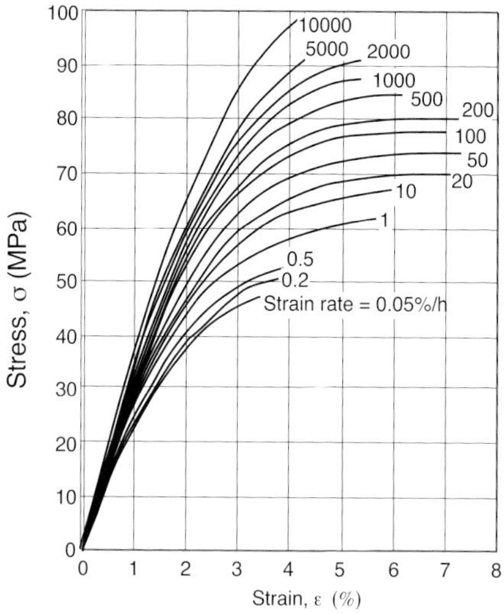

Figure 9.9 Stress-strain behavior of PMMA at various strain rates.

It can be shown that for small strains, or in linear viscoelasticity, the *secant modulus*, described by

$$E_s = \frac{\sigma}{\varepsilon} \tag{9.22}$$

and the tangent modulus, defined by

$$E_t = \frac{d\sigma}{d\varepsilon} \tag{9.23}$$

are independent of strain rate and are functions only of time and temperature. This is demonstrated in Fig. 9.10 [13]. The figure shows two stress-strain responses: one at a slow elongational strain rate, $\dot{\varepsilon}_1$, and one at twice the speed, defined by $\dot{\varepsilon}_2$. The tangent modulus at ε_1 in the curve with $\dot{\varepsilon}_1$ is identical to the tangent modulus at ε_2 in the curve with $\dot{\varepsilon}_2$, where ε_1 and ε_2 occurred at the same time. For small strains the tangent modulus, E_t, is identical to the relaxation modulus, E_r, measured with a stress relaxation test. This is important since the complex stress relaxation test can be replaced by the relatively simple short-term tensile test by plotting the tangent modulus versus time.

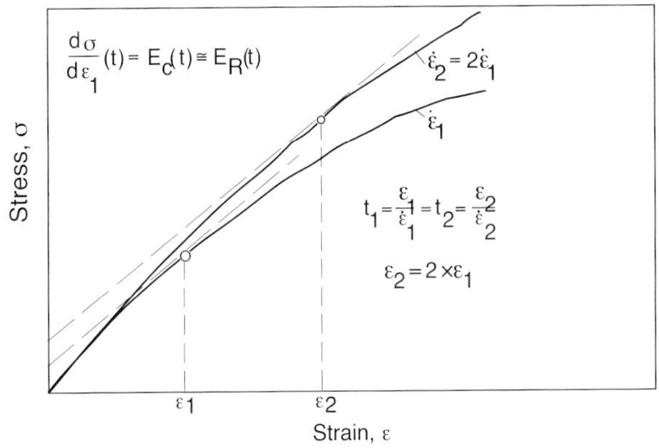

Figure 9.10 Schematic of the stress-strain behavior of a viscoelastic material at two rates of deformation.

Generic stress-strain curves and stiffness and compliance plots for amorphous and semi-crystalline thermoplastics are shown in Fig. 9.11 [14]. For amorphous thermoplastics one can usually approximate the stress-strain behavior in the curves of Fig. 9.11 by

$$\sigma(T,t) = E_0(T,t)\left(1 - D_1(T,t)\varepsilon\right)\varepsilon \tag{9.24}$$

and in a short-term test a semi-crystalline polymer would behave more like

$$\sigma(T,t) = E_0(T,t)\frac{\varepsilon}{1 + D_2(T,t)\varepsilon} \tag{9.25}$$

where E_0, D_1 and D_2 are time- and temperature-dependent material properties. However, below the glass transition temperature, the stress-strain curve of an amorphous polymer has a long and much steeper rise, with less relaxation effects, as shown in Fig. 9.10 and Eq. 9.24. In the stress-strain response of semi-crystalline polymers, on the other hand, the amorphous regions make themselves visible in long-term tensile tests. Hence, Eqs. 9.24 and 9.25 can be written in a more general form as

$$\sigma = E_0\varepsilon\frac{1 - D_1\varepsilon}{1 + D_2\varepsilon} \tag{9.26}$$

The coefficients in Eq. 9.26 can be determined for various rates of deformation. For example, the curves in Fig. 9.12 [15] show the coefficient E_0 for an amorphous unplasticized PVC measured at various strain rates, $\dot{\varepsilon}$, and temperatures. The curves in the figure suggest that there is a direct relationship between temperature and strain rate or time. It can be seen that the curves, separated by equal temperature differences, are shifted from each other at equal distances, $\log(a_T)$, where

$$a_T = \frac{\dot{\varepsilon}_{ref}}{\dot{\varepsilon}} \tag{9.27}$$

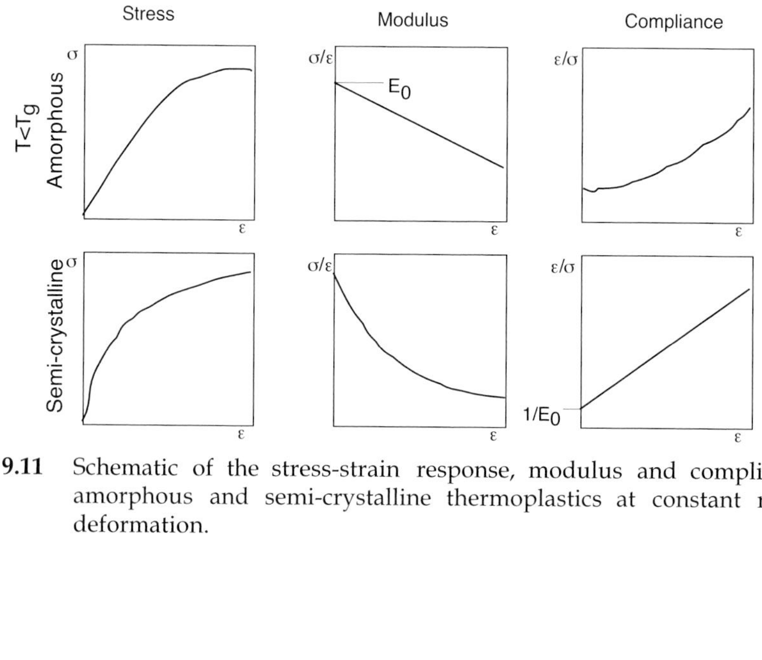

Figure 9.11 Schematic of the stress-strain response, modulus and compliance of amorphous and semi-crystalline thermoplastics at constant rates of deformation.

Since strain rate is directly related to time, one can make use of Arrhenius' relation between relaxation time, λ, and a reference relaxation time, λ_{ref}, stated by

$$\frac{\lambda}{\lambda_{ref}} = e^{-A/KT} \tag{9.28}$$

where A is the activation energy, T temperature and K a material property. The modified form of the Arrhenius equation for shifting the data can be written as

$$\log(a_T) = K\left(\frac{1}{T} - \frac{1}{T_{ref}}\right) \tag{9.29}$$

where T_{ref} is the reference temperature. The material constant K can be calculated by using data from Fig. 9.12, as shown in the sample graphical shift displayed in the figure. The coefficient K, which can be solved for by using

$$K = \frac{\log(\dot{\varepsilon}_{ref}/\dot{\varepsilon})}{1/T - 1/T_{ref}} \tag{9.30}$$

turns out to be 10,000 for the conditions shown in Fig. 9.12. This is true unless the test temperature is above the glass transition temperature, at which point the shift factor, a_T, and the coefficient K become functions of strain rate, as well as time and temperature. This is demonstrated for unplasticized PVC in Fig. 9.13 [15]. For the temperature range below T_g, displayed in Fig. 9.12, the data can easily be shifted, allowing the generation of a *master curve* at the reference temperature, T_{ref}. Figure 9.14 [15] shows such master curves for the three coefficients E_0, D_1 and D_2 in Eq. 9.26 for the amorphous PVC shown in Figs. 9.12 and 9.13. For comparison, Fig. 9.15 [15] shows E_0 and D_2 for a high density polyethylene at 23 °C as a function of strain rate.

The values of E_0, D_1 and D_2 can be easily calculated for each strain rate from the stress-strain diagram [16]. The modulus E_0 simply corresponds to the tangent modulus at small deformations where

$$\sigma \approx E_0\varepsilon \tag{9.31}$$

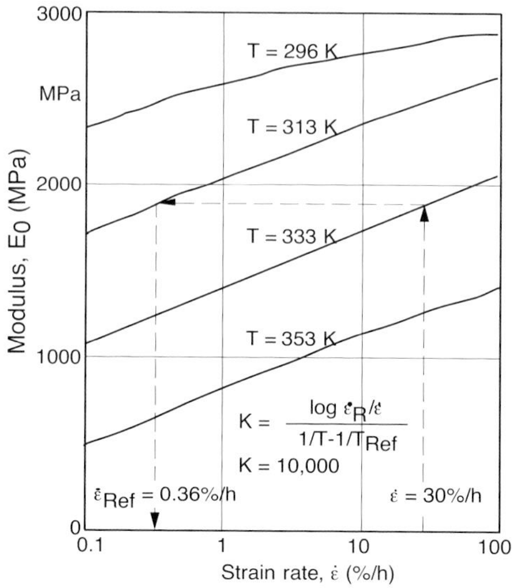

Figure 9.12 Plot of the elastic property E_0 and determining strain rate shift for an unplasticized PVC.

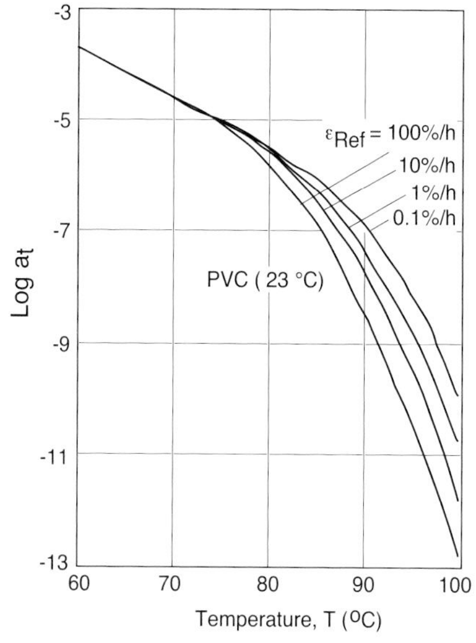

Figure 9.13 Time-temperature shift factor for an unplasticized PVC at several rates of deformation.

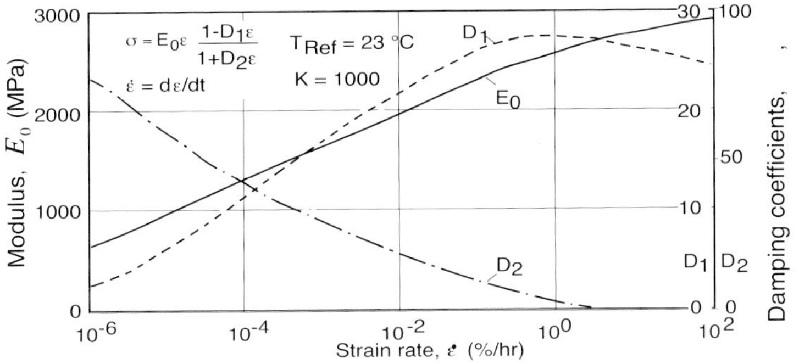

Figure 9.14 Master curves for E_0, D_1, and D_2 for an amorphous unplasticized PVC.

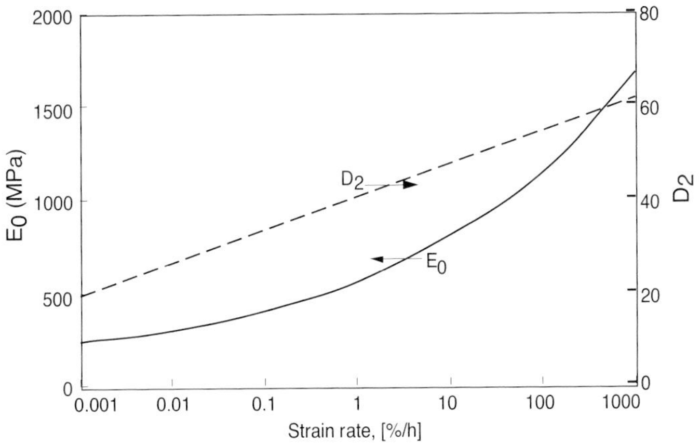

Figure 9.15 Coefficients E_0 and D_2 for a high density polyethylene at 23 °C.

Assuming that for amorphous thermoplastics $D_2 \approx 0$ when $T << T_g$ and for semi-crystalline thermoplastics $D_1 \approx 0$ when $T >> T_g$ we can compute D_1 from

$$D_1 = \frac{\sigma_2\varepsilon_1 - \sigma_1\varepsilon_2}{\sigma_2\varepsilon_1^2 - \sigma_1\varepsilon_2^2} \quad \text{and} \quad D_2 \quad \text{from} \tag{9.32}$$

$$D_2 = \frac{\sigma_1\varepsilon_2 - \sigma_2\varepsilon_1}{\varepsilon_1\varepsilon_2(\sigma_2 - \sigma_1)} \tag{9.33}$$

By introducing the values of E_0, D_1, and D_2, plotted in Fig. 9.14, into Eq. 9.26 one can generate long-term behavior curves as shown in the isochronous plots in Fig. 9.16 [15]. Here, the stress-strain behavior of unplasticized PVC is presented at constant times from 0.1 to 1000 hours of loading time. This is done by simply determining which strain rate results in a certain value of strain for a specific isochronous curve (time), then reading the values of E_0, D_1, and D_2 from the graphs and computing the corresponding stress using Eq. 9.26. When one uses this technique and furthermore applies the time-temperature superposition on short-term tests, one can approximate the long-term behavior of polymers which is usually measured using time consuming creep or relaxation tests.

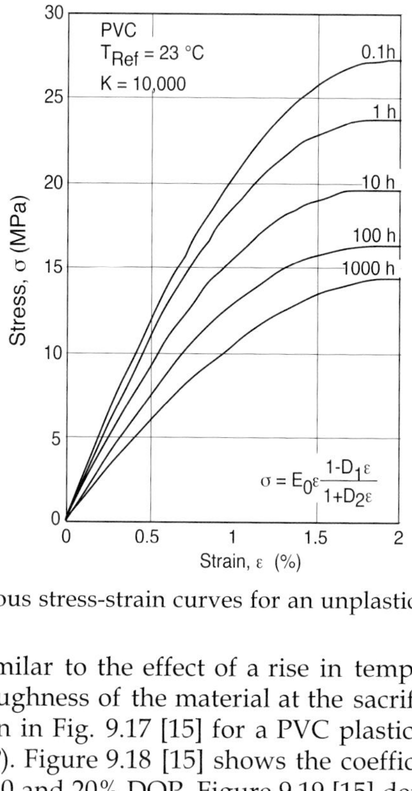

Figure 9.16 Isochronous stress-strain curves for an unplasticized PVC.

As a final note, similar to the effect of a rise in temperature, a solvent can increase the overall toughness of the material at the sacrifice of its strength and stiffness. This is shown in Fig. 9.17 [15] for a PVC plasticized with 10 and 20% dioctylphthalate (DOP). Figure 9.18 [15] shows the coefficients E_0 and D_2 for a PVC plasticized with 10 and 20% DOP. Figure 9.19 [15] demonstrates the similar effects of temperature and plasticizer.

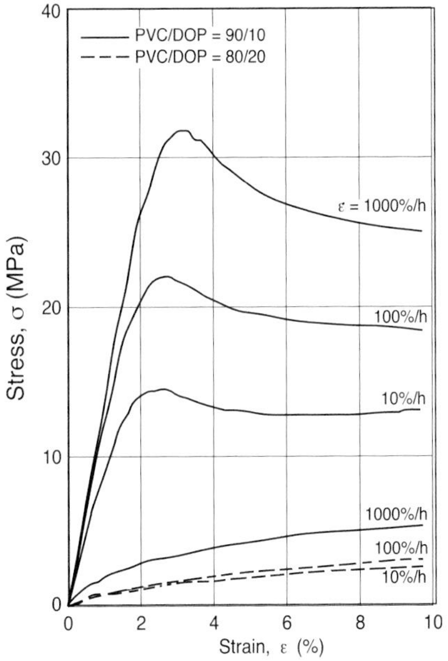

Figure 9.17 Stress-strain responses at various rates of deformation for a plasticized PVC with two plasticizer (DOP) concentrations.

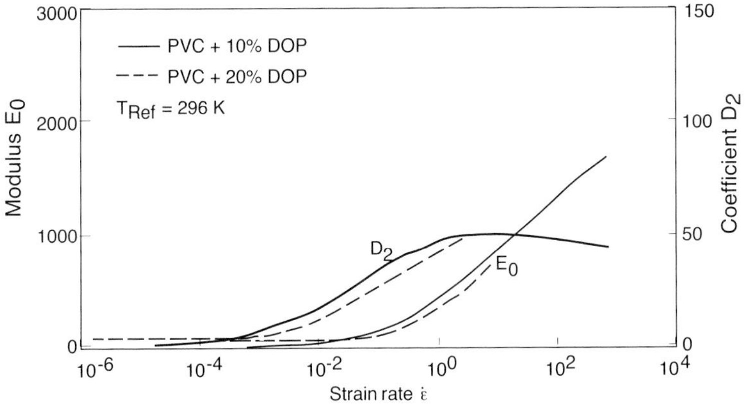

Figure 9.18 Coefficients E_0 and D_2 for polyvinyl chloride plasticized with 10% and 20% DOP.

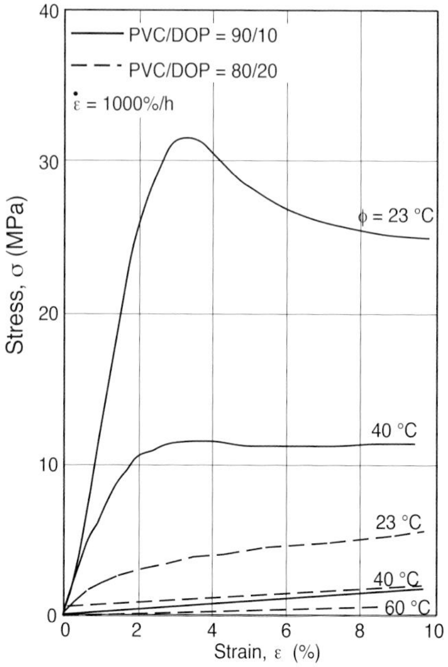

Figure 9.19 Stress-strain responses at various temperatures for a plasticized PVC with two plasticizer (DOP) concentrations.

9.3 Long-Term Tests

The stress relaxation and the creep test are well-known long-term tests. The stress relaxation test, discussed in Chapter 2, is difficult to perform and is therefore often approximated by data acquired through the more commonly used *creep test*. The stress relaxation of a polymer is often thought of as the inverse of creep.

The creep test, which can be performed either in shear, compression, or tension, measures the flow of a polymer component under a constant load. It is a common test that measures the strain, ε, as a function of stress, time, and temperature. Standard creep tests such as DIN 53 444 and ASTM D2990 can be used. Creep tests are performed at a constant temperature using a range of applied stress, as shown in Fig. 9.20 [17], where the creep responses of a polypropylene copolymer are presented for a range of stresses in a graph with a log scale for time. When plotting creep data in a log-log graph, in the majority of the cases, the creep curves reduce to straight lines as shown for polypropylene in Fig. 9.21 [1]. Hence, the creep behavior of most polymers can be approximated with a power-law model represented by

$$\varepsilon(t) = M(\sigma, T)t^n \tag{9.34}$$

where M and n are material-dependent properties.

Similar to the stress relaxation test, the creep behavior of a polymer depends heavily on the material temperature during testing, having the highest rates of deformation around the glass transition temperature. This is demonstrated in Fig. 9.22 [18], which presents the creep compliance of plasticized PVC.

Creep data is very often presented in terms of creep modulus, E_c, defined by

$$E_c = \frac{\sigma_0}{\varepsilon(t)} \tag{9.35}$$

Figure 9.23 [19] presents the creep modulus for various materials as a function of time.

Depending on the time scale of the experiment, a property that also varies considerably during testing is Poisson's ratio, ν. Figure 9.24 [14] shows Poisson's ratio for PMMA deformed at rates (%/h) between 10^{-2} (creep) and 10^3 (impact). The limits are ν = 0.5 (fluid) for high temperatures or very slow deformation speeds and ν = 0.33 (solid) at low temperatures or high deformation speeds.

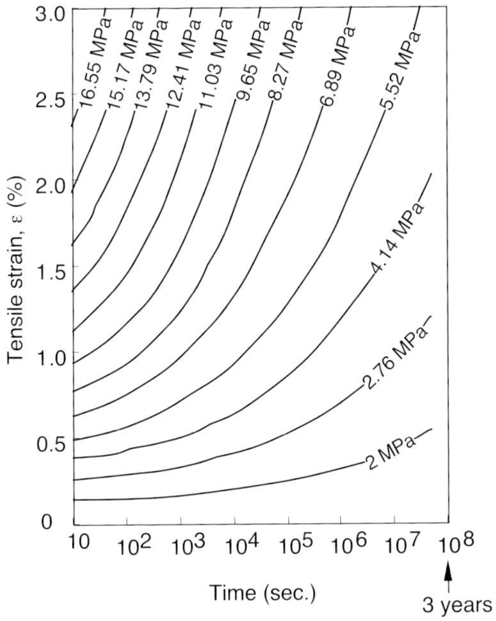

Figure 9.20 Creep response of a propylene-ethylene copolymer at 20 °C.

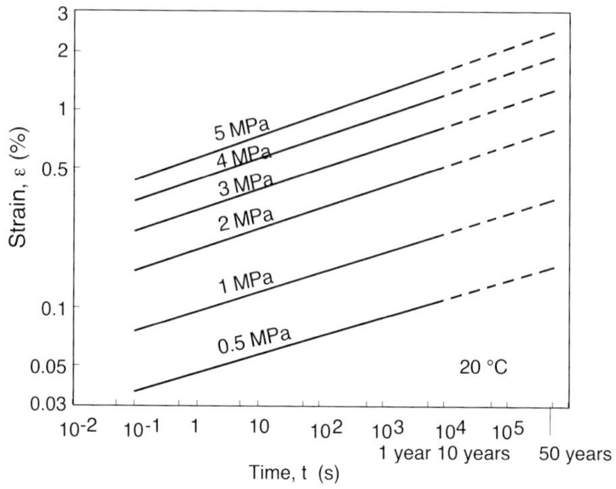

Figure 9.21 Creep response of a polypropylene plotted on a log-log scale.

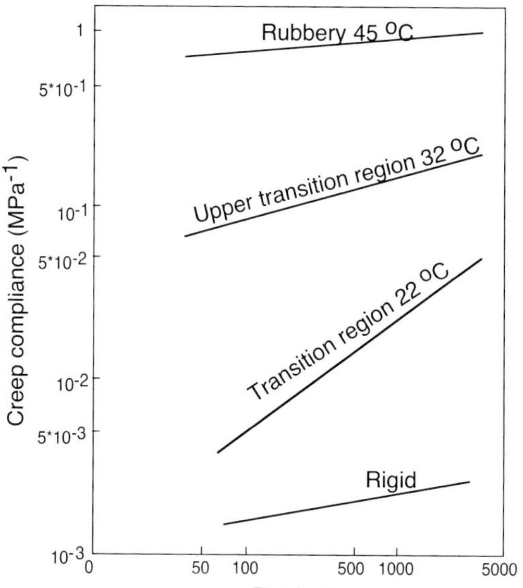

Figure 9.22 Creep compliance of a plasticized PVC at different temperatures.

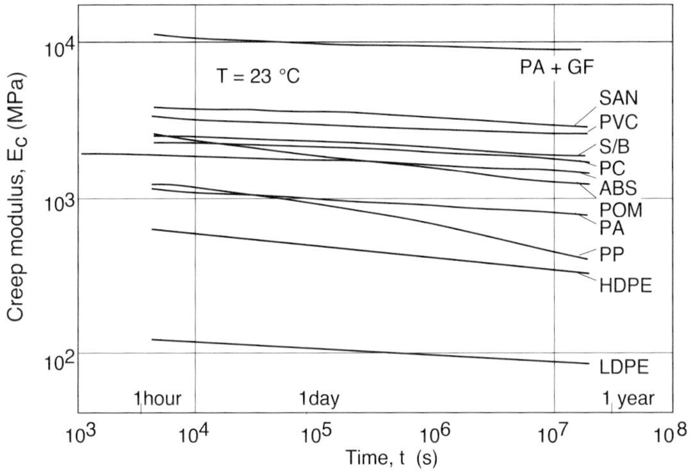

Figure 9.23 Time dependence of creep moduli for several polymer.

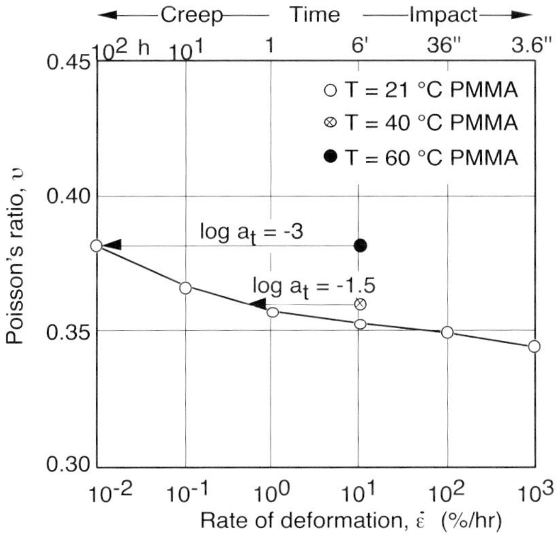

Figure 9.24 Poisson's ratio as a function of rate of deformation for PMMA at various temperatures.

9.3.1 Isochronous and Isometric Creep Plots

Typical creep test data, as shown in Fig. 9.20, can be manipulated to be displayed as short-term stress-strain tests or as stress relaxation tests. These manipulated creep-test-data curves are called *isochronous* and *isometric* graphs.

An isochronous plot of the creep data is generated by cutting sections through the creep curves at constant times and plotting the stress as a function of strain. The isochronous curves of the creep data displayed in Fig. 9.20 are presented in Fig. 9.25 [17]. Similar curves can also be generated by performing a series of *short creep tests*, where a specimen is loaded at a specific stress for a short period of time, typically around 100 s [20]. The load is then removed, and the specimen is allowed to relax for a period of 4 times greater than the time of the creep test. The specimen is then reloaded at a different stress, and the test is repeated until a sufficient number of points exists to plot an isochronous graph. This procedure is less time-consuming than the regular creep test and is often used to predict the short-term behavior of polymers. However, it should be pointed out that the short-term tests described in Section 9.2.2 are more accurate and are less time consuming and cheaper to perform.

The isometric or "equal size" plots of the creep data are generated by taking constant strain sections of the creep curves and by plotting the stress as a function of time. Isometric curves of the polypropylene creep data presented in Fig. 9.20 are shown in Fig. 9.26 [17]. This plot resembles the stress relaxation test results and is often used in the same manner. When we divide the stress axis by the strain, we can also plot the modulus versus time.

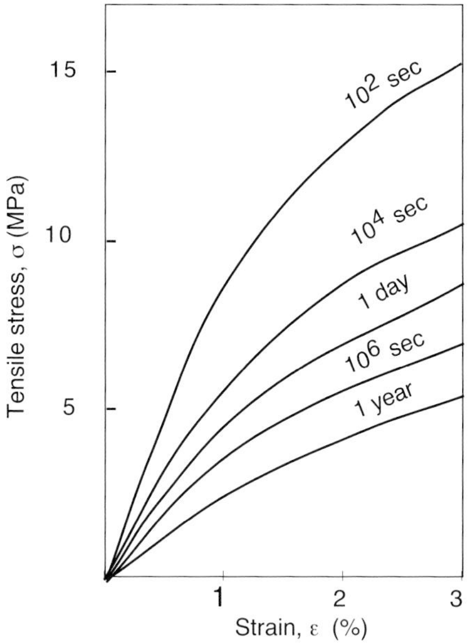

Figure 9.25 Isochronous stress-strain curves for the propylene-ethylene copolymer creep responses shown in Fig. 9.20.

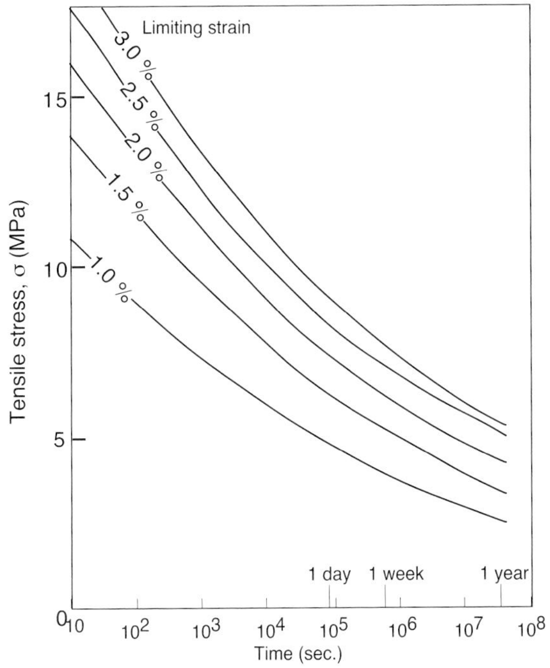

Figure 9.26 Isometric stress-time curves for the propylene-ethylene copolymer creep responses shown in Fig. 9.20.

9.4 Dynamic Mechanical Tests

9.4.1 Torsion Pendulum

The simplest dynamic mechanical test is the torsion pendulum. The standard procedure for the torsional pendulum, schematically shown in Fig. 9.27 [21], is described in DIN 53445 and ASTM D2236. The technique is applicable to virtually all plastics, through a wide range of temperatures; from the temperature of liquid nitrogen, -180 °C to 50-80 °C above the glass transition temperature in amorphous thermoplastics and up to the melting temperature in semi-crystalline thermoplastics. With thermoset polymers one can apply torsional tests up to the degradation temperatures of the material.

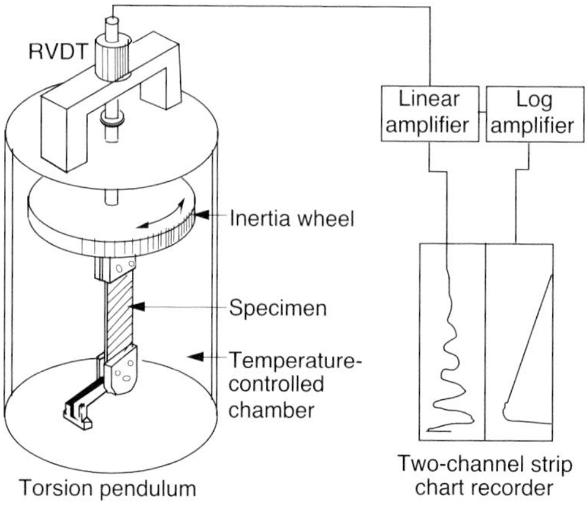

Figure 9.27 Schematic diagram of the torsion pendulum test equipment.

The torsion pendulum apparatus is made up of an inertia wheel, grips, and the specimen contained in a temperature-controlled chamber. The rectangular test specimen can be cut from a polymer sheet or part, or it can be made by injection molding. To execute the test the inertia wheel is deflected, then released and allowed to oscillate freely. The angular displacement or twist of the specimen is recorded over time. The frequency of the oscillations is directly related to the elastic shear modulus of the specimen, G', and the decay of the amplitude is related to the damping or *logarithmic decrement*, Δ, of the material. The elastic shear modulus (Pascals) can be computed using the relation[2]

$$G' = \frac{6.4\pi^2 ILf^2}{\mu bt^3} \tag{9.36}$$

where I is the polar moment of inertia (g/cm^2), L the specimen length (cm), f the frequency (Hz), b the width of the specimen, t the thickness of the specimen, and μ a shape factor which depends on the width-to-thickness ratio. Typical values of μ are listed in Table 9.2 [18]. The logarithmic decrement can be computed using

$$\Delta = \ln\left(\frac{A_n}{A_{n+1}}\right) \tag{9.37}$$

where A_n represents the amplitude of the nth oscillation.[3] Although the elastic shear modulus, G', and the logarithmic decrement, Δ, are sufficient to characterize a material, one can also compute the loss modulus G'' by using

2 For more detail, please consult ASTM D2236.

3 When $\Delta > 1$, a correction factor must be used to compute G'. See ASTM D2236.

$$G'' = \left(\frac{G'\Delta}{\pi} \right) \tag{9.38}$$

Table 9.2 Shape Factor μ for Various Rectangular Cross-Sections

Ratio of specimen width to thickness	μ
1.0	2.249
1.2	2.658
1.4	2.990
1.6	3.250
1.8	3.479
2.0	3.659
2.5	3.990
3.0	4.213
4.0	4.493
5.0	4.662
10.0	4.997
50.0	5.266
∞	5.333

The logarithmic decrement can also be written in terms of *loss tangent*, tan δ, where δ is the out-of-phase angle between the strain and stress responses. The loss tangent is defined as

$$\tan\delta = \frac{G''}{G'} = \frac{\Delta}{\pi} \tag{9.39}$$

Figures 9.28 [1] and 9.29 [1] show the elastic shear modulus and the loss tangent for high impact polystyrene, and various polypropylene grades, respectively. In the graph for high impact polystyrene, the glass transition temperatures for polystyrene at 120 °C and for butadiene at -50 °C, are visible. For the polypropylene grades, the glass transition temperatures and the melting temperatures can be seen. The vertical scale in plots such as Figs. 9.28 and 9.29 is usually a logarithmic scale. However, a linear scale better describes the mechanical behavior of polymers in design aspects. Figure 9.30 [1] presents the elastic shear modulus on a linear scale for several thermoplastic polymers as a function of temperature.

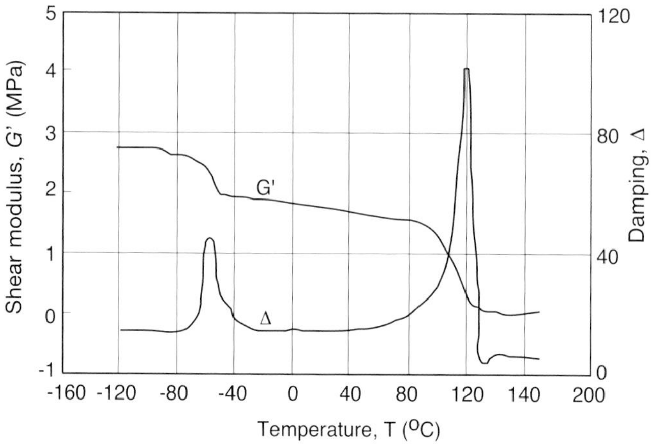

Figure 9.28 Elastic shear modulus and loss tangent for HIPS.

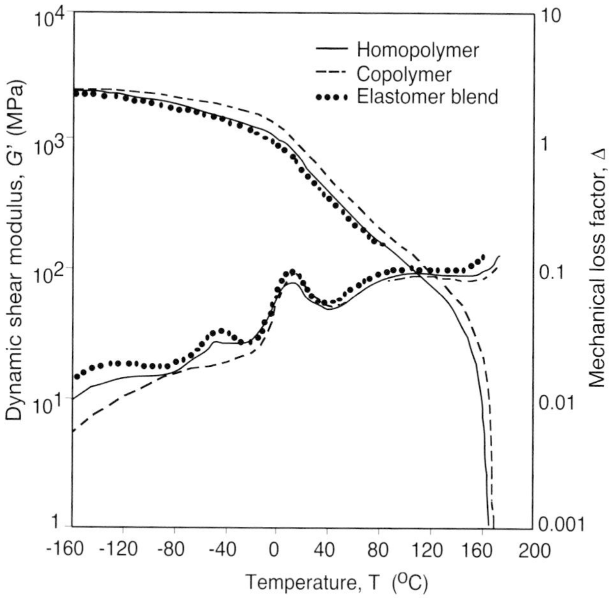

Figure 9.29 Elastic shear modulus and loss tangent for various polypropylene grades.

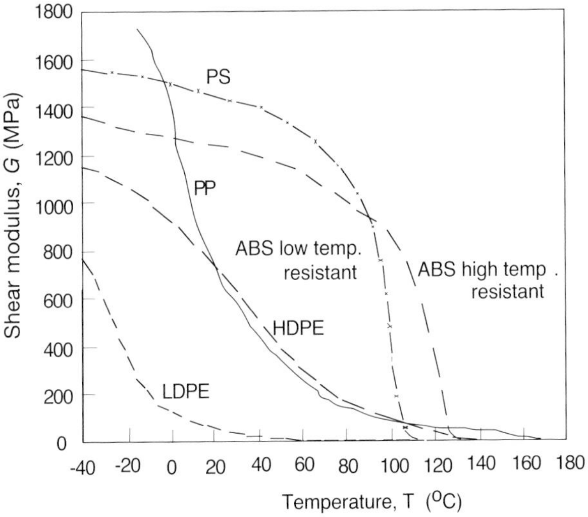

Figure 9.30 Elastic shear modulus for several thermoplastics.

9.4.2 Sinusoidal Oscillatory Test

In the sinusoidal oscillatory test, a specimen is excited with a low frequency stress input which is recorded along with the strain response. The shapes of the test specimen and the testing procedure vary significantly from test to test. The various tests and their corresponding specimens are described by ASTM D4065 and the terminology, such as the one already used in Eqs. 9.36–9.39, is described by ASTM D4092.

If the test specimen in a sinusoidal oscillatory test is perfectly elastic, the stress input and strain response would be as follows:

$$\tau(t) = \tau_0 \cos \omega t \tag{9.40}$$

$$\gamma(t) = \gamma_0 \cos \omega t \tag{9.41}$$

For an ideally viscous test specimen, the strain response would lag $\pi/2$ radians behind the stress input:

$$\tau(t) = \tau_0 \cos \omega t \tag{9.42}$$

$$\gamma(t) = \gamma_0 \cos\left(\omega t - \frac{\pi}{2}\right) \tag{9.43}$$

Polymers behave somewhere in between the perfectly elastic and the perfectly viscous materials and their response is described by

$$\tau(t) = \tau_0 \cos \omega t \quad \text{and} \tag{9.44}$$

$$\gamma(t) = \gamma_0 \cos(\omega t - \delta) \tag{9.45}$$

The shear modulus takes a complex form of

$$G^* = \frac{\tau(t)}{\gamma(t)} = \frac{\tau_0 e^{i\delta}}{\gamma_0} = \frac{\tau_0}{\gamma_0}(\cos\delta + \mathbf{i}\sin\delta) = G' + G'' \tag{9.46}$$

which is graphically represented in Fig. 9.31. G' is usually referred to as *storage modulus* and G'' as *loss modulus*. The ratio of loss modulus to storage modulus is referred to as *loss tangent:*

$$\tan\delta = \frac{G''}{G'} \tag{9.47}$$

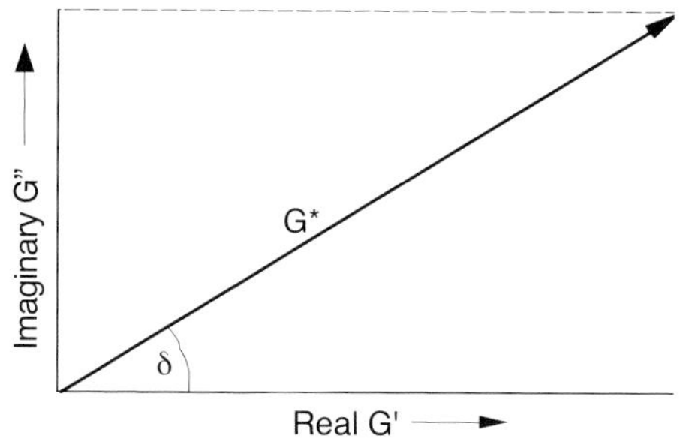

Figure 9.31 Vector representation of the complex shear modulus.

9.5 Viscoelastic Behavior of Polymers

Linear viscoelastic models were already introduced in Chapter 2. The main assumptions made in linear viscoelasticity is that the deformations must be small and that various loadings at different times are simply superimposed on one another as stated by *Boltzmann's superposition principle.* Several models exist to simulate the linear viscoelastic behavior of polymers. These physical models are generally composed of one or several elements such as dashpots, springs, or friction elements that represent viscous, elastic, or yielding properties, respectively. All models must satisfy the momentum balance and continuity or deformation equation, along with the appropriate constitutive laws. The most commonly used constitutive equations are the viscous or *Newtonian model*, which is written as

$$\sigma = \eta \dot{\varepsilon} \tag{9.48}$$

and the linear elastic or *Hookean model*, which is represented by

$$\sigma = E\varepsilon \tag{9.49}$$

where η and E are the viscosity and Young's modulus, respectively.

9.5.1 Kelvin Model

The Kelvin model, sometimes also called the Kelvin-Voigt model, is shown in Fig. 9.32. It is the simplest model that can be used to represent the behavior of a solid polymer component at the beginning of loading.

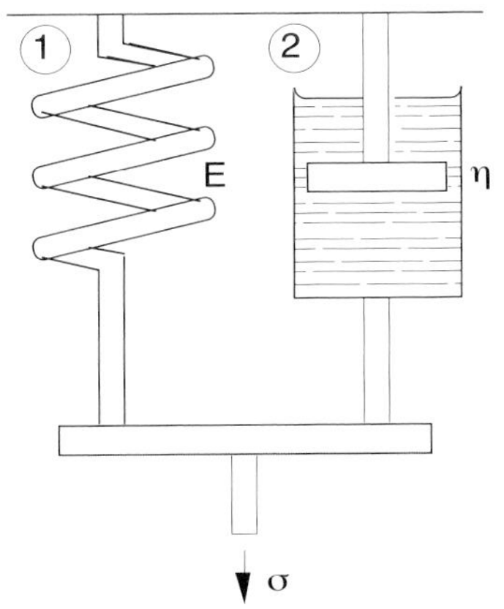

Figure 9.32 Schematic diagram of the Kelvin model.

The momentum balance for the Kelvin model is stated as

$$\sigma = \sigma_1 + \sigma_2 \tag{9.50}$$

and the continuity equation is represented by

$$\varepsilon = \varepsilon_1 = \varepsilon_2 \tag{9.51}$$

Using Eq. 9.51 with the constitutive relations in Eqs. 9.48 and 9.49, the governing equation, Eq. 9.50, can be rewritten as

$$\sigma = E\varepsilon + \eta\dot{\varepsilon} \tag{9.52}$$

9.5.1.1 Kelvin Model Creep Response

Using Eq. 9.52, the strain in a creep test in the Kelvin model can be solved for as

$$\varepsilon(t) = \frac{\sigma_0}{E}\left(-e^{-t/\lambda}\right) \tag{9.53}$$

where λ, (E/η), is the relaxation time. The creep modulus is defined as

$$E_c(t) = \frac{E}{\left(1 - e^{-t/\lambda}\right)} \tag{9.54}$$

The creep response of the Kelvin model is shown in Fig. 9.33.

9.5.1.2 Kelvin Model Stress Relaxation

In the Kelvin model the stress does not relax and remains constant at

$$\sigma = E\varepsilon_0 \tag{9.55}$$

This is shown in Fig. 9.33.

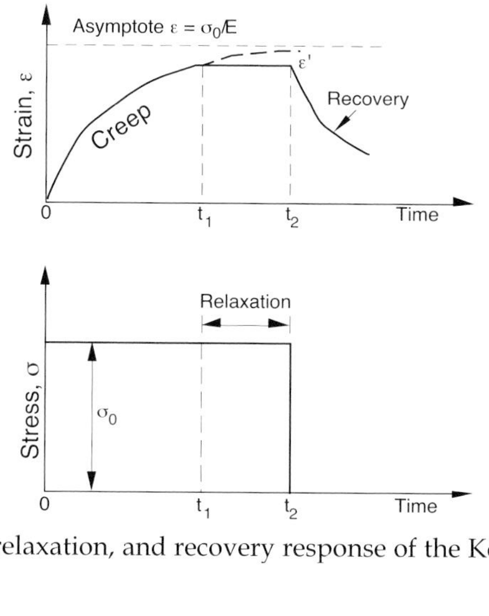

Figure 9.33 Creep, relaxation, and recovery response of the Kelvin model.

9.5.1.3 Kelvin Model Strain Recovery

Since the stresses do not relax in a Kelvin model, the full shape of the original component or specimen can be recovered. The strain recovery response can be written as

$$\varepsilon(t) = \varepsilon_0 e^{-t/\lambda} \tag{9.56}$$

and is represented in Fig. 9.33.

9.5.1.4 Kelvin Model Dynamic Response

We can also consider the response of a Kelvin model subjected to a sinusoidal strain given by

$$\varepsilon(t) = \varepsilon_0 \sin(\omega t) \tag{9.57}$$

where ε_0 is the strain amplitude and ω is the frequency. Differentiating Eq. 9.57 and substituting into Eq. 9.52 results in

$$\sigma(t) = E\varepsilon_0 \sin(\omega t) + \eta\omega\varepsilon_0 \cos(\omega t) \tag{9.58}$$

Dividing Eq. 9.58 by the strain amplitude results in the *complex modulus*, which is formed by the *storage modulus* defined by

$$E' = E \tag{9.59}$$

and the loss modulus given by

$$E'' = \eta\omega \tag{9.60}$$

9.5.2 Jeffrey Model

As shown in Fig. 9.34, the Jeffrey model is a Kelvin model with a dashpot. This extra feature adds the missing long-term creep to the Kelvin model.

The momentum balance of the Jeffrey model is represented by two equations as

$$\sigma = \sigma_3 \quad \text{and} \tag{9.61}$$

$$\sigma = \sigma_1 + \sigma_2 \tag{9.62}$$

as is the continuity equation by

$$\varepsilon_1 = \varepsilon_2 \quad \text{and} \tag{9.63}$$

$$\varepsilon = \varepsilon_2 + \varepsilon_3 \tag{9.64}$$

Combining Eq. 9.61–9.64 and applying the constitutive equations gives

$$\sigma + \left(\frac{\eta_1 + \eta_3}{E_2}\right)\dot{\sigma} = \eta_3\dot{\varepsilon} + \left(\frac{\eta_3\eta_1}{E_2}\right)\ddot{\varepsilon} \tag{9.65}$$

which is sometimes written as

$$\sigma + \lambda_1\dot{\sigma} = \eta_0\left(\dot{\varepsilon} + \lambda_2\ddot{\varepsilon}\right) \tag{9.66}$$

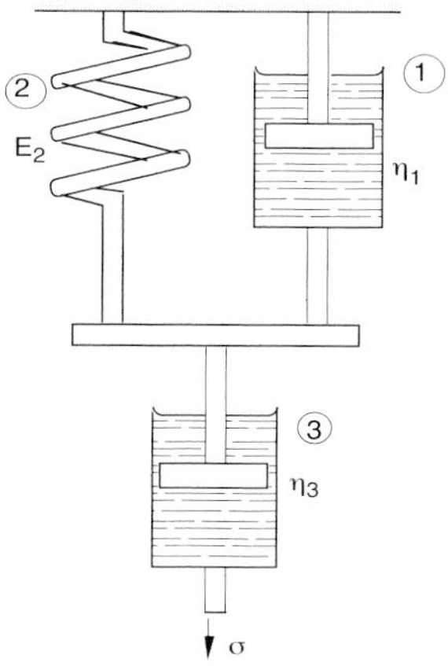

Figure 9.34 Schematic diagram of the Jeffrey model.

9.5.2.1 Jeffrey Model Creep Response

Using Eq. 9.66, the strain in a creep test in the Jeffrey model can be solved for as

$$\varepsilon(t) = \frac{\sigma_0}{E}\left(1 - e^{-t/\lambda_2}\right) + \frac{\sigma_0}{\eta_0}t \tag{9.67}$$

and is depicted in Fig. 9.35. The creep modulus of the Jeffrey model is written as

$$E_c(t) = \left(\frac{\left(1 - e^{-t/\lambda_2}\right)}{E} + \frac{t}{\eta_3}\right)^{-1} \tag{9.68}$$

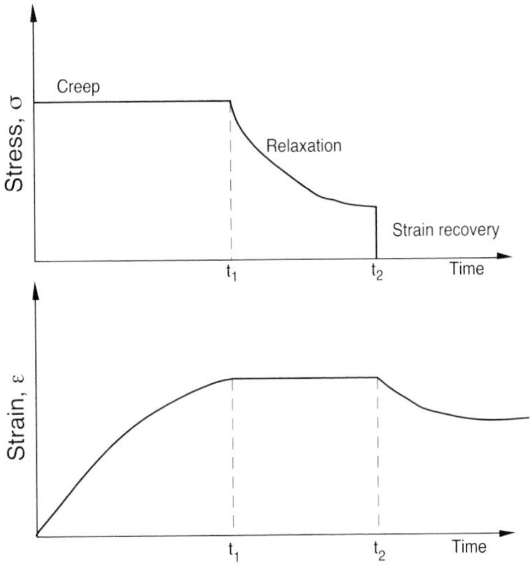

Figure 9.35 Creep, relaxation and recovery response of the Jeffrey model.

9.5.2.2 Jeffrey Model Stress Relaxation

The stress relaxation of the Jeffrey model is derived from the governing equation, Eq. 9.66 as

$$\sigma = \sigma_0 e^{-t/\lambda_1} \tag{9.69}$$

and is also represented in Fig. 9.35.

9.5.2.3 Jeffrey Model Strain Recovery

The unrelaxed stress is recovered in the same way as in the Kelvin model

$$\varepsilon(t) = \varepsilon_0 e^{-t/\lambda} \tag{9.70}$$

9.5.3 Standard Linear Solid Model

The standard linear solid model, shown in Fig. 9.36, is a commonly used model to simulate the short-term behavior of solid polymer components. The momentum balance of the standard linear solid model is expressed with two equations as

$$\sigma = \sigma_1 + \sigma_2 \tag{9.71}$$

and

$$\sigma_1 = \sigma_3 \tag{9.72}$$

Continuity or deformation is represented with

$$\varepsilon = \varepsilon_1 + \varepsilon_3 \tag{9.73}$$

and

$$\varepsilon = \varepsilon_2 \tag{9.74}$$

When we combine Eqs. 9.71–9.74 and use the constitutive equations for the spring and dashpot elements, we get the governing equation for the standard linear solid model:

$$\eta\dot{\sigma} + E_1\sigma = \eta(E_1 + E_2)\dot{\varepsilon} + E_1E_2\varepsilon \tag{9.75}$$

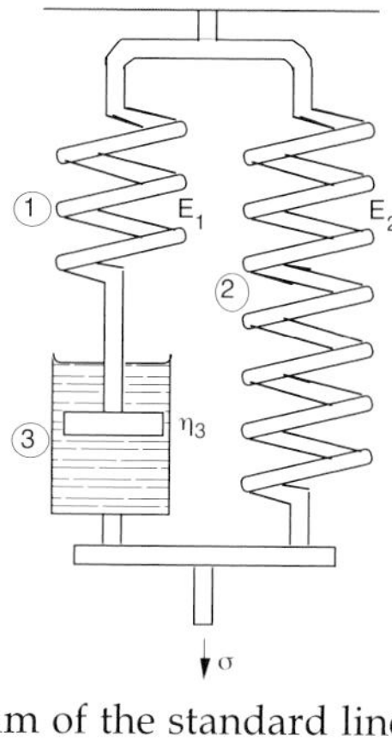

Figure 9.36 Schematic diagram of the standard linear solid model.

9.5.3.1 Standard Linear Solid Model Creep Response

Using Eq. 9.75, the strain in a creep test in the standard linear solid model can be solved for as

$$\varepsilon = \frac{\sigma_0}{E_2} + \left(\frac{\sigma_0}{E_1 + E_2} - \frac{\sigma_0}{E_2}\right)e^{-(E_1E_2/\eta(E_1+E_2))t} \tag{9.76}$$

and is plotted in Fig. 9.37.

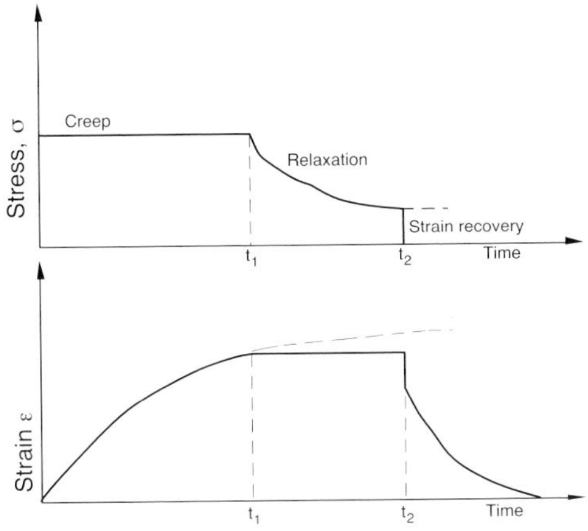

Figure 9.37 Creep, relaxation, and recovery response of the standard linear solid model.

9.5.3.2 Standard Linear Solid Model Stress Relaxation

The stress relaxation of the standard linear solid model can be derived by integrating Eq. 9.75 and is represented by

$$\sigma = \varepsilon_0 \left(E_2 + E_1\right) e^{-(E_1/\eta)t} \tag{9.77}$$

9.5.4 Maxwell-Wiechert Model

The Maxwell-Wiechert model, shown in Fig. 9.38, is a generalized model which consists of an arbitrary number of Maxwell models connected in parallel.

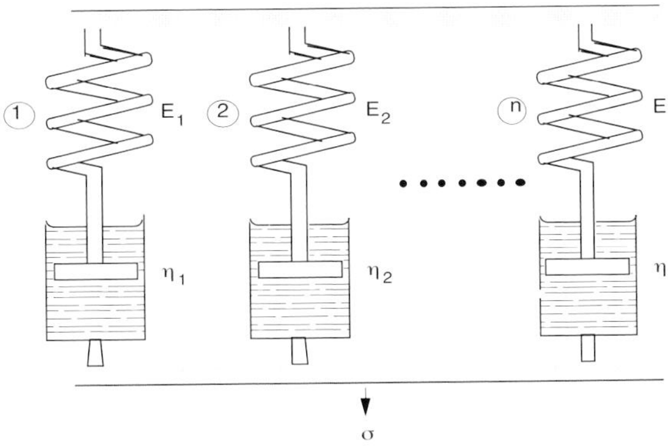

Figure 9.38 Schematic diagram of the Maxwell-Wiechert model.

The momentum balance in the ith Maxwell element of the Maxwell-Wiechert model is expressed as

$$\sigma_1 = \sigma_{i1} + \sigma_{i2} \tag{9.78}$$

and the full momentum balance for a model with n elements is written as

$$\sigma = \sum_{i=1}^{n} \sigma_i \tag{9.79}$$

Continuity or deformation for the ith Maxwell element is expressed as

$$\varepsilon_i = \varepsilon_{i1} + \varepsilon_{i2} \tag{9.80}$$

and for the full model

$$\varepsilon = \varepsilon_1 = \varepsilon_2 = \varepsilon_i ... \tag{9.81}$$

The governing equation for the Maxwell-Wiechert model is written as

$$\dot{\varepsilon} = \frac{\dot{\sigma}_i}{E_i} + \frac{\sigma_i}{\eta_i} \tag{9.82}$$

9.5.4.1 Maxwell-Wiechert Model Stress Relaxation

The stress relaxation of the Maxwell-Wiechert model can be derived by integrating Eq. 9.82 and substituting the resulting stress into Eq. 9.79. Dividing by the applied strain ε_0 results in an expression for the relaxation model which is written as

$$E(t) = \sum_{i=1}^{n} E_i e^{-(1/\lambda_i)t} \tag{9.83}$$

which represents a model with n relaxation times and where $\lambda_i = \eta_i / E_i$. As an example, we can approximate the relaxation behavior of the polyisobutylene shown in Fig. 2.29 by using a Maxwell-Wiechert model having two Maxwell elements with $\lambda_1 = 10^{-8}$ h and λ_2=100 h, and E_1 = 3x10^9 Pa and E_2=10^6 Pa. Figure 9.39 compares the experimental relaxation modulus with the model. One can see that although there are big differences between the two curves, the model, with its two relaxation times, does at least qualitatively represent the experimental values.

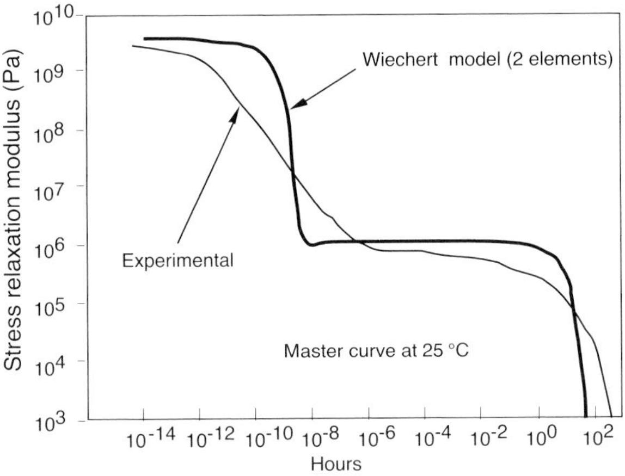

Figure 9.39 Comparison of the experimental stress relaxation for polyisobutylene to a two-component Maxwell-Wiechert model response.

9.5.4.2 Maxwell-Wiechert Model Dynamic Response

We can also consider the response of a Maxwell-Wiechert model subjected to a sinusoidal strain given by Eq. 9.57. In a similar analysis to that presented for the Kelvin model and for the Maxwell model in Chapter 2, the *storage modulus* is given by

$$E' = \sum_{i=1}^{n} \left(\frac{E_i(\omega_i \lambda_i)^2}{1 + (\omega_i \lambda_i)^2} \right) \tag{9.84}$$

and the viscous term or the *loss modulus*, is given by

$$E'' = \sum_{i=1}^{n} \left(\frac{E_i \omega_i \lambda_i}{1 + (\omega_i \lambda_i)^2} \right) \tag{9.85}$$

9.6 Effects of Structure and Composition on Mechanical Properties

The shear modulus versus temperature diagram is a very useful description of the mechanical behavior of certain materials. It is possible to generate a general shear modulus versus temperature diagram for all polymers by using a reduced temperature described by

$$T_{red} = \frac{293K}{T_g} \tag{9.86}$$

Fig. 9.40 [22] shows this diagram with the shear modulus of several polymers. The upper left side of the curve represents the stiff and brittle cross-linked materials, and the upper right side represents the semi-crystalline thermoplastics whose glass transition temperature is below room temperature. The lower right side of the curve represents elastomers, positioned accordingly on the curve depending on their degree of cross-linkage.

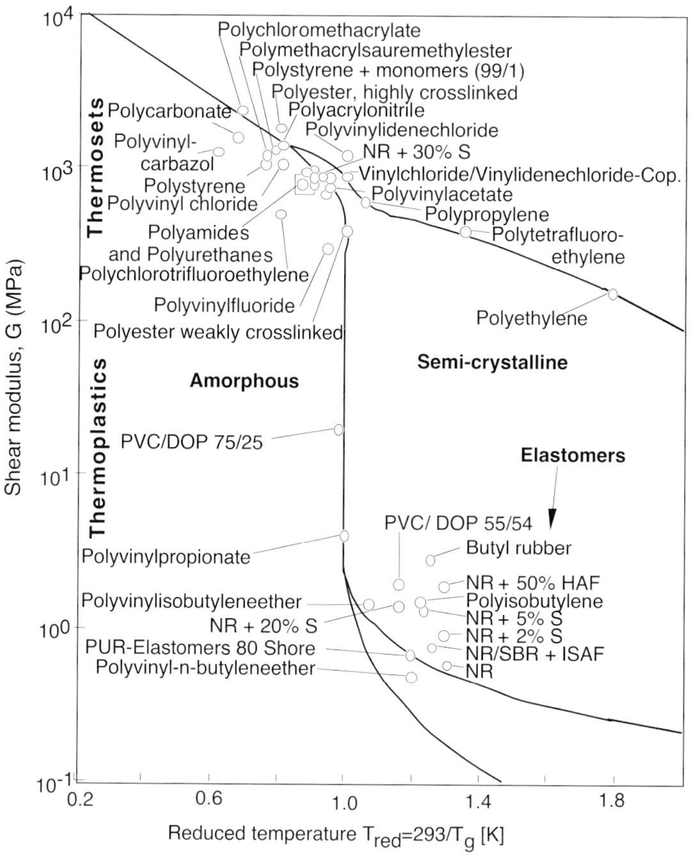

Figure 9.40 Shear modulus of several polymers as a function of reduced glass transition temperature.

9.6.1 Amorphous Thermoplastics

Amorphous thermoplastics exhibit "useful" behavior below their glass transition temperature. Figure 9.41 shows the shear modulus of an unplasticized PVC with respect to temperature. In this figure we observe that the material solidifies at the glass transition temperature, between 80 and 90 °C. We note that one cannot exactly pinpoint T_g but only a range within which it will occur. In fact, already at 60 °C the stiffness dramatically drops as the U-PVC starts to soften. Below -10 °C, the U-PVC becomes very stiff and brittle, making it useful only between -10 °C and 60 °C for most applications. As mentioned before, the properties of thermoplastics can be modified by adding plasticizing agents. This is shown for PVC in Fig. 9.42 where the shear modulus drops at much lower temperatures when a plasticizing agent is added.

Often the tensile stress and strain at failure are plotted as a function of temperature. Figure 9.43 shows this for a typical amorphous thermoplastic. The figure shows how the material is brittle below the glass transition temperature and, therefore, fails at low strains. As the temperature increases, the strength of the amorphous thermoplastic decreases, since it becomes leathery in texture and is able to withstand larger deformations. Above T_g, the strength decreases significantly, as the maximum strain continues to increase, until the flow properties have been reached at which point the mechanical strength is negligible. This occurs around the "flow temperature" marked as T_f in the diagram. If the temperature is further increased, the material will eventually thermally degrade at the degradation temperature, T_d.

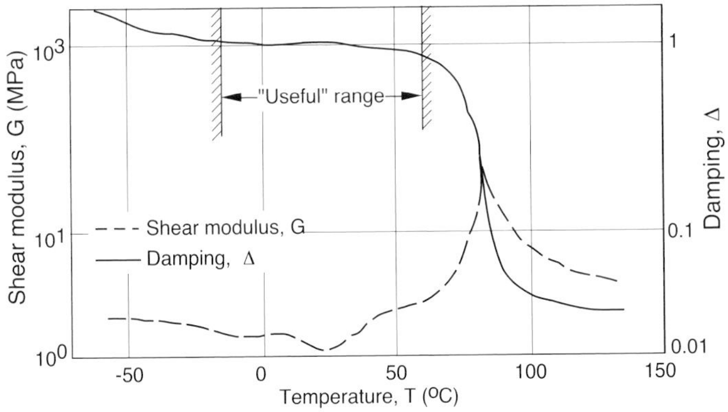

Figure 9.41 Shear modulus and mechanical damping for an unplasticized PVC.

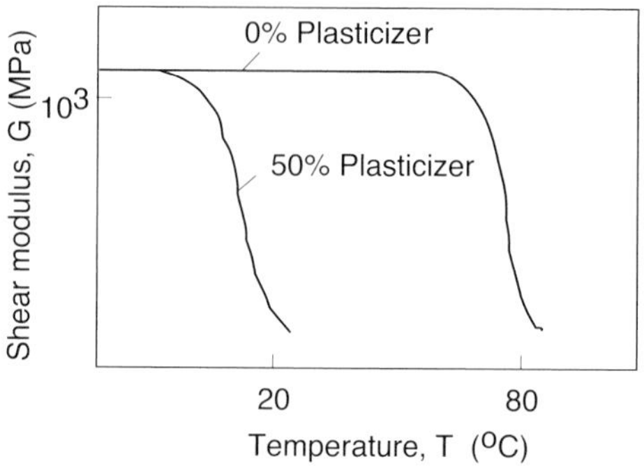

Figure 9.42 Shear modulus as a function of temperature for a PVC with and without a plasticizing agent.

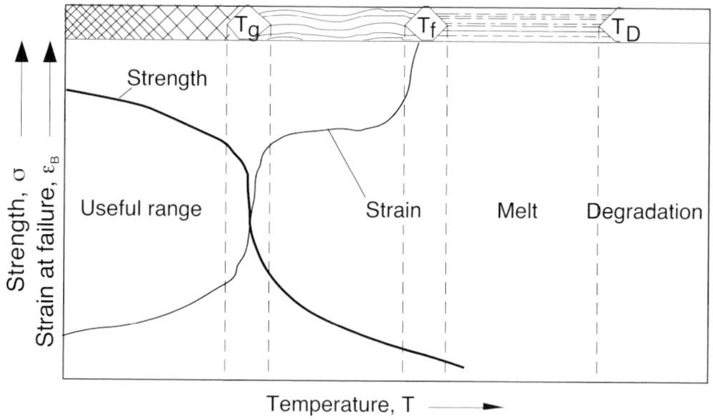

Figure 9.43 Tensile strength and strain at failure as a function of temperature for an amorphous thermoplastic.

Some amorphous thermoplastics can be made high impact resistant (less brittle) through copolymerization. The most common example is acrylonitrile-butadiene-styrene, also known as ABS. Since butadiene chains vitrify at temperatures below -50 °C, ABS is very tough at room temperature in contrast to polystyrene and acrylics by themselves. Due to the different glass transition temperatures present in the materials which form the blend, ABS shows two general transition regions, one around -50 °C and the other at 110 °C, visible in both the logarithmic decrement and the shear modulus.

9.6.2 Semi-Crystalline Thermoplastics

The properties of semi-crystalline thermoplastics can also be analyzed by plotting mechanical properties with respect to temperature. An interesting example is shown in Fig. 9.44, which presents plots of shear modulus versus temperature of a polystyrene with different molecular structures after having gone through different stereo-specific polymerization techniques: low molecular weight PS (A), a high molecular weight PS (B), a semi-crystalline PS (C), and a cross-linked PS (D). In Fig. 9.44 we can see that the low molecular weight material flows before the high molecular weight one, simply due to the fact that the shorter chains can slide past each other more easily— reflected in the lower viscosity of the low molecular weight polymer. The semi-crystalline PS shows a certain amount of stiffness between its glass transition temperature at around 100 °C and its melting temperature at 230 °C. Since a semi-crystalline polystyrene is still brittle at room temperature, it is not very useful to the polymer industry. Figure 9.44 also demonstrates that a cross-linked polystyrene will not melt.

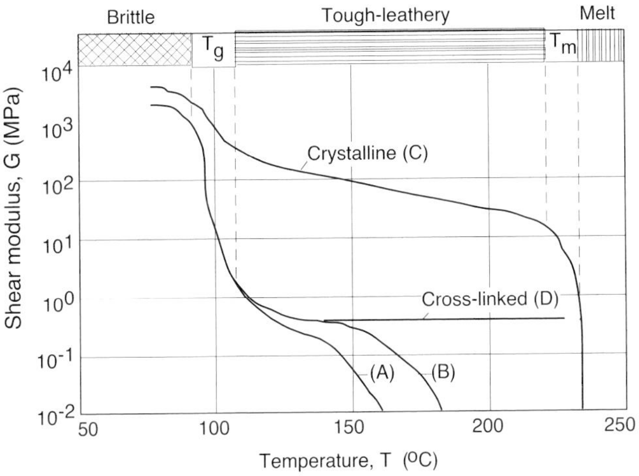

Figure 9.44 Shear modulus curves for amorphous, semi-crystalline and cross-linked polystyrene. (A) low molecular weight amorphous, (B) high molecular weight amorphous, (C) semi-crystalline, (D) cross-linked.

Semi-crystalline thermoplastics are leathery and tough at room temperature since their atactic and amorphous regions vitrify at much lower temperatures. Figure 9.45 shows the shear modulus plotted versus temperature for polypropylene at various degrees of crystallinity. In each case the amorphous regions "solidify" at around 0 °C, whereas the melting temperature goes up significantly with increasing degree of crystallinity. The brittle behavior of polypropylene at 0 °C can sometimes pose a problem in design. This problem can be mitigated through copolymerization— in this case PP copolymerized with ethylene or in elastomers such as ethylene-propylene-diene terpolymer (EPDM).

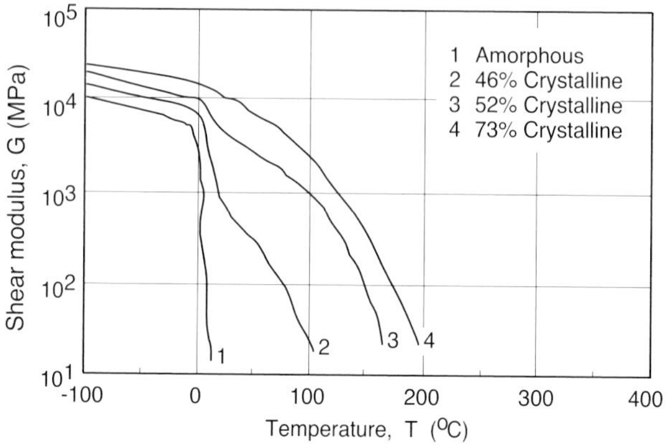

Figure 9.45 Shear modulus for polypropylene with various degrees of crystallinity.

The tensile stress and the strain at failure for a common semi-crystalline thermoplastic is shown in Fig. 9.46. The figure shows an increase in toughness between the glass transition temperature and the melting temperature. The range between T_g and T_m applies to most semi-crystalline thermoplastics.

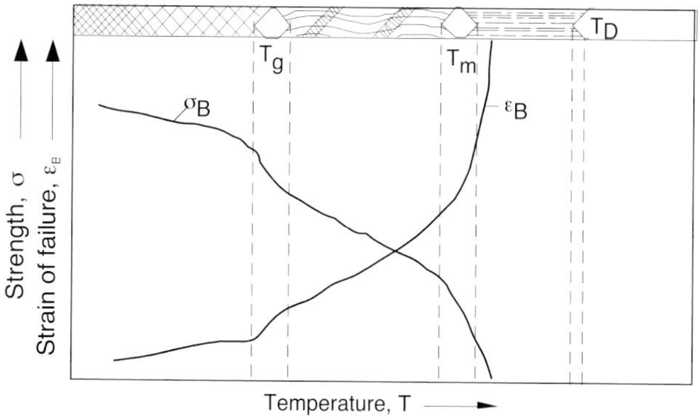

Figure 9.46 Tensile strength and strain at failure as a function of temperature for a semi-crystalline thermoplastic.

9.6.3 Oriented Thermoplastics

If a thermoplastic is deformed at a temperature high enough so that the polymer chains can slide past each other but low enough such that the relaxation time is much longer than the time it takes to stretch the material, then the orientation generated during stretching is retained within the polymer component. We note that the amount of stretching, L/L_0, is not always proportional to the degree of orientation within the component; for example, if the temperature is too high during stretching, the molecules may have a chance to fully relax, resulting in a component with little or no orientation. Any degree of orientation results in property variations within thermoplastic polymers. Figure 9.47 [23] shows the influence stretching has on various properties of common amorphous thermoplastics. The stretching will lead to decreased strength and stiffness properties perpendicular to the orientation and increased properties parallel to the direction of deformation. In addition, highly oriented materials tend to split along the orientation direction under small loads.

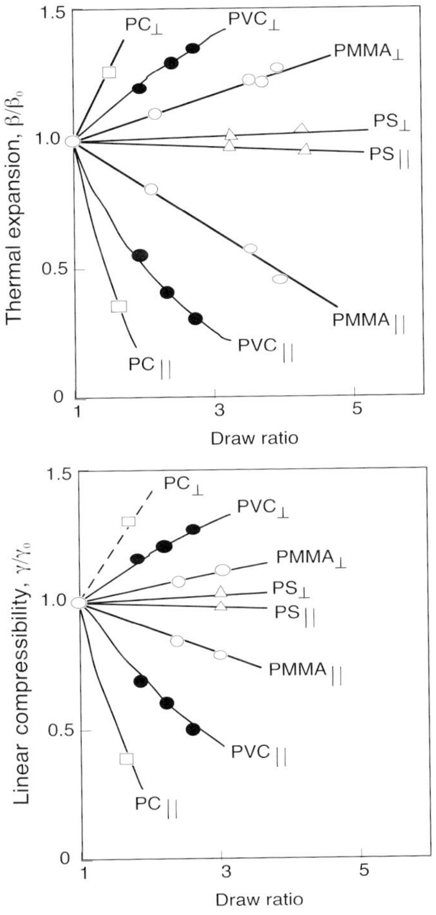

Figure 9.47 Influence of stretch on different properties of amorphous thermoplastics: (top) Thermal expansion, (bottom) Linear compressibility *(continued)*.

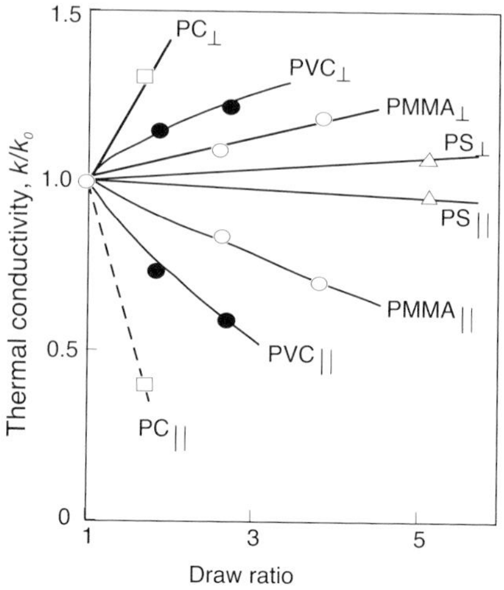

Figure 9.47 Influence of stretch on different properties of amorphous thermoplastics *(continued):* Thermal conductivity.

In amorphous thermoplastics the stretching that leads to permanent property changes occurs between 20 and 40 °C above the glass transition temperature, T_g, whereas with semi-crystalline thermoplastics they occur between 10 and 20 °C below the melting temperature, T_m. After having stretched a semi-crystalline polymer, one must anneal it at temperatures high enough that the amorphous regions relax. During stretching the spherulites break up as whole blocks of lamellae slide out, shown schematically in Fig. 9.48 [24]. Whole lamellae can also rotate such that by sufficiently high stretching, all molecules are oriented in the same direction. The lamellae blocks are now interconnected by what is generally called *tie molecules*. If this material is annealed in a fixed position, a very regular, oriented structure can be generated. This highly oriented material becomes dimensionally stable at elevated temperatures, including temperatures slightly below the annealing or fixing temperature. However, if the component is not fixed during the annealing process, the structure before stretching would be recovered. Figure 9.49 has stress-strain plots for polyethylene with various morphological structures. If the material is stretched such that a needle-like or fibrilic morphological structure results, the resulting stiffness of the material is very high. Obviously, a more realistic structure that would result from stretching would lead to a stacked platelike structure with lower stiffness and ultimate strength. An unstretched morphological structure would be composed of spherulites and exhibit much lower stiffness and ultimate strength. The strength of fibrilic structures is taken advantage of when making synthetic fibers. Figure 9.50 shows theoretical and achievable elastic moduli of various synthetic fiber materials.

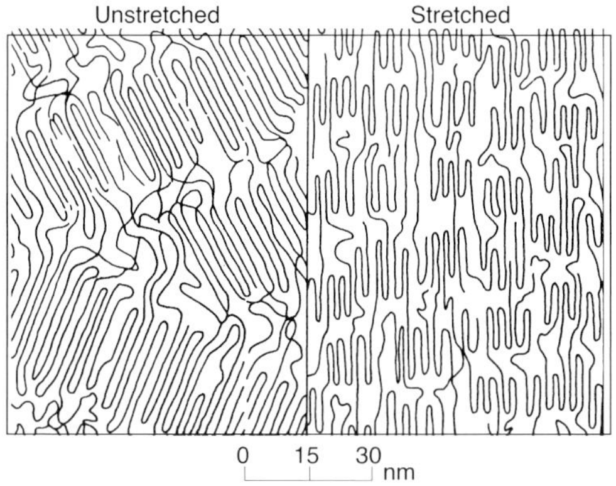

Figure 9.48 Schematic of the sliding and re-orientation of crystalline blocks in semi-crystalline thermoplastics.

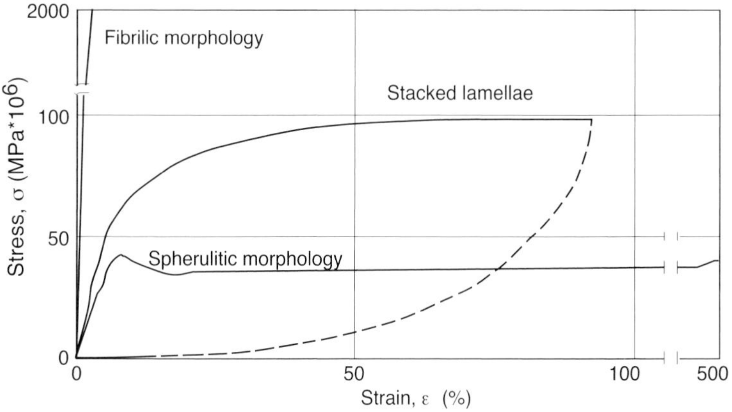

Figure 9.49 Stress-strain behavior of polyethylene with various morphologies.

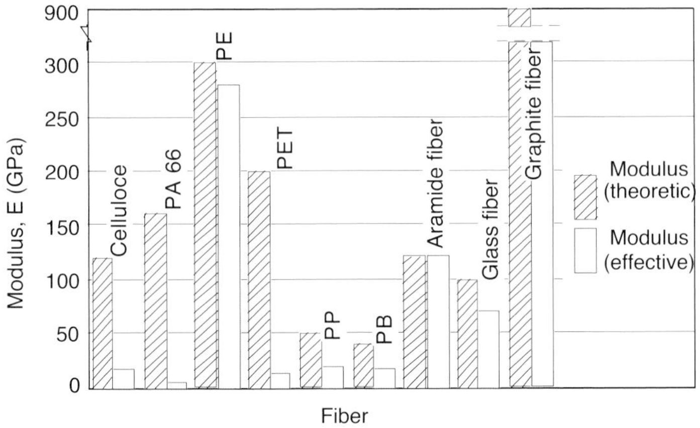

Figure 9.50 Tensile modulus for various fibers.

High-stiffness and high-strength synthetic fibers are becoming increasingly important for lightweight high-strength applications. Extended-chain ultra-high molecular weight polyethylene fibers have only been available commercially since the mid 1980s. The fibers are manufactured by drawing or extending fibers of small diameters at temperatures below the melting point. The modulus and strength of the fiber increase with the drawing ratio or stretch. Due to intermolecular entanglement, the natural draw ratio of high molecular weight high-density polyethylene is only 5[4]. To increase the draw ratio by a factor of 10 or 100, polyethylene must be processed in a solvent such as paraffin oil or paraffin wax.

Figure 9.51 [25] presents the tensile modulus of super-drawn ultra-high molecular weight high-density polyethylene fibers as a function of draw ratio. It can be seen that at draw ratios of 250, a maximum modulus of 200 GPa is reached. In addition to amorphous and semi-crystalline thermoplastics, there is a whole family of thermoplastic materials whose molecules do not relax and, thus, retain their orientation even in the melt state. This class of thermoplastics is the *liquid crystalline polymers*. One such material is the aramid fiber, most commonly known by its tradename, Kevlar®, which has been available in the market for several years. To demonstrate the structure of liquid crystalline polymers, successive enlargement of an aramid pellet is shown in Fig. 9.52 [26]. For comparison, Table 9.3 presents mechanical properties of aramid and polyethylene fibers and other materials.

4 It is interesting that a semi-crystalline thermoplastic stretches more at low molecular weights than at high molecular weights. This contradicts what we expect from theory that longer molecules allow the component to stretch following the relation $\lambda_{max} \approx M^{0.5}$. An explanation for this may be the *trapped entanglements* found in high molecular weight, semi-crystalline polymers that act as semi-permanent cross-links which rip at smaller deformations.

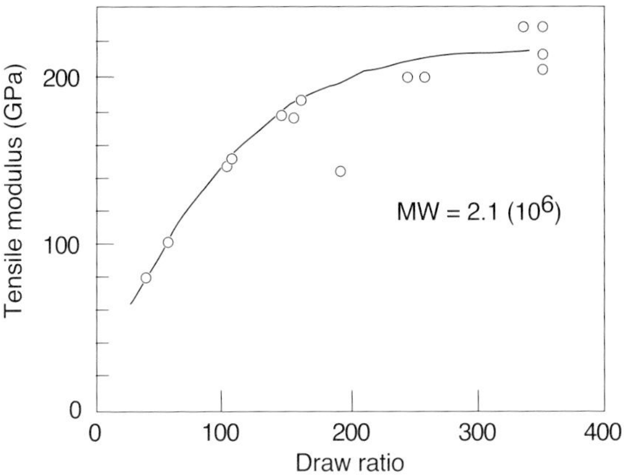

Figure 9.51 Tensile modulus as a function of draw ratio for a UHMWPE (M_W~$2x10^6$).

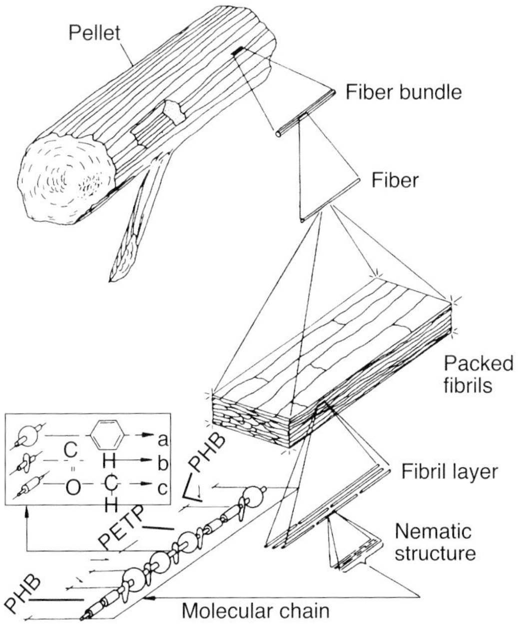

Figure 9.52 Schematic of the structure of a LC-PET.

Table 9.3 Mechanical Properties of Selected Fibers

Fiber	Tensile strength (MPA)	Tensile modulus (GPa)	Elongation to break (%)	Specific gravity
Polyethylene	3000	172	2.7	0.97
Aramid	2760	124	2.5	1.44
Graphite	2410	379	0.6	1.81
S-glass	4585	90	2.75	2.50

The anisotropy of the oriented material can be approximated by assuming that there are covalent bonds joining the molecular chains along the orientation direction, whereas only van der Waals forces act in the two directions perpendicular to the main orientation. With this so called 1:2 rule, we can write

$$a_{\parallel} + 2a_{\perp} = 3a_o \quad \text{and} \tag{9.87}$$

$$a_1 + 2a_2 = 3a_o \tag{9.88}$$

where a_0 is the property for the isotropic material, a_1 and a_2 are the values for the fully oriented material, and $a_{\parallel}$ and $a_{\perp}$ are the values that correspond to the parallel and perpendicular directions with respect to the main orientation. The actual degree of orientation in the above analysis is unknown. The 1:2 rule can be applied to strength properties such as elastic modulus and Poisson's ratio as well as to thermal expansion coefficient and thermal diffusivity. For the elastic modulus, we can write

$$\frac{1}{E_{\parallel}} + \frac{2}{E_{\perp}} = \frac{3}{E_0} \tag{9.89}$$

This rule can be used to approximate mechanical properties of synthetic fibers. For example, a polypropylene fiber will have a tensile elastic modulus of 700 MPa compared to an elastic modulus of 30 MPa for the isotropic material.

9.6.4 Cross-Linked Polymers

Cross-linked polymers, such as thermosets and elastomers, behave completely differently than their counterparts, thermoplastic polymers. In cross-linked systems, the mechanical behavior is also best reflected by the plot of the shear modulus versus temperature. Figure 9.53 compares the shear modulus between highly cross-liked, cross-linked, and uncross-linked polymers. The coarse cross-linked system, typical of elastomers, has a low modulus above the glass transition temperature. The glass transition temperature of these materials is usually below -50 °C, so they are soft and flexible at room temperature. On the other hand, highly cross-linked systems, typical in thermosets, show a smaller decrease in stiffness as the material is raised above the glass transition temperature; the decrease in properties becomes smaller as the degree of cross-linking increases. Figure 9.54 shows ultimate tensile strength and strain curves plotted versus temperature. It is clear that the strength remains fairly constant up to the thermal degradation temperature of the material.

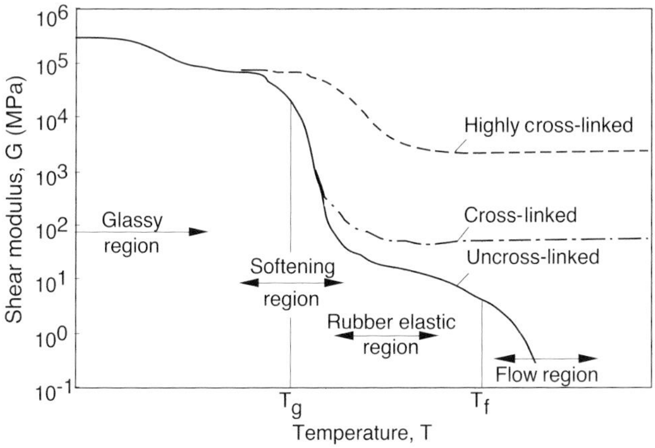

Figure 9.53 Shear modulus and behavior of cross-linked and uncross-linked polymers.

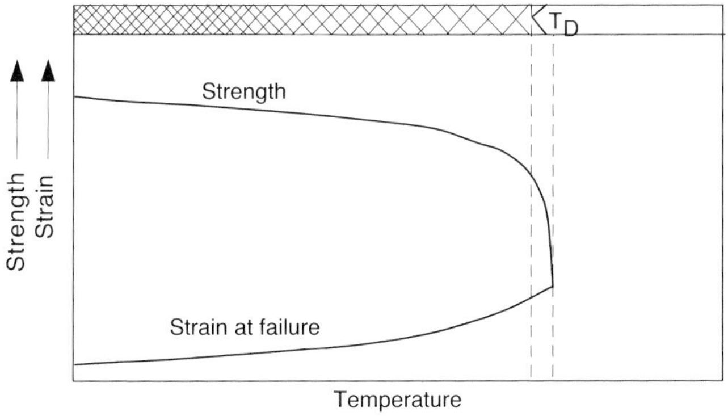

Figure 9.54 Tensile strength and strain at failure as a function of temperature for typical thermosets.

9.7 Mechanical Behavior of Filled and Reinforced Polymers

When we talk about fillers, we refer to materials that are intentionally placed in polymers to make them stronger, lighter, electrically conductive, or cheaper. Any filler will affect the mechanical behavior of a polymeric material. For example, long fibers will make it stiffer but usually denser, whereas foaming will make it more compliant but much lighter. On the other hand, a filler such as calcium carbonate will decrease the polymer's toughness while making it considerably cheaper. Figure 9.55 [27] shows a schematic plot of the change in stiffness as a function of volume fraction for several types of filler materials.

Figure 9.56 shows the increase in dynamic shear modulus for polybutylene terephthalate with 10 and 30% glass fiber content. However, fillers often decrease the strength properties of polymers— this is discussed in more detail in the next chapter.

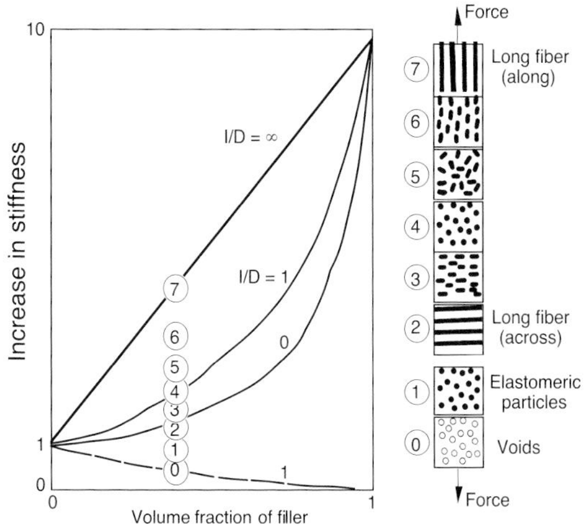

Figure 9.55 Relation between stiffness and filler type and orientation in polymeric materials.

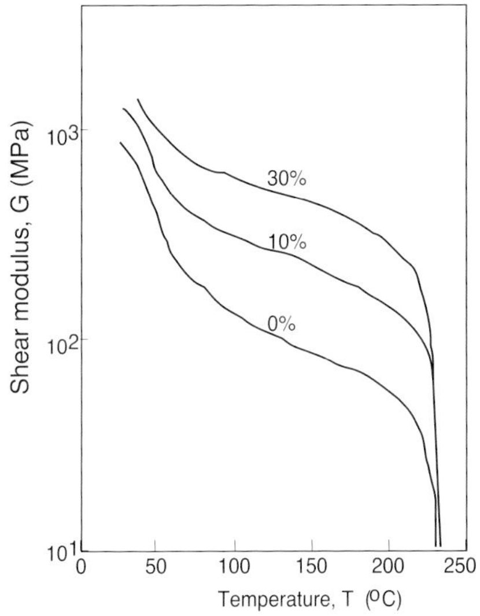

Figure 9.56 Shear modulus for a polybutylene terephthalate with various degrees of glass fiber content by weight.

However, when we refer to reinforced plastics, we talk about polymers (matrix) whose properties have been enhanced by introducing a reinforcement (fibers) of higher stiffness and strength. Such a material is usually called a *fiber reinforced polymer* (FRP) or a *fiber reinforced composite* (FRC). The purpose of introducing a fiber into a matrix is to transfer the load from the weaker material to the stronger one. This load transfer occurs over the length of the fiber as schematically represented in Fig. 9.57. The length it takes to complete the load transfer from the matrix to the fiber, without fiber or matrix fracture, is usually referred to as critical length, L_C. For the specific case where there is perfect adhesion between fiber and matrix, the critical length can be computed as

$$L_c = D\frac{\sigma_{uf}}{2\tau_{um}} \tag{9.90}$$

where D is the fiber diameter, σ_{uf} is the tensile strength of the fiber and τ_{um} is the shear strength of the matrix. Although Eq. 9.90 predicts L/D as low as 10, experimental evidence suggests that aspect ratios of 100 or higher are required to achieve maximum strength [28]. If composites have fibers that are shorter than their critical length they are referred to as *short fiber composites*, and if the fibers are longer they are referred to as *long fiber composites* [29].

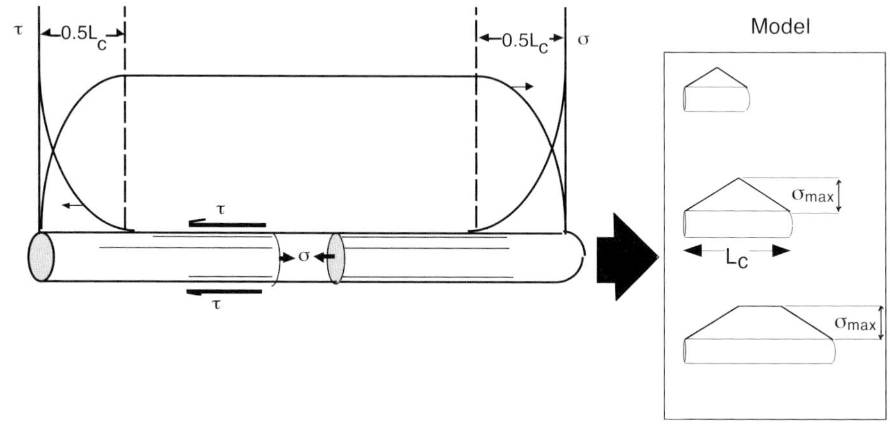

Figure 9.57 Schematic diagram of load transfer from matrix to fiber in a composite.

9.7.1 Anisotropic Strain-Stress Relation

As discussed in Chapter 6, filled polymers are often anisotropic, and the relations presented in Eqs. 9.1–9.15 are not valid. The three-dimensional anisotropic strain-stress relation where, for simplicity, x, y, and z have been replaced by 1, 2, and 3, respectively, is often written as

$$\varepsilon_{11} = \frac{1}{E_{11}}\sigma_{11} - \frac{\nu_{21}}{E_{22}}\sigma_{22} - \frac{\nu_{31}}{E_{33}}\sigma_{33} \tag{9.91}$$

$$\varepsilon_{22} = -\frac{\nu_{12}}{E_{11}}\sigma_{11} + \frac{1}{E_{22}}\sigma_{22} - \frac{\nu_{32}}{E_{33}}\sigma_{33} \tag{9.92}$$

$$\varepsilon_{33} = -\frac{\nu_{13}}{E_{11}}\sigma_{11} - \frac{\nu_{23}}{E_{22}}\sigma_{22} + \frac{1}{E_{33}}\sigma_{33} \tag{9.93}$$

$$\gamma_{12} = \frac{1}{G_{12}}\tau_{12} \tag{9.94}$$

$$\gamma_{23} = \frac{1}{G_{23}}\tau_{23} \tag{9.95}$$

$$\gamma_{31} = \frac{1}{G_{31}} \tau_{31} \tag{9.96}$$

and in matrix form for the more general case:

$$\begin{Bmatrix} \varepsilon_{11} \\ \varepsilon_{22} \\ \varepsilon_{33} \\ \gamma_{12} \\ \gamma_{23} \\ \gamma_{31} \end{Bmatrix} = \begin{bmatrix} S_{11} & S_{12} & S_{13} & S_{14} & S_{15} & S_{16} \\ S_{21} & S_{22} & S_{23} & S_{24} & S_{25} & S_{26} \\ S_{31} & S_{32} & S_{33} & S_{34} & S_{35} & S_{36} \\ S_{41} & S_{42} & S_{43} & S_{44} & S_{45} & S_{46} \\ S_{51} & S_{52} & S_{53} & S_{54} & S_{55} & S_{56} \\ S_{61} & S_{62} & S_{63} & S_{64} & S_{65} & S_{66} \end{bmatrix} \begin{Bmatrix} \sigma_{11} \\ \sigma_{22} \\ \sigma_{33} \\ \tau_{12} \\ \tau_{23} \\ \tau_{31} \end{Bmatrix} \tag{9.97}$$

where coupling between the shear terms and the elongational terms can be introduced.

9.7.2 Aligned Fiber Reinforced Composite Laminates

The most often applied form of the above equations is the two-dimensional model used to analyze the behavior of aligned fiber reinforced laminates, such as that shown schematically in Fig. 9.58. For this simplified case, Eqs. 9.91–9.96 reduce to

$$\varepsilon_L = \frac{1}{E_L} \sigma_L - \frac{\nu_{TL}}{E_T} \sigma_T \tag{9.98}$$

$$\varepsilon_T = -\frac{\nu_{LT}}{E_L} \sigma_L + \frac{1}{E_T} \sigma_T \tag{9.99}$$

$$\gamma_{LT} = \frac{1}{G_{LT}} \tau_{LT} \tag{9.100}$$

which can also be written as

$$\{\varepsilon_{LT}\} = [S_{LT}]\{\sigma_{LT}\} \tag{9.101}$$

where the subscripts L and T define the longitudinal and transverse directions, respectively, as described in Fig. 9.58, and $[S_{LT}]$ is referred to as the compliance matrix.

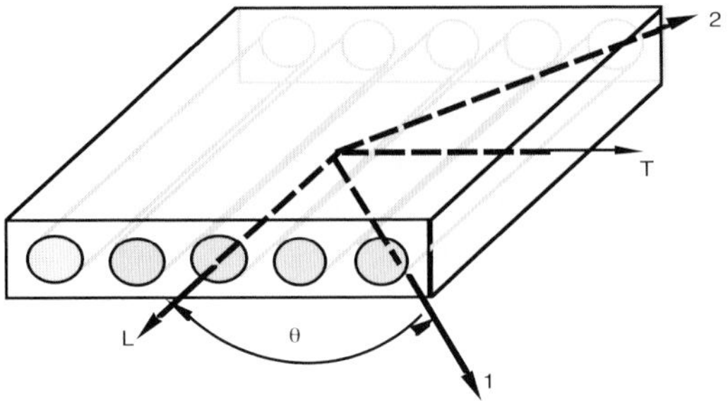

Figure 9.58 Schematic diagram of unidirectional continuous fiber reinforced laminated structure.

The longitudinal and transverse properties can be calculated using the widely used Halpin-Tsai model [30] as

$$E_L = E_m\left(\frac{1+\xi\eta\phi}{1-\eta\phi}\right) \tag{9.102}$$

$$E_T = E_m\left(\frac{1+\eta\phi}{1-\eta\phi}\right) \tag{9.103}$$

$$G_{LT} = G_m\left(\frac{1+\lambda\phi}{1-\lambda\phi}\right) = G_m\frac{v_{LT}}{v_m} \tag{9.104}$$

where

$$\eta = \frac{\left(\dfrac{E_f}{E_m}-1\right)}{\left(\dfrac{E_f}{E_m}-\xi\right)} \tag{9.105}$$

$$\lambda = \frac{\left(\dfrac{G_f}{G_m}-1\right)}{\left(\dfrac{G_f}{G_m}+1\right)} \tag{9.106}$$

$$\xi = 2\left(\frac{L}{D}\right) \tag{9.107}$$

Here, the subscripts *t* and *m* represent the fiber and matrix, respectively; *L* the fiber length; *D* the fiber diameter; ϕ the volume fiber fraction which can be expressed in terms of weight fraction, ψ, as

$$\phi = \frac{\psi}{\psi + (1+\psi)(\rho_f / \rho_m)} \tag{9.108}$$

It should be pointed out that, in addition to the Halpin-Tsai model, there are several other models in use today to predict the elastic properties of aligned fiber reinforced laminates [31, 33]. Most models predict the longitudinal modulus quite accurately as shown in Fig. 9.59 [31], which compares measured values to computed values using the *mixing rule*. This comes as no surprise, since experimental evidence clearly shows that longitudinal modulus is directly proportional to the fiber content for composites with unidirectional reinforcement. However, differences do exist between the models when predicting the transverse modulus, as shown in Fig. 9.60 [31].

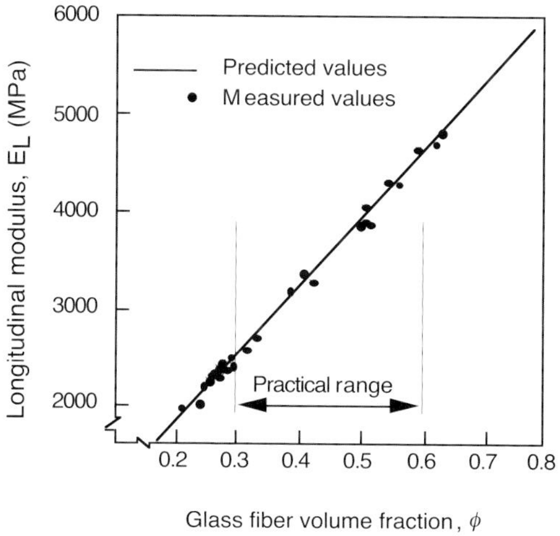

Figure 9.59 Measured and predicted longitudinal modulus for an unsaturated polyester/aligned glass fiber composite laminate as a function of volume fraction of glass content.

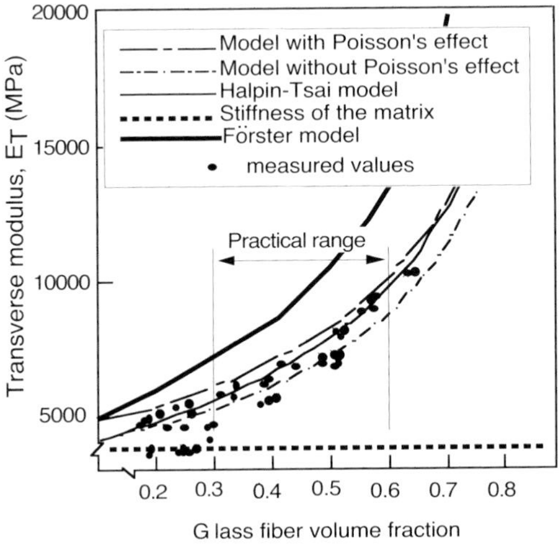

Figure 9.60 Measured and predicted transverse modulus for an unsaturated polyester/aligned glass fiber composite laminate as a function of volume fraction of glass content.

9.7.3 Transformation of Fiber Reinforced Composite Laminate Properties

The loading in a laminated structure is not always aligned with the transverse and longitudinal directions of the reinforcement. Hence, it is often necessary to rotate the laminate and its properties by an angle θ. Figure 9.58 depicts the laminate's material coordinate system L-T and a rotated arbitrary coordinate system 1-2. If we rotate the axes from the 1-2 system to the L-T system, we can transform the stress components using

$$\begin{Bmatrix} \sigma_L \\ \sigma_T \\ \tau_{LT} \end{Bmatrix} = \begin{bmatrix} c^2 & s^2 & 2sc \\ s^2 & c^2 & -2sc \\ -sc & sc & \left(c^2 - s^2\right) \end{bmatrix} \begin{Bmatrix} \sigma_{11} \\ \sigma_{22} \\ \tau_{12} \end{Bmatrix} \quad \text{or} \tag{9.109}$$

$$\{\sigma_{LT}\} = [T_\sigma]\{\sigma_{12}\} \tag{9.110}$$

The transformation of the strain components carry an extra 1/2 term for the shear strains and is written as

$$\begin{Bmatrix} \varepsilon_L \\ \varepsilon_T \\ \gamma_{LT} \end{Bmatrix} = \begin{bmatrix} c^2 & s^2 & sc \\ s^2 & c^2 & -sc \\ -2sc & 2sc & \left(c^2 - s^2\right) \end{bmatrix} \begin{Bmatrix} \varepsilon_{11} \\ \varepsilon_{22} \\ \gamma_{12} \end{Bmatrix} \quad \text{or} \tag{9.111}$$

$$\{\varepsilon_{LT}\} = [T_\varepsilon]\{\varepsilon_{12}\} \tag{9.112}$$

Combining Eq. 9.101 with the above transformations, we can write

$$[T_\varepsilon]\{\varepsilon_{12}\} = [S_{LT}][T\sigma]\{\sigma_{12}\} \tag{9.113}$$

or

$$\{\varepsilon_{12}\} = [T_\varepsilon]^{-1}[S_{LT}][T\sigma]\{\sigma_{12}\} \tag{9.114}$$

The compliance matrix in the L-T coordinate system has four independent components and the 1-2 system has six. The inverse of $[T_\varepsilon]$ is equivalent to rotating the coordinates back by $-\theta$. This leads to

$$\begin{Bmatrix} S_{11} \\ S_{22} \\ S_{12} \\ S_{44} \\ S_{14} \\ S_{24} \end{Bmatrix} = \begin{bmatrix} c^4 & s^4 & 2c^2s^2 & c^2s^2 \\ s^4 & c^4 & 2s^2c^2 & c^2s^2 \\ c^2s^2 & c^2s^2 & (c^4+s^4) & -c^2s^2 \\ 4c^2s^2 & 4c^2s^2 & -8c^2s^2 & (c^2-s^2)^2 \\ 2c^3s & -2cs^3 & 2(cs^3-c^3s) & (cs^3-c^3s) \\ 2cs^3 & -2c^3s & 2(c^3s-cs^3) & (c^3s-cs^3) \end{bmatrix} \begin{Bmatrix} S_{LL} \\ S_{TT} \\ S_{LT} \\ S_{SS} \end{Bmatrix} \tag{9.115}$$

or

$$\{S_{12}\} = [R(\theta)]\{S_{LT}\} \tag{9.116}$$

The engineering elastic constants in the 1-2 system can easily be computed:

$$E_{11} = \frac{1}{S_{11}} \tag{9.117}$$

$$E_{22} = \frac{1}{S_{22}} \tag{9.118}$$

$$G_{12} = \frac{1}{S_{44}} \tag{9.119}$$

$$\nu_{12} = -\frac{S_{12}}{S_{11}} \tag{9.120}$$

$$\nu_{21} = -\frac{S_{12}}{S_{22}} \tag{9.121}$$

$$\eta_{14} = \frac{S_{14}}{S_{11}} \tag{9.122}$$

$$\eta_{24} = \frac{S_{24}}{S_{22}} \tag{9.123}$$

$$\eta_{41} = \frac{S_{41}}{S_{44}} \tag{9.124}$$

$$\eta_{42} = \frac{S_{42}}{S_{44}} \tag{9.125}$$

Figure 9.61 [33] shows how the stiffness decreases as one rotates away from the longitudinal axis for an aligned fiber reinforced composite with different volume fraction fiber content. From the figure it is evident that for high volume fraction fiber contents only a slight misalignment of the fibers from the loading direction results in a drastic reduction of the properties. Along with the predicted stiffness properties, the figure also presents the stiffness for a composite with 0.56 volume fraction of fibers measured at various angles from the longitudinal axis of the composite. The measured and the predicted values agree quite well.

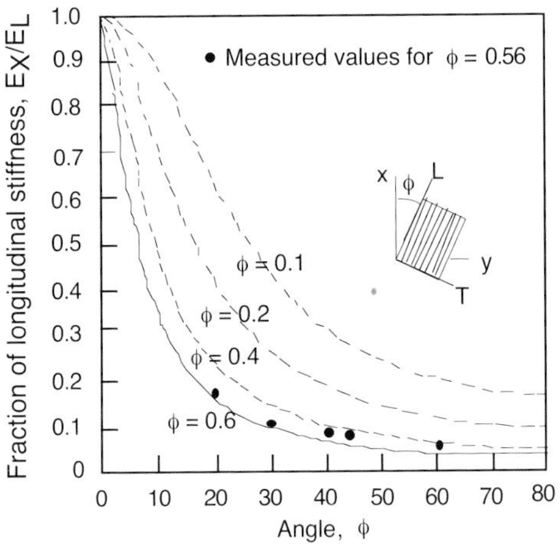

Figure 9.61 Measured and predicted elastic modulus in a unidirectional fiber reinforced laminate as a function of angle between loading and fiber direction.

9.7.4 Reinforced Composite Laminates with a Fiber Orientation Distribution Function

The above transformation can be used to compute the properties of planar systems with a fiber orientation distribution function. This is done by superposing aligned fiber laminates rotated away from the principal 1-2 coordinate system by an angle θ and with a volume fiber fraction given by $\psi(\theta)$. The transformation is written as

$$\{S_{12}\} = \int_{-\pi/2}^{\pi/2} \left([R(\theta)]\{S_{LT}\}\psi(\theta)\right) d\theta \tag{9.126}$$

which can be written in discrete terms to be used with fiber orientation distribution function attained from computer simulation:

$$\{S_{12}\} = \sum_{i=1}^{N} \left([R(\theta_i)]\{S_{LT}\}\psi(\theta_i)\right)\Delta\theta \tag{9.127}$$

Using Eq. 9.126, one can easily predict the stiffness properties of a part with randomly oriented fibers, where $\psi(\theta) = 1/\pi$, using[5]

$$\frac{1}{E_{11}} = \frac{1}{E_{22}} = \frac{1}{E_{\text{random}}} = \frac{3}{8}\frac{1}{E_L} + \frac{3}{8}\frac{1}{E_T} - \frac{2}{8}\frac{\nu_{LT}}{E_L} + \frac{1}{8}\frac{1}{G_{LT}} \tag{9.128}$$

9.8 Strength Stability Under Heat

As mentioned earlier, polymers soften and eventually flow as they are heated. It is, therefore, important to know what the limiting temperatures are at which a polymer component can still be loaded with moderate deformations. Three tests are commonly performed on polymer specimens to determine this limiting temperature for that material. They are the *Vicat temperature test* (DIN 53460), shown in Fig. 9.62, the *Martens temperature test* (DIN 53458 or 53462), and the *heat-distortion temperature* (HDT) test (ASTM D 648-72) shown in Fig. 9.63[6].

5 The incorrect expression $E_{Random} = 3/8E_L + 5/8E_T$ is often successfully used, for low fiber content, to approximate the stiffness of the composite with randomly oriented fibers. However, using this equation for composites with large differences between E_L and E_T, the stiffness can be overestimated by 50%.

6 Courtesy of BASF.

In the Vicat temperature test, a needle loaded with weights is pushed against a plastic specimen inside a glycol bath. This is shown schematically in Fig. 9.62. The uniformly heated glycol bath rises in temperature during the test. The *Vicat number* or Vicat temperature is measured when the needle has penetrated the polymer by 1 mm. The advantage of this test method is that the test results are not influenced by the part geometry or manufacturing technique. The practical limit for thermoplastics, such that the finished part does not deform under its own weight, lies around 15 K below the Vicat temperature.

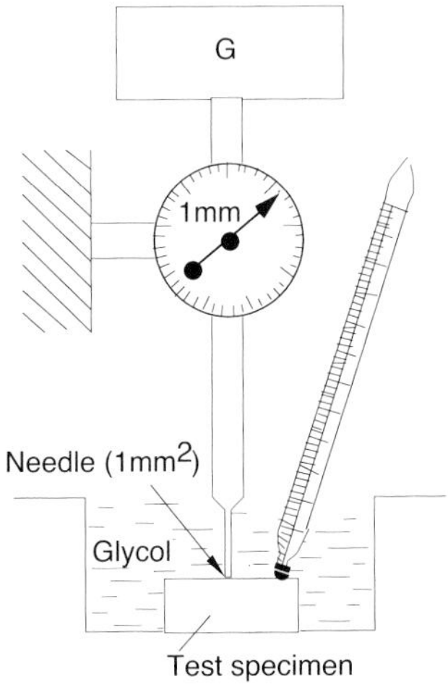

Figure 9.62 Apparatus to determine a material's shape stability under heat [BASF] using the Vicat temperature test.

To determine the heat distortion temperature, the standard specimen lies in a fluid bath on two knife edges separated by a 10 cm distance. A bending force is applied on the center of the specimen. Similar to the Vicat temperature test, the bath's temperature is increased during the test. The HDT is the temperature at which the rod has bent 0.2 mm to 0.3 mm (see Fig. 9.63). The Vicat temperature is relatively independent of the shape and type of part, whereas the heat-distortion-data are influenced by the shaping and pretreatment of the test sample. Table 9.4 shows the heat distortion temperature for selected thermoplastics measured using ASTM D648.

In the Martens temperature test, the temperature at which a cantilevered beam has bent 6 mm is recorded. The test sample is placed in a convection oven with a constantly rising temperature. In Europe, the HDT test has replaced the *Martens temperature test.*

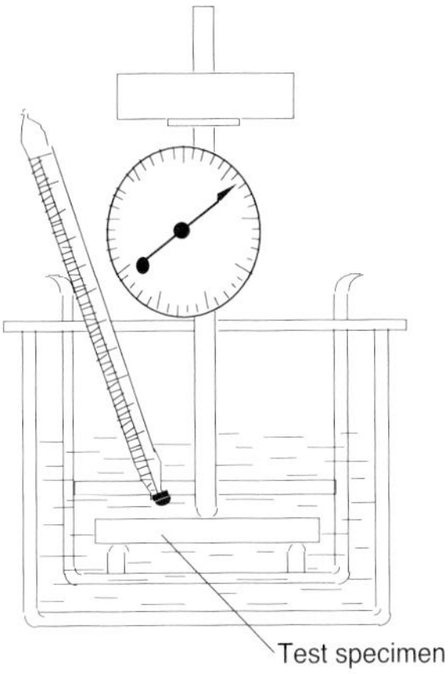

Figure 9.63 Apparatus to determine a material's shape stability under heat using the heat-distortion-temperature test (HDT).

It is important to point out that these test methods do not give enough information to determine the allowable operating temperature of molded plastic components subjected to a stress. Heat distortion data is excellent when comparing the performance of different materials and should only be used as a reference not as a direct design criterion.

Table 9.4 Heat Distorsion Temperature for Selected Thermoplastics

Material	HDT (°C) 1.86 MPa	0.45 MPa
HDPE	50	50
PP	45	120
uPVC	60	82
PMMA	60	100
PA66	105	200
PC	130	145

Examples

9.1 A loaded high-density polyethylene component is not to exceed 4% strain after 60 days of loading at room temperature. Using the short term data presented in Fig. 9.64 estimate the maximum allowable stress within the component.

Figure 9.64 presents constants E_0 and D_2 as a function of rate of the deformation $\dot{\varepsilon}$. The time given in this design is 60 days (1440 h) and the strain is 4%. Therefore the average rate of deformation is $4\%/1440\text{h} = 0.0028\%/\text{h}$. At that rate, we read $E_0 = 250\text{MPa}$ and $D_2 = 28$. We can now use Eq. 9.25 to compute the allowable stress

$$\sigma = \frac{E_0 \varepsilon}{1 + D_2 \varepsilon} = \frac{250\text{MPa}(0.04)}{1 + 28(0.04)} = 4.72\text{MPa}$$

It should be noted that the prescribed strain of 4% is quite high. As a rule of thumb, a component that is subjected to a long term load should not have strains larger than 2%, to avoid creep rupture. Hence, if we reduce our allowable strain to 2%, the rate of deformation during the 60 days should be $\dot{\varepsilon} = 2\% / 1440\text{h} = 0.0014\% / \text{h}$. At the new rate we read $E_0 = 245\text{MPa}$ and $D_2 = 26$ from Fig. 9.64. Using the new constants, the maximum allowable stress for a 2% strain in 60 days is 3.22 MPa.

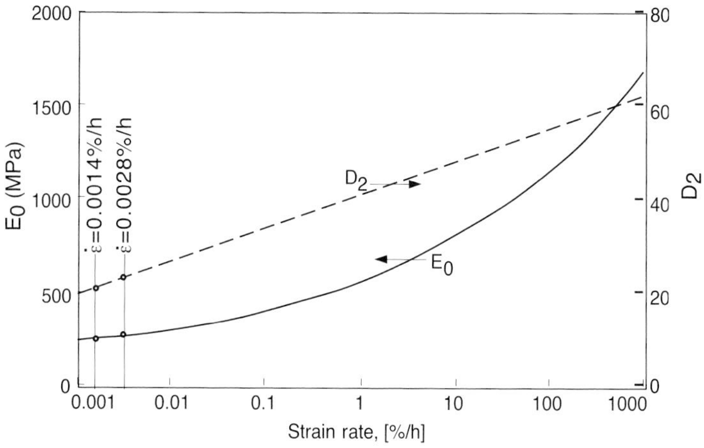

Figure 9.64 Coefficients for PE-HD short term data

9.2 In a special laboratory experiment, a PMMA pipe is used to cap a tank that is pressurized at 2 MPa, as shown in Fig. 9.65. The 3 mm thick pipe has a 50 mm internal diameter and is 300 mm long. Estimate the diameter change of the pipe after one year of testing? Use the creep data given below [34].

To solve this problem we can use the thin pressure vessel approximation, working with an average diameter, $\overline{D} = 53\text{mm}$. This is a case of biaxial stress, composed of a hoop stress, σ_H, and an axial stress, σ_A, defined by

$$\sigma_H = \frac{p\overline{D}}{2h} = 17.67\ \text{MPa and}\ \sigma_A = \frac{p\overline{D}}{4h} = 8.83\ \text{MPa}$$

where, p is the pressure and h is the thickness. These stresses are constant and will cause the pipe to creep. Using the PMMA creep data we can generate a 1 year isochronous curve, which is presented in Fig.9.67. The strains that correspond to stresses of 17.67 MPa and 8.83 MPa are $\varepsilon_H = 1.32\%$ and $\varepsilon_A = 0.6\%$, respectively. Since this is a biaxial case we must correct the hoop strain using Poisson's effect, before computing the diameter change. For this we use

$$\varepsilon_H' = \varepsilon_H - \nu\varepsilon_A$$

However, since we were not given Poisson's ration, ν, we assume a value of 1/3. Thus,

$$\varepsilon_H' = 1.32 - \left(\tfrac{1}{3}\right)0.6 = 1.12\%$$

To compute the diameter change we use

$$\varepsilon_H' = \Delta D / D$$

Hence,

$$\Delta D = (53\text{mm})(0.0112) = 0.594\text{mm}$$

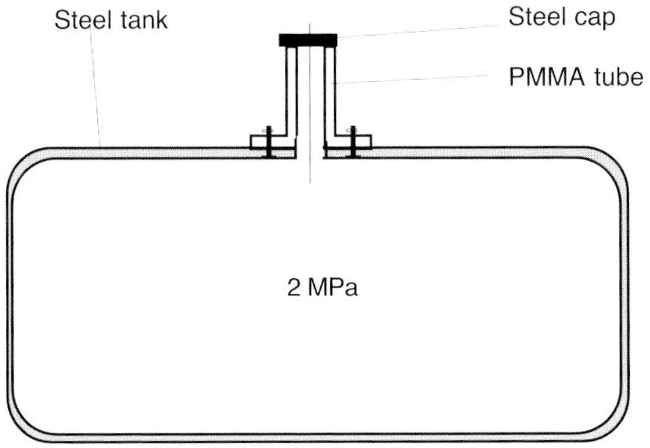

Figure 9.65 Schematic of the apparatus described in Example 9.2.

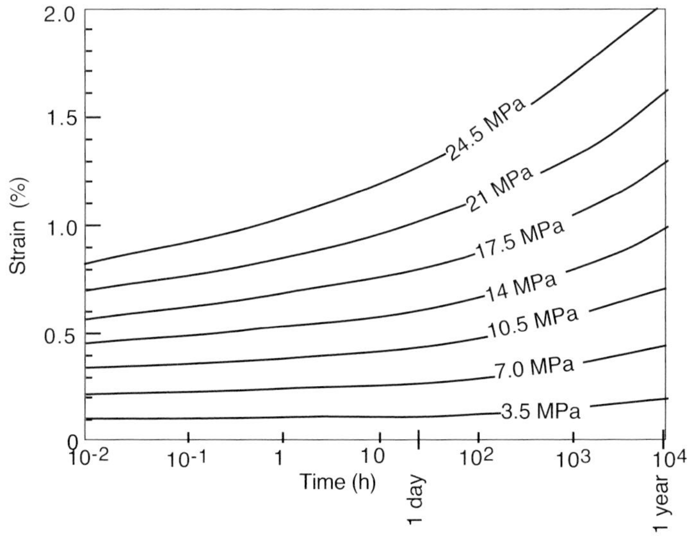

Figure 9.66 PMMA creep data.

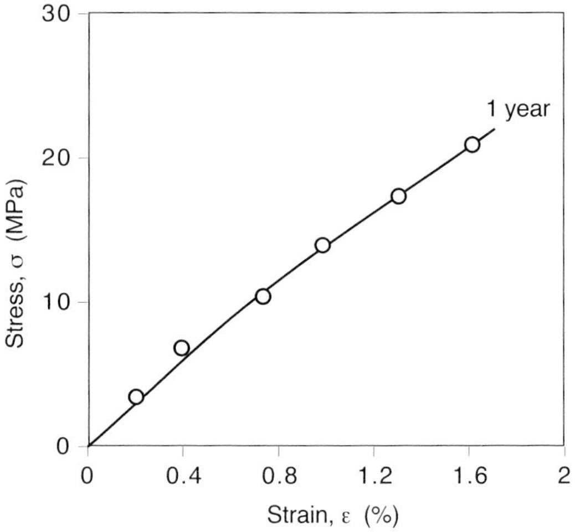

Figure 9.67 A 1 year isochronous curve for PMMA.

9.3 In the assembly shown in Fig. 9.68, a tubular polypropylene feature is pressed on a 15 mm long metal stud. The inner diameter of the 1 mm thick PP tubular element is 10 mm. The metal stud is slightly oversized with a diameter of 10.15 mm. With a coefficient of friction $\mu = 0.3$ estimate the force required to disassemble the parts shortly after assembly, and after one year. Use the creep data given in Fig. 9.69.

This is a classic constant strain, ε_0, stress relaxation problem. The initial hoop stress that holds the assembly together can be quite high. However, as time passes the hoop stress relaxes and it becomes easier to disassemble the two components. The strain in the system after assembly is computed using

$$\varepsilon_0 = \Delta D / D = 0.16\,\text{mm} / 11\,\text{mm} = 0.0145 \longrightarrow 1.45\%$$

In order to follow the hoop stress history after assembly we generate a 1.45% isometric curve, which is shown below (Fig. 9.70).

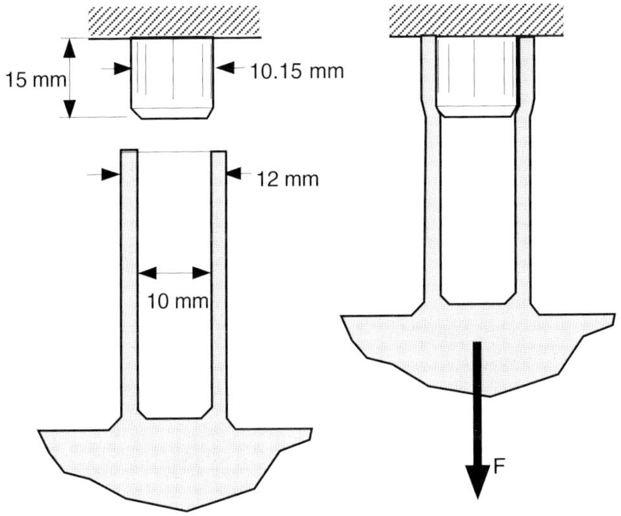

Figure 9.68 Assembly for Example 9.3.

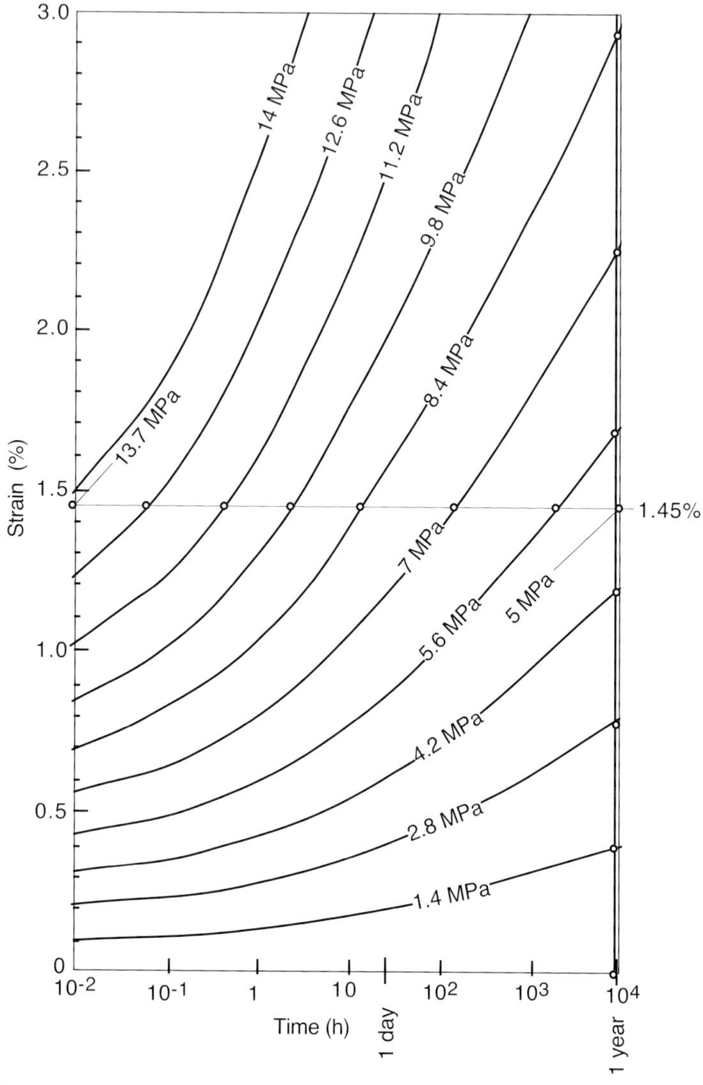

Figure 9.69 PP creep data.

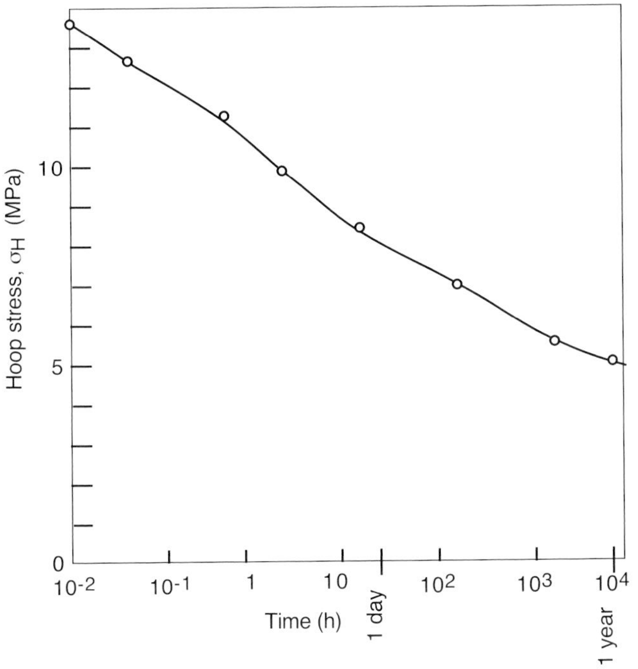

Figure 9.70 Isochronous curve for PP at 1.45% strain.

From the isochronous curve we can deduce that the hoop stress, σ_H, is 13.7 MPa shortly after assembly and about 5 MPa one year after assembly. The pressure acting on the metal stud, due to the hoop stress, can be computed using

$$p = {}^{2h\sigma_H}\!\big/_{D}$$

which gives $p = 2.49\,\text{MPa}$ right after assembly and $p = 0.91\,\text{MPa}$ after one year. From the pressure and the friction we can calculate the disassembly force with

$$F = \mu p(\pi D_i L)$$

where $L = 15\,\text{mm}$ is the length of the stud. Using the above equation, the computed force necessary to pull the two components apart is 358 N (80 lb) after assembly and 130 N (29 lb) after one year.

9.4 You are asked to design a polypropylene bracket of circular cross-section, as depicted in Fig. 9.71, to hold a 10 N load for a one year period. The maximum strain you would like the bracket to feel is 2%. Calculate the diameter of the bracket using the PP creep data given in Example 9.3.

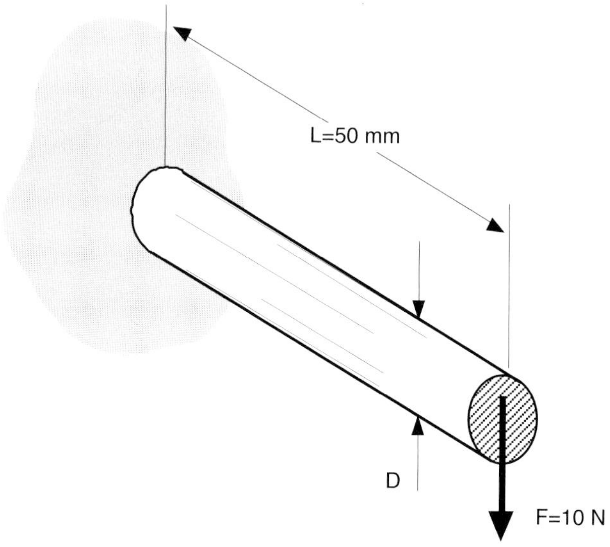

Figure 9.71 Cantilevered bracket of circular cross-section.

In this design, the instant the bracket is loaded it will begin to creep. Since the life of the system is designed to be one year, with a maximum strain of 2%, we will work with a 1 year isochronous curve(Fig.9.72), generated using Fig.9.69.

From the isochronous curve we find that 6.5 MPa is the stress that corresponds to a 2% strain. In a cantilevered beam the maximum stresses occur in the upper (tensile) and lower (compressive) points where the bracket joins the wall. For a cantilevered system we use

$$\sigma = \frac{Mc}{I}, \text{ where } \begin{cases} c = D/2 \\ M = FL \\ I = \dfrac{\pi D^4}{64} \end{cases}$$

Hence, we can solve for D using

$$D = \sqrt[3]{\frac{32FL}{\pi\sigma}} = \sqrt[3]{\frac{32(10\,\text{N})(0.05\,\text{m})}{\pi(6.5\times10^{6}\,\text{Pa})}} = 9.2\,\text{mm}$$

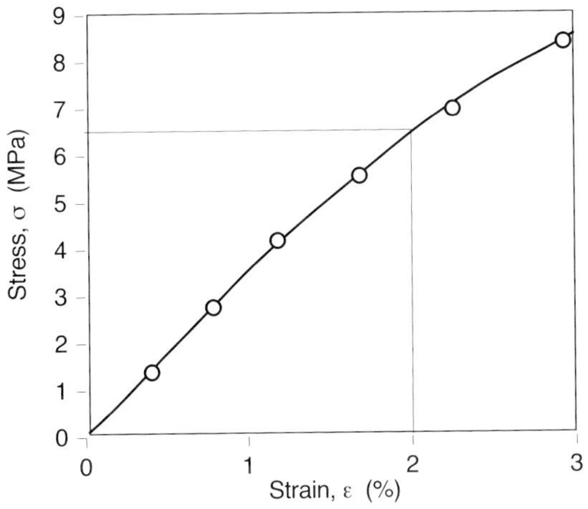

Figure 9.72 A 1 year isochronous curve for PP.

At this point it is important to mention that this design has many flaws. First a solid cross-section such as this will result in a long cycle time during manufacturing due to the slow cooling. In addition, thick cross-sections result in large amounts of shrinkage, especially with a semi-crystalline material such as polypropylene. Finally, the bending strength of a solid circular cross-section is only a fraction (lower than 20%) of the bending strength that an I-beam could deliver. This means that by changing the cross-sectional geometry of the beam, we would only need 20% or less of the material required for the present design.

Problems

9.1 Using the coefficients E_0 and D_2 for the high-density polyethylene presented in Fig. 9.73 compute the stress for 0%, 2%, 4%, and 6% strain and reconstruct the stress-strain curve for a specimen tested at 23 °C with a 100%/h deformation rate.

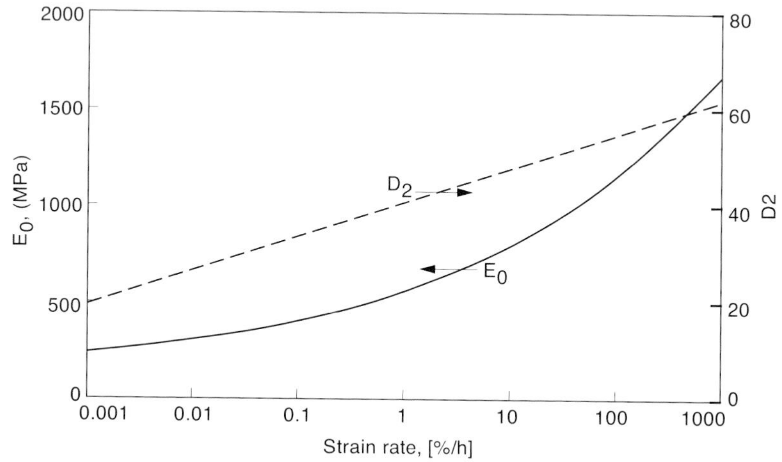

Figure 9.73 Coefficients E_0 and D_2 for a high density polyethylene at 23 °C.

9.2 For the above material calculate the 1% strain secant modulus for 6 min, 1 h, 10 h, and 100 h. Plot the data and comment on your results.

9.3 Using the time-temperature superposition principle described below, plot the stress-strain curve of Problem 9.1 for T = 40 °C.

$$\mathrm{Log}\left(\dot{\varepsilon}_{ref} / \dot{\varepsilon}\right) = k\left(1/T - 1/T_{ref}\right), \quad \text{with} \quad k = 10400\,\mathrm{K}$$

9.4 Generate a 2 year isochronous curve for the polypropylene curve presented in Fig. 9.21. Using this curve estimate the thickness of a 50 mm diameter polypropylene pipe whose maximum allowable strain in a 2 year period should be 2%, while subjected to a 200 kPa internal pressure. Note: neglect effect of biaxial state of stress.

9.5 After a continuos 1 year operation, what is the diameter change of a cast acrylic tube which is part of a fluids experimental set-up that transports air at a pressure of 10 bar. Use the following

- Tube inside diameter = 50 mm,
- Tube thickness = 3 mm,
- Tube length = 300 mm.

Use the 20 °C creep curves given below.

9.6 You are to re-design the 50 mm long solid polypropylene circular bracket described in Example 9.4 using a hollow circular cross-section with a 2 mm thickness. How much material are you saving?

9.7 A cast iron pump housing has a polypropylene cover. The cover is tightened with steel bolts causing a compressive strain of 1.5% on the edge of the plate. When will the pump start leaking if it generates internal pressures of up to 4 MPa. Use the creep data given in Fig. 9.20.

9.8 Rework Example 9.3 using polyacetal data given in Fig. 9.74.

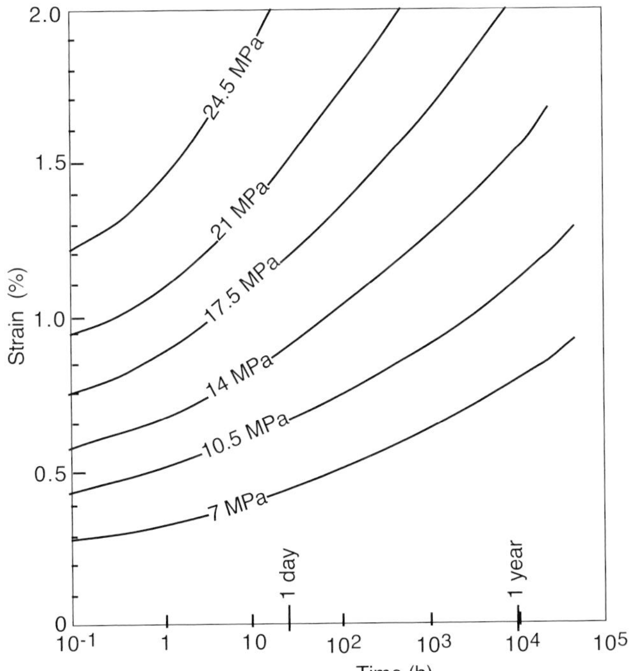

Figure 9.74 Creep data for polyacetal.

9.9 Two metal pipes are to be connected using a polypropylene fitting as shown in Fig. 9.75. The tubes are inserted onto the fitting causing a 1% hoop strain.

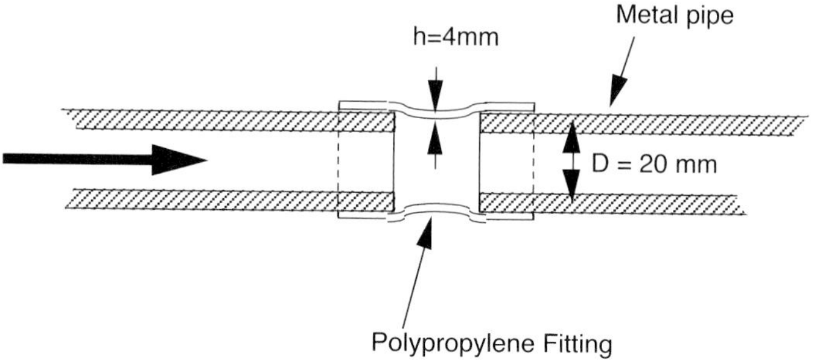

Figure 9.75 Polypropylene snap-fit assembly.

Using the creep curves given in Fig. 9.69, estimate the initial hoop stress in the polypropylene fitting.

The pressure of the water inside the tubes is 5 MPa. Estimate how long will this design be water-proof?

What initial hoop strain do you recommend to make the design water proof for 1 year?

Draw a 1 day isochronous stress-strain curve.

9.10 Derive the equation given for Young's modulus, E_{random}, for a plate with randomly oriented fibers. Assume that the fiber length is much larger than the thickness of the plate.

9.11 Similar to the Halpin-Tsai model, the mixing rule is another common technique to compute the properties of a unidirectional fiber reinforced composite part. The mixing rule for the longitudinal modulus is written as

$$E_L = \phi E_f + (1-\phi) E_m$$

Plot E_L as a function of ϕ for $E_f/E_m = 2, 10$ and 20.

9.12 You are to design two long fiber reinforced cylindrical structural components. Component A is to be loaded axially. Component B is to be loaded in torsion.

- What is the ideal orientation of the fibers for each component?
- If you were to choose between pultrusion, filament winding, and injection molding, which process would you choose to manufacture your components. Why?

9.13 You are asked to design a polyacetal snap-fit as depicted in Fig. 9.76. Find the maximum force between the two components shortly after assembly, and after one year. Use the creep data presented in Fig. 9.74 [20].

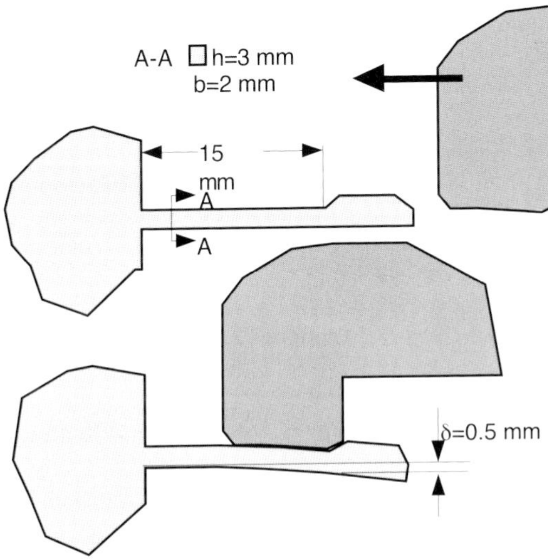

Figure 9.76 Polyacetal snap-fit assembly.

9.14 In a certain design, in which creep plays a significant role, you need to decide between a polyacetal and a polypropylene component. Your maximum allowable strain is 2% after one year. Which one would you choose if cost plays a major role in your design decision. To help you solve this problem, use the creep data presented in Figs. 9.76 and 9.69, for POM and PP, respectively, and the cost data presented in Chapter 1.

9.15 In Problem 9.13, how would your decision be if weight played a higher role than cost?

References

1. Domininghaus, H., *Plastics for Engineers*, Hanser Publishers, Munich, (1993).
2. Treloar, L.R.G., *The Physics of Rubber Elasticity*, 3rd. Ed., Clarendon Press, Oxford, (1975).
3. Courtesy ICIPC, Medellín, Colombia.
4. Ward, I.M., and Hadley, D.W., *An Introduction to Mechanical Properties of Solid Polymers*, John Wiley & Sons, Chichester, (1993).
5. Mooney, M., *J. Appl. Phys., 11*, 582, (1940).
6. Rivlin, R.S., and Saunders, D.W., *Phil. Trans. Roy. Soc., A243*, 251, (1951).
7. Gumbrell, S.M., Mullins, L., and Rivlin, R.S., *Trans. Faraday Soc., 49*, 1495, (1953).
8. Guth, E., and Simha, R., *Kolloid-Zeitschrift, 74*, 266, (1936).
9. Guth, E., *Proceedings of the American Physical Society*, (1937): *Physical Review, 53*, 321, (1938).
10. Smallwood, H.M., *J. Appl. Phys., 15*, 758, (1944).
11. Mullins, L., and Tobin, N.R. N.R., *J. Appl. Polym. Sci., 9*, 2993, (1965).
12. Knausenberger, R., and Menges, G., *SPE Technical Papers*, 39th ANTEC, 240, (1981).
13. Retting, W., *Rhelogica Acta, 8*, 258, (1969).
14. Schmachtenberg, E., Ph.D. Thesis, IKV, RWTH-Aachen, Germany, (1985).
15. Krämer, S., Ph.D. Thesis, IKV, RWTH-Aachen, Germany, (1987).
16. Weng, M., Ph.D. Thesis, IKV, RWTH-Aachen, Germany, (1988).
17. ASTM, Plastics (II), 08.02, ASTM, Philadelphia, (1994).
18. Nielsen, L.E., *Mechanical Properties of Polymers*, Van Nostrand Reinhold, New York, (1962).
19. Menges, G., (after Retting, W. and Lörtsch),*Werkstoffkunde Kunststoffe*, 3rd. Ed., Hanser Publishers, Munich, (1989),
20. Crawford, R.J., *Plastics Engineering*, 2nd ed., 47, Pergamon Press, Oxford, (1987).
21. O'Toole, J.L., *Modern Plastics Encyclopedia*, McGraw Hill, New York, (1983).
22. Thimm, Th., *Plastomere, Elastomere, Duromere, Kautschuk und Gummi, 14*, 8, 233, (1961).
23. Henning, F., *Kunststoffe, 65*, 401, (1975).
24. Hosemann, R., *Kristall und Technik, 11*, 1139, (1976).
25. Zachariades, A.E., and Kanamoto, T., *High Modulus Polymers*, A.E. Zachariades, and R.S. Porter, Eds., Marcel Dekker, Inc., New York, (1988).
26. Becker, H., Diploma Thesis, IKV, RWTH-Aachen, Germany, (1986).
27. Wende, A., *Glasfaserverstärkte Plaste*, VEB Deutscher Verlag für die Grundstoff-Industrie, Leipzig, (1969).
28. Nielsen, L.E., and Landel, R.F., *Mechanical Properties of Polymers and Composites*, 2nd Ed., Marcel Dekker, Inc., New York, (1994).

29. Krishnamachari, S.I., *Applied Stress Analysis of Plastics*, Van Nostrand Reinhold, New York, (1993).
30. Tsai, S.W., Halpin, J.C., and Pagano, N.J., *Composite Materials Workshop*, Technomic Publishing Co., Stamford, (1968).
31. Brintrup, H., Ph.D. Thesis, IKV, RWTH-Aachen, Germany (1974).
32. Ehrenstein, G.W., *Faserverbundkunststoffe*, Hanser Verlag, München, (1992).
33. Menges, G., and Bintrup, H., Chapter 4 in *Polymere Wrkstoffe*, Georg Thieme Verlag Stuttgart, (1984).
34. Powell, P.C., "*Engineering with Polymers*," Chapman and Hall, London, (1983).

10

Failure and Damage of Polymers

Failure of a component is a designer and engineer's biggest headache. The field of study that analyzes failed products or predicts failure is very large and complex. With polymers, failure is caused by mechanical, thermal, chemical, or other environmental influences. This chapter begins with a small overview of fracture mechanics. Here, simple models are presented that can be used to quantify the strength of polymers. The subsequent sections cover short term behavior, impact strength, creep rupture, fatigue, wear, and environmental effects.

10.1 Fracture Mechanics

A common and relatively simple approach to analyzing failure of polymer components is derived from linear elastic fracture mechanics (LEFM). The main assumption when using LEFM is that the material under consideration behaves like a linear elastic solid. The technique has been found to work well even for those materials for which the region near the crack tip behaves inelastically but everywhere the materials shows elastic behavior. When a polymer component is loaded at relatively high speeds, its behavior can be considered elastic, justifying the usage of linear elastic fracture mechanics to analyze its failure. In fact, polymer components made of a ductile material will often undergo a brittle failure when subjected to impact.

However, LEFM is not appropriate to model the fracture behavior in viscoelastic media or where extensive plasticity is present during deformation. For these cases the *J-integrals* can be used to determine energy change during fracture. LEFM does not apply to long-term tests such as creep rupture since the mechanical properties of the component or test specimen are viscoelastic. If the time scale of the loaded component is carried to the extreme, say on the order of a few months or years, the material behavior is viscous.

The next sections will discuss linear elastic as well as linear viscoelastic fracture mechanics, applying them to quantitatively analyze strength properties of cracked polymer components or parts with sharp notches or stress concentrators.

10.1.1 Fracture Predictions Based on the Stress Intensity Factor

From the basic three crack growth modes of failure, shown in Fig. 10.1, mode I most resembles an internal crack growing under a tensile load. For the analysis, consider the cracked body displayed in Fig. 10.2 with a crack length of $2a$ and an applied stress σ. Near but not at the crack tip, the stress in the direction of loading can be approximated by

$$\sigma_y = \frac{K_{Ic}}{\sqrt{2\pi r}} \tag{10.1}$$

where r is the distance from the crack tip and K_{Ic} is the *critical stress intensity factor* needed for mode *I* crack growth and failure, defined by

$$K_{Ic} = \sigma\sqrt{\pi a} \tag{10.2}$$

This stress intensity factor depends on crack size relative to component size, crack shape, boundary conditions, etc. Hence, a more general form of Eq. 10.2 is given by

$$K_{Ic} = Y\sigma\sqrt{\pi a} \tag{10.3}$$

where Y is a dimensionless correction factor tabulated for various geometries in Table 10.1. In a linear elastic solid, fracture occurs instantaneously at a combination of stress and crack size that results in a critical stress intensity factor, K_{Ic}. The critical stress intensity factor is a material property that depends on the temperature, grade of polymer, orientation, etc., where a large value implies a tough material. Table 10.2 [1, 2] shows stress intensity factors, also called *fracture toughness*, for common polymers.

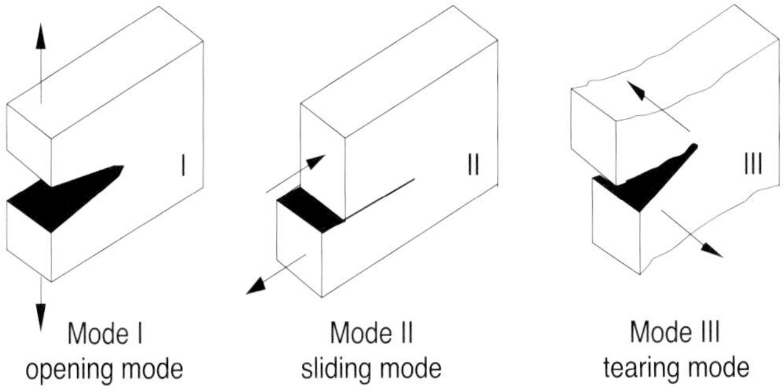

Figure 10.1 Three modes of crack loadings.

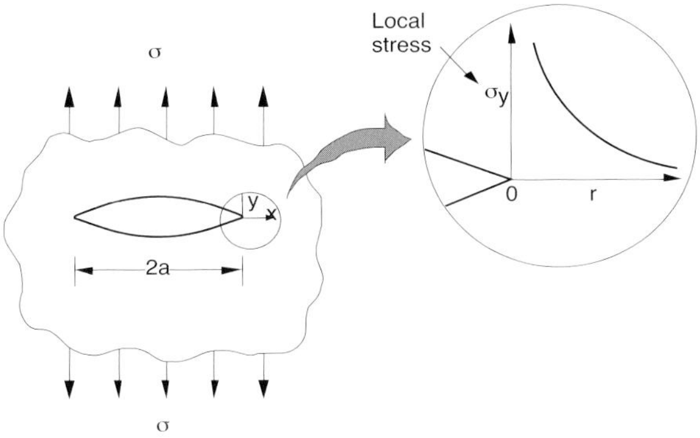

Figure 10.2 Stress near a crack tip in an infinite plate.

Table 10.1 Values for Geometry Factor Y For Various Crack Configurations

Central crack	$Y = \sqrt{\frac{W}{\pi a}\tan\left(\frac{\pi a}{W}\right)}$
Double edge crack	$Y = \sqrt{\frac{W}{\pi a}\tan\left(\frac{\pi a}{W}\right) + \frac{0.2W}{\pi a}\sin\left(\frac{\pi a}{W}\right)}$
Single edge crack	$Y = 1.12 - 0.23\left(\frac{a}{w}\right) + 10.6\left(\frac{a}{w}\right)^2 - 21.7\left(\frac{a}{w}\right)^3 + 30.4\left(\frac{a}{w}\right)^4$

Table 10.2 Values of Plane Stress Intensity Factor and Strain Toughness for Various Materials

Material	$K_{Ic}(MN/m^{3/2})$	$G_{Ic}(kJ/m^2)$
ABS	2-4	5
Acetal	4	1.2-2
Epoxy	0.3-0.5	0.1-0.3
LDPE	1	6.5
MDPE-HDPE	0.5-5	3.5-6.5
Polyamide 66	3	0.25-4
Polycarbonate	1-2.6	5
Polyester-glass reinforced	5-7	5-7
Polypropylene copolymer	3-4.5	8
Polystyrene	0.7-1.1	0.3-0.8
PMMA	1.1	1.3
uPVC	1-4	1.3-1.4
Aluminum-alloy	37	20
Glass	0.75	0.01-0.02
Steel-mild	50	12
Steel-alloy	150	107
Wood	0.5	0.12

10.1.2 Fracture Predictions Based on an Energy Balance

To analyze a mode I crack growth case using an energy balance and LEFM, consider the cracked body used in the previous analysis, where the actual forces are used instead of stresses, as displayed in Fig. 10.3. The crack width is also $2a$, and the body is subjected to a load F. The load-displacement behavior of the cracked body is described by the solid line in Fig. 10.4. The elastic energy stored in the loaded component is given by the area under the curve:

$$U_e = \frac{1}{2}FX \tag{10.4}$$

As the crack grows by an amount Δa, the stiffness of the component decreases as shown by the dashed line in Fig. 10.4, and the elastic energy changes to

$$U'_e = \frac{1}{2}(F + \Delta F)(X + \Delta X) \tag{10.5}$$

The change in stored elastic energy for the body with the growing crack is described by

$$\Delta U = U_e - U_e' \tag{10.6}$$

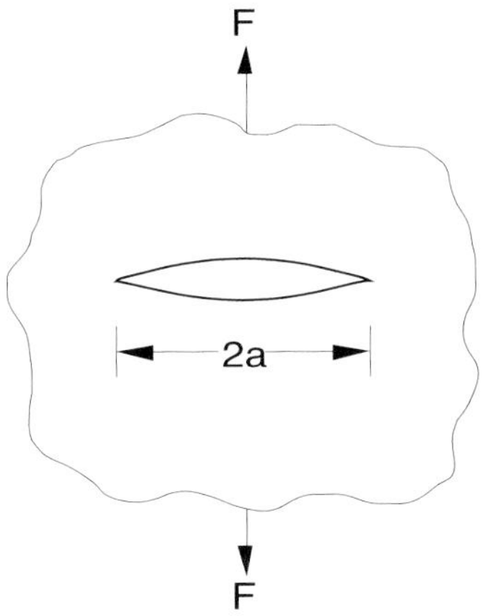

Figure 10.3 Load applied on a cracked specimen.

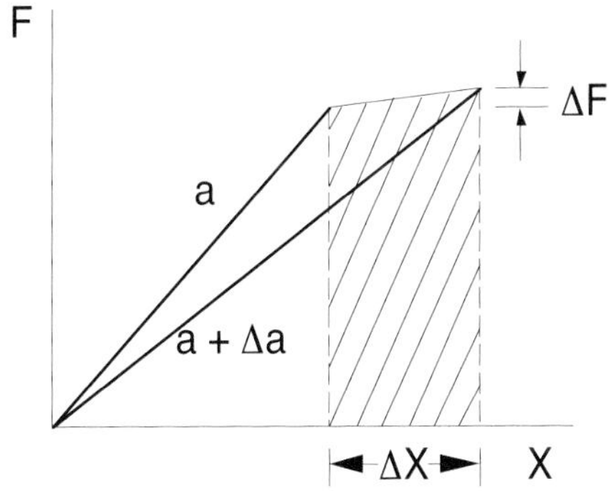

Figure 10.4 Linear elastic behavior of a cracked specimen.

The external work done as the crack grows the distance Δa is defined by the shaded area in Fig. 10.4 and can be written as

$$\Delta U_c = \left(F + \frac{\Delta F}{2}\right)\Delta X \tag{10.7}$$

By subtracting Eq. 10.6 from Eq. 10.7 we can compute the energy required to generate new surfaces during fracture:

$$\Delta U_c - \Delta U = \frac{1}{2}\left(F\Delta X - X\Delta F\right) \tag{10.8}$$

Griffith's hypothesis [3], commonly used as a fracture criterion, states that for a crack to increase in size, the rate of stored elastic energy decrease must be larger or equal to the rate at which surface energy is created during crack growth. Griffith's hypothesis can be related to the expression in Eqs. 10.6 and 10.7 as

$$\frac{1}{t}\frac{\partial(U_c - U)}{\partial a} \geq G_{Ic} \tag{10.9}$$

where G_{Ic} defines the energy required to increase a crack by a unit length in a component of unit width and is usually referred to as the *elastic energy release rate*, the *toughness* or the *critical energy release rate.* Equations 10.6, 10.7, and 10.9 can be combined to give

$$G_{Ic} = \frac{1}{2}\left(F\frac{\partial x}{\partial a} - X\frac{\partial F}{\partial a}\right) \tag{10.10}$$

Making use of the compliance defined by

$$J = \frac{X}{F} \tag{10.11}$$

and defining the force at the onset of crack propagation by F_c, we can rewrite Eq. 10.10 as

$$G_{Ic} = \frac{F_c^2}{2t}\frac{\partial J}{\partial a} \tag{10.12}$$

This equation describes the fundamental material property, G_{Ic}, as a function of applied force at fracture and the rate at which compliance changes with respect to crack size. Equation 10.12 is more useful if it is written in terms of stress as

$$G_{Ic} = \frac{\pi\sigma_c^2 a}{E} \tag{10.13}$$

This equation only applies for plane stress and must be redefined for the plane strain case as

$$G_{Ic} = \frac{\pi \sigma_c^2 a}{E}\left(1 - \nu^2\right) \tag{10.14}$$

Table 10.2 also displays typical values of toughness for common materials.

By substituting Eq. 10.3 into Eqs. 10.13 and 10.14, we can relate the stress intensity factor K_{Ic} and the toughness G_{Ic} with

$$G_{Ic} = \frac{K_{Ic}^2}{Y^2 E} \tag{10.15}$$

for the plane stress case and

$$G_{Ic} = \frac{K_{Ic}^2}{Y^2 E}\left(1 - \nu^2\right) \tag{10.16}$$

for the plane strain case.

10.1.3 Linear Viscoelastic Fracture Predictions Based on J-Integrals

Because of the presence of viscous components and the relatively large crack-tip plastic zone during deformation, the elastic energy release rate, G_{Ic}, is not an appropriate measure of the energy release or toughness during fracture of many thermoplastic polymers. The J-integral concept was developed by Rice [4] to describe the strain energy transfer into the crack tip region and, using the notation in Fig. 10.5, is defined by

$$J_{Ic} = \int_{\Gamma} W dy - \int_{\Gamma} \underline{\mathbf{T}} \frac{\partial \underline{\mathbf{u}}}{\partial \mathbf{x}} ds \tag{10.17}$$

where W represents the strain energy density, $\underline{\mathbf{T}}$ the traction vector, and $\underline{\mathbf{u}}$ the displacement vector. As with G_{Ic}, J_{Ic} is a measure of the energy release rate, thus, analogous to Fig. 10.4. We can consider Fig. 10.6 to determine the energy release in a viscoelastic material with a growing crack. Again, the shaded area in the curve represents the energy release rate which, apart from second order effects, is the same for constant load and constant deformation and for a specimen of thickness t can be computed by

$$\Delta U = J_{Ic} t \Delta a \tag{10.18}$$

which gives

$$J_{Ic} = -\frac{1}{t}\left(\frac{\partial U}{\partial a}\right) \tag{10.19}$$

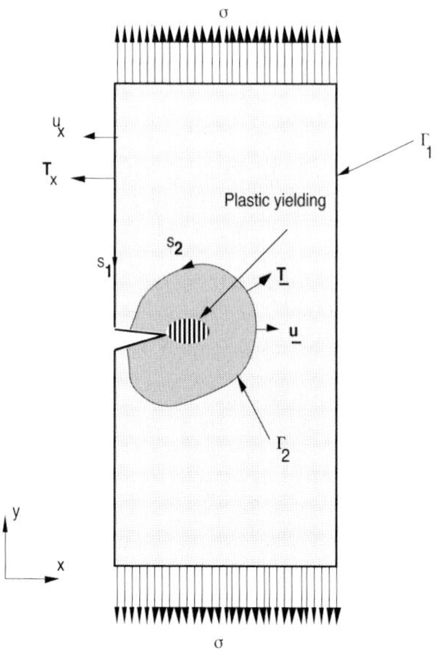

Figure 10.5 Strain energy transfer into the crack tip region of a component.

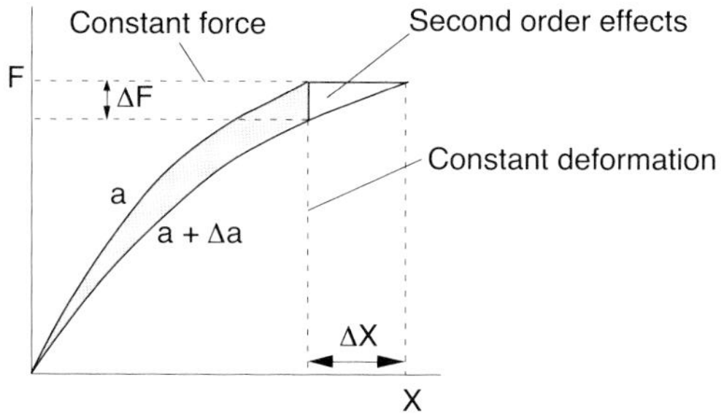

Figure 10.6 Non-linear behavior of a cracked specimen.

At fracture, J_{Ic} can be related to the crack opening displacement, δ, and the yield stress, σ_y, by

$$J_{Ic} = \sigma_y \delta \tag{10.20}$$

One important feature of the J-integral as defined in Eq. 10.17 is that it is path independent, so any convenient path can be chosen where stresses and displacements are known. A good choice for a path is one that completely encloses the crack tip with the plastic region but which lies within the elastic region of the body. Another property of the J-integral is that for linear elastic cases, $J_{Ic} = G_{Ic}$.

10.2 Short-Term Tensile Strength

As discussed in the previous chapter, short-term tensile tests DIN 53457 and ASTM D638 are available to test the stress-strain behavior of polymeric materials at room temperature (23 °C)[1] . The resulting data is plotted in a stress-strain curve that reflects the type of material and the mode of failure associated with the test polymer. When regarding polymeric materials from the failure mode point of view, there are two general types of fracture: *brittle* and *ductile*. However, depending on the temperature, environment, and whether a component is notched or not, a material can fail either way. Figure 10.7 shows a diagram developed by Vincent [5] which helps the engineer to distinguish between the different modes of failure of various thermoplastics. The figure represents a plot of brittle stress at about -180 °C (σ_B) versus yield stress at -20 °C (Δ) and 20 °C (O). The data points to the right of line A represent polymers that are brittle unnotched. The points that are between lines A and B are those polymers that are ductile unnotched but brittle notched, and those to the left of line B represent polymers that are ductile even when notched.

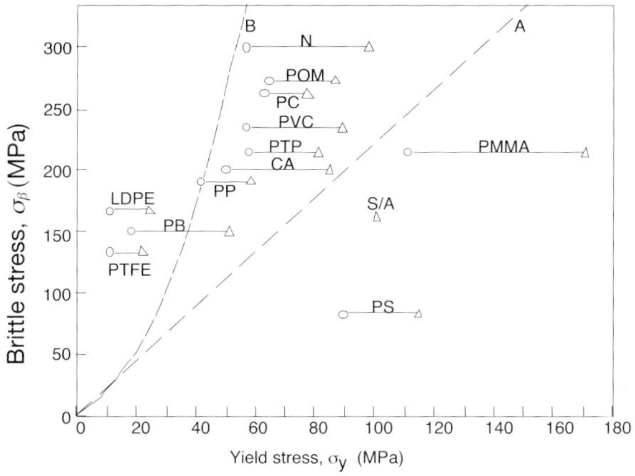

Figure 10.7 Brittle stress at -180 °C versus yield stress at -20 °C (Δ) and 20 °C (o) for various polymers.

1 We believe that it is imperative that the rate of deformation within the tensile specimen be maintained constant. In Chapter 9 we presented the Knausenberger-Menges technique that maintains this condition using optical techniques.

10.2.1 Brittle Failure

Brittle failure usually occurs with thermoplastics below their glass transition temperature, T_g, and with highly cross-linked polymers. However, as discussed later in this chapter, brittle failure also occurs in creep rupture and fatigue tests performed at temperatures above the glass transition temperature of the polymer. Typically, it occurs at very small strains— perhaps 1% or less— and it is generally associated with amorphous thermoplastics below their glass transition temperature. Figure 10.8 shows the stress-strain curve for the injection molded polystyrene ASTM D638M type I test specimen shown in Fig. 10.9. The stress strain curve has a constant slope until the point where small microcracks form just before failure. These small microcracks form in the plane perpendicular to the maximum principal stress and are nucleated at points of high stress concentrations such as scratches, dust particles in the material and material inhomogeneities. These cracks, which are more commonly known as*crazes*, impair clarity and reflect light, which makes them particularly obvious in transparent materials.

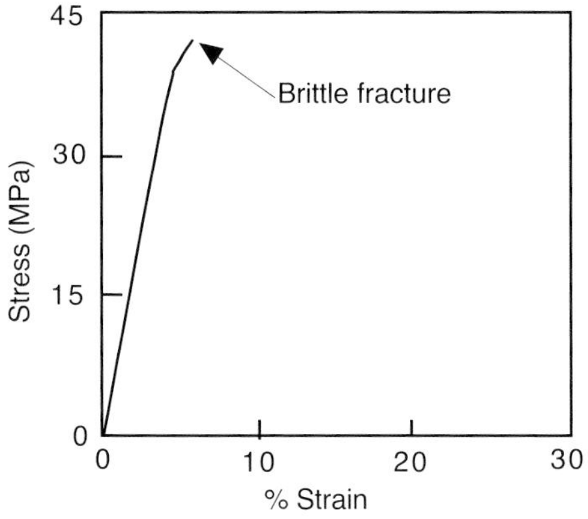

Figure 10.8 Stress-strain curve for a polystyrene at room temperature.

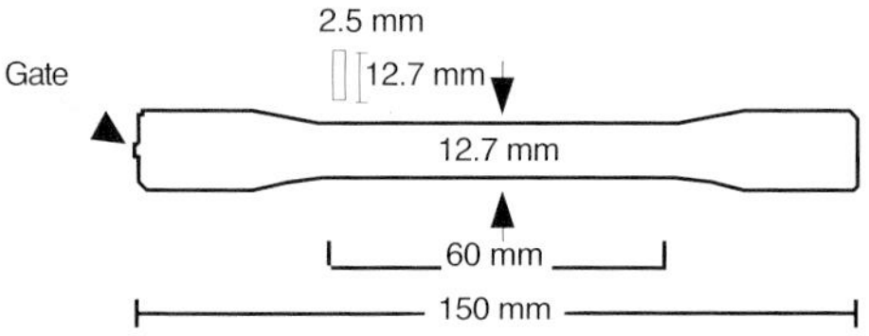

Figure 10.9 ASTM Type I test specimen.

Figure 10.10 [6] shows electron micrographs of the center and edge of a craze in a polystyrene specimen. As can be seen in the micrograph, the craze boundaries are connected with load bearing fibrils which make them less dangerous than actual cracks. Crazing is directly related to the speed at which the component or test specimen is being deformed. At high deformation speeds, the crazes are small and form shortly before failure, which makes them difficult to detect. At slow rates of deformation, crazes tend to be large and occur early on during loading. A typical craze is about 0.5 μm wide at the center and 200 μm long.[2] However, the length and width of a craze is material-dependent.

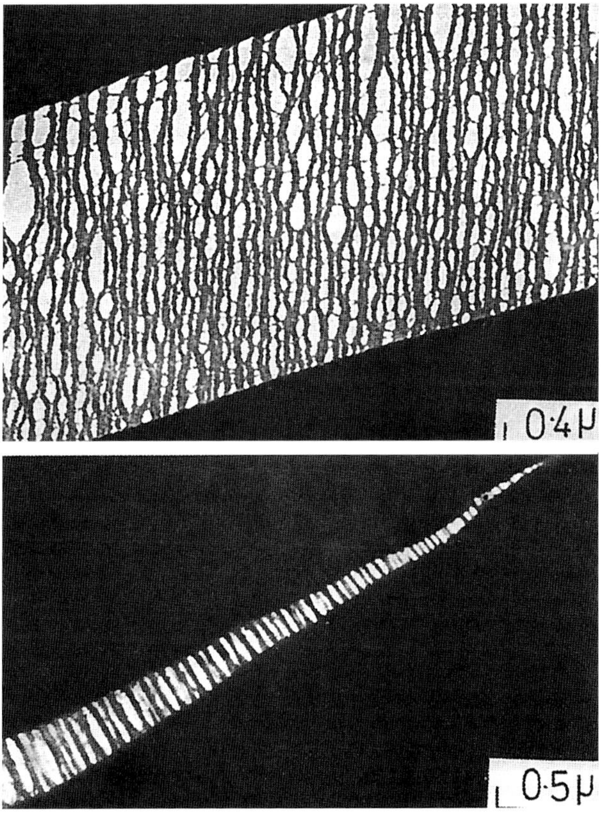

Figure 10.10 Electron micrographs of the center and edge of a craze in a polystyrene specimen.

2 The length of a craze will vary between 10 μm and 1000 μm.

Figure 10.11 [7] shows the relation between stress, strain, craze size formation and failure. By extrapolating the craze formation line, one can see that at high stress crazes form at the same time as failure. When crazes form under static loading, they do not pose immediate danger to the polymer component. However, crazes are irreversible, and they imply a permanent damage within the material. It should also be noted that once crazes and microcracks have formed, the material no longer obeys the laws of linear viscoelasticity. The limit strain at which microcracks will form is sometimes depicted by ε_{Fx}. Complex models exist, beyond the scope of this book, which relate the critical strain to the surface energy within a craze [8].

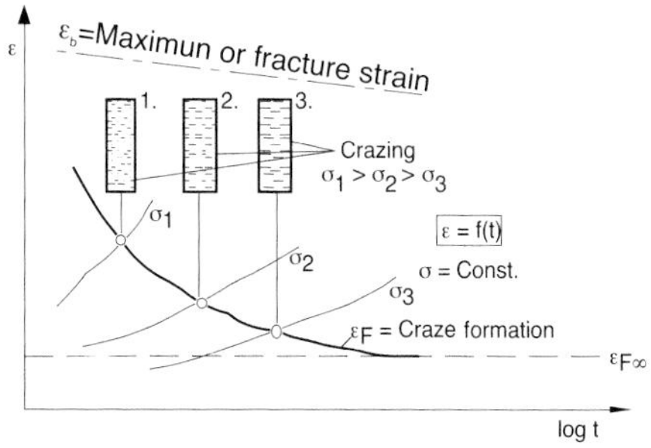

Figure 10.11 Relation between stress, strain, time and craze formation.

Figure 10.12 [7] shows the relationship between strain, time, and damage for PMMA. The bottom of the figure shows the time-temperature superposition. For example, the damage that occurs at 10 hours for 23 °C will occur at 10^5 hours for a component at 60 °C. It is interesting to see how the critical strain, ε_{Fx}, is delayed as the temperature of the polymer component rises.

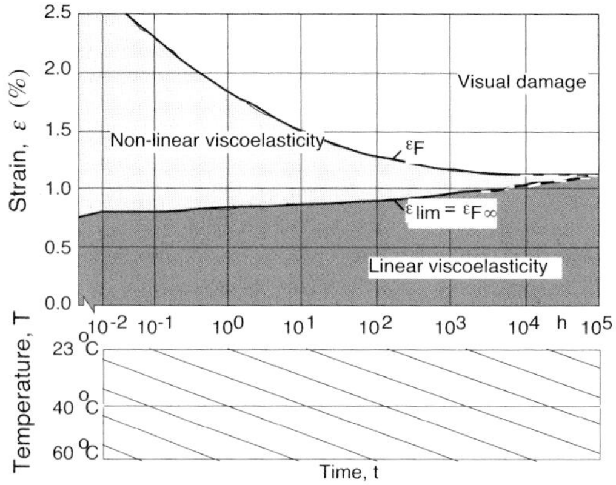

Figure 10.12 Strain limits for linear viscoelasticity.

Figure 10.13 [7] shows a plot of critical strain versus temperature for an impact resistant polystyrene and compares the damage behavior curve to the shear modulus. The two curves are almost a mirror images of each other and demonstrate that the formation of microcracks is inversely proportional to the stiffness of the material. The figure also demonstrates the influence of orientation on the onset of microcracks. As expected, in a component loaded perpendicular to the orientation, or across the polymer chains, the microcracks form earlier than in one loaded parallel to the orientation direction.

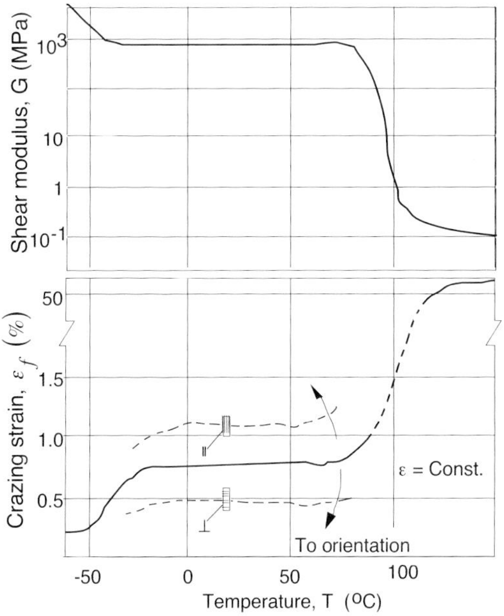

Figure 10.13 Shear modulus and crazing strain as a function of temperature for a high impact polystyrene.

10.2.2 Ductile Failure

A ductile failure takes place with semi-crystalline thermoplastics at temperatures between the glass transition temperature, T_g, and the melting temperature, T_m. A ductile failure is a succession of several events, as is clearly shown in the stress-strain diagram for polypropylene, shown in Fig. 10.14 and explained in the following paragraphs.

At first, the semi-crystalline polymer behaves like an elastic solid whose deformation is reversible. For the polypropylene sample test results shown in Fig. 10.14, this linear elastic behavior holds for deformations of up to 0.5%. This behavior takes place when the component's load is applied and released fairly quickly, without causing a permanent damage to the material and allowing the component to return to its original shape. This is graphically depicted in Fig. 10.15 [7].

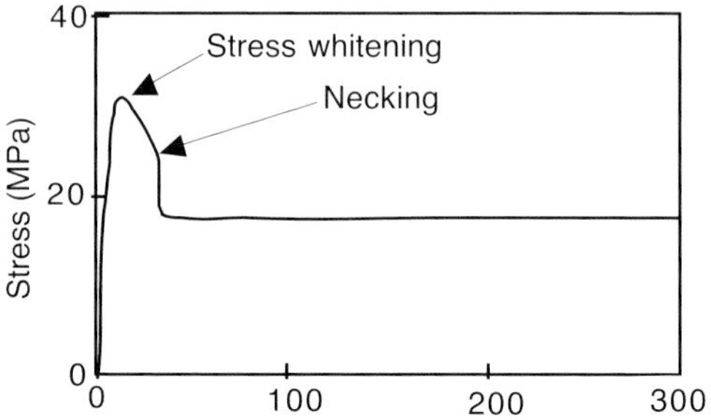

Figure 10.14 Stress-strain curve for a polypropylene at room temperature.

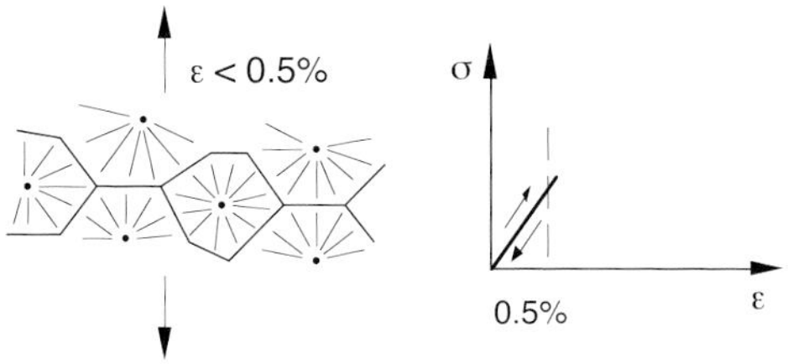

Figure 10.15 Elastic deformation within a spherulitic structure.

If the load is increased or the process is slowed, the stress-strain curve becomes non-linear, reflected by the reduction of rate of stress increase. At this point, microcracks form in the interface between neighboring spherulites, as shown in the photograph of Fig. 10.16 [9] and schematically in Fig. 10.17 [7].

Figure 10.16 Micrograph of crack formation at inter-spherulitic boundaries (polypropylene).

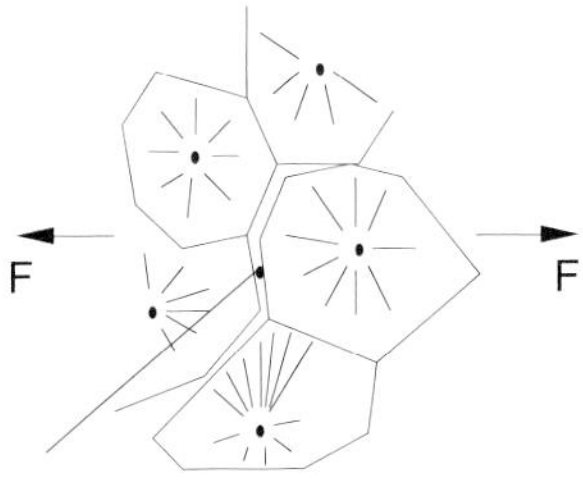

Figure 10.17 Schematic of crack formation at inter-spherulitic boundaries.

The formation of such microcracks, also called *stress whitening*, is an irreversible process, causing a permanent deformation in the polymer component. Other than the white coloration that makes itself noticeable in the stressed component, the cracks are not visible to the naked eye. These microcracks are fairly constant in length, about the size of the spherulites. Their formation and growth, and their relation to the stress-strain behavior of a semicrystalline polymer, are depicted in Fig. 10.18 [7].

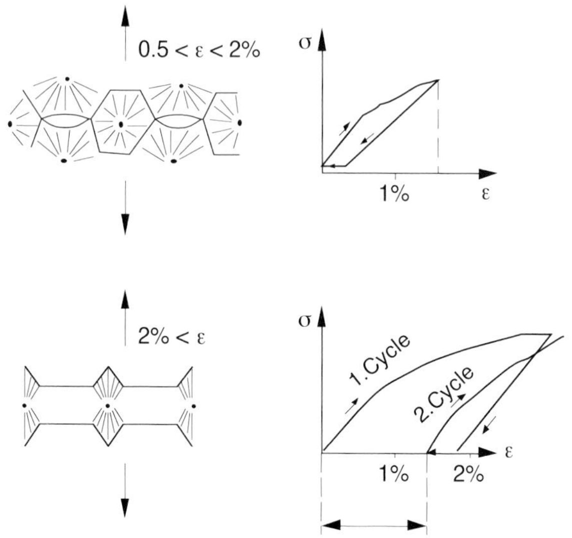

Figure 10.18 Relation between microcrack formation and the stress-strain behavior of a semi-crystalline polymer.

Figure 10.14 shows that by further deforming the specimen, the stress-strain curve reaches a maxima called the *yield point* or *yield strength*. Beyond the yield point the stress drops, an event that is followed by *necking*, a localized reduction in cross-sectional area. Once necking has occurred, the specimen or component continues a long cold drawing process where the spherulitic structure is first deformed and then broken up, creating highly oriented regions within the polymer component. Figure 10.19 shows the progression of the necked region during tensile tests of polypropylene samples. This mode of failure is common in semi-crystalline components, such as the one shown in Fig. 10.20 for a polyethylene automotive fan blade.

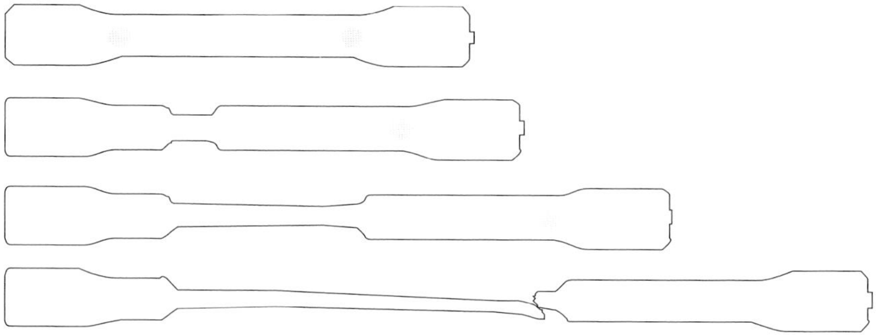

Figure 10.19 Necking progression of a polypropylene specimen during a tensile test.

Figure 10.20 Necking of a polyethylene automotive fan blade [11].

Necking and cold drawing can be explained with the molecular model [10] shown in Fig. 10.21. Once the amorphous ties between lamellae have been completely extended, a slip-tilting of the lamellae is induced. As deformation continues, lamellae fragments get aligned in the direction of draw, forming fibrils of alternating crystal blocks and stretched amorphous regions.

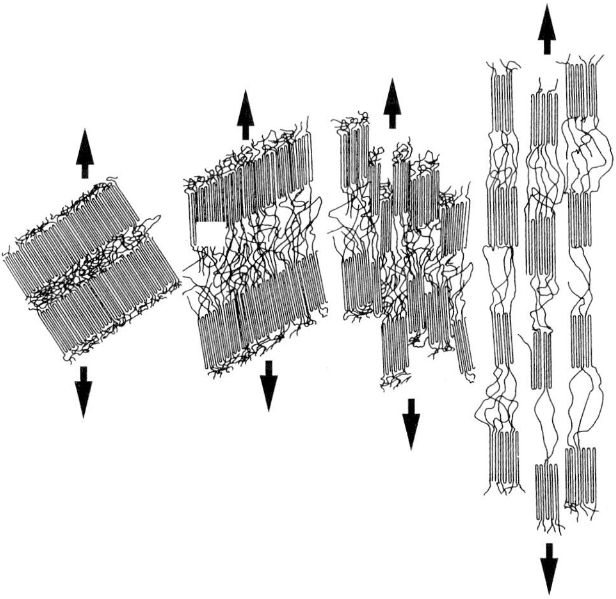

Figure 10.21 Molecular model used to represent necking and drawing during a tensile test.

10.2.3 Failure of Highly Filled Systems or Composites

A polymer that usually fails under a brittle fracture can be toughened by adding filler particles. The most common examples for this effect are high impact polystyrene (HIPS) and ABS. In both these systems, brittle polymers, acrylic, and polystyrene, were toughened with the inclusion of rubbery particles into the material as shown in the schematic of the structure of HIPS in Fig. 10.22 [12]. This increase in toughness is reflected in the stress-strain behavior of HIPS shown in Fig. 10.23, where the rubbery elastic behavior of the rubber particles lowered the stiffness and ultimate strength of the material but increased its toughness. The rubber particles halt the propagation of a growing craze. Characteristic lengths of crazes that form in such systems are only as large as the characteristic gap between filler particles. This creates a system that has a large number of small crazes instead of the small number of large crazes present in the unfilled polymer. Table 10.3 [2] presents the effect of volume fraction of rubber particles on the mechanical and fracture properties of rubber-modified polystyrene. The impact strength and the fracture strain are maximized at a rubber particle volume fraction of about 20%.

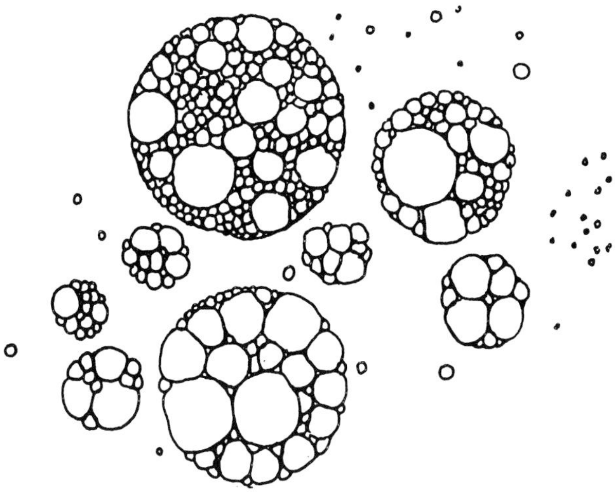

Figure 10.22 Schematic of the structure of a high impact polystyrene.

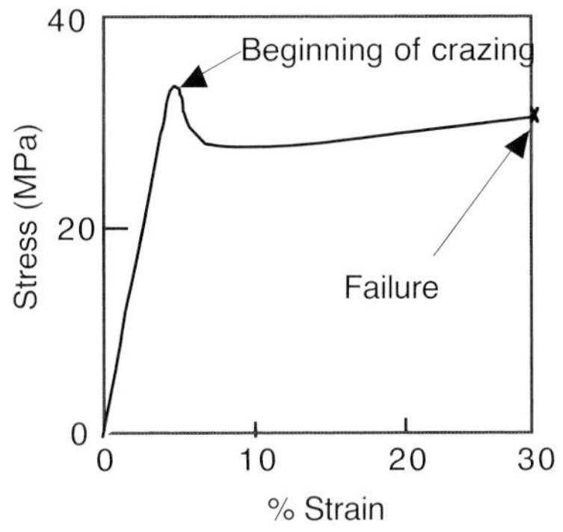

Figure 10.23 Stress-strain curve of a high impact polystyrene.

Table 10.3 Effect of Rubber Particle Volume Fraction, ϕ, on the Properties of Rubber Modified Polystyrene

Volume fraction ϕ	Tensile modulus GPa	Impact strength MJ/m^3	Fracture strain %
0.06	2.8	0.42	3
0.12	2.4	1.90	20
0.22	1.9	11.6	45
0.30	1.0	5.6	34
0.78	0.55	1.2	8

This increase in toughness can be observed even if the filler material is also brittle. Electron micrographs of such systems have shown that cracks propagate until they hit a filler particle, which often stops the propagation [2]. In thermosetting polymers, this effect is commonly referred to as *crack pinning*. Figure 10.24 compares plots of impact-absorbed energy as a function of specimen size for unfilled epoxy and epoxies filled with irregular-shaped silica with weight percents of 55% and 64%.

The failure of a fiber-filled material begins at the interface between filler or reinforcement and the matrix, as shown in the electron micrograph presented in Fig. 10.25 [13] . The micrograph was taken when a glass fiber filled polyester specimen was placed under loading and shows the breakage of the adhesion between imbedded glass fibers and their matrix. This breakage is generally referred to as *debonding*. This initial microcrack formation is reflected in a stress-strain curve by the deviation from the linear range of the elastic constants. In fact, the failure is analogous to the microcracks that form between spherulites when a semi-crystalline polymer is deformed.

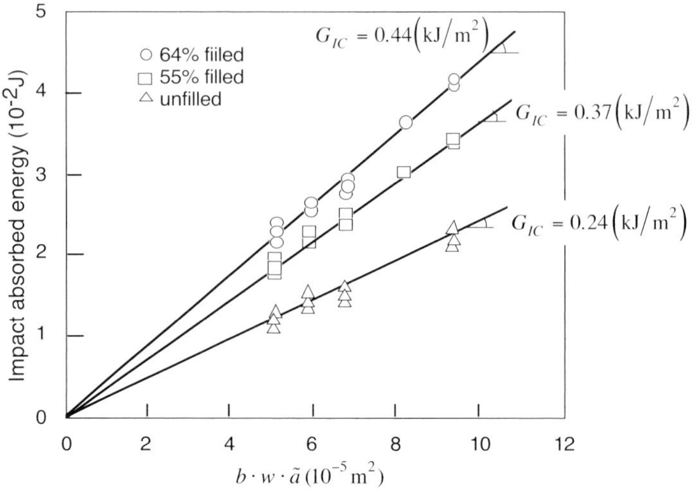

Figure 10.24 Impact absorbed energy as a function of specimen size for unfilled epoxy and epoxies filled with irregular-shaped silica with weight percents of 55% and 64%.

Figure 10.25 Electron micrograph of crack formation between polyester matrix and a glass fiber.

10.3 Impact Strength[3]

In practice, nearly all polymer components are subjected to impact loads. Since many polymers are tough and ductile, they are often well suited for this type of loading. However, under specific conditions even the most ductile materials, such as polypropylene, can fail in a brittle manner at very low strains. These types of failure are prone to occur at low temperatures and at very high deformation rates.

According to several researchers [14, 15] a significantly high rate of deformation leads to complete embrittlement of polymers which results in a lower threshold of elongation at break. Menges and Boden designed a special high-speed elongational testing device that was used to measure the minimum work required to break the specimens. The minimum strain, ε_{min}, which can be measured with such a device, is a safe value to use in calculations for design purposes. One should always assume that if this minimum strain value is exceeded at any point in the component, initial fracture has already occurred. Table 10.4 [16, 17] presents minimum elongation at break values for selected thermoplastics on impact loading.

On the other hand, the stiffness and the stress at break of the material under consideration increase with the rate of deformation. Table 10.5 [16] presents data for the stress at break, σ_{min}, for selected thermoplastics on impact loading. This stress corresponds to the point where the minimum elongation at break has just been reached.

Table 10.4 Minimum Elongation at Break on Impact Loading

Polymer	ε_{min} (%)
HMW-PMMA	2.2
PA6 + 25% SFR	1.8
PP	1.8
uPVC	2.0
POM	4.0
PC + 20% SFR	4.0
PC	6.0

3 The term "impact strength" is widely misused, since one is actually referring to energy absorbed before failure.

Table 10.5 Minimum Stress at Break on Impact Loading

Polymer	σ_{min} (MPa)
HMW-PMMA	135
PA6 + 25% SFR	175
uPVC	125
POM	>130
PC + 20% SFR	>110
PC	>70

Figure 10.26 summarizes the stress-strain and fracture behavior of a HMW-PMMA tested at various rates of deformation. The area under the stress-strain curves represents the *volume-specific energy to fracture* (w). For impact, the elongation at break of 2.2% and the stress at break of 135 MPa represent a minimum of volume-specific energy because the stress increases with higher rates of deformation, but the elongation at break remains constant. Hence, if we assume a linear behavior, the *minimum volume-specific energy absorption* up to fracture can be calculated using

$$w_{min} = 0.5\sigma_{max}\varepsilon_{min} \tag{10.21}$$

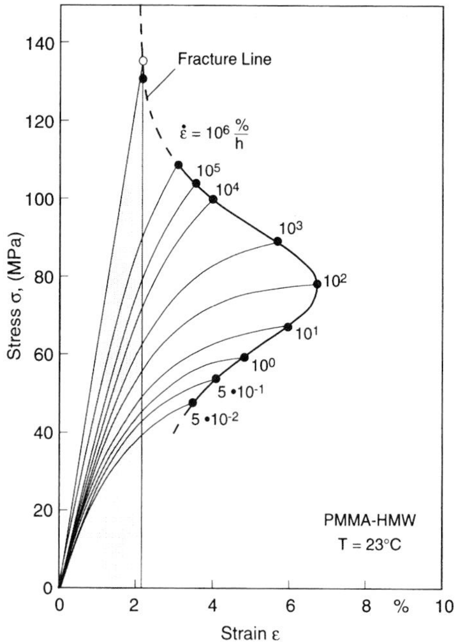

Figure 10.26 Stress-strain behavior of HMW-PMMA at various rates of deformation.

If the stress-strain distribution in the polymer component is known, one can estimate the minimum energy absorption capacity using w_{min}. It can be assumed that failure occurs if w_{min} is exceeded in any part of the loaded component. This minimum volume-specific energy absorption, w_{min}, can be used as a design parameter. It was also used by Rest [18] for fiber-reinforced polymeric materials.

Glass fiber reinforced thermoplastics generally undergo brittle failure. Figure 10.27 [16] shows how the impact resistance of a polyamide 6 material was dramatically reduced with the addition of a glass reinforcement. Interesting to note is that the impact resistance of polyamide 6 with 6% and 30% glass reinforcement is essentially the same when compared to the unfilled material. However, a specimen with a sharp notch that resembles a crack will have a higher impact resistance if it is glass reinforced. Figure 10.28 [16] illustrates this by showing a plot of izod test data for polyamide 6 specimens and polyamide 6 specimens filled with 30% glass reinforcement as a function of notch radius. Here, the data for a small notch radius reflect the energy it takes to propagate the crack through the specimen. The data for a large notch radius approach the energy it takes to both, initiate and propagate a crack. In a filled polymer, a filler can sometimes increases the impact resistance of the component. For example, the volume fraction of crazes in a rubber particle filled polystyrene increases with increasing particle content. Figure 10.29 shows the volume fraction of craze voids as a function of rubber particle content in a high impact polystyrene. The figure also schematically depicts the relation between rubber particle content and craze geometry.

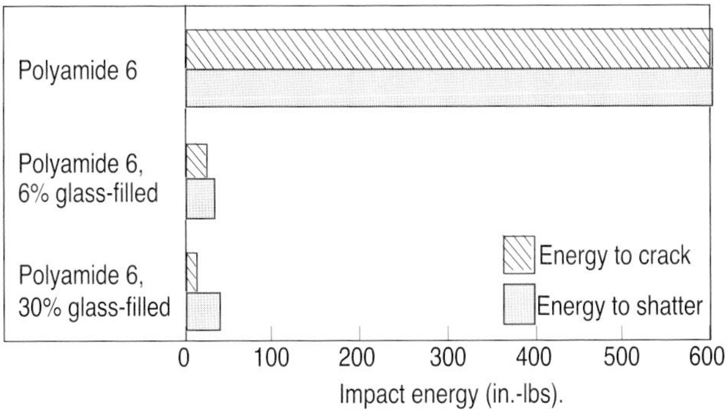

Figure 10.27 Effect of glass fiber content on energy to crack and shatter for polyamide 6.

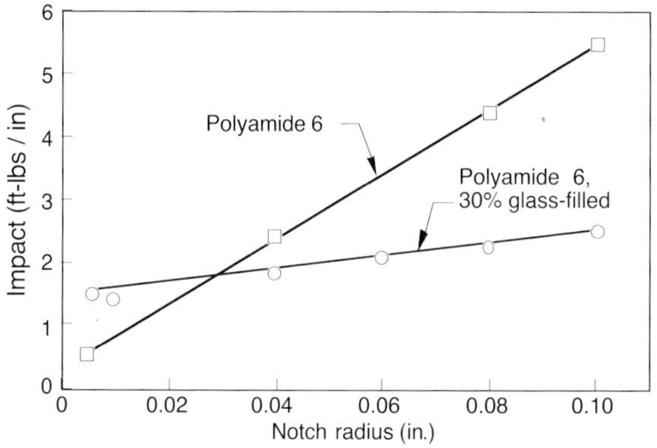

Figure 10.28 Impact strength as a function of notch tip radius for a filled and unfilled polyamide 6.

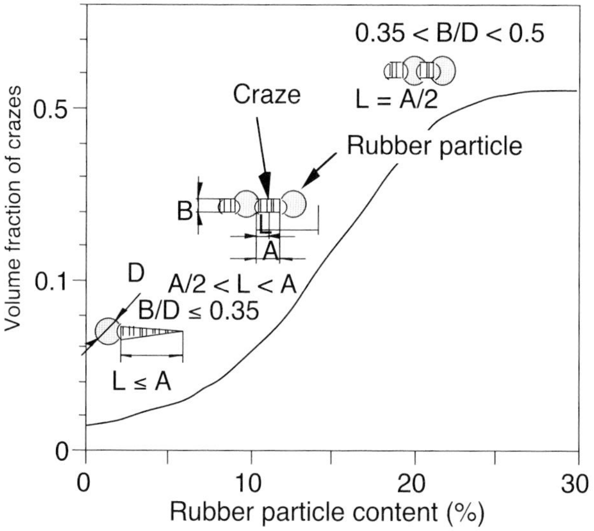

Figure 10.29 Volume fraction of craze voids as a function of rubber particle content in a high impact polystyrene.

The impact strength of a copolymer and polymer blend of the same materials can be quite different, as shown in Fig. 10.30. From the figure it is clear that the propylene-ethylene copolymer, which is an elastomer, has a much higher impact resistance than the basic polypropylene-polyethylene blend. It should be pointed out here that elastomers usually fail by ripping. The ripping or tear strength of elastomers can be tested using the ASTM D1004, ASTM D1938, or DIN 53507 test methods. The latter two methods make use of rectangular test specimens with clean slits cut along the center. A typical tear propagation test for an elastomer composed of 75 parts of natural rubber (NR) and 25 parts of styrene butadiene rubber (SBR) is presented in Fig. 10.31.[4] The tear strength of elastomers can be increased by introducing certain types of particulate fillers. For example, a well dispersed carbon black filler can double the ripping strength of a typical elastomer. Figure 10.32 [7] shows the effect that different types of fillers have on the ripping strength of a polychloroprene elastomer.

In general, one can say if the filler particles are well dispersed and have diameters between 20 nm and 80 nm, they will reinforce the matrix. Larger particles will act as microscopic stress concentrators and will lower the strength of the polymer component. A case where the filler adversely affects the polymer matrix is presented in Fig. 10.33 [7], where the strength of PVC is lowered with the addition of a calcium carbonate powder.

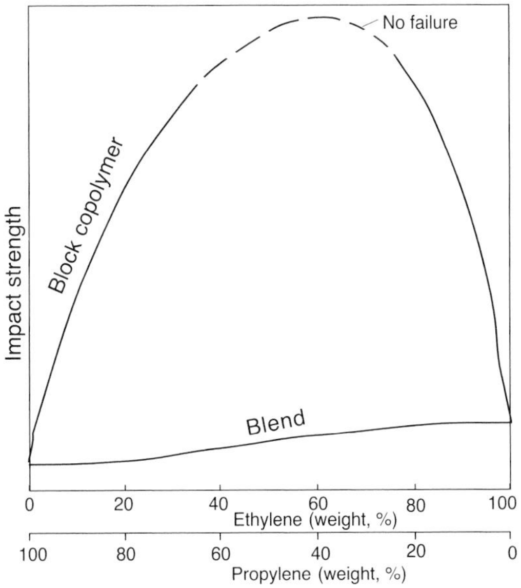

Figure 10.30 Impact strength of a propylene-ethylene copolymer and a polypropylene -polyethylene polymer blend.

4 Courtesy of ICIPC, Medellín, Colombia.

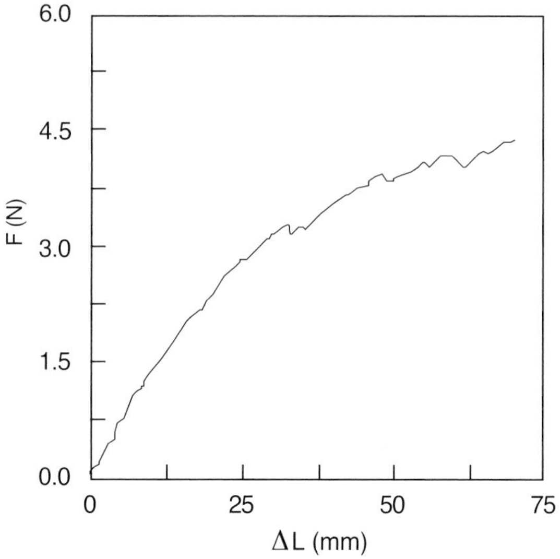

Figure 10.31 Tear propagation test for an elastomer composed of 75 parts of natural rubber (NR) and 25 parts of styrene butadiene rubber (SBR).

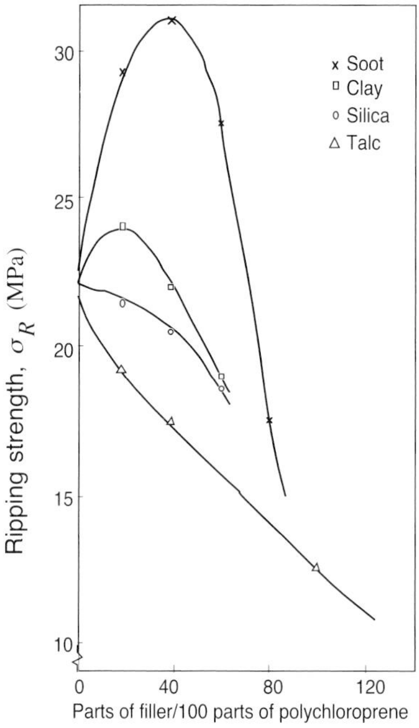

Figure 10.32 Ripping strength of a polychloroprene elastomer as a function of filler content for different types of fillers.

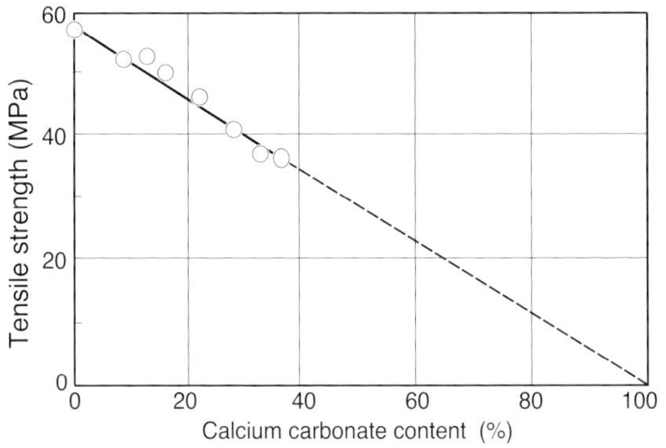

Figure 10.33 Tensile strength of PVC as a function of calcium carbonate content.

10.3.1 Impact Test Methods

Impact tests are widely used to evaluate a material's capability to withstand high velocity impact loadings. The most common impact tests are the *Izod* and the *Charpy* tests. The Izod test evaluates the impact resistance of a cantilevered notched bending specimen as it is struck by a swinging hammer. Figure 10.34 [19] shows a typical Izod-type impact machine, and Fig. 10.35 [19] shows a detailed view of the specimen, the clamp, and the striking hammer. The pendulum or hammer is released from a predetermined height and after striking the specimen, travels to a recorded height. The energy absorbed by the breaking specimen is computed from the difference between the two heights. The standard test method that describes the Izod impact test is the ASTM-D 256 test. There are several variations of the Izod test. These variations include positioning the test specimen such that the stresses in the notch are tensile or compressive by having the notch face away or toward the swinging pendulum, respectively. In addition, the clamping force exerted on the test specimen can have a great effect on the test results. The Charpy test evaluates the bending impact strength of a small notched or unnotched simply supported specimen that is struck by a hammer similar to the Izod impact tester [19]. The notched Charpy test is done such that the notch faces away from the swinging hammer creating tensile stresses within the notch, see Fig. 10.36. Both, the standard ASTM D256 and DIN 53453 tests describe the Charpy impact test.

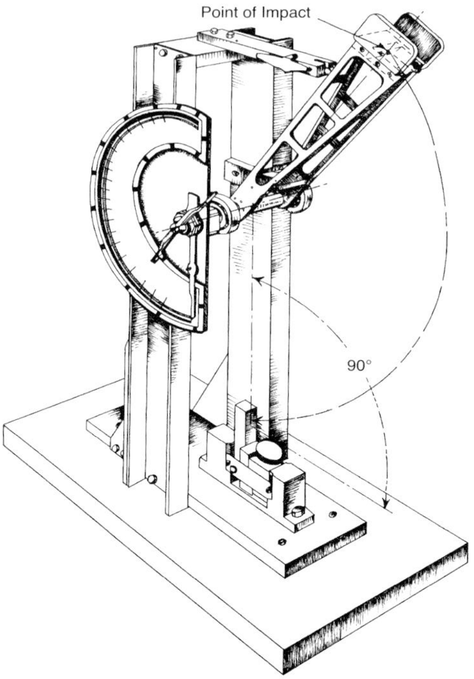

Figure 10.34 Cantilever beam Izod impact machine.

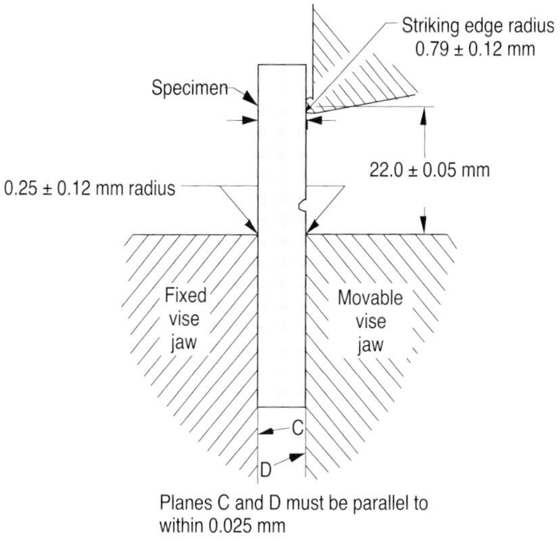

Figure 10.35 Schematic of the clamp, specimen and striking hammer in an izod impact test.

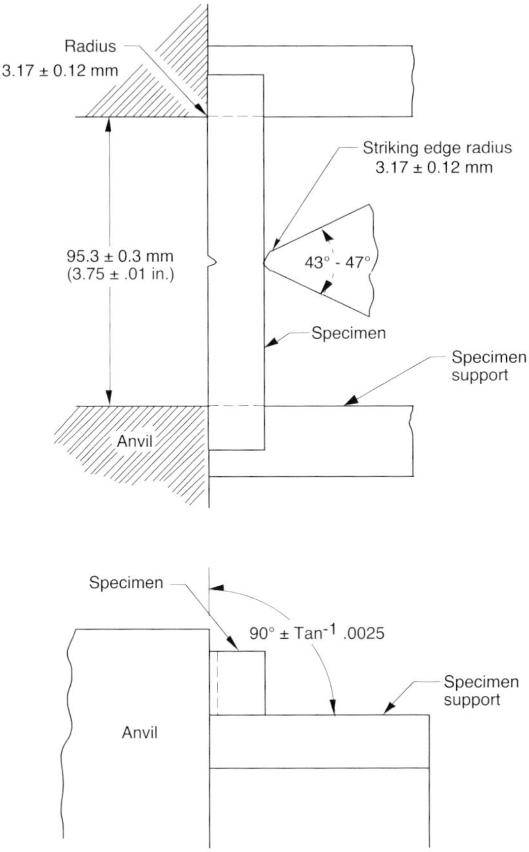

Figure 10.36 Schematic of the clamp, specimen, and striking hammer in a Charpy impact test.

A variable of both tests is the notch tip radius. Depending on the type of material, the notch tip radius may significantly influence the impact resistance of the specimen. Figure 10.37 [2] presents impact strength for various thermoplastics as a function of notch tip radius. As expected, impact strength is significantly reduced with decreasing notch radius. Another factor that influences the impact resistance of polymeric materials is the temperature. This is clearly demonstrated in Fig. 10.38 [2], in which PVC specimens with several notch radii are tested at various temperatures. In addition, the impact test sometimes brings out brittle failure in materials that undergo a ductile breakage in a short-term tensile test. The brittle behavior is sometimes developed by lowering the temperature of the specimen or by decreasing the notch tip radius. Figure 10.39 [1] shows the brittle to ductile behavior regimes as a function of temperature for several thermoplastic polymers.

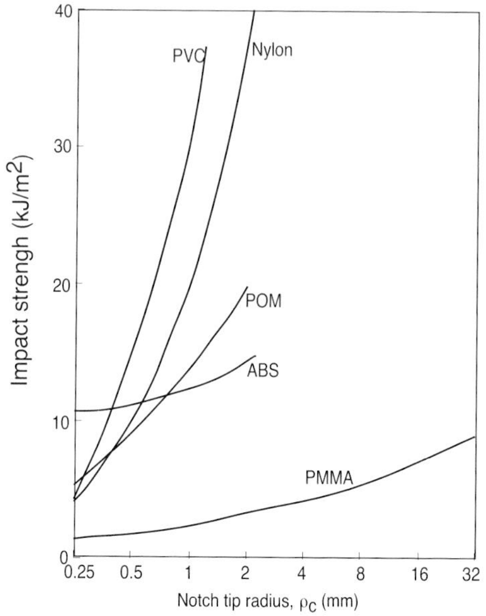

Figure 10.37 Impact strength as a function of notch tip radius for various polymers.

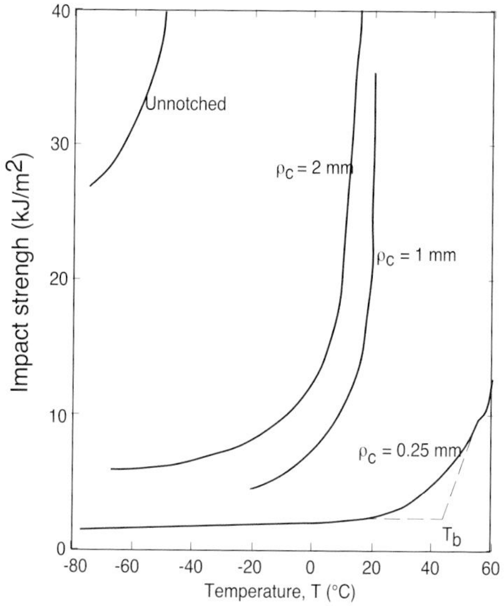

Figure 10.38 Impact strength of PVC as a function of temperature for various notch tip radii.

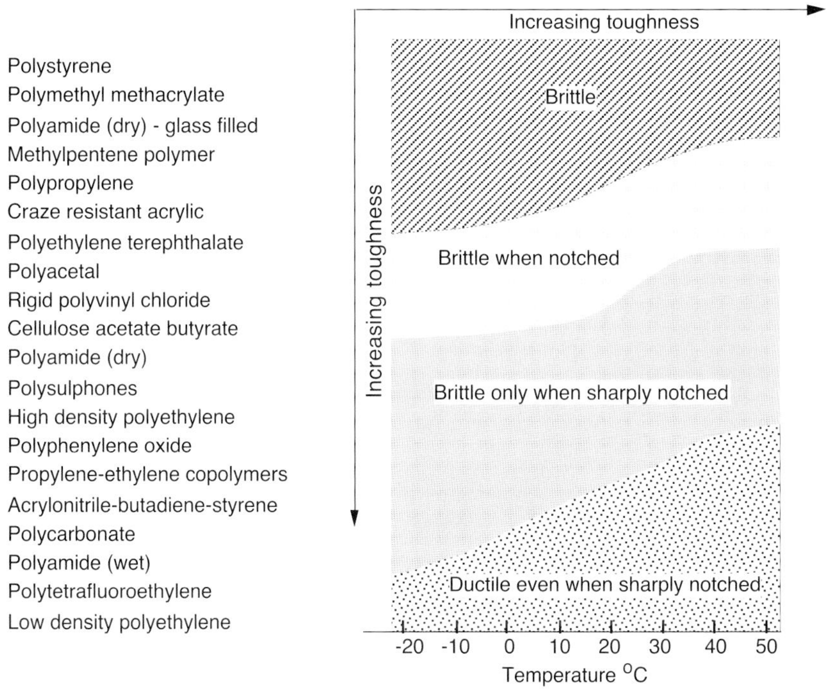

Figure 10.39 Brittle to ductile behavior regimes as a function of temperature for several thermoplastic polymers.

Another impact test worth mentioning is the *falling dart* test. This test is well suited for specimens that are too thin or flexible to be tested using the Charpy and Izod tests. This test, which is described by the ASTM 3029 and DIN 53 453 standard methods, also works well when the fracture toughness of a finished product with large surfaces is sought. Figure 10.40 shows a schematic of a typical falling dart test set-up [19]. The test consists of dropping a *tup*, with a spherical tip and otherwise variable shape and weight on a usually circular test specimen that is clamped around the periphery. The weight of the tup and the height from which it is dropped are the test variables. The energy needed to fracture the test specimen is directly computed from the potential energy of the tup before it is released; written as

$$U_p = mgh \tag{10.22}$$

where m is the mass of the tup, h the height from which it is dropped, and g gravity. It is assumed that the potential energy is fully transformed into kinetic energy and, in turn, into fracture energy. The test itself is much simpler than the Charpy and Izod tests, since Eq. 10.22 can be used to interpret the results directly. However, a large number of tests are required to determine the energy required for fracture.

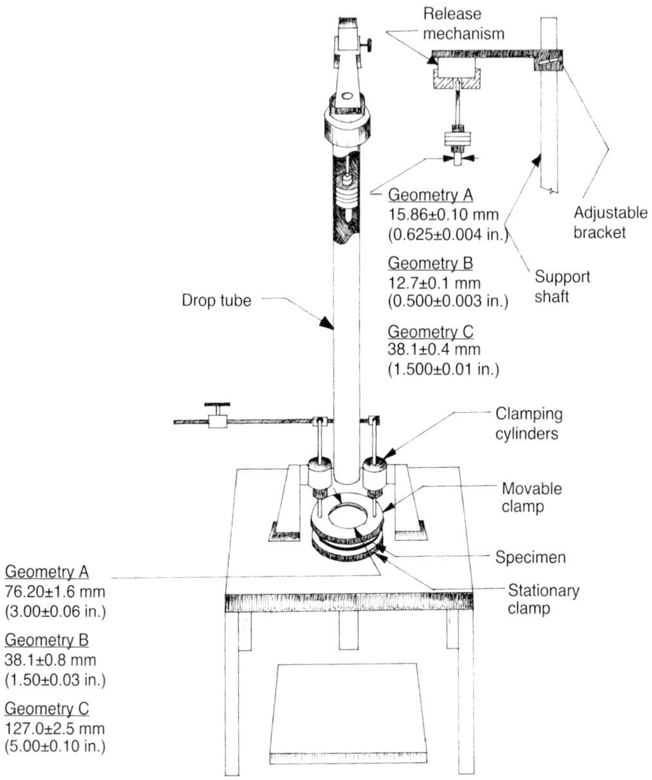

Figure 10.40 Schematic of a drop weight impact tester.

10.3.2 Fracture Mechanics Analysis of Impact Failure

Although the most common interpretation of impact tests is qualitative, it is possible to use linear elastic fracture mechanics to quantitatively evaluate impact test results. Using LEFM, it is common to compute the material's fracture toughness G_{Ic} from impact test results. Obviously, LEFM is only valid if the Izod or Charpy test specimen is assumed to follow linear elastic behavior and contains a sharp notch. At the onset of crack propagation, Eq. 10.4, which gives the elastic energy stored in the loaded test specimen, can be rewritten in terms of compliance, J, as

$$U_e = \frac{1}{2} F_c^2 J \tag{10.23}$$

Solving for F_c in Eq. 10.12 and substituting into Eq. 10.23 results in

$$U_e = G_{Ic} t \left(\frac{J}{\partial J / \partial a} \right) \tag{10.24}$$

Introducing the test specimen's width, w, and a geometrical crack factor, $\tilde{a}$, given by

$$\tilde{a} = \left(\frac{1}{J} \frac{\partial J}{\partial a} \right)^{-1} \tag{10.25}$$

we can write Eq. 10.24 as

$$U_e = G_{Ic} w t \tilde{a} \tag{10.26}$$

The parameter $\tilde{a}$ is found in Table 10.6 [20] for various Charpy impact test specimens and in Table 10.7 [20] for various Izod impact test specimens. The elastic energy absorbed by the test specimen during fracture, U_e, can also be represented with energy lost by the pendulum, U_L. This allows the test engineer to relate impact test results with the fracture toughness G_{Ic} of a material. When plotting U_e versus $tw\tilde{a}$, the kinetic effects lead to a positive intercept on the U_e axis. This can be corrected by subtracting the kinetic effects, U_k, from U_e. The kinetic effects can be computed using [20]

$$U_k = (1+e)\left(1 + \frac{1}{2}\frac{m}{M}(1-e)\right) m \left(\frac{M}{m+M} \right) V^2 \tag{10.27}$$

where m is the mass of the specimen, M the mass of the tup, V the velocity, and e the coefficient of restitution.

Figure 10.41 contains both Charpy and Izod test result data for a medium density polyethylene [20] as plots of U_e versus $tw\tilde{a}$ with kinetic energy corrections. We can now calculate G_{Ic} from the slope of the curve (Eq. 10.26).

However, as mentioned earlier for polymers that undergo significant plastic deformation before failure, Eq. 10.26 does not apply and the J-integral must be used. Here, by taking the Charpy or Izod specimen and assuming full yield, having a plastic hinge, we can calculate the energy by using

$$U_e = \frac{\delta}{2} \sigma_y t (w - a) \tag{10.28}$$

Table 10.6 Charpy Impact Test Geometric Crack Factors $\tilde{a}$

a/D	$2L/D$=4	$2L/D$=6	$2L/D$=8	$2L/D$=10	$2L/D$=12
0.04	1.681	2.456	3.197	3.904	4.580
0.06	1.183	1.715	2..220	2.700	3.155
0.08	0.933	1.340	1.725	2.089	2.432
0.10	0.781	1.112	1.423	1.716	1.990
0.12	0.680	0.957	1.217	1.461	1.688
0.14	0.605	0.844	1.067	1.274	1.467
0.16	0.550	0.757	0.950	1.130	1.297
0.18	0.505	0.688	0.858	1.015	1.161
0.20	0.468	0.631	0.781	0.921	1.050
0.22	0.438	0.584	0.718	0.842	0.956
0.24	0.413	0.543	0.664	0.775	0.877
0.26	0.391	0.508	0.616	0.716	0.808
0.28	0.371	0.477	0.575	0.665	0.748
0.30	0.354	0.450	0.538	0.619	0.694
0.32	0.339	0.425	0.505	0.578	0.647
0.34	0.324	0.403	0.475	0.542	0.603
0.36	0.311	0.382	0.447	0.508	0.564
0.38	0.299	0.363	0.422	0.477	0.527
0.42	0.276	0.328	0.376	0.421	0.462
0.44	0.265	0.311	0.355	0.395	0.433
0.46	0.254	0.296	0.335	0.371	0.405
0.48	0.244	0.281	0.316	0.349	0.379
0.50	0.233	0.267	0.298	0.327	0.355
0.52	0.224	0.253	0.281	0.307	0.332
0.54	0.214	0.240	0.265	0.88	0.310
0.56	0.205	0.228	0.249	0.270	0.290
0.58	0.196	0.216	0.235	0.253	0.271
0.60	0.187	0.205	0.222	0.238	0.253

Table 10.7 Izod Impact Test Geometric Crack Factors $\tilde{a}$

a/D	$2L/D$=4	$2L/D$=6	$2L/D$=8	$2L/D$=10	$2L/D$=12
0.06	1.540	1.744	1.850	2.040	-------
0.08	1.273	1.400	1.485	1.675	1.906
0.10	1.060	1.165	1.230	1.360	1.570
0.12	0.911	1.008	1.056	1.153	1.294
0.14	0.795	0.890	0.932	1.010	1.114
0.16	0.708	0.788	0.830	0.900	0.990
0.18	0.650	0.706	0.741	0.809	0.890
0.20	0.600	0.642	0.670	0.730	0.810
0.22	0.560	0.595	0.614	0.669	0.750
0.24	0.529	0.555	0.572	0.617	0.697
0.26	0.500	0.525	0.538	0.577	0.656
0.28	0.473	0.500	0.510	0.545	0.618
0.30	0.452	0.480	0.489	0.519	0.587
0.32	0.434	0.463	0.470	0.500	0.561
0.34	0.420	0.446	0.454	0.481	0.538
0.36	0.410	0.432	0.440	0.468	0.514
0.38	0.397	0.420	0.430	0.454	0.494
0.40	0.387	0.410	0.420	0.441	0.478
0.42	0.380	0.400	0.411	0.431	0.460
0.44	0.375	0.396	0.402	0.423	0.454
0.46	0.369	0.390	0.395	0.415	0.434
0.48	0.364	0.385	0.390	0.408	0.422
0.50	0.360	0.379	0.385	0.399	0.411

Using the relation in Eq. 10.20 with Eq. 10.28 we can write

$$J_{Ic} = \frac{2U_e}{A} \tag{10.29}$$

where $t(w-a)$ was replaced by A or the cross-sectional area of the specimen where fracture occurs. Figure 10.42 gives results for U_e as a function of A for high-impact polystyrene. The results show close agreement between the Charpy and Izod test methods and a linear correlation exists, as predicted with Eq. 10.29.

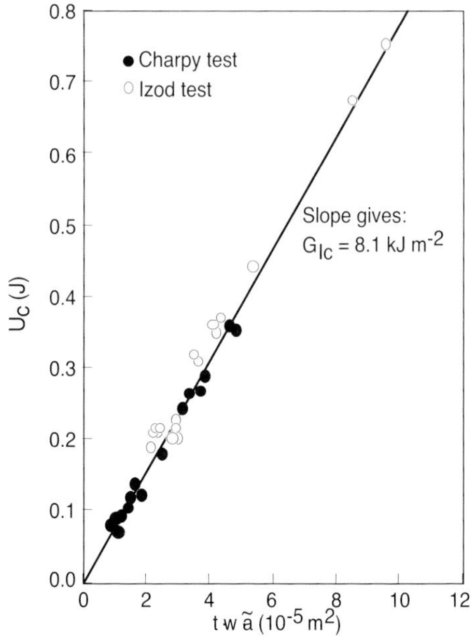

Figure 10.41 Elastic energy absorbed at impact fracture as a function of test specimen cross-sectional geometry for a medium-density polyethylene.

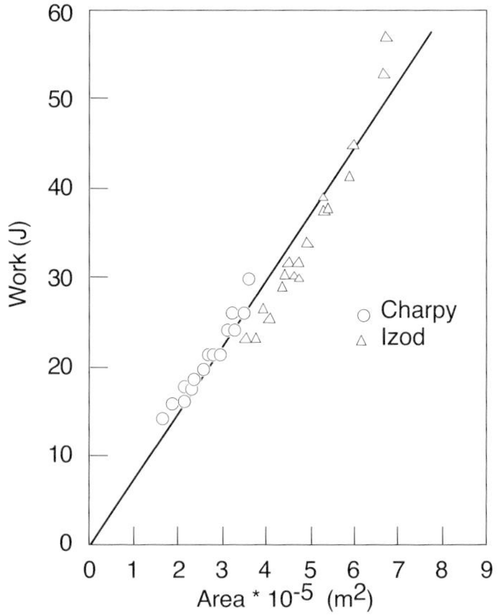

Figure 10.42 Elastic energy absorbed as a function of cross-sectional area for a high impact polystyrene test specimen.

10.4 Creep Rupture

During creep, a loaded polymer component will gradually increase in length until fracture or failure occurs. This phenomenon is usually referred to as *creep rupture* or, sometimes, as *static fatigue*. During creep, a component is loaded under a constant stress, constantly straining until the material cannot withstand further deformation, causing it to rupture. At high stresses, the rupture occurs sooner than at lower stresses. However, at low enough stresses, failure may never occur. The time it takes for a component or test specimen to fail depends on temperature, load, manufacturing process, environment, etc. It is important to point out that damage is often present and visible before creep rupture occurs. This is clearly demonstrated in Fig. 10.43 [7], which presents isochronous creep curves for polymethyl methacrylate at three different temperatures. The regions of linear and non-linear viscoelasticity and of visual damage are highlighted in the figure.

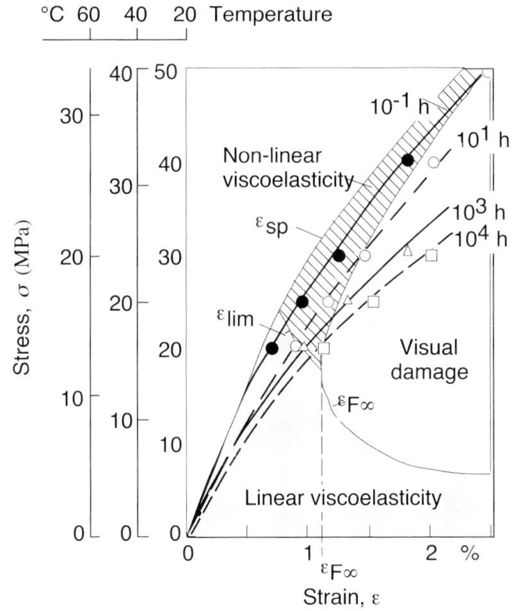

Figure 10.43 Isochronous creep curves for PMMA at three different temperatures.

10.4.1 Creep Rupture Tests

The standard test to measure creep rupture is the same as the creep test discussed in the previous chapter. Results from creep rupture tests are usually presented in graphs of applied stress versus the logarithm of time to rupture. A sample of creep rupture behavior for several thermoplastics is presented in Fig. 10.44 [21]. As the scale in the figure suggests, the tests were carried out with loadings that cause the material to fail within a few weeks. An example of a creep rupture test that ran for 10 years is shown in Fig. 10.45 [22, 23]. Here, the creep rupture of high density polyethylene pipes under internal pressures was tested at different temperatures. Two general regions with different slopes become obvious in the plots. The points to the left of the knee represent pipes that underwent a ductile failure, whereas those points to the right represent the pipes that had a brittle failure. As pointed out, generating a graph such as the one presented in Fig. 10.45, is an extremely involved and lengthy task, that takes several years of testing.[5] Usually, these types of tests are carried out to 1,000 h (6 weeks) or 10,000 h (60 weeks) as shown in Fig. 10.46[6] for a polyamide 6 with 30% glass fibers tested at different temperatures. Once the steeper slope, which is typical of the brittle fracture has been reached, the line can be extrapolated with some degree of confidence to estimate values of creep rupture at future times.

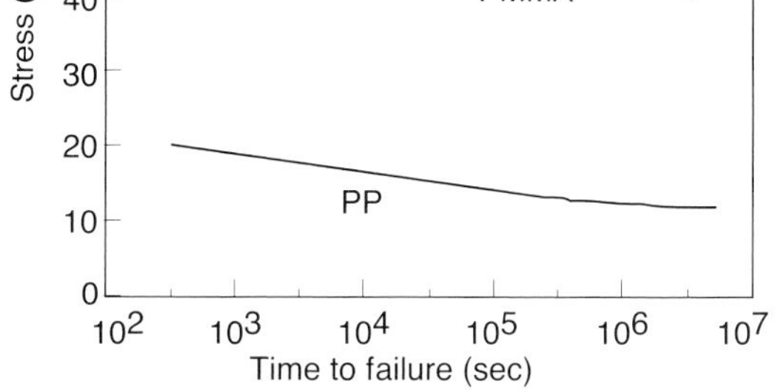

Figure 10.44 Creep rupture behavior for various thermoplastics.

5 These tests where done between 1958 and 1968 at Hoechst AG, Germany.

6 Courtesy Bayer AG.

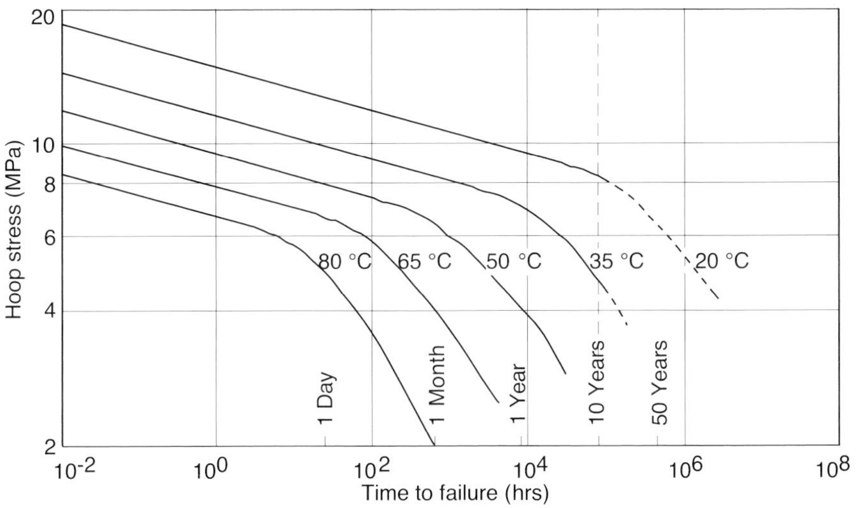

Figure 10.45 Creep rupture behavior as a function of temperature for a high density polyethylene.

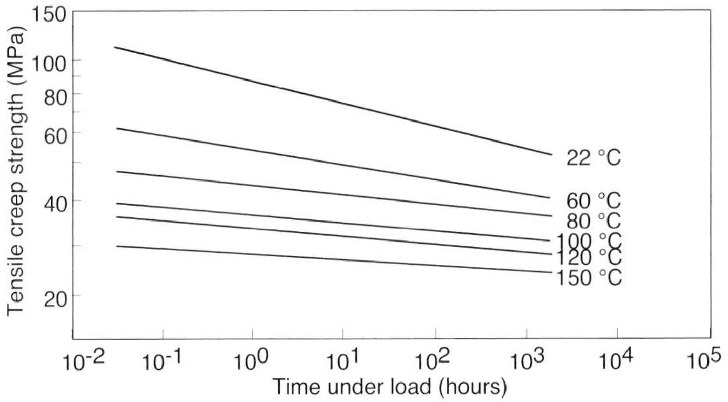

Figure 10.46 Creep rupture behavior as a function of temperature for a polyamide 6 with 30% glass fibers (Durethan BKV 30).

Although the creep test is considered a long term test, in principle it is difficult to actually distinguish it from monotonic stress strain tests or even impact tests. In fact, one can plot the full behavior of the material, from impact to creep, on the same graph as shown for PMMA under tensile loads at room temperature in Fig. 10.47 [7].

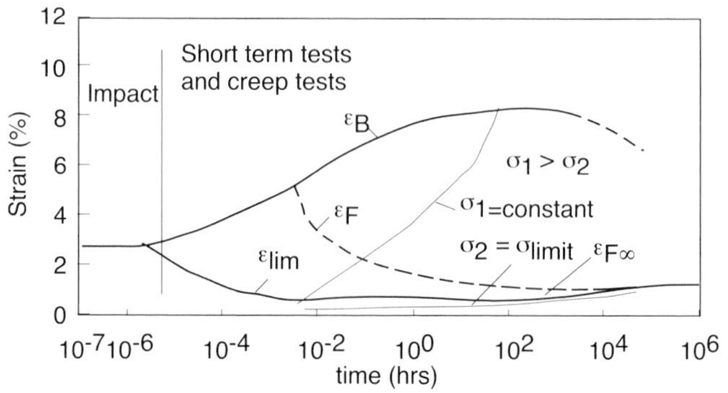

Figure 10.47 Plot of material behavior at room temperature, from impact to creep, for a PMMA under tensile loads.

The figure represents strain as a function of the logarithm of time. The strain line that represents rupture is denoted by ε_B. This line represents the maximum attainable strain before failure as a function of time. Obviously, a material tested under an impact tensile loading will strain much less than the same material tested in a creep test. Of interest in Fig. 10.47 are the two constant stress lines denoted by σ_1 and σ_2. The following example will help the reader interpret Fig. 10.47. It can be seen that a PMMA specimen loaded to a hypothetical stress of σ_1 will behave as a linear viscoelastic material up to a strain of 1%, at which point the first microcracks start forming or the craze nucleation begins. The crazing appears a little later after the specimen's deformation is slightly over 2%. The test specimen continues to strain for the next 100 h until it ruptures at a strain of about 8%. From the figure it can be deduced that the first signs of crazing can occur days and perhaps months or years before the material actually fractures. The stress line denoted by σ_2, where $\sigma_1 > \sigma_2$, is a limiting stress under which the component will not craze. Figure 10.47 also demonstrates that a component loaded at high speeds (i.e., impact) will craze and fail at the same strain. A limiting strain of 2.2% is shown.

Since these tests take a long time to perform, it is often useful to test the material at higher temperatures, where a similar behavior occurs in a shorter period of time. Figure 10.48 [7] shows tests performed on PMMA samples at five different temperatures. When comparing the results in Fig. 10.48 to the curve presented in Fig. 10.47, a clear time-temperature superposition becomes visible. In the applied stress versus logarithm of time to rupture curves, such as the one shown in Fig. 10.45, the time-temperature superposition is also evident.

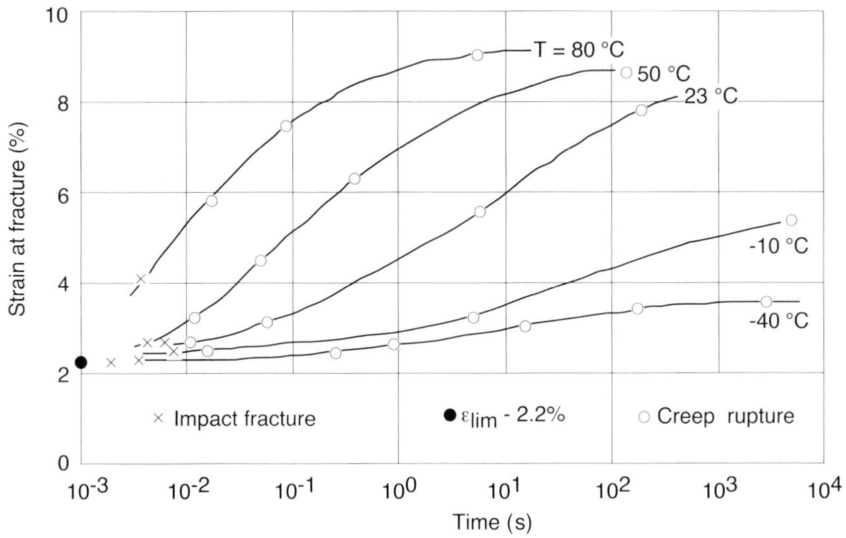

Figure 10.48 Strain at fracture for a PMMA in creep tests at various temperatures.

10.4.2 Fracture Mechanics Analysis of Creep Rupture

Crack growth rates during static tests have been found to have a power law relation with the stress intensity factor K_{Ic}, as

$$\frac{da}{dt} = AK_{Ic}^{m} \tag{10.30}$$

where A and m are material properties and K_{Ic} can be computed using Eq. 10.3 which results in

$$\frac{da}{dt} = A(Y\sigma)^{m}(\pi a)^{m/2} \tag{10.31}$$

By ignoring the time it takes for crack initiation, this equation may be used to predict a conservative time for creep rupture of a polymer component. If we integrate Eq. 10.31, we can predict the time it takes for a crack to grow from a length a_1 to a length a_2 by using

$$t = \frac{2\left(a_1^{1-m/2} - a_2^{1-m/2}\right)}{\left[(m-2)C_1\left(Y\sigma\sqrt{\pi}\right)^{m}\right]} \tag{10.32}$$

Experimental evidence shows that for slow crack growth, the value of m is large, and ranges between 7 and 25. Hence, the time it takes for a crack to grow between a_1 and a_2 is dominated by the initial crack length, because a small crack will grow much slower than a large one.

10.5 Fatigue

Dynamic loading of any material that leads to failure after a certain number of cycles is called *fatigue* or *dynamic fatigue*. Dynamic fatigue is of extreme importance since a cyclic or fluctuating load will cause a component to fail at much lower stresses than it does under monotonic loads.

10.5.1 Fatigue Test Methods

The standard fatigue tests for polymeric materials are the ASTM-D 671 test and the DIN 50100 test. In the ASTM test, a cantilever beam, shown in Fig. 10.49 [19], is held in a vise and bent at the other end by a yoke which is attached to a rotating variably eccentric shaft. A constant stress throughout the test region in the specimen is achieved by its triangular shape.

Fatigue testing results are plotted as stress amplitude versus number of cycles to failure. These graphs are usually called *S-N curves*, a term inherited from metal fatigue testing [24]. Figure 10.50 [25] presents S-N curves for several thermoplastic and thermoset polymers tested at a 30-Hz frequency and about a zero mean stress, σ_m.

We must point out here, that most fatigue data presented in the literature and in resin supplier data sheets does not present the frequency, specimen geometry or environmental conditions at which the tests were performed. Hence, such data is not suitable for use in design. The data we present in this section is only intended to illustrate the various problems that arise when measuring fatigue life of a polymer. The information should also serve to reflect trends and as a comparison between various materials and conditions.

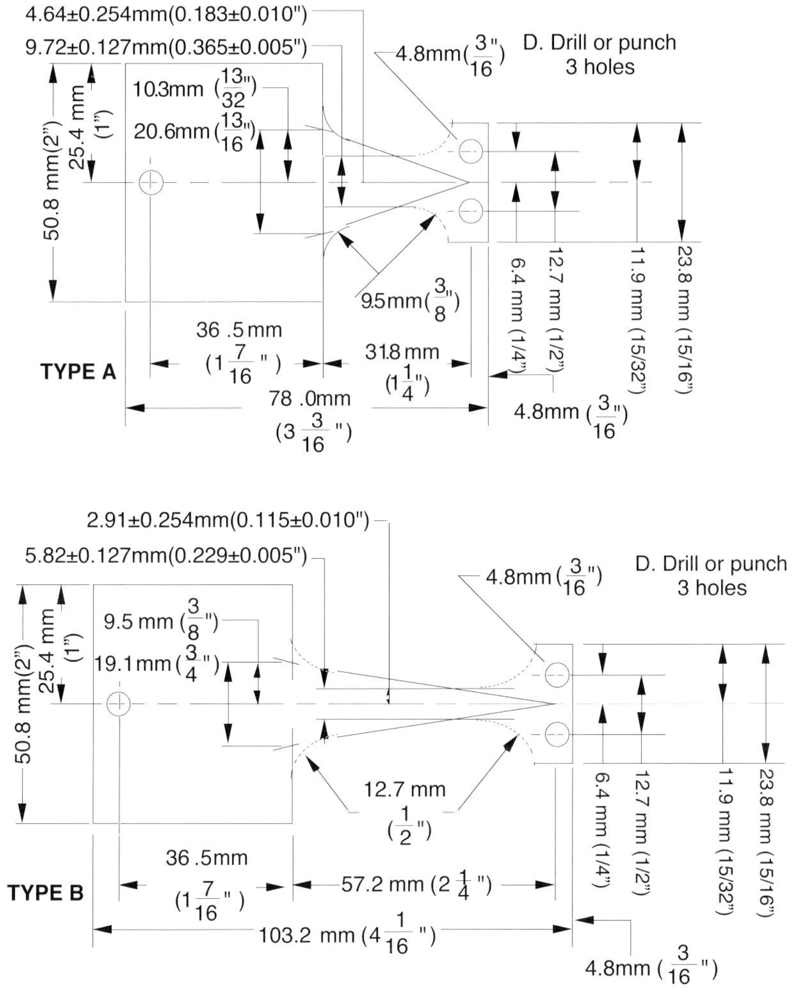

Figure 10.49 ASTM constant force fatigue test specimens.

Fatigue in plastics is strongly dependent on the environment, the temperature, the frequency of loading, the surface, etc. For example, due to surface irregularities and scratches, crack initiation at the surface is more likely in a polymer component that has been machined than in one that was injection molded. As mentioned in Chapter 5, an injection molded article is formed by several layers of different orientation. In such parts the outer layers act as a protective skin that inhibits crack initiation. In an injection molded article, cracks are more likely to initiate inside the component by defects such as weld lines and filler particles. The gate region is also a prime initiator of fatigue cracks. Corrosive environments also accelerate crack initiation and failure via fatigue. Corrosive environments and weathering will be discussed in more detail later in this chapter.

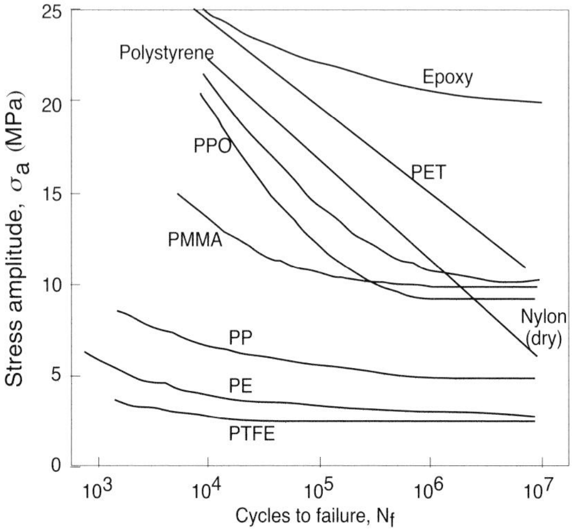

Figure 10.50 Stress-life (S-N) curves for several thermoplastic and thermoset polymers tested at a 30-Hz frequency about a zero mean stress.

It is interesting to point out in Fig. 10.50 that thermoset polymers show a higher fatigue strength than thermoplastics. An obvious cause for this is their greater rigidity. However, more important is the lower internal damping or friction, which reduces temperature rise during testing. Temperature rise during testing is one of the main factors that lead to failure when experimentally testing thermoplastic polymers under cyclic loads. The heat generation during testing is caused by the combination of internal frictional or hysteretic heating and low thermal conductivity. At a low frequency and low stress level, the temperature inside the polymer specimen will rise and eventually reach thermal equilibrium when the heat generated by hysteretic heating equals the heat removed from the specimen by conduction. As the frequency is increased, viscous heat is generated faster, causing the temperature to rise even further. This phenomenon is shown in Fig. 10.51 [21] in which the temperature rise during uniaxial cyclic testing of polyacetal is plotted. After thermal equilibrium has been reached, a specimen eventually fails by conventional brittle fatigue, assuming the stress is above the endurance limit.

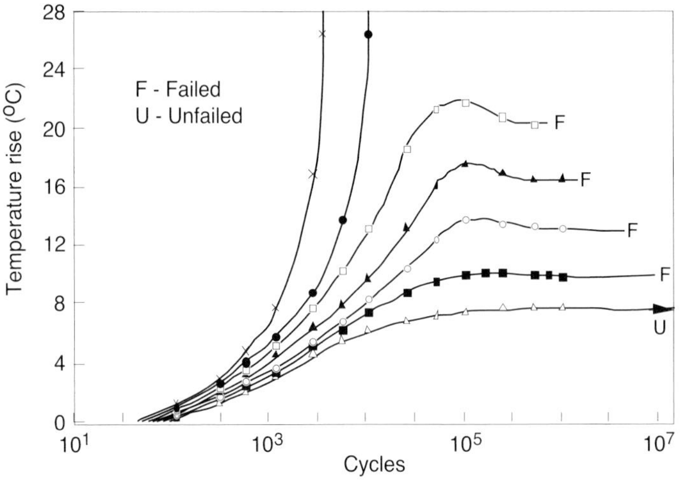

Figure 10.51 Temperature rise during uniaxial cycling under various stresses at 5-Hz.

However, if the frequency or stress level is increased even further, the temperature will rise to the point at which the test specimen softens and ruptures before reaching thermal equilibrium. This mode of failure is usually referred to as *thermal fatigue*. This effect is clearly demonstrated in Fig. 10.52 [21]. The points marked T denote those specimens that failed due to thermal fatigue. The other points represent the specimens that failed by conventional mechanical fatigue. A better picture of how frequency plays a significant role in fatigue testing of polymeric materials is generated by plotting results such as those shown in Fig. 10.52 [21] for several frequencies (Fig. 10.53 [21]). The temperature rise in the component depends on the geometry and size of test specimen. For example, thicker specimens will cool slower and are less likely to reach thermal equilibrium. Similarly, material around a stress concentrator will be subjected to higher stresses which will result in temperatures that are higher than the rest of the specimen leading to crack initiation caused by localized thermal fatigue. To neglect the effect of thermal fatigue, cyclic tests with polymers must be performed at very low frequencies that make them much lengthier than those performed with metals and other materials which have high thermal conductivity.

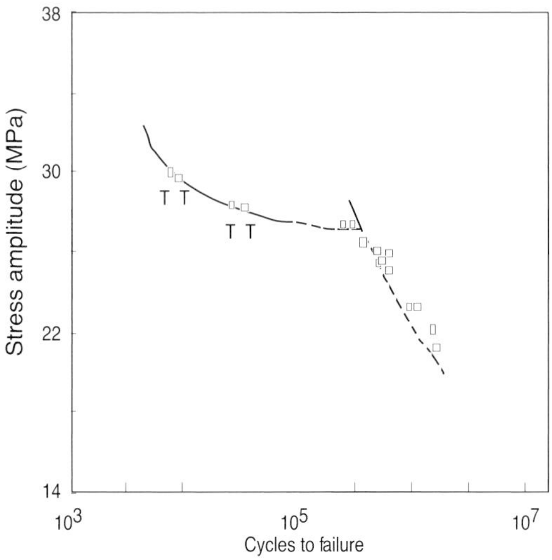

Figure 10.52 Fatigue and thermal failures in acetal tested at 1.67 Hz.

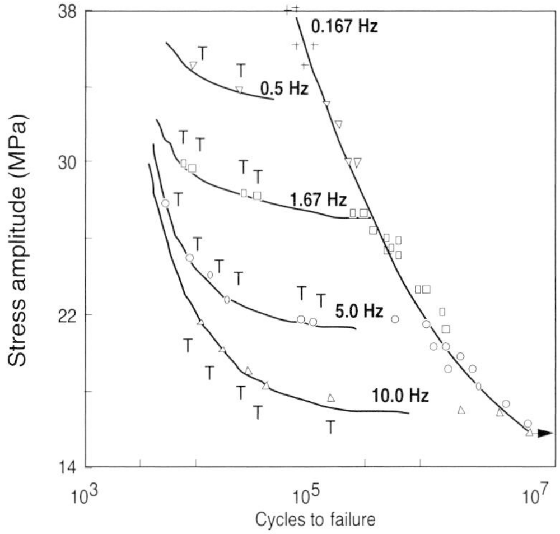

Figure 10.53 Fatigue and thermal failures in acetal tested at various frequencies.

As mentioned earlier, stress concentrations have a great impact on the fatigue life of a component. Figures 10.54[7] and 10.55[6] compare S-N curves for uPVC and polyamide 66, respectively, for specimens with and without a 3 mm

7 All courtesy of Bayer AG, Germany.

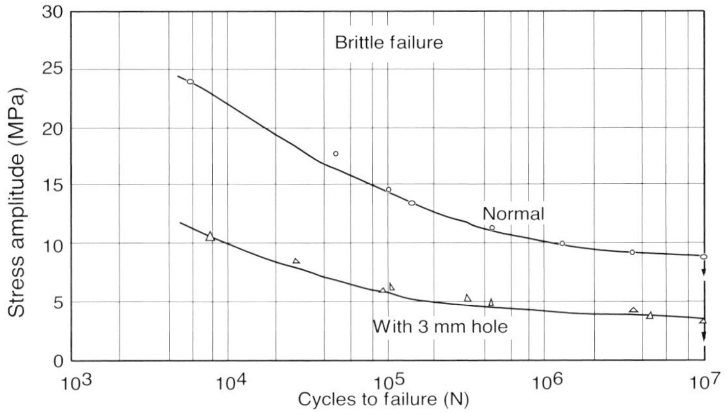

Figure 10.54 Fatigue curves for an uPVC using specimens with and without 3 mm hole stress concentrators tested at 23 °C and 7-Hz with a zero mean stress.

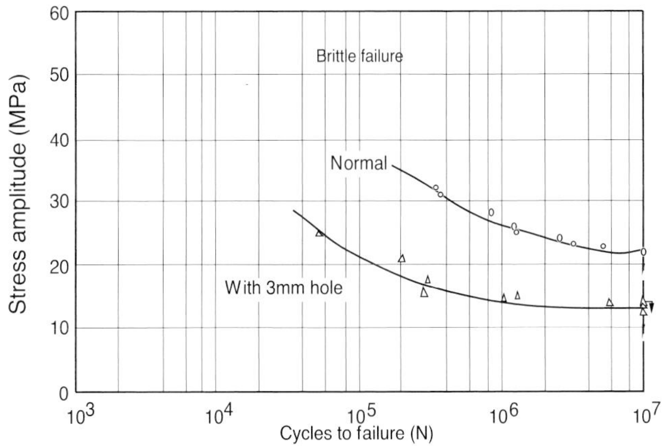

Figure 10.55 Fatigue curves for polyamide 66 (Durethan A30S) using specimens with and without 3 mm hole stress concentrators tested at 23 °C and 7-Hz with a zero mean stress.

circular hole acting as a stress concentrator. Material irregularities caused by filler particles or by weld lines also affect the fatigue of a component. Figures 10.56[6] and 10.57[6] compare S-N curves for regular PC and ABS test specimens to fatigue behavior of specimens with a weld line and specimens with a 3-mm circular hole.

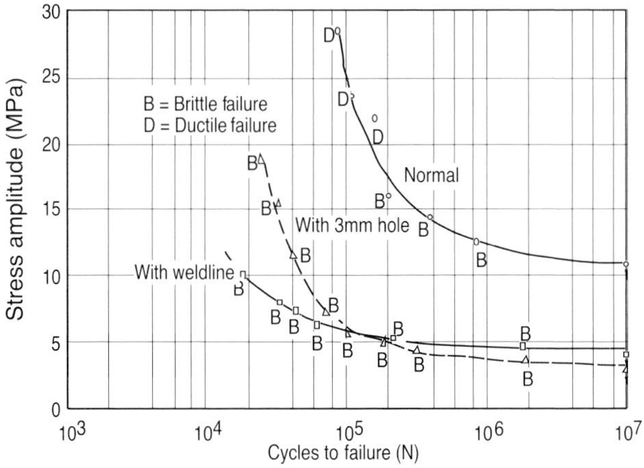

Figure 10.56 Fatigue curves for polycarbonate (Makrolon 2800) using regular specimens and specimens with 3 mm hole stress concentrators and weldlines tested at 23 °C and 7-Hz with a zero mean stress.

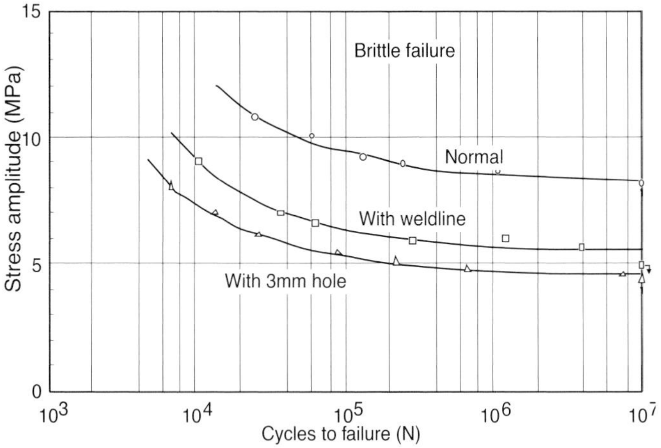

Figure 10.57 Fatigue curves for ABS (Novodur PH/AT) using regular specimens and specimens with 3 mm hole stress concentrators and weldlines tested at 23 °C and 7-Hz with a zero mean stress.

Up to this point, we assumed a zero mean stress, σ_m. However, many polymer components that are subjected to cyclic loading have other loads and stresses applied to them, leading to non-zero mean stress values. This superposition of two types of loading will lead to a combination of creep, caused by the mean stress, and fatigue, caused by the cyclic stress, σ_a. Test results from experiments with cyclic loading and non-zero mean stresses are complicated by the fact that some specimens fail due to creep and others due to conventional brittle fatigue. Figure 10.58 illustrates this phenomenon for both cases with and without thermal fatigue, comparing them to experiments in which a simple static loading is applied. For cases with two or more dynamic loadings with different stress or strain amplitudes, a similar strain deformation progression is observed. Figure 10.59 [26] presents the strain progression in polyacetal specimens in which two stress amplitudes, one above and one below the linear viscoelastic range of the material, are applied. The strain progression, $\Delta\varepsilon$, is the added creep per cycle caused by different loadings, similar to *ratcheting* effects in metal components where different loadings are combined.

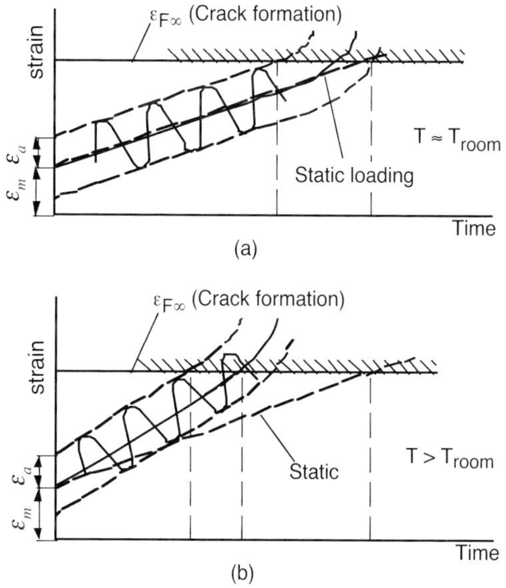

Figure 10.58 Creep and thermal fatigue effects during cyclic loading.

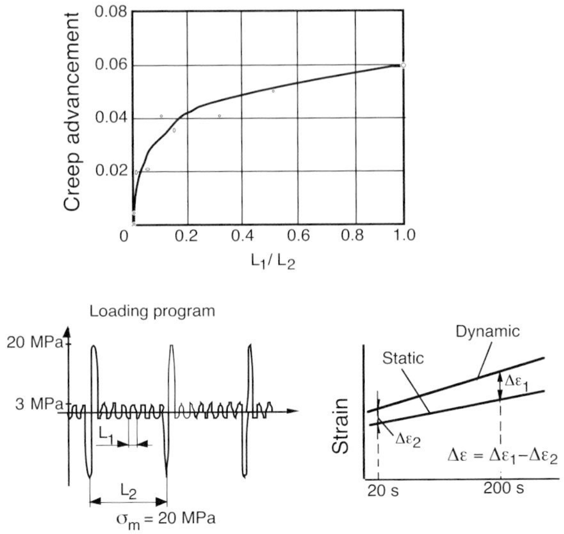

Figure 10.59 Strain progression in polyacetal specimens during fatigue tests with two stress amplitudes.

Fiber-reinforced composite polymers are stiffer and less susceptible to fatigue failure. Reinforced plastics have also been found to have lower hysteretic heating effects, making them less likely to fail by thermal fatigue. Figure 10.60 [27] presents the flexural fatigue behavior for glass fiber filled and unfilled polyamide 66 tested at 20 °C and a 0.5 Hz frequency with a zero mean stress. Parallel to the fiber orientation, the fatigue life was greater than the life of the specimens tested perpendicular to the orientation direction and the unfilled material specimens. The fatigue life of the unfilled specimen and the behavior perpendicular to the orientation direction were similar. However, the unfilled material failed by thermal fatigue at high stresses, whereas both the specimens tested perpendicular and parallel to the orientation direction failed by conventional fatigue at high stress levels. Fiber reinforced systems generally follow a sequence of events during failure consisting of debonding, cracking, and separation [28]. Figure 10.61 [29] clearly demonstrates this sequence of events with a glass-filled polyester mat tested at 20 °C and a frequency of 1.67 Hz. In most composites, debonding occurs after just a few cycles. It should be pointed out that often reinforced polymer composites do not exhibit an endurance limit, making it necessary to use factors of safety between 3 and 4. The fracture by fatigue is generally preceded by cracking of the matrix material, which gives a visual warning of imminent failure. It is important to mention that the fatigue life of thermoset composites is also affected by temperature. Figure 10.62 [30] shows the tensile strength versus

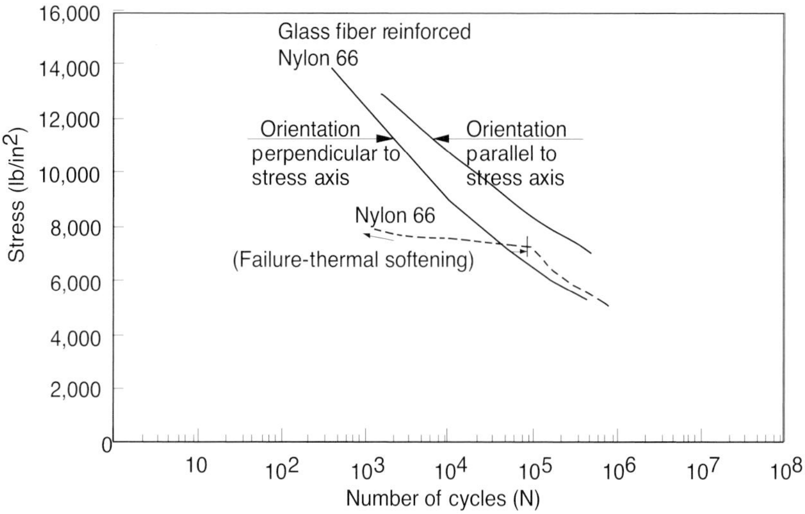

Figure 10.60 Flexural fatigue curves for a polyamide 66 and a glass fiber filled polyamide 66 tested at 20 °C and 0.5 Hz with a zero mean stress.

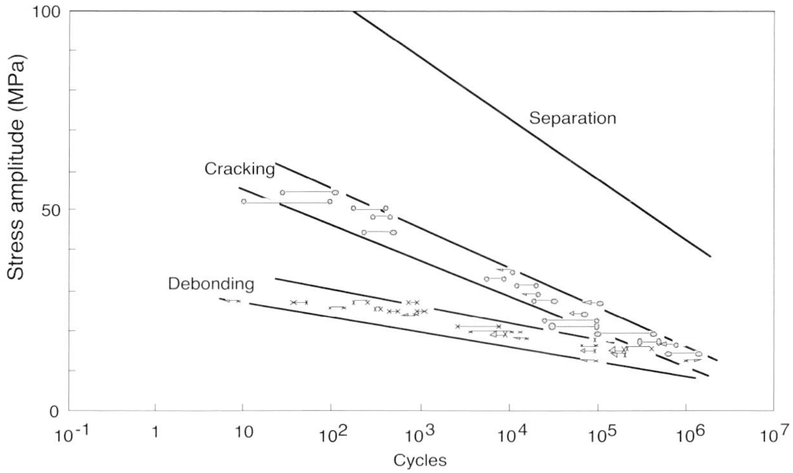

Figure 10.61 Fatigue curves for a glass filled polyester mat tested at 20 °C and a frequency of 1.67 Hz.

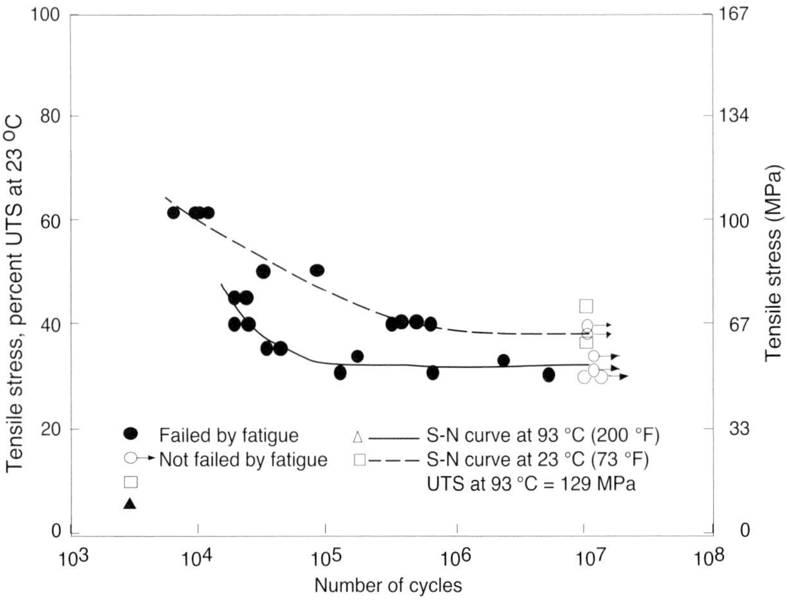

Figure 10.62 Fatigue curves for a 50% by weight glass fiber reinforced polyester resin sheet molding compound tested at 23 °C and 93 °C and 10 Hz.

number of cycles to failure for a 50% glass fiber filled unsaturated polyester tested at 23 °C and 93 °C. At ambient temperature, the material exhibits an endurance limit of about 65 MPa, which is reduced to 52 MPa at 93 °C.

10.5.2 Fracture Mechanics Analysis of Fatigue Failure

Crack growth rates during cyclic fatigue tests are related to the stress intensity factor difference, ΔK_{Ic},

$$\frac{da}{dt} = B(\Delta K_{Ic})^n \tag{10.33}$$

where B and n are material properties and $\Delta K_{Ic} = K_{Ic\,\max} - K_{Ic\,\min}$ can be computed using Eq. 10.3 with the maximum and minimum alternating stresses. Crack growth behavior for several polymers is shown in Fig. 10.63 [29, 31]. Hertzberg and Manson [29] also show that for some materials, the crack growth rate is reduced somewhat with increasing test frequency.

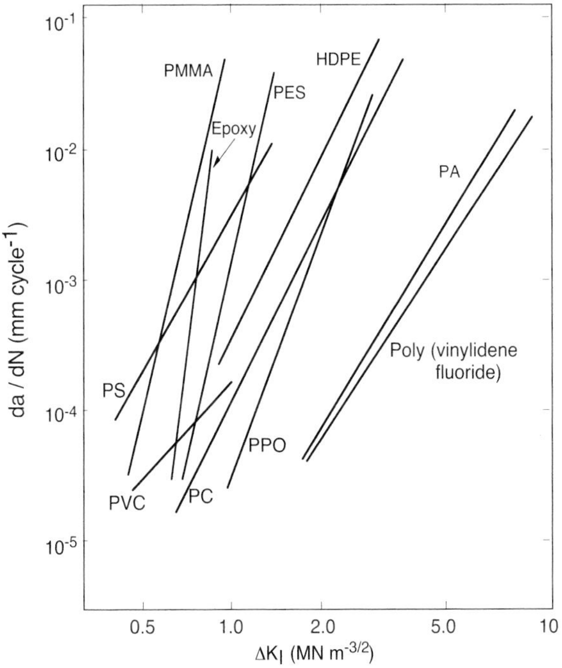

Figure 10.63 Crack growth rate during fatigue for various polymers.

10.6 Friction and Wear

Friction is the resistance that two surfaces experience as they slide or try to slide past each other. Friction can be dry (i.e., direct surface-surface interaction) or lubricated, where the surfaces are separated by a thin film of a lubricating fluid.

The force that arises in a dry friction environment can be computed using *Coulomb's law of friction* as

$$F = \mu N \tag{10.34}$$

where F is the force in surface or sliding direction, N the normal force, and μ the coefficient of friction. Coefficients of friction between several polymers and different surfaces are listed in Table 10.8 [7]. However, when dealing with polymers, the process of two surfaces sliding past each other is complicated by the fact that enormous amounts of frictional heat can be generated and stored near the surface due to the low thermal conductivity of the material. The analysis of friction between polymer surfaces is complicated further by environmental effects such as relative humidity and by the likeliness of a polymer surface to deform when stressed, such as shown in Fig. 10.64 [7]. The top two figures illustrate metal-metal friction, wheareas the bottom figures illustrate metal-polymer friction.

Table 10.8 Coefficient of Friction for Various Polymers

Specimen	Partner	Velocity (mm/s)					
		0.03	0.1	0.4	0.8	3.0	10.6
Dry friction							
PP^i	PP^s	0.54	0.65	0.71	0.77	0.77	0.71
PA^i	PA^i	0.63	-	0.69	0.70	0.70	0.65
PP^s	PP^s	0.26	0.29	0.22	0.21	0.31	0.27
PA^m	PA^m	0.42	-	0.44	0.46	0.46	0.47
Steel	PP^s	0.24	0.26	0.27	0.29	0.30	0.31
Steel	PA^m	0.33	-	0.33	0.33	0.30	0.30
PP^s	Steel	0.33	0.34	0.37	0.37	0.38	0.38
PA^m	Steel	0.39	-	0.41	0.41	0.40	0.40
Water lubricated							
PP^s	PP^s	0.25	0.26	0.29	0.30	0.28	0.31
PA^m	PA^m	0.27	-	0.24	0.22	0.21	0.19
Steel	PP^s	0.23	0.25	0.26	0.26	0.26	0.22
PP^s	Steel	0.25	0.25	0.26	0.26	0.25	0.25
PA^m	Steel	0.20	-	0.23	0.23	0.22	0.18
Oil lubricated							
PP^s	PP^s	0.29	0.26	0.24	0.25	0.22	0.21
PA^m	PA^m	0.22	-	0.15	0.13	0.11	0.08
Steel	PP^s	0.17	0.17	0.16	0.16	0.14	0.14
Steel	PA^m	0.16	-	0.11	0.09	0.08	0.08
PP^s	Steel	0.31	0.30	0.30	0.29	0.27	0.25
PA^m	Steel	0.26	-	0.15	0.12	0.07	0.04

Note ^{i} = injection molded, ^{s} = sandblasted, ^{m} = machined.

Temperature plays a significant role on the coefficient of friction μ as clearly demonstrated in Fig. 10.65 for polyamide 66 and polyethylene. In the case of polyethylene, the friction first decreases with temperature. At 100 ^oC the friction increases because the polymer surface becomes tacky. The friction coefficient starts to drop as the melt temperature is approached. A similar behavior is encountered with the polyamide curve.

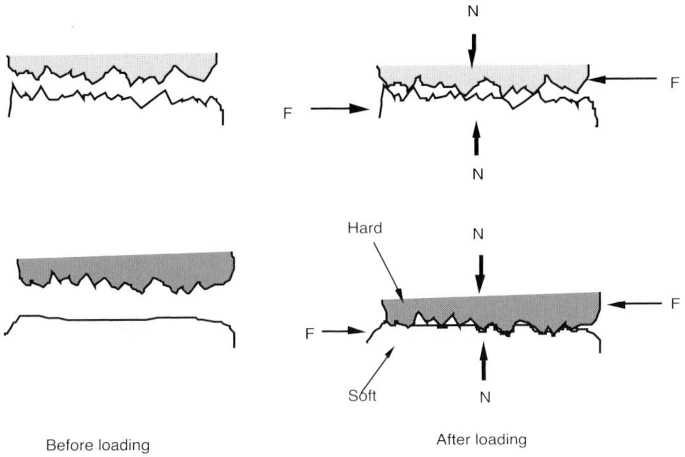

Figure 10.64 Effect of surface finish and hardness on frictional force build-up.

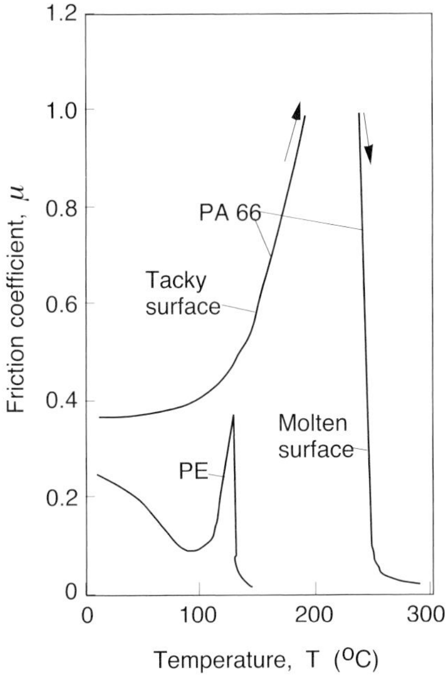

Figure 10.65 Temperature effect on coefficient of friction for a polyamide 66 and a high density polyethylene.

As mentioned earlier, temperature increases can be caused by the energy released by the frictional forces. A temperature increase in time, due to friction between surfaces of the same material, can be estimated using

$$\Delta T = \frac{2\dot{Q}\sqrt{t}}{\sqrt{\pi}\sqrt{k\rho C_p}} \tag{10.35}$$

where k is the thermal conductivity of the polymer, ρ the density, C_p the specific heat and the rate of energy created by the frictional forces which can be computed using

$$\dot{Q} = Fv \tag{10.36}$$

where v is speed between the sliding surfaces.

Wear is also affected by the temperature of the environment. Figure 10.66[8] shows how wear rates increase dramatically as the surface temperature of the polymer increases, causing it to become tacky.

Table 10.9 [32] presents relative volumetric wear values for selected polymers and beechwood using the volumetric wear of steel St 37 as a reference.

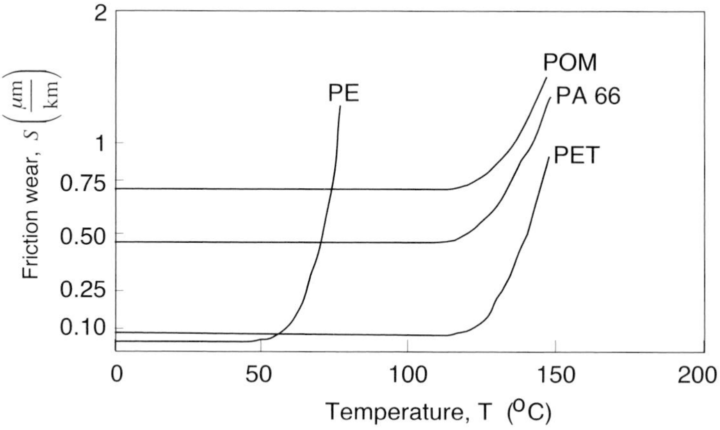

Figure 10.66 Wear as a function of temperature for various thermoplastics.

8 Courtesy of BASF.

Table 10.9 Relative Volumetric Wear

Polymer	Density (g/cm^3)	$Wear/Wear_{steel}$
Steel	7.45	1.0
Beechwood	0.83	17.9
PMMA	1.31	11.2
PVC-U	1.33	5.8
HDPE	0.92	3.8
PP	0.90	2.8
HDPE	0.95	2.1
PA 66	1.13	1.0
UHMW-HDPE	0.94	0.6

10.7 Stability of Polymer Structures

The failure of certain structures is often determined by the stiffness of the structural element and not by the strength of the material. Such a failure is commonly referred to as *buckling*. The common example of buckling is the failure of a slender column under compressive loads. Slender columns with hinged ends have a critical load, F_{crit}, defined by

$$F_{crit} = \frac{\pi^2 EI}{L^2} \tag{10.37}$$

where E is the modulus, I the area moment of inertia and L the length of the column. If we use the relation $I = Ar^2$, where A is the cross-sectional area of the column and r is the radius of gyration, we can rewrite Eq. 10.37 as

$$\frac{F_{crit}}{A} = \frac{\pi^2 EI}{\left(L/r\right)^2} = \frac{\pi^2 EI}{\lambda^2} \tag{10.38}$$

where F_{crit}/A is often referred to as critical unit load and λ the slenderness ratio. For a square cross-section of dimensions $a \times a$ the slenderness ratio is defined by

$$\lambda = \frac{L}{a}\sqrt{12} \tag{10.39}$$

Menges and Dolfen performed a series of compression tests on glass fiber reinforced unsaturated polyester specimens of square cross-section. Figure 10.67 [33] presents the unit loads and compressive strains at failure of the composite columns as a function of slenderness ratio. Their experiments demonstrated that at slenderness ratios smaller than 20 the columns failed by a shear delamination mode, while columns with slenderness ratios larger than 20 failed by buckling. Similarly, Gaube and Menges [34] performed experiments with various shell structures. Figure 10.68 presents the critical strain as a function of slenderness ratio. We can see here that in all cases buckling of polymer structures occurred below the value predicted by classical stability theory.

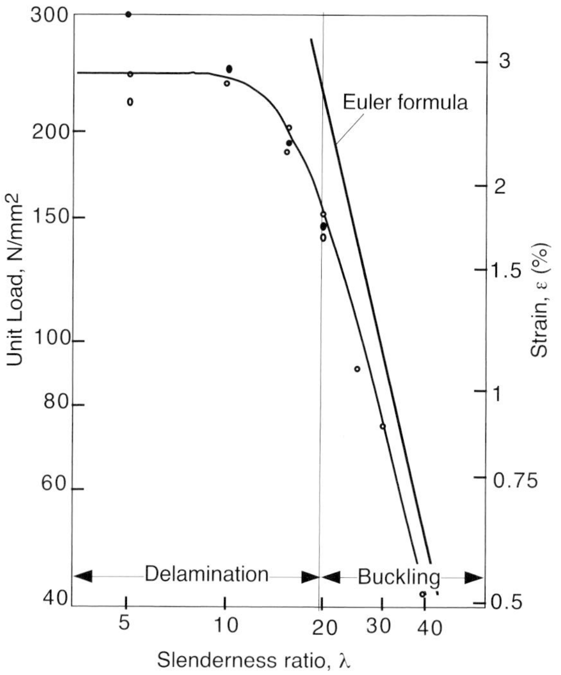

Figure 10.67 Delamination and buckling of glass reinforced UPE columns.

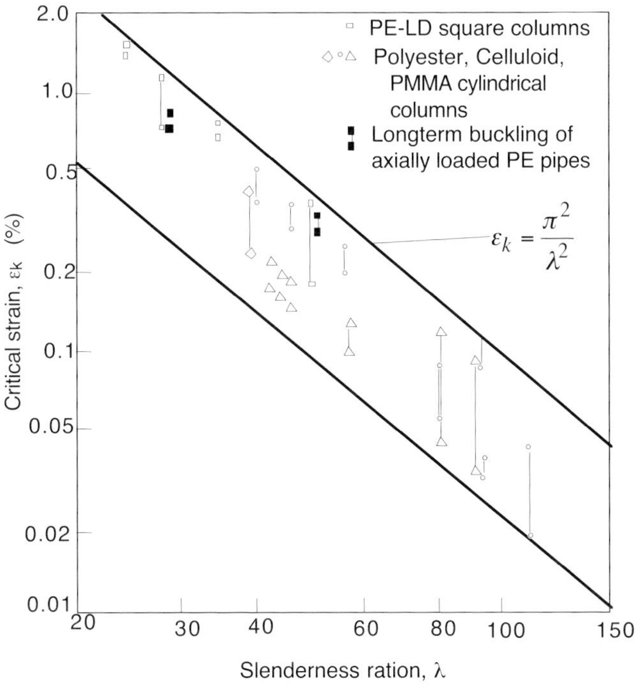

Figure 10.68 Buckling of columns and tubes.

10.8 Environmental Effects on Polymer Failure

The environment or the media in contact with a loaded or unloaded component plays a significant role on its properties, life span, and mode of failure. The environment can be a natural one, such as rain, hail, solar ultra-violet radiation, extreme temperatures, etc., or an artificially created one, such as solvents, oils, detergents, high temperature environments, etc. Damage in a polymer component due to natural environmental influences is usually referred to as *weathering*.

10.8.1 Weathering

When exposed to the elements, polymeric materials begin to exhibit environmental cracks, which lead to early failure at stress levels significantly lower than those in the absence of these environments. Figure 10.69 [35] shows an electron micrograph of the surface of a high-density polyethylene beer crate after nine years of use and exposure to weathering. The surface of the HDPE exhibits brittle cracks, which resulted from ultra violet rays, moisture, and extremes in temperature.

Figure 10.69 Electron micrograph of the surface of a high density polyethylene beer crate after nine years of use and exposure to weathering.

Standard tests such as the DIN 53486 test are available to evaluate effects of weathering on properties of polymeric materials. It is often unclear which weathering aspects or what combination of them influence material decay the most. Hence, laboratory tests are often done to isolate individual weathering factors such as ultra-violet radiation. For example, Fig. 10.70 shows the surface of a polyoxymethylene specimen irradiated with ultra violet light for 100 h in a laboratory environment. The DIN 53487 xenotest is a standard test to expose polymer test specimens to UV radiation in a controlled environment. Figure 10.71 is a plot of impact strength of notched PMMA specimens as a function of hours of UV radiation exposure in a controlled DIN 53487 test and years of weathering under standard DIN 53486 conditions. The correlation between the two tests is clear. The ASTM-D 4674 test evaluates the color stability of polymer specimens exposed to ultra violet radiation. Standard tests also exist to test materials for specific applications such as the ASTM-D 2561 test, which evaluates the environmental stress cracking resistance of blow molded polyethylene containers.

Figure 10.70 Surface of a polyoxymethylene specimen irradiated with ultra violet light for 100 h in a laboratory environment.

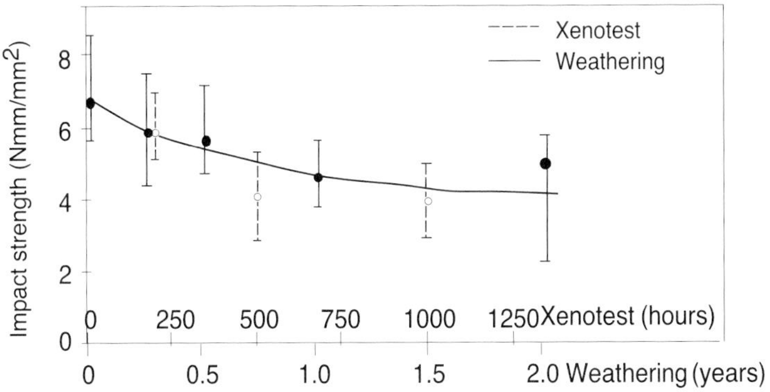

Figure 10.71 Impact strength of notched PMMA specimens as a function of hours of UV radiation exposure in a controlled test and weathering exposure time.

As can be seen, the effect of ultra violet radiation, moisture, and extreme temperature is detrimental to the mechanical properties of plastic parts. One example in which weathering completely destroys the strength properties of a material is shown for PVC in Fig. 10.72. The figure presents the decay of the impact strength of PVC pipes exposed to weathering in the United Kingdom [36]. As can be seen, the impact strength rapidly decreases in the first six months and is only 11% of its original value after only two years. The location and climate of a region can play a significant role on the weathering of polymer components. Figure 10.73 [36] shows the decrease in impact strength of rigid PVC as a function of time at five different sites. After five years of weathering, the PVC exposed in Germany still has 95% of its original impact strength, whereas the samples exposed in Singapore have less than 5% of their initial strength. The degradation in PVC samples is also accompanied by discoloration as presented in Fig. 10.74 [36]. The figure shows discoloration of white PVC as a function of time at various locations. The samples exposed in Arizona showed significantly higher discoloration than those exposed in Pennsylvania and Florida.

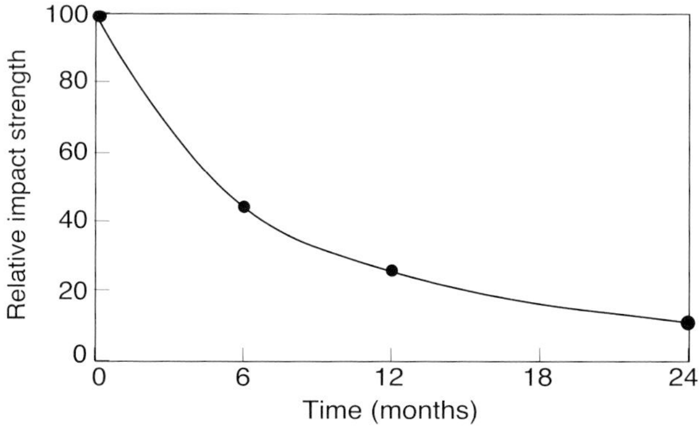

Figure 10.72 Impact strength of PVC pipe as a function weathering exposure time in the United Kingdom.

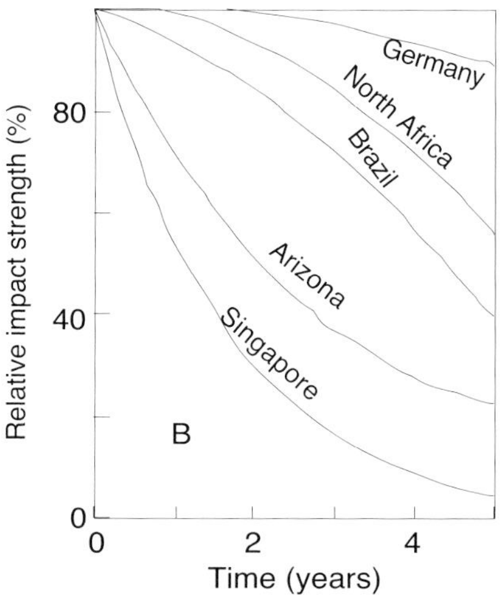

Figure 10.73 Impact strength as a function of weathering time of uPVC exposed in different geographic locations.

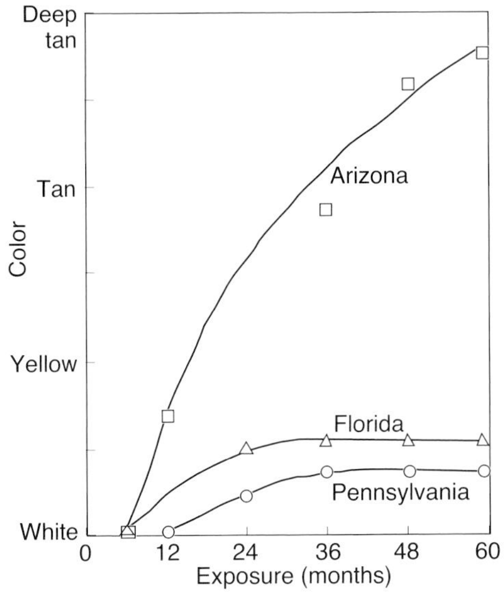

Figure 10.74 Discoloration as a function of weathering time of white PVC exposed in different geographic locations.

The strength losses and discoloration in a weathering process are mainly attributed to the ultra-violet rays received from sunshine. This can be demonstrated by plotting properties as a function of actual sunshine received instead of time exposed. Figure 10.75 [37] is a plot of percent of initial impact strength for an ABS as a function of total hours of exposure to sun light in three different locations: Florida, Arizona, and West Virginia. The curve reveals the fact that by "normalizing" the curves with respect to exposure to actual sunshine, the three different sites with three completely different weather conditions[9] lead to the same relation between impact strength and total sunshine.

9 Florida has a subtropical coastal climate with a yearly rainfall of 952 mm and sunshine of 2750 hours. Arizona has a hot dry climate with 116 mm of rainfall and 3850 hours of sunshine. West Virginia has a milder climate with 992 mm of rainfall and 2150 hours of sunshine [28].

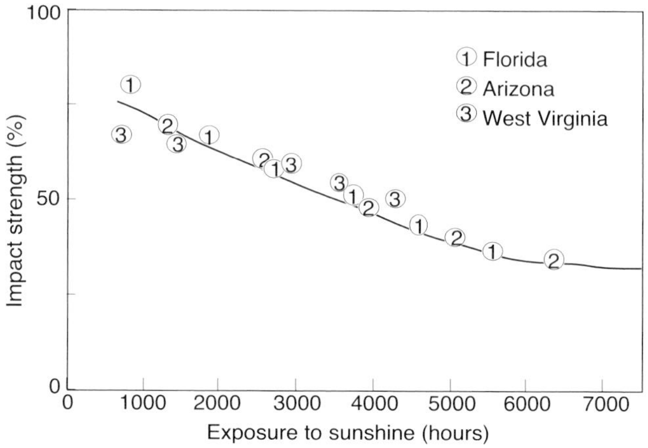

Figure 10.75 Impact strength of an ABS as a function of hours to actual sunshine exposure.

The effect of weathering can often be mitigated with the use of pigments, such as TiO_2 or soot, which absorb ultra violet radiation, making it more difficult to penetrate the surface of a polymer component. The most important one is soot. For example, ABS with white and black pigments exhibit a noticeable improvement in properties after exposure to ultra violet radiation. Figure 10.76 [37] shows the reduction of impact strength in ABS samples as a function of exposure time to sunshine for four pigment concentrations: 0.5%, 0.7%, 1%, and 2%. It is clear that the optimal pigment concentration is around 1%. Beyond 1% of pigmentation there is little improvement in the properties.

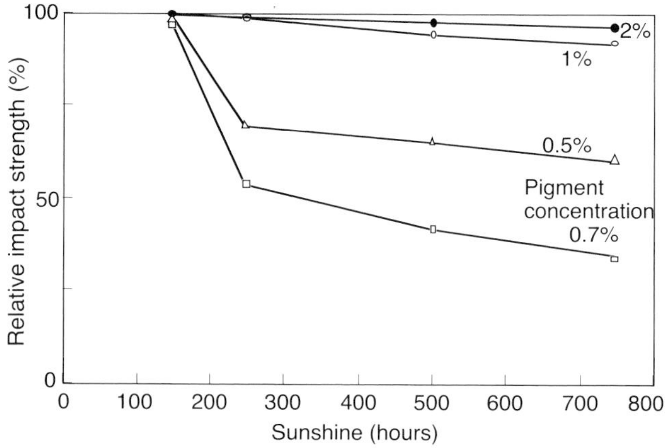

Figure 10.76 Influence of pigment concentration on the impact strength reduction of ABS specimens exposed to weathering.

10.8.2 Chemical Degradation

Liquid environments can have positive and negative effects on the properties of polymeric materials. Some chemicals or solvents can have detrimental effects on a polymer component. Figure 10.77 [25] shows results of creep rupture tests done on PVC tubes as a function of the hoop stress. It can be seen that the life span of the tubes in contact with the iso-octane and isopropanol has been significantly reduced as compared to the tubes in contact with water. The measured data for the pipes that contained iso-octane clearly show a slope reduction with a visible endurance limit, making it possible to do long-life predictions. On the other hand, the isopropanol samples do not exhibit such a slope reduction, suggesting that isopropanol is a harmful environment which acts as a solving agent and leads to gradual degradation of the PVC surface.

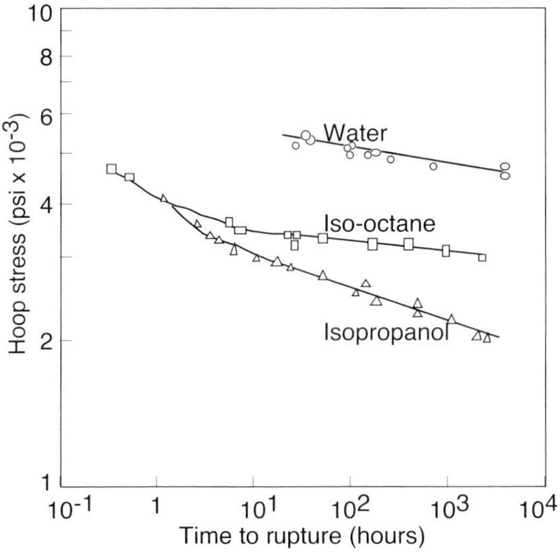

Figure 10.77 Effect of different environments on the stress rupture life of PVC pipe at 23 °C.

The question, of whether a chemical is harmful to a specific polymeric material is one that needs to be addressed if the polymer component is to be placed in a possibly threatening environment. Similar to polymer solutions, a chemical reaction between a polymer and another substance is governed by the Gibbs free energy equation, as discussed in Chapter 5. If the change in enthalpy, ΔH, is negative then a chemical reaction will occur between the polymer and the solvent.

The effect of the solubility parameter of several solvents on the fatigue response of polystyrene samples is presented in Fig. 10.78 [31]. Here, the relation in Eq. 6.24 becomes evident; as the absolute difference between the solubility parameter of polystyrene, which is 9.1 $(cal/cm^3)^{1/2}$, and the solubility parameter of the solvent decreases, the fatigue life drops significantly.

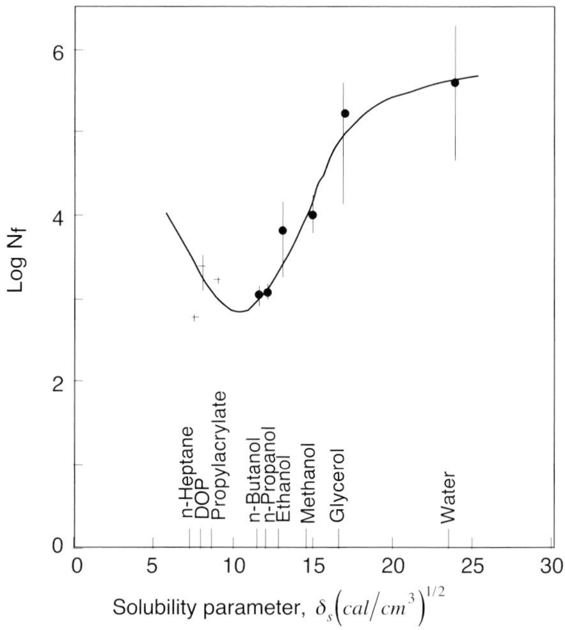

Figure 10.78 Effect of solubility parameter of the surrounding media on the fatigue life of polystyrene specimens.

It should be pointed out again that some substances are more likely to be absorbed by the polymer than others[10]. A polymer that is in a soluble environment is more likely to generate stress cracks and fail. This is illustrated in Fig. 10.79 [38], which shows the strain for crack formation in polyphenylene oxide samples as a function of solubility parameter[11] of various solutions. The specimens in solutions that were ± 1 $(cal/cm^3)^{1/2}$ away from the solubility parameter of the polymer generated cracks at fairly low strains, whereas those specimens in solutions with a solubility parameter further away from the solubility of the polymer formed crazes at much higher strains.

10 Please refer to Chapter 13.

11 Please refer to Chapter 6.

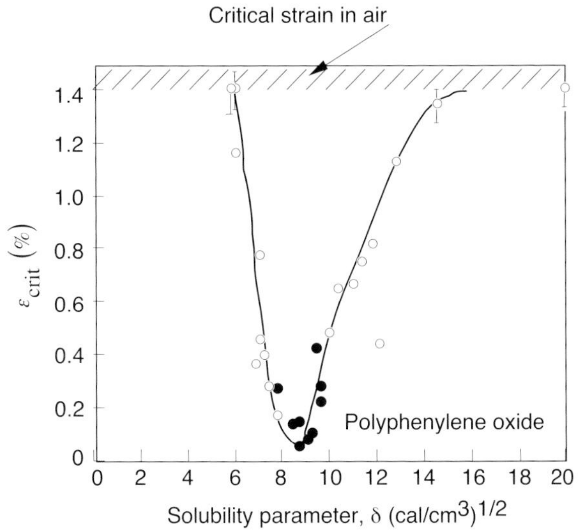

Figure 10.79 Strains at failure as a function of solubility parameter for polyphenylene oxide specimens: (•) cracking, (O) crazing.

Environmental stress cracking or stress corrosion in a polymer component only occurs if crazes or microcracks exist. Hence, stress corrosion in a hostile environment can be avoided if the strain within the component is below the critical strain, ε_{fx}.

10.8.3 Thermal Degradation of Polymers

Because plastics are organic materials, they are threatened by chain breaking, splitting off of substituents, and oxidation. This degradation generally follows a reaction which can be described by the Arrhenius principle. The period of dwell or residence time permitted before thermal degradation occurs is given by

$$t_{\text{permitted}} \sim \exp\left(\frac{\Delta}{RT}\right) \tag{10.40}$$

where Δ is the activation energy of the polymer, R the gas constant and T the absolute temperature.

A material that is especially sensitive to thermal degradation is PVC; furthermore, the hydrogen chloride that results during degradation attacks metal parts. Ferrous metals act as a catalyzer and accelerate degradation.

An easy method for determining the flash point of molding batches is by burning the hydrocarbons which are released at certain temperatures. This is shown schematically in Fig. 10.80[12] . For PVC one should use a vial with soda lye, instead of a flame, to determine the conversion of chlorine.

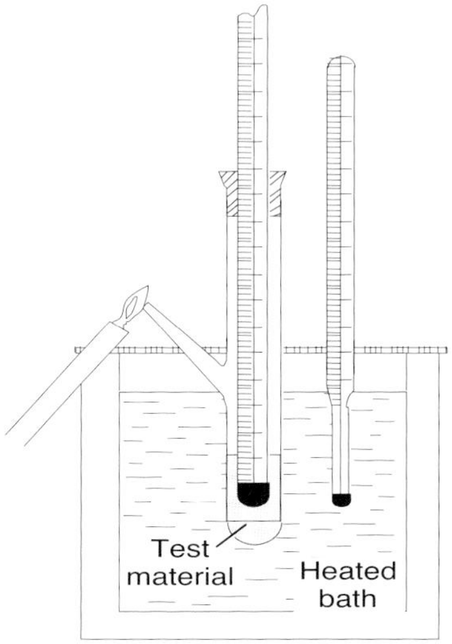

Figure 10.80 Test procedure to determine flash point of polymers.

Thermogravimetry (TGA) is another widely used method to determine the resistance to decomposition or degradation of polymers at high temperatures. For this purpose, the test sample is warmed up in air or in a protective gas while placed on a highly sensitive scale. The change in weight of the test sample is then observed and recorded (see Chapter 3). It is also very useful to observe color changes in a sample while they are heated inside an oven. For example, to analyze the effect of processing additives, polymers are kneaded for different amounts of time, pressed onto a plate, and placed inside a heated oven. The time when a color change occurs is recorded to signify when degradation occurs.

12 Courtesy of BASF.

Examples

10.1 Charpy Impact tests were performed on PVC samples with crack lengths of 1, 2, 3, 4, and 5 mm. The energy absorbed by the specimen during impact was 0.1, 0.062, 0.046, 0.037 and 0.031 J. The specimen's cross-section was 10 mm x 10 mm, and its length 40 mm. For a PVC modulus of 2 GPa what is the material's fracture toughness ?

We must first compute a/D and read $\tilde{a}$ from Table 10.6, where $D = 10$ mm and $L = 40$ mm. Next we can calculate $tw\tilde{a}$ ($Dw\tilde{a}$ for the notation used in Table 10.6), where $w = 10$ mm. Table 10.10 tabulates all the data.

We can now plot U versus $Dw\tilde{a}$ and estimate the fracture toughness from the slope, as shown in Fig. 10.82.

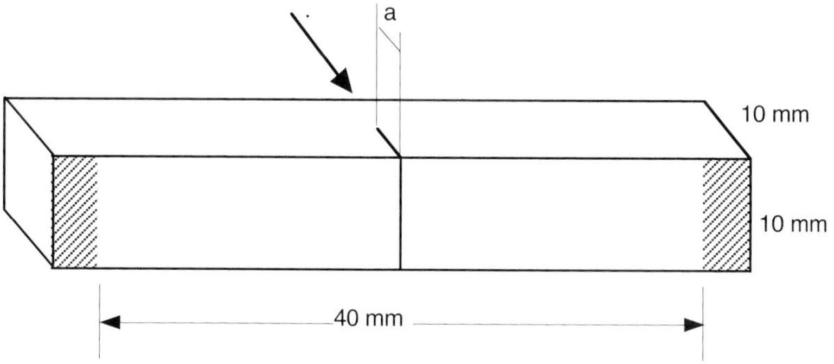

Figure 10.81 Sample geometry for Example 10.1.

Table 10.10 Tabulated data for Example 10.1

a (mm)	a/D	$\tilde{a}$	$Dw\tilde{a}$ (m^2)	U (J)
1	0.1	0.781	78.1×10^{-6}	0.1
2	0.2	0.468	46.8×10^{-6}	0.062
3	0.3	0.354	35.4×10^{-6}	0.0465
4	0.4	0.287	28.7×10^{-6}	0.037
5	0.5	0.233	23.3×10^{-6}	0.031

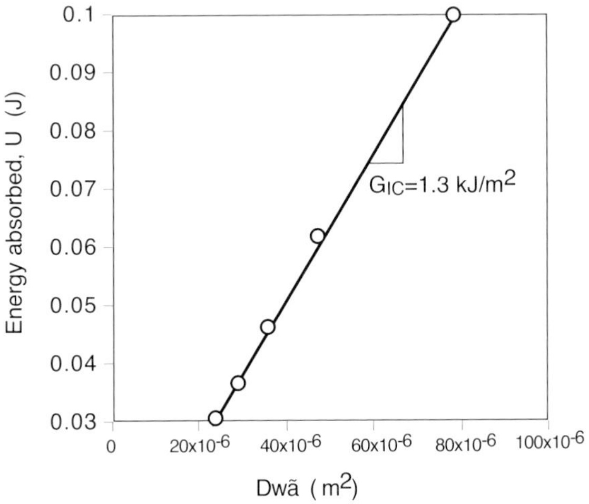

Figure 10.82 Fracture toughness plot for Example 10.1.

We can also compute the critical stress intensity factor K_{IC} using

$$K_{IC} = \sqrt{EG_{IC}} = 1.61 MNm^{-3/2}$$

10.2 You are asked to design a cylindrical 200 mm diameter, 5 mm thick polycarbonate pressure vessel, as depicted in Fig. 10.83. To attach a fixture, the pressure vessel has a series of 0.5 mm deep grooves along its length. During its life, the pressure vessel will experience occasional pressure surges. Using the data given below, estimate the critical pressure inside the vessel that will result in fracture.

$$G_{IC} = 5\,\text{kJ}/\text{m}^2, \quad E = 2.1\,\text{GPa}, \quad \nu = 0.33$$

This problem can be solved using Eq. 10.14

$$G_{Ic} = \frac{\pi\sigma_c^2 a}{E}\left(1-\nu^2\right)$$

where, σ_c is the hoop stress given by

$$\sigma_c = \frac{p\overline{D}}{2h}$$

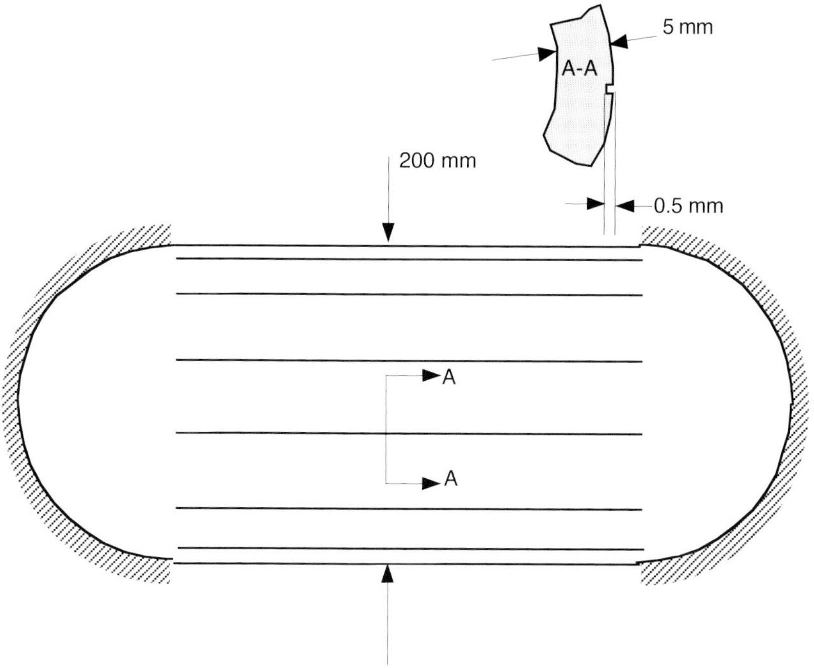

Figure 10.83 Pressure vessel geometry for Example 10.2.

Note that since the vessel is axially restricted, the axial stress can be neglected. We can now solve for the pressure that will lead to failure using

$$p = \left[\frac{G_{IC}}{\pi a} \frac{4Eh^2}{\left(1-\nu^2\right)\bar{D}^2} \right]^{1/2} = \left[\frac{5\ \mathrm{kJ/m^2}}{\pi(0.5\ \mathrm{mm})} \frac{4(2.1\ \mathrm{GPa})(5\ \mathrm{mm})^2}{(1-.33^2)(197.5\ \mathrm{mm})^2} \right]^{1/2} = 4.39\mathrm{MPa}\ (645\ \mathrm{psi})$$

Problems

10.1 Is polystyrene an appropriate material to use for safety goggles? Why?

10.2 A high impact strength polycarbonate has a stress intensity factor, K_{Ic}, of 2.6 MN/m$^{3/2}$, and a strain toughness, G_{Ic} of 5.0 kJ/m^2. The polycarbonate will be used to assemble a hockey face guard. The face guard will fit onto the helmet using several snap-fits located on the helmet. This requires

1.5 mm deep grooves on the polycarbonate face guard. What maximum stresses will you be able to subject the face guard in the region of the grooves.

10.3 Someone in your company designed high-density polyethylene water pipes to transport cold and hot water at pressures of 1.5 bar. The hot water will be at 50 ^{o}C, and you can assume the cold water to be at room temperature. The pipe's inner diameter is 50 mm with a wall thickness of 1 mm. Would you approve this product with a 5 year warranty for the above specifications? Why?

10.4 A PS-HI test specimen with a 20 mm x 1 mm cross-section and 2 mm long central crack is carrying a 100 N load. Plot the stress distribution near, but not at the crack tip.

10.5 A PS component with 0.5 mm long crazes is being stressed between -5 MPa and +5MPa. At what rate is the length of the crazes growing? Estimate the number of cycles to failure.

10.6 A creeping PS component loaded to a stress of 2 MPa is showing the formation of small crazes that measure approximately 200 μm. Estimate the time it takes for the crazes to grow to a length of 500 μm. Use $m = 7$ and $m = 25$.

References

1. Crawford, R.J., *Plastics Engineering*, 2nd, Ed., Pergamon Press, (1987).
2. Kinloch, A.J., Young, R.J., *Fracture Behavior of Polymers*, Applied Science Publishers, London, (1983).
3. Griffith, A.A., *Phil. Trans. Roy. Soc., A221*, 163, (1920).
4. Rice, J.R., *J. Appl. Mech., 35*, 379, (1965).
5. Vincent, P.I., *Plastics, 29*, 79 (1964).
6. Beahan, P., Bevis, M., and Hull, D., *Phil. Mag., 24*, 1267, (1971).
7. Menges, G., *Werkstoffkunde der Kunststoffe*, 2nd Ed., Hanser Publishers, Munich, (1984).
8. Nicolay, Ph.D. Thesis, IKV, RWTH-Aachen, (1976).
9. Menges, G., and Alf, E., *Kunststoffe, 62*, 259, (1972).
10. Schultz, J.M., *Polymer Materials Science*, Prentice-Hall, Englewood Cliffs, N.J., (1974).
11. Ehrenstein, G.W., *Polymeric Materials*, Hanser Publishers, Munich, (2001).
12. Engel, L., Klingele, H., Ehrenstein, G.W., and Schaper, H., *An Atlas of Polymer Damage*, Hanser Publishers, Munich, (1978).
13. Roskothen, H.J., Ph.D. Thesis, IKV, RWTH-Aachen, (1974).
14. Boyer, R.F., *Polymer Eng. Sci., 8*, 161, (1968).

15. Boden, H.E., Ph.D. Thesis, IKV, RWTH-Aachen, Germany, (1983).
16. Menges, G., and Boden, H.-E., *Failure of Plastics*, Chapter 9, Eds. W. Brostow, and R.D. Corneliussen, Hanser Publishers, Munich, (1986).
17. Andrews, E.H., *Fracture in Polymers*, Oliver and Body, London, (1968).
18. Rest, H., Ph.D. Thesis, IKV-Aachen, (1984).
19. ASTM, Vol .08.01, Plastics (I): C 177-D 1600, ASTM-D 256, 58-74, (1991).
20. Plati, E. and Williams, J.G., *Polym. Eng. Sci., 15*, 470, (1975).
21. Crawford, R.J., and Benham, P.P., *Polymer, 16*, 908, (1975).
22. Richard, K., Gaube, E., and Diedrich, G., *Kunststoffe, 49*, 516, (1959).
23. Gaube, E. and Kausch, H.H., *Kunststoffe, 63*, 391, (1973).
24. Bannantine, J.A., Comer, J.J., and Handrock, J.L., *Fundamentals of Metal Fatigue Analysis*, Prentice Hall, Englewood Cliffs, (1990).
25. Riddell, M.N., *Plast. Eng., 40*, 4, 71, (1974).
26. Kleinemeier, B., Ph.D. Thesis, IKW-Aachen, Germany, (1979).
27. Bucknall, C.B., Gotham, K.V., and Vincent, P.I., *Polymer Science. A Materials Handbook*, Ed. A.D. Jenkins, Vol. 1, Chapter 10, American Elsevier, New York, (1972).
28. Owen, M.J., Smith, T.R., and Dukes, R., *Plast.Polym., 37*, 227, (1969).
29. Hertzberg, R.W., and Manson, J.A., *Fatigue of Engineering Plastics*, Academic Press,New York, (1980).
30. Denton, D.L., *The Mechanical Properties of an SMC-R50 Composite*, Owens-Corning Fiberglas Corp., (1979).
31. Hertzberg, R.W., Manson, J.A., and Skibo, M.D., *Polym. Eng. Sci., 15*, 252, (1975).
32. Domininghaus, H., *Plastics for Engineers*, Hanser Publishers, Munich, (1993).
33. Dolfen, E., Ph.D. Thesis, IKV, RWTH-Aachen, (1978).
34. Menges, G., *Wekstoffkunde Kunststoffe*, Hanser Publishers, Munich, (1990).
35. Engel, L., Klingele, H., Ehrenstein, G.W., and Schaper, H., *An Atlas of Polymer Damage*, Hanser Publishers, Munich, (1978).
36. Davis, A. and Sims, D., *Weathering of Polymers*, Applied Science Publishers, London, (1983).
37. Ruhnke, G.M. and Biritz, L.F., *Kunststoffe*, 62, 250, (1972).
38. Bernier, G.A., and Kambour, R.P., *Macromolecules*,1,393, (1968).

11

Electrical Properties of Polymers

In contrast to metals, common polymers are poor electron conductors. Similar to mechanical properties, their electric properties depend to a great extend on the flexibility of the polymer's molecular blocks. The intent of this chapter is to familiarize the reader with electrical properties of polymers by discussing the dielectric, conductive, and magnetic properties.

11.1 Dielectric Behavior[1]

11.1.1 Dielectric Coefficient

The most commonly used electrical property is the dielectric coefficient, ε_r. Let us begin the discussion on dielectricity by looking at a disk condenser charged by the circuit shown in Fig 11.1. The accumulated charge, Q, is proportional to the consumed voltage, U:

$$Q = CU \tag{11.1}$$

where the proportionality constant, C, is called *capacitance*. The capacitance for the disk condenser in a vacuum, also valid for air, is defined by

$$C_0 = \varepsilon_0 \frac{A}{d} \tag{11.2}$$

where ε_0 is the vacuum's dielectric coefficient, A the disk's area and d the separation between the plates. Hence, the condenser's charge is given by

$$Q_0 = C_0 U \tag{11.3}$$

1 Parts of this chapter are based on the lecture notes of Prof. H. Hersping at the RWTH-Aachen, Germany, (1972).

If we replace the vacuum or air between the disks of the condenser by a real dielectric, the charge increases, for the same voltage U, by the factor ε_r represented by

$$Q = \varepsilon_r Q_0 \tag{11.4}$$

and the capacity changes to

$$C = \varepsilon_r C_0 = \varepsilon_r \varepsilon_0 \frac{A}{d} \tag{11.5}$$

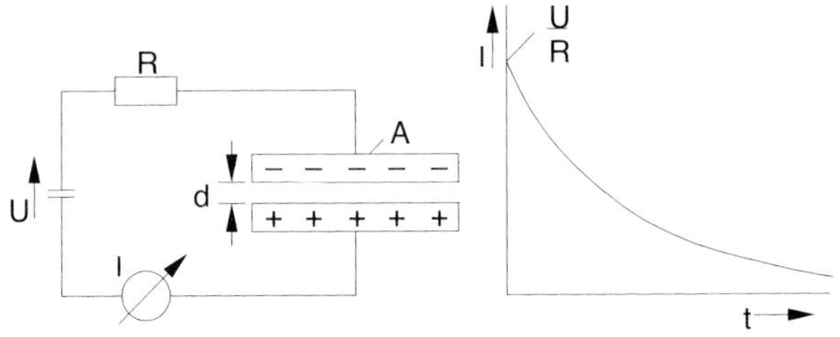

Figure 11.1 Condenser circuit used to measure capacitance properties.

The constant ε_r is often called the *relative dielectric coefficient*. It is dimensionless, and it is dependent on the material, temperature, and frequency. However, the charge changes when a dielectric material is inserted between the plates. This change in charge is due to the influence of the electric field developing polarization charges in the dielectric. This is more clearly represented in Fig. 11.2. The new charges that develop between the condenser's metal disks are called Q_p. Hence, the total charge becomes

$$Q = Q_0 + Q_p \tag{11.6}$$

In general terms, the charge is expressed per unit area as

$$\frac{Q}{A} = \frac{Q_0}{A} + \frac{Q_p}{A} = D \tag{11.7}$$

where D is the total charge per unit area. Introducing the *electric field intensity*, E, Eq. 11.2 can be rewritten as

$$E = \frac{U}{d} = \frac{1}{\varepsilon_0}\frac{Q_0}{A} = \frac{1}{\varepsilon_0 \varepsilon_r}\frac{Q}{A} \tag{11.8}$$

which results in

$$\frac{Q_0}{A} = \varepsilon_0 E \tag{11.9}$$

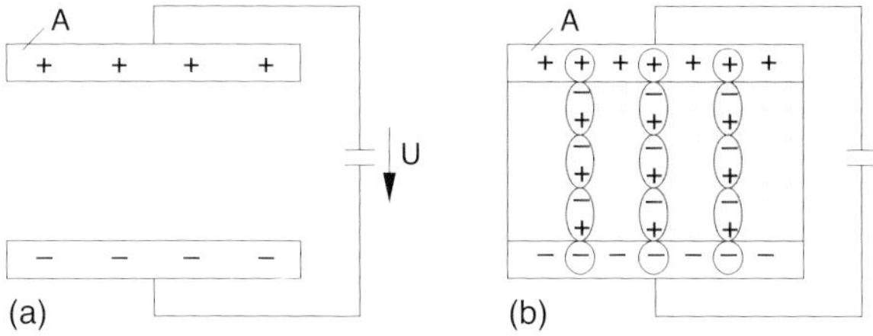

Figure 11.2 Polarization charges a) without a dielectric, b) with a dielectric.

If the charged condenser is separated from the voltage source beforehand, the voltage of the condenser will decrease with insertion of a dielectric. Thus, the inserted dielectric also increases the capacity of the condenser. Let us define the charge density of the polarization surface as

$$P = \frac{Q_p}{A} \tag{11.10}$$

This causes the total charge per unit area to become

$$D = \varepsilon_0 E + P \tag{11.11}$$

which can be rewritten as

$$P = D - \varepsilon_0 E \tag{11.12}$$

No field can develop within metallic condenser plates because of the high electric conductivity. Therefore, using the dielectric charge density per unit area, Eqs. 11.7 and 11.8 can be combined to give

$$D = \varepsilon_0 \varepsilon_r E \tag{11.13}$$

Substituting this result into Eq. 11.12, we get

$$P = \varepsilon_0 \varepsilon_r E - \varepsilon_0 E \tag{11.14}$$

or

$$P = \varepsilon_0 \left[\varepsilon_r - 1 \right] E = \varepsilon_0 \chi E \tag{11.15}$$

The factor χ is generally referred to as *dielectric susceptibility*. It is a measurement of the ability of a material to be a polarizer.

Table 11.1 lists the relative dielectric coefficients of important polymers. The measurements were conducted using the standard test DIN 53 483 in condensers of different geometries which, in turn depended on the sample type. The ASTM standard test is described by ASTM D150. Figures 11.3 [1] and 11.4 [1] present the dielectric coefficient for selected polymers as a function of temperature and frequency, respectively.

Table 11.1 Relative Dielectric Coefficient, ε_r, of Various Polymers [Hersping]

Polymer	Relative dielectric coefficient, ε_r	
	800 Hz	10^6 Hz
Expanded polystyrene	1.05	1.05
Polytetrafluoroethylene	2.05	2.05
Polyethylene (density dependent)	2.3–2.4	2.3–2.4
Polystyrene	2.5	2.5
Polypropylene	2.3	2.3
Polyphenyleneether	2.7	2.7
Polycarbonate	3.0	3.0
Polyethyleneterephthalate	3.0–4.0	3.0–4.0
ABS	4.6	3.4
Celluloseacetate, type 433	5.3	4.6
Polyamide 6 (moisture content dependent)	3.7–7.0	
Polyamide 66 (moisture content dependent)	3.6–5.0	
Epoxy resin (unfilled)		2.5–5.4
Phenolic type 31.5	6.0–9.0	6.0
Phenol type 74	6.0–10.0	4.0–7.0
Urea type 131.5	6.0–7.0	6.0–8.0
Melamine type 154	5.0	10.0

11.1.2 Mechanisms of Dielectrical Polarization

The two most important molecular types for the polarization of a dielectric in an electric field are *displacement polarization* and *orientation polarization*.

Under the influence of an electric field, the charges deform in field direction by aligning with the atomic nucleus (electron polarization) or with the ions (ionic polarization). This is usually called *displacement polarization* and is clearly demonstrated in Fig. 11.5.

Because of their structure, some molecules possess a dipole moment in the spaces that are free of an electric field. Hence, when these molecules enter an electric field, they will orient according to the strength of the field. This is generally referred to as *orientation polarization* and is schematically shown in Fig. 11.5.

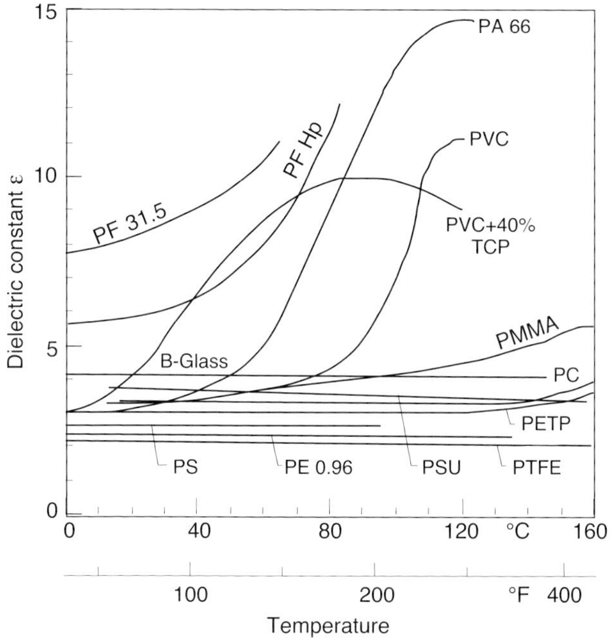

Figure 11.3 Dielectric constant as a function of temperature for various polymers.

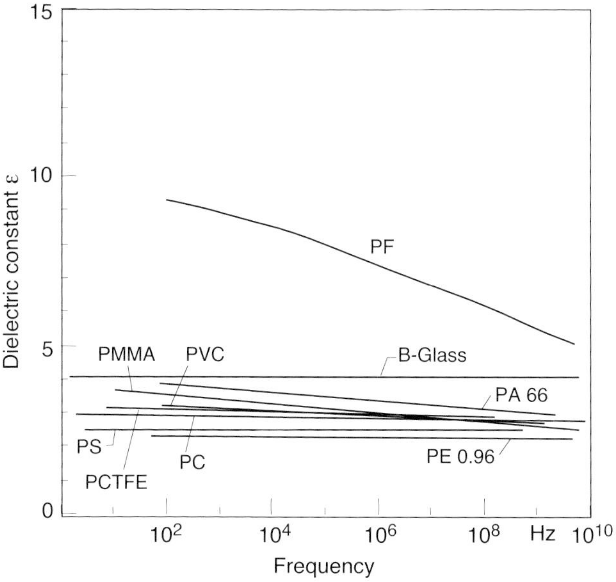

Figure 11.4 Dielectric constant as a function of frequency for various polymers.

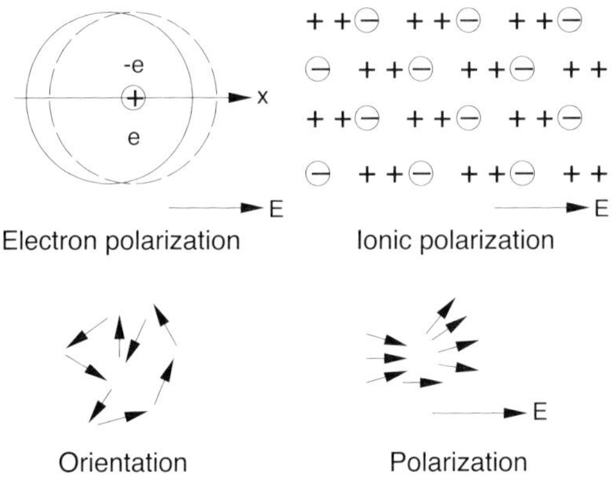

Figure 11.5 Polarization processes.

It takes some time to displace or deform the molecular dipoles in the field direction and even longer time for the orientation polarization. The more viscous the surrounding medium is, the longer it takes. In alternating fields of high frequency, the dipole movement can lag behind at certain frequencies. This is called dielectric relaxation, which leads to dielectric losses that appear as dielectric heating of the polar molecules.

In contrast to this, the changes in the displacement polarization happen so quickly that it can even follow a lightwave. Hence, the refractive index, n, of light is determined by the displacement contribution, ε_v, of the dielectric constant[2]. The relation between n and ε_v is given by

$$n = \sqrt{\varepsilon_v} \tag{11.16}$$

Hence, we have a way of measuring polarization properties since the polarization of electrons determines the refractive index of polymers. It should be noted that ion or molecular segments of polymers are mainly stimulated in the middle of the infrared spectrum.

A number of polymers have permanent dipoles. The best known polar polymer is polyvinyl chloride, and C = O groups also represent a permanent dipole. Therefore, polymers with that kind of building block suffer dielectric losses in alternating fields of certain frequencies. For example, Fig. 11.6 shows the frequency dependence of susceptibility.

2 For a more in-depth coverage of optical properties the reader is referred to Chapter 11 of this book.

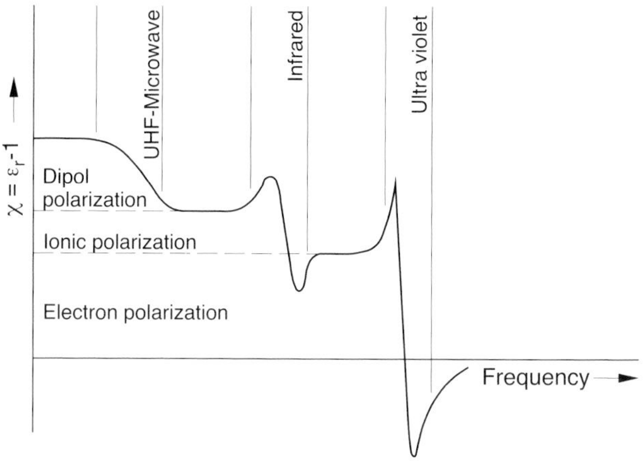

Figure 11.6 Frequency dependence of different polarization cases.

In addition, the influence of fillers on the relative dielectric coefficient is of considerable practical interest. The rule of mixtures can be used to calculate the effective dielectric coefficient of a matrix with assumingly spherically shaped fillers as

$$\varepsilon_{\text{eff}} = \varepsilon_{\text{matrix}}\left(1 - 3\phi\frac{\varepsilon_{\text{matrix}} - \varepsilon_{\text{filler}}}{2\varepsilon_{\text{matrix}} + \varepsilon_{\text{filler}}}\right) \tag{11.17}$$

Materials with air entrapments such as foams, have a filler dielectric coefficient of $\varepsilon_{air} = 1$; thus, the effective dielectric coefficient of the material reduces to

$$\varepsilon_{\text{foam}} = \varepsilon_{\text{matrix}}\left(1 - 3\phi\frac{\varepsilon_{\text{matrix}} - 1}{2\varepsilon_{\text{matrix}} + 1}\right) \tag{11.18}$$

and, for metal fillers where $\varepsilon_{metal} = \infty$, it can be written as

$$\varepsilon_{\text{eff}} = \varepsilon_{\text{matrix}}(1 + 3\phi) \tag{11.19}$$

Whether a molecule is stimulated to its resonant frequency in alternating fields or not depends on its relaxation time. The relaxation time, in turn, depends on viscosity, η, temperature, T, and radius, r, of the molecule. The following relationship can be used:

$$\lambda_m \sim \frac{\eta r^3}{T} \tag{11.20}$$

The parameter λ_m is the time a molecule needs to move back to its original shape after a small deformation. Hence, the resonance frequency, f_m, can be computed using

$$f_m = \frac{\omega_m}{2\pi} = \frac{1}{2\pi\lambda_m} \tag{11.21}$$

11.1.3 Dielectric Dissipation Factor

The movement of molecules, for example, during dipole polarization or ion polarization in an alternating electric field, leads to internal friction and, therefore, to the heating of the dielectric. The equivalent circuit shown in Fig. 11.7 is used here to explain this phenomenon. Assume an alternating current is passing through this circuit, with the effective value of U volts and an angular frequency ω defined by

$$\omega = 2\pi f \tag{11.22}$$

where f is the frequency in Hz. Through such a system, a complex current I^* will flow, composed of a resistive or loss component, I_r, and of a capacitive component, I_c. The vector diagram in Fig. 11.8 shows that with

$$I_r = U/R \tag{11.23}$$

and

$$I_c = \omega CU \tag{11.24}$$

we can write

$$I^* = U/R + i\omega CU \tag{11.25}$$

Here, i represents an imaginary component oriented in the imaginary axis of the vector diagram in Fig. 11.8. An alternating current applied to a condenser free of any current loss components would result in

$$\frac{I_r}{I_c} = \tan\delta \rightarrow 0 \tag{11.26}$$

In such a case, the condenser current is purely capacitive, which leads to no losses at all. This results in a voltage that is lagging the current by 90°, as demonstrated in Fig. 11.9. Accordingly, capacitance also consists of a real component and an imaginary component.

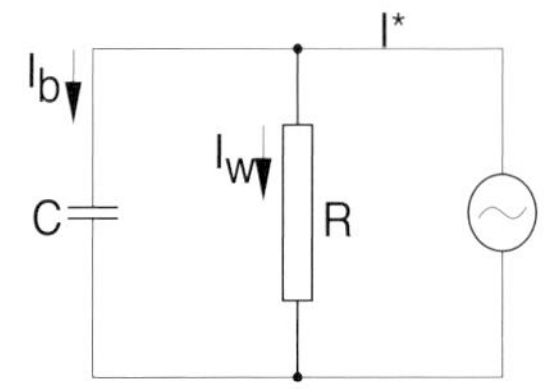

Figure 11.7 Equivalent circuit diagram for the losses in a dielectric.

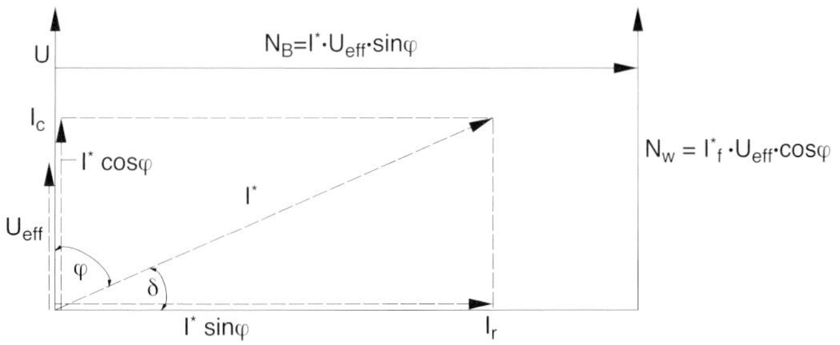

Figure 11.8 Current-voltage diagram or power-indicator diagram of electric alternating currents.

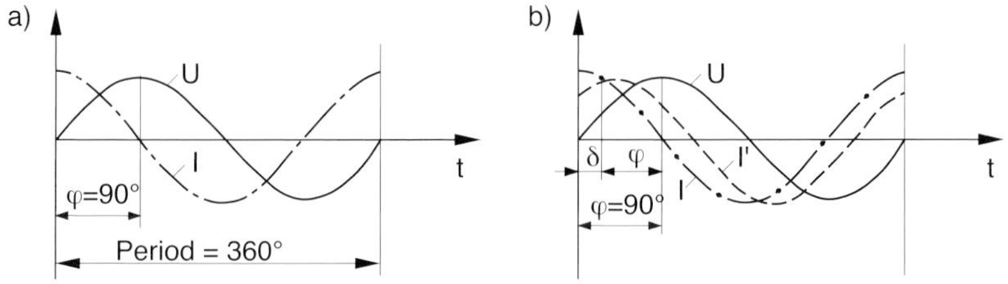

Figure 11.9 Current and voltage in a condenser. I ≡ current, U ≡ voltage, t ≡ time a) Without dielectric losses (ideal condition), current and voltage are displaced by the phase angle $\varphi = 90^\circ$ or $\pi/2$; b) With dielectric loss, the current curve I' is delayed by the loss angle δ.

If the condenser has losses, when $\tan\delta > 0$, a resistive current I_r is formed which leads to a heating energy rate in the dielectric of

$$E_h = \frac{1}{2} U I_{\text{eff}} \tan\delta \tag{11.27}$$

where I_{eff} represents the total current or the magnitude of the vector in Fig. 11.8. Using Eq. 11.25 for capacitance leads to

$$C^* = C' - \frac{1}{R\omega} = C' - iC'' \tag{11.28}$$

where C^* is the complex capacitance, with C' as the real component defined by

$$C' = \varepsilon_0 \varepsilon_r' \frac{A}{d} \tag{11.29}$$

and C'' as the imaginary component described by

$$C'' = \frac{1}{R\omega} = \varepsilon_0 \varepsilon_r'' \frac{A}{d}. \tag{11.30}$$

Using the relationship in Eq. 11.5 we can write

$$C^* = C_0\left(\varepsilon_r' - i\varepsilon_r''\right) = C_0 \varepsilon_r^* \tag{11.31}$$

where ε_r^* is called the *complex dielectric coefficient*. According to Eqs. 11.25 and 11.31, the phase angle difference or *dielectric dissipation factor* can be defined by

$$\tan\delta = \frac{I_r}{I_c} = \frac{\varepsilon_r''}{\varepsilon_r'} \tag{11.32}$$

If we furthermore consider that electric conductivity is determined by

$$\sigma = \frac{1}{R}\frac{d}{A} \tag{11.33}$$

then the imaginary component of the complex dielectric coefficient can be rewritten as

$$\varepsilon_r'' = \frac{\sigma}{\omega \varepsilon_0} = \varepsilon_r' \tan\delta \tag{11.34}$$

Typical ranges for the dielectric dissipation factor of various polymer groups are shown in Table 11.2. Figures 11.10 [1] and 11.11 [1] present the dissipation factor tan δ as a function of temperature and frequency, respectively.

Table 11.2 Dielectric Dissipation Factor ($\tan\delta$) for Varoius Polymers

Material	$\tan\delta$
Non-polar polymers (PS, PE, PTFE)	< 0.0005
Polar polymers (PVC and others)	0.001-0.02
Thermoset resins filled with glass, paper, cellulose	0.02-0.5

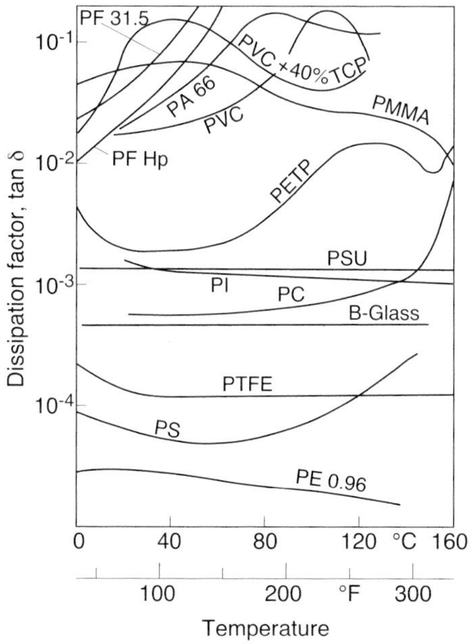

Figure 11.10 Dielectric dissipation factor as a function of temperature for various polymers.

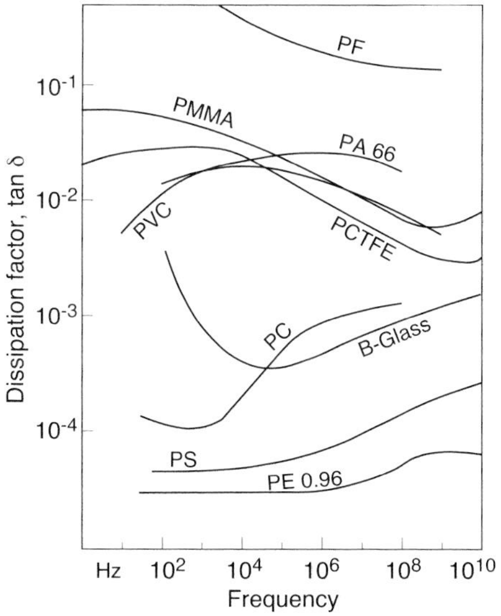

Figure 11.11 Dielectric dissipation factor as a function of frequency for various polymers.

11.1.4 Implications of Electrical and Thermal Loss in a Dielectric

The electric losses through wire insulation running high frequency currents must be kept as small as possible. Insulators are encountered in transmission lines or in high-frequency fields such as the housings of radar antennas. Hence, we would select materials which have low electrical losses for these types of applications.

On the other hand, in some cases we want to generate heat at high frequencies. Heat sealing of polar polymers at high frequencies is an important technique used in the manufacturing of soft PVC sheets such as the ones encountered in automobile vinyl seat covers.

To assess whether a material is suitable for either application, one must know the loss properties of the material and calculate the actual electrical loss. To do this, we can rewrite Eq. 11.27 as

$$E_h = U^2 \omega C \tan\delta \tag{11.35}$$

or as

$$E_h = 2\pi f U^2 d^2 \varepsilon_0 \varepsilon_r' \tan\delta C_0 \tag{11.36}$$

The factor that is dependent on the material and indicates the loss is the *loss factor* $\varepsilon_r' \tan\delta$, called ε_r'' in Eq. 11.34. As a rule, the following should be used:

$\varepsilon_r' \tan\delta < 10^{-3}$ for high-frequency insulation applications, and

$\varepsilon_r' \tan\delta > 10^{-2}$ for heating applications.

In fact, polyethylene and polystyrene are perfectly suitable as insulators in high-frequency applications. To measure the necessary properties of the dielectric, the standard DIN 53 483 and ASTM D 150 tests are recommended.

11.2 Electric Conductivity

11.2.1 Electric Resistance

The current flow resistance, R, in a plate-shaped sample in a direct voltage field is defined by Ohm's law as

$$R = \frac{U}{I} \tag{11.37}$$

or by

$$R = \frac{1}{\sigma}\frac{d}{A} \tag{11.38}$$

where σ is known as the conductivity and d and A are the sample's thickness and surface area, respectively. The resistance is often described as the inverse of the conductance, G,

$$R = \frac{1}{G} \tag{11.39}$$

and the conductivity as the inverse of the specific resistance, ρ,

$$\sigma = \frac{1}{\rho} \tag{11.40}$$

The simple relationship found in Eq. 11.37-39 is seldom encountered since the voltage, U, is rarely steady and usually varies in cyclic fashion between 10^{-1} to 10^{11} Hz [2].

Current flow resistance is called *volume conductivity* and is measured one minute after direct voltage has been applied using the DIN 53 482 standard test. The time definition is necessary, because the resistance decreases with polarization. For some polymers we still do not know the final values of resistance. However, this has no practical impact, since we only need relative values for comparison. Figure 11.12 compares the specific resistance, ρ, of various polymers and shows its dependence on temperature. Here, we can see that similar to other polymer properties, such as the relaxation modulus, the specific resistance not only decreases with time but also with temperature.

The surface of polymer parts often shows different electric direct-current resistance values than their volume. The main cause is from surface contamination (e.g., dust and moisture). We therefore have to measure the surface resistance using a different technique. One common test is DIN 53 482, which uses a contacting sample. Another test often used to measure surface resistance is DIN 53 480. With this technique, the surface resistance is tested between electrodes placed on the surface. During the test, a saline solution is dripped on the electrodes causing the surface to become conductive, thus heating up the surface and causing the water to evaporate. This leads not only to an increased artificial contamination but also to the decomposition of the polymer surface. If during this process conductive derivatives such as carbon form, the conductivity quickly increases to eventually create a short circuit. Polymers that develop only small traces of conductive derivatives are considered resistant. Such polymers are polyethylene, fluoropolymers, and melamines.

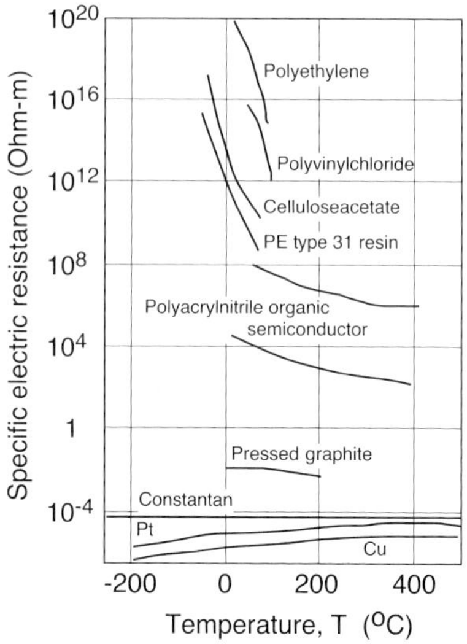

Figure 11.12 Specific electric resistance of polymers and metals as a function of temperature.

11.2.2 Physical Causes of Volume Conductivity

Polymers with a homopolar atomic bond, which leads to pairing of electrons, do not have free electrons and are not considered to be conductive. Conductive polymers–still in the state of development–in contrast, allow for movement of electrons along the molecular cluster, since they are polymer salts. The classification of these polymers with different materials is given in Fig. 11.13.

Potential uses of electric conductive polymers in electrical engineering include flexible electric conductors of low density, strip heaters, anti-static equipment, high-frequency shields, and housings. In semi-conductor engineering, some applications include semi-conductor devices (Schottky-Barriers) and solar cells. In electrochemistry, applications include batteries with high energy and power density, electrodes for electrochemical processes, and electrochrome instruments.

Because of their structure, polymers cannot be expected to conduct ions. Yet the extremely weak electric conductivity of polymers at room temperature and the fast decrease of conductivity with increasing temperatures is an indication that ions do move. They move because engineering polymers always contain a certain amount of added low-molecular constituents which act as moveable charge carriers. This is a diffusion process which acts in field direction and across the field. The ions "jump" from potential hole to potential hole as activated by higher temperatures (Fig. 11.12). At the same time, the lower density speeds up this diffusion process. The strong decrease of specific resistance with the absorption of moisture is caused by ion conductivity.

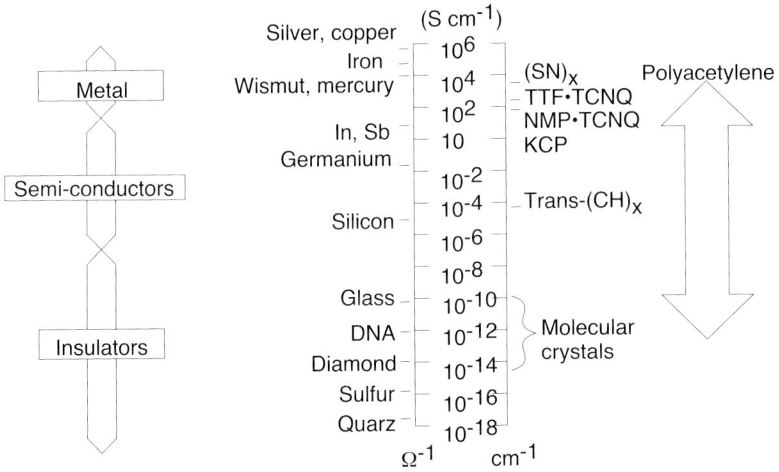

Figure 11.13 Electric conductivity of polyacetylene (*trans*-$(CH)_X$) in comparison to other materials.

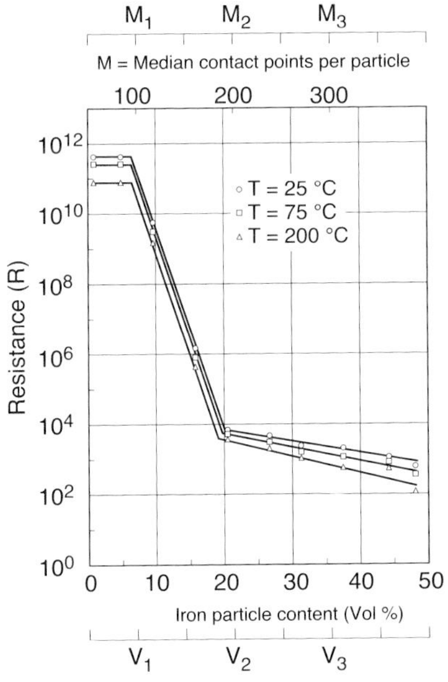

Figure 11.14 Resistance R of a polymer filled with metal powder (iron).

Conducting polymers are useful for certain purposes. When we insulate high-energy cables, for example, as a first transition layer we use a polyethylene filled with conductive filler particles such as soot. Figure 11.14 demonstrates the relationship between filler content and resistance. When contact tracks develop, resistance drops spontaneously. The number of inter-particle contacts, M, determines the resistance of a composite. At M_1 or M=1 there is one contact per particle. At this point, the resistance starts dropping. When two contacts per particle exist, practically all particles participate in setting up contact and the resistance levels off. The sudden drop in the resistance curve indicates why it is difficult to obtain a medium specific resistance by filling a polymer.

Figure 11.15 [3] presents the resistance in metal flakes or powder filled epoxy resins. The figure shows how the *critical volume concentration* for the epoxy systems filled with copper or nickel flakes is about 7% concentration of filler, and the critical volume concentration for the epoxy filled with steel powder is around 15%.

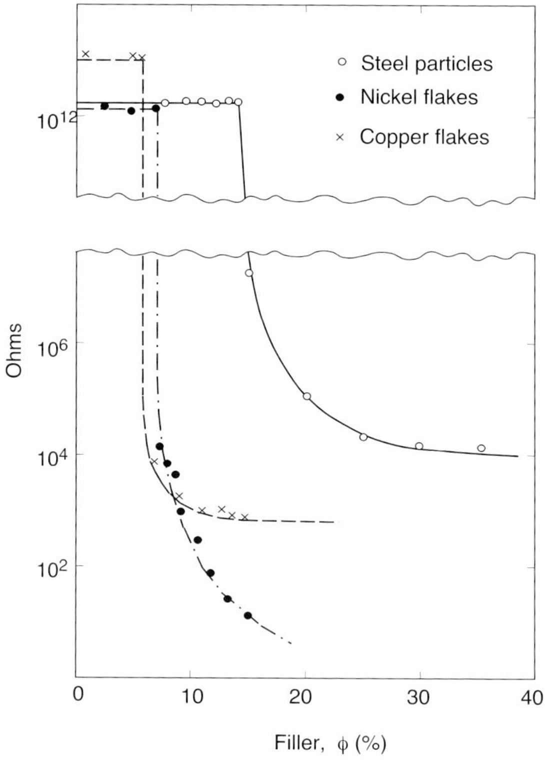

Figure 11.15 Resistance in metal flakes and powder filled epoxy resins.

11.3 Application Problems

11.3.1 Electric Breakdown

Since the electric breakdown of insulation may lead to failure of an electric component or may endanger people handling the component, it must be prevented. Hence, we have to know the critical load of the insulating material to design the insulation for long continuous use and with a great degree of confidence. One of the standard tests used to generate this important material property data for plate or block-shaped specimens is DIN 53 481. This test neglects the effect of material structure and of processing conditions. From the properties already described, we know that the *electric breakdown resistance* or *dielectric strength* must depend on time, temperature, material condition, load application rate, and frequency. It is furthermore dependent on electrode shape and sample thickness. In practice, however, it is very important that the upper limits measured on the experimental specimens in the laboratory are never reached. The rule of thumb is to use long-term load values of only 10% of the short-term laboratory data. Experimental evidence shows that the *dielectric strength* decreases as soon as crazes form in a specimen under strain and continues to decrease with increasing strain. This is demonstrated in Fig 11.16 [4].

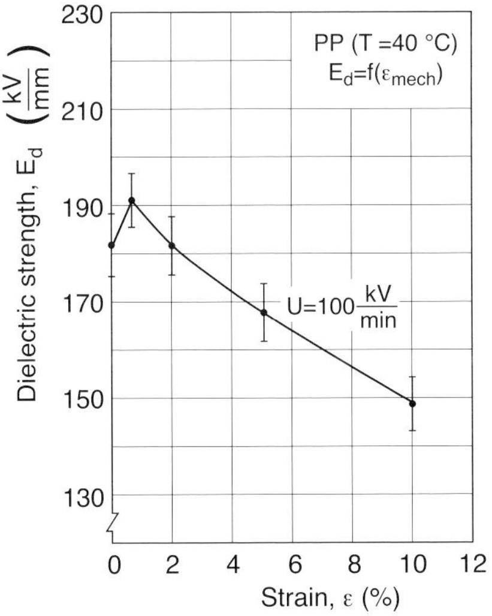

Figure 11.16 Drop of the dielectric strength of PP films with increasing strain.

On the other hand, Fig. 11.17 [4] demonstrates how the *dielectric dissipation factor*, tan δ, rises with strain. Hence, one can easily determine the beginning of the viscoelastic region (begin of crazing) by noting the starting point of the change in tan δ. It is also known that amorphous polymers act more favorably to electric breakdown resistance than partly crystalline polymers. Semi-crystalline polymers are more susceptible to electric breakdown as a result of breakdown along inter-spherulitic boundaries as shown in Fig. 11.18 [5]. Long-term breakdown of semi-crystalline polymers is either linked to "treeing," as shown in Fig. 11.19, or occurs as a heat breakdown, burning a hole into the insulation, such as the one in Fig. 11.18. In general, with rising temperature and frequency, the dielectric strength continuously drops.

Insulation materials— mostly LDPE— are especially pure and contain voltage stabilizers. These stabilizers are low-molecular cyclic aromatic hydrocarbons. Presumably, they diffuse into small imperfections or failures, fill the empty space and thereby protect them from breakdown.

Table 11.3 [6] gives dielectric strength and resistivity for selected polymeric materials.

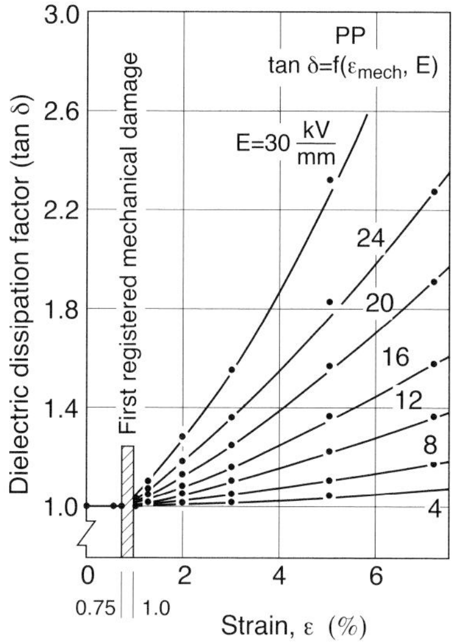

Figure 11.17 Increase of dielectric dissipation with increased strain in PP foils.

Figure 11.18 Breakdown channel around a polypropylene spherulitic boundary.

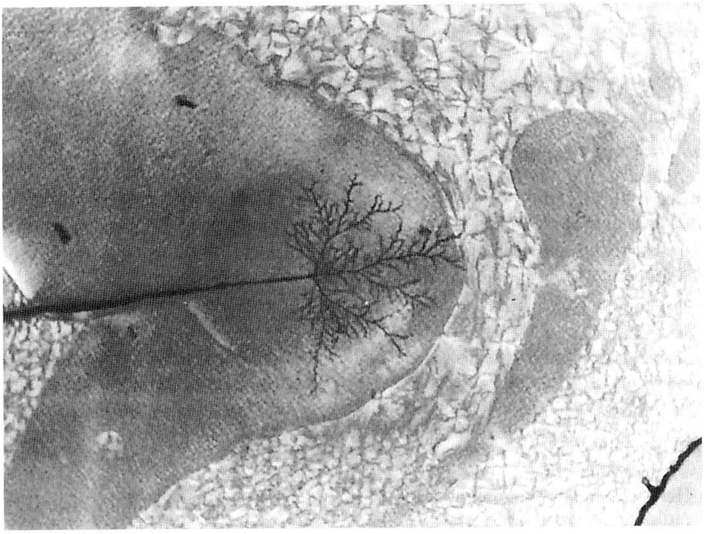

Figure 11.19 Breakdown channel in structureless finely crystalline zone of polypropylene.

Table 11.3 Dielectric Strength and Resistivity for Selected Polymers

Polymer	Dielectric strength MV/m	Resistivity Ohm-m
ABS	25	10^{14}
Acetal (homopolymer)	20	10^{13}
Acetal (copolymer) acrylic	20	10^{13}
Acrylic	11	10^{13}
Cellulose acetate	11	10^{9}
CAB	10	10^{9}
Epoxy	16	10^{13}
Modified PPO	22	10^{15}
Nylon 66	8	10^{13}
Nylon 66 + 30% GF	15	10^{12}
PEEK	19	10^{14}
PET	17	10^{13}
PET + 36% GF	50	10^{14}
Phenolic (mineral filled)	12	10^{9}
Polycarbonate	23	10^{15}
Polypropylene	28	10^{15}
Polystyrene	20	10^{14}
LDPE	27	10^{14}
HDPE	22	10^{15}
PTFE	45	10^{16}
uPVC	14	10^{12}
pPVC	30	10^{11}
SAN	25	10^{14}

11.3.2 Electrostatic Charge

An electrostatic charge is often a result of the excellent insulation properties of polymers— the very high surface resistance and current-flow resistance. Since polymers are bad conductors, the charge displacement of rubbing bodies, which develops with mechanical friction, cannot equalize. This charge displacement results from a surplus of electrons on one surface and a lack of electrons on the other. Electrons are charged positively or negatively up to hundreds of volts. They release their surface charge only when they touch another conductive body or a body which is inversely charged. Often the discharge occurs without contact, as the charge arches through the air to the close-by conductive or inversely charged body, as demonstrated in Fig. 11.20.

The currents of these breakdowns are low. For example, there is no danger when a person suffers an electric shock caused by a charge from friction of synthetic carpets or vinyls. There is danger of explosion, though, when the sparks ignite flammable liquids or gases.

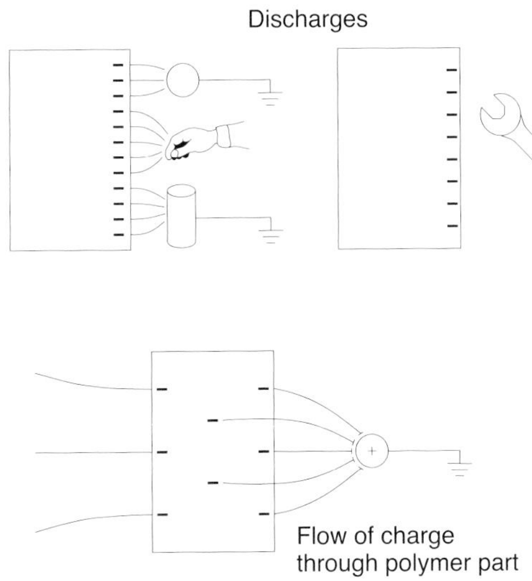

Figure 11.20 Electrostatic charges in polymers.

As the current-flow resistance of air is generally about 10^9 Ωcm, charges and flashovers only occur if the polymer has a current-flow resistance of $>10^9$ to 10^{10} Ωcm. Another effect of electrostatic charges is that they attract dust particles on polymer surfaces.

Electrostatic charges can be reduced or prevented by the following means:

- Reduce current-flow resistance to values of $<10^9$ Ωcm, for example, by using conductive fillers such as graphite.
- Make the surfaces conductive by using hygroscopic fillers that are incompatible with the polymer and surface. It can also be achieved by mixing-in hygroscopic materials such as strong soap solutions. In both cases, the water absorbed from the air acts as a conductive layer. It should be pointed out that this treatment loses its effect over time. Especially, the rubbing in of hygroscopic materials has to be repeated over time.
- Reduce air resistance by ionization through discharge or radioactive radiation.

11.3.3 Electrets

An electret is a solid dielectric body that exhibits permanent dielectric polarization. One can manufacture electrets out of some polymers when they are solidified under the influence of an electric field, when bombarded by electrons, or sometimes through mechanical forming processes.

Applications include films for condensers (polyester, polycarbonate, or fluoropolymers).

11.3.4 Electromagnetic Interference Shielding (EMI Shielding)

Electric fields surge through polymers as shown schematically in Fig. 11.20. Since we always have to deal with the influence of interference fields, signal-sensitive equipment such as computers cannot operate in polymer housings. Such housings must therefore have the function of Faradayic shields. Preferably, a multilayered structure is used— the simplest solution is to use one metallic layer. Figure 11.21 classifies several materials on a scale of resistances. We need at least 10^2 Ωcm for a material to fulfill the shielding purpose. With carbon fibers or nitrate coated carbon fibers used as a filler, one achieves the best protective properties. The shielding properties are determined using the standard ASTM ES 7-83 test.

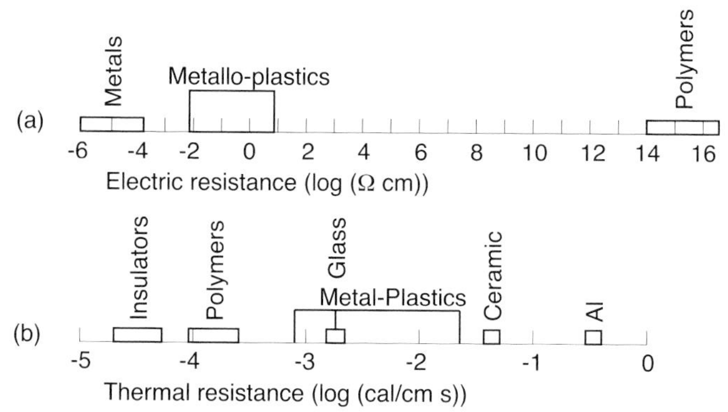

Figure 11.21 Comparison of conductive polymers with other materials: a) Electric resistance ρ of metal-plastics compared to resistance of metals and polymers b) Thermal resistance λ of metal-plastics compared to other materials

11.4 Magnetic Properties

External magnetic fields have an impact on substances which are subordinate to them because the external field interacts with the internal fields of electrons and atomic nuclei.

11.4.1 Magnetizability

Pure polymers are diamagnetic; that is, the external magnetic field induces magnetic moments. However, permanent magnetic moments, which are induced on ferromagnetic or paramagnetic substances, do not exist in polymers. This magnetizability M of a substance in a magnetic field with a field intensity H is computed with the *magnetic susceptibility*, χ as

$$M = \chi H \tag{11.41}$$

The susceptibility of pure polymers as *diamagnetic substances* has a very small and negative value. However, in some cases, we make use of the fact that fillers can alter the magnetic character of a polymer completely. The magnetic properties of polymers are often changed using magnetic fillers. Well-known applications are injection molded or extruded magnets or magnetic profiles, and all forms of electronic storage such as recording tape, floppy or magnetic disks.

11.4.2 Magnetic Resonance

Magnetic resonance occurs when a substance, in a permanent magnetic field, absorbs energy from an oscillating magnetic field. This absorption develops as a result of small paramagnetic molecular particles stimulated to vibration. We use this phenomenon to a great extent to clarify structures in physical chemistry. Methods to achieve this include *electron spinning resonance* (ESR) and, above all, *nuclear magnetic resonance* (NMR) spectroscopy.

Electron spinning resonance becomes noticeable when the field intensity of a static magnetic field is altered and the microwaves in a high frequency alternating field are absorbed. Since we can only detect unpaired electrons using this method, we use it to determine radical molecule groups.

When atoms have an odd number of nuclei, protons and neutrons, the magnetic fields which are caused by self-motivated spin cannot equalize. The alignment of nuclear spins in an external magnetic field leads to a magnetization vector which can be measured macroscopically as is schematically demonstrated in Fig. 11.22 [7]. This method is of great importance for the polymer physicist to learn more about molecular structures.

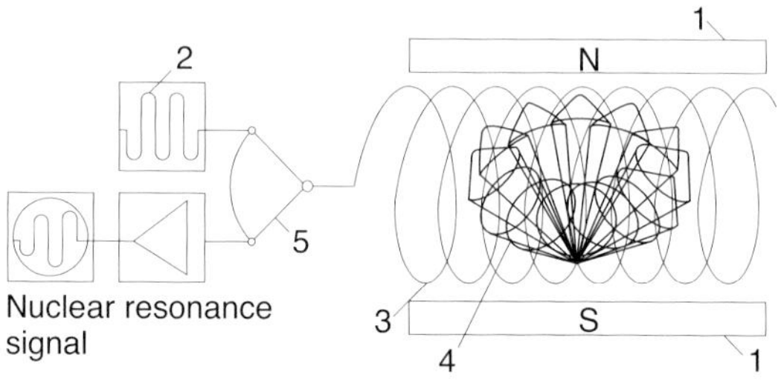

Figure 11.22 Schematic of the operating method of a nuclear spin tomograph: 1) magnet which produces a high steady magnetic field 2) radio wave generator 3) high-frequency field, produced by 2, when switch 5 is in upper position 4) processing nucleus, simulated by high frequency field 5) switch; in this position the decrease of relaxation of the nucleus' vibrations is measured.

References

1. Domininghaus, H., Plastics for Engineers, Hanser Publishers, Munich, (1993).
2. Baer, E., *Engineering Design for Plastics*, Robert E. Krieger Publishing Company, (1975).
3. Reboul, J.-P., *Thermoplastic Polymer Additives*, Chapter 6, J.T. Lutz, Jr., Ed., Marcel Dekker, Inc., New York, (1989).
4. Berg, H., Ph .D Thesis, IKV, RWTH-Aachen, Germany, (1976).
5. Wagner, H., Internal report, AEG, Kassel, Germany, (1974).
6. Crawford, R.J., *Plastics Engineering*, 2nd Ed., Pergamon Press, (1987).
7. From *Bild der Wissenschaft*.

12

Optical Properties of Polymers

Since some polymers have excellent optical properties and are easy to mold and form into any shape, they are often used to replace transparent materials such as inorganic glass. Polymers have been introduced into a variety of applications such as automotive headlights, signal light covers, optical fibers, imitation jewelry, chandeliers, toys, and home appliances. Organic materials such as polymers are also an excellent choice for high-impact applications where inorganic materials such as glass would easily shatter. However, due to the difficulties encountered in maintaining dimensional stability, they are not apt for precision optical applications. Other drawbacks include lower scratch resistance, when compared to inorganic glasses, making them still impractical for applications such as automotive windshields.

In this section, we will discuss basic optical properties which include the index of refraction, birefringence, transparency, transmittance, gloss, color, and behavior of polymers in the infrared spectrum.

12.1 Index of Refraction

As rays of light pass through one material into another, the rays are bent due to the change in the speed of light from one media to the other. The fundamental material property that controls the bending of the light rays is the *index of refraction,* N. The index of refraction for a specific material is defined as the ratio between the speed of light in a vacuum to the speed of light through the material under consideration

$$N = \frac{c}{v} \tag{12.1}$$

where c and v are the speeds of light through a vacuum and transparent media, respectively. In more practical terms, the refractive index can also be computed as a function of the angle of incidence, θ_i, and the angle of refraction, θ_r, as follows:

$$N = \frac{\sin\theta_i}{\sin\theta_r} \tag{12.2}$$

where θ_i and θ_r are defined in Fig. 12.1.

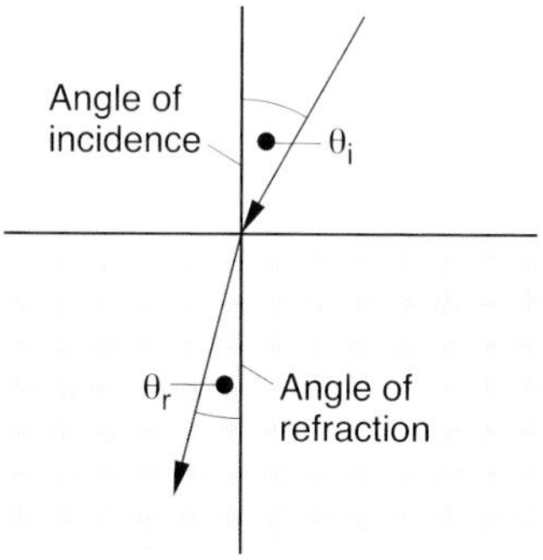

Figure 12.1 Schematic of light refraction.

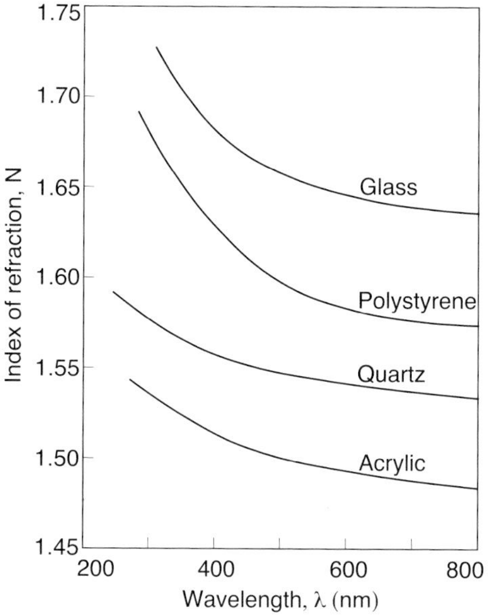

Figure 12.2 Index of refraction as a function of wavelength for various materials.

The index of refraction for organic plastic materials can be measured using the standard ASTM D 542 test. It is important to mention that the index of refraction is dependent on the wavelength of the light under which it is being measured. Figure 12.2 shows plots of the refractive index for various organic and inorganic materials as a function of wavelength. One of the significant points of this plot is that acrylic materials and polystyrene have similar refractive properties as inorganic glasses.

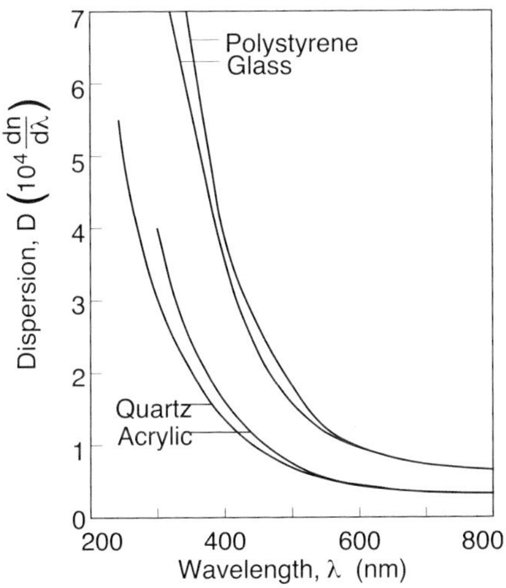

Figure 12.3 Dispersion as a function of wavelength for various materials.

An important quantity that can be deduced from the light's wavelength dependence on the refractive index is the dispersion, *D*, which is defined by

$$D = \frac{dN}{d\lambda} \tag{12.3}$$

Figure 12.3 shows plots of dispersion as a function of wavelength for the same materials shown in Fig. 12.2. The plots show that polystyrene and glass have a high dispersion in the ultra-violet light domain.

It is also important to mention that since the index of refraction is a function of density, it is indirectly affected by temperature. Figure 12.4 shows how the refractive index of PMMA changes with temperature. A closer look at the plot reveals the glass transition temperature.

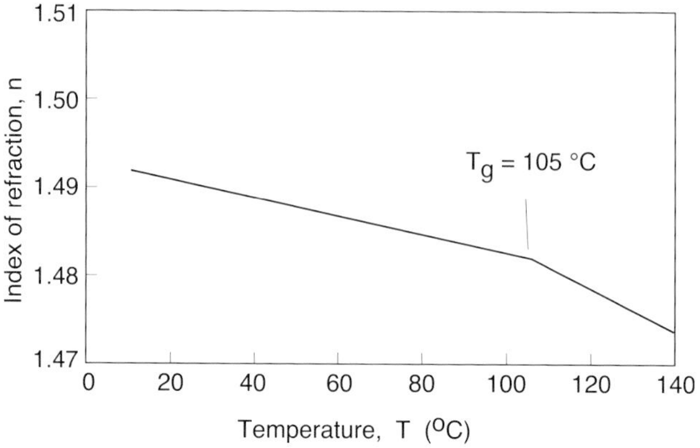

Figure 12.4 Index of refraction as a function of temperature for PMMA (λ = 589.3 nm).

12.2 Photoelasticity and Birefringence

Photoelasticity and flow birefringence are applications of the optical anisotropy of transparent media. When a transparent material is subjected to a strain field or a molecular orientation, the index of refraction becomes directional; the principal strains ε_1 and $\varepsilon_2$are associated with principal indices of refraction N_1 and N_2 in a two-dimensional system. The difference between the two principal indices of refraction (*birefringence*) can be related to the difference of the principal strains using the *strain-optical coefficient, k,* as

$$N_1 - N_2 = k(\varepsilon_1 - \varepsilon_2) \tag{12.4}$$

or, in terms of principal stress, and

$$N_1 - N_2 = C(\sigma_1 - \sigma_2) \tag{12.5}$$

where C is the stress-optical coefficient.

Double refractance in a material is caused when a beam of light travels through a transparent media in a direction perpendicular to the plane that contains the principal directions of strain or refraction index, as shown schematically in Fig. 12.5 [1]. The incoming light waves split into two waves that oscillate along the two principal directions. These two waves are out of phase by a distance δ, defined by

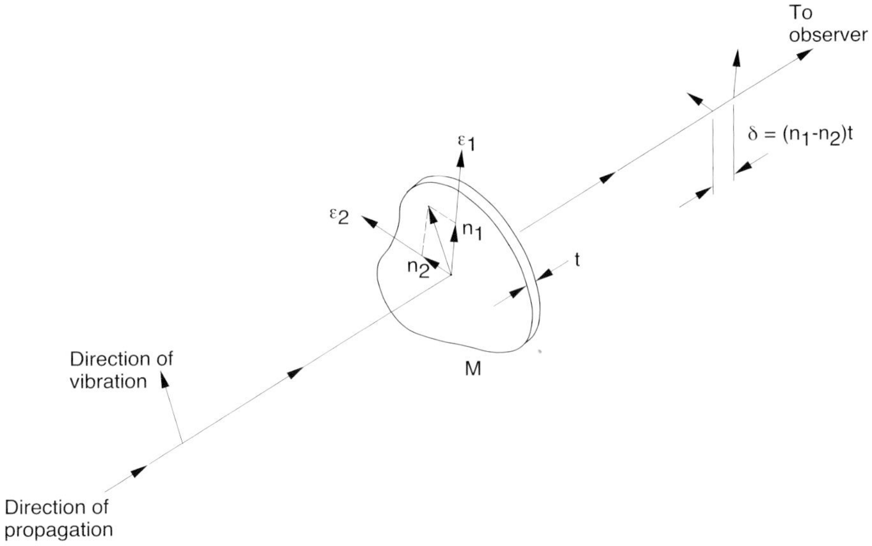

Figure 12.5 Propagation of light in a strained transparent media.

$$\delta = (N_1 - N_2)d \tag{12.6}$$

where d is the thickness of the transparent body. The out-of-phase distance, δ, between the oscillating light waves is usually referred to as the *retardation*.

In photoelastic analysis, one measures the direction of the principal stresses or strains and the retardation to determine the magnitude of the stresses. The technique and apparatus used to performed such measurements is described in the ASTM D 4093 test. Figure 12.6 shows a schematic of such a set-up, composed of a narrow wavelength band light source, two polarizers, two quaterwave plates, a compensator, and a monochromatic filter. The polarizers and quaterwave plates must be perpendicular to each other (90°). The compensator is used for measuring retardation, and the monochromatic filter is needed when white light is not sufficient to perform the photoelastic measurement. The set-up presented in Fig. 12.6 is generally called a *polariscope*.

The parameter used to quantify the strain field in a specimen observed through a polariscope is the color. The retardation in a strained specimen is associated to a specific color. The sequence of colors and their respective retardation values and fringe order are shown in Table 12.1 [1]. The retardation and color can also be associated to a *fringe order* using

$$\text{fringe order} = \frac{\delta}{\lambda} \tag{12.7}$$

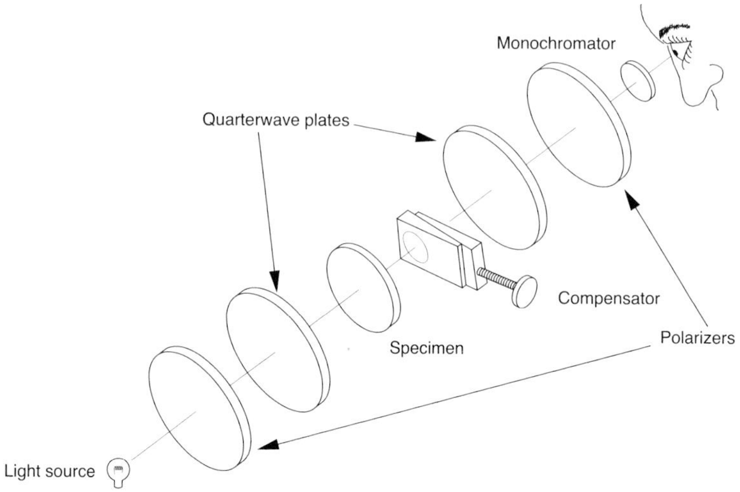

Figure 12.6 Schematic diagram of a polariscope.

Table 12.1 Retardation and Fringe Order Produced in a Polariscope

Color	Retardation (nm)	Fringe order
Black	0	0
Gray	160	0.28
White	260	0.45
Yellow	350	0.60
Orange	460	0.79
Red	520	0.90
Tint of passage	577	1.00
Blue	620	1.06
Blue-green	700	1.20
Green-yellow	800	1.38
Orange	940	1.62
Red	1050	1.81
Tint of passage	1150	2.00
Green	1350	2.33
Green-yellow	1450	2.50
Pink	1550	2.67
Tint of passage	1730	3.00
Green	1800	3.10
Pink	2100	3.60
Tint of passage	2300	4.00
Green	2400	4.13

A black body (fringe order zero) represents a strain free body, and closely spaced color bands represent a component with high strain gradients. The color bands are generally called the *isochromatics*. Figure 12.7 shows the isochromatic fringe pattern in a stressed notched bar. The fringe pattern can also be a result of molecular orientation and residual stresses in a molded transparent polymer component. Figure 12.8 shows the orientation induced fringe pattern in a molded part. The residual stress-induced birefringence is usually smaller than the orientation-induced pattern, making them more difficult to measure.

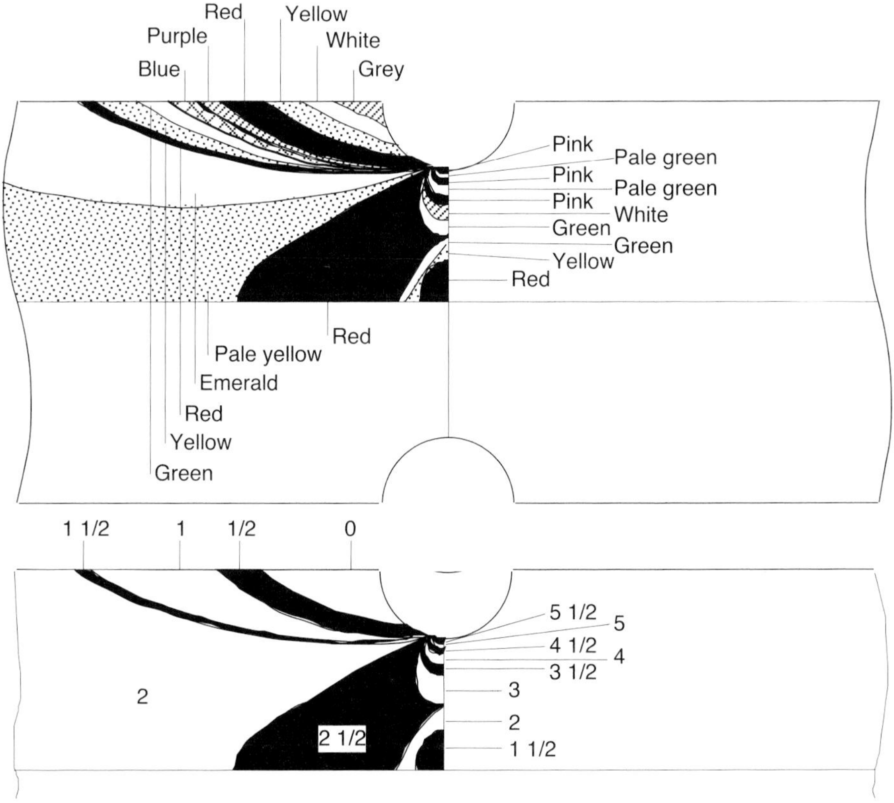

Figure 12.7 Fringe pattern on a notched bar under tension.

Flow induced birefringence is an area explored by several researchers [2–4]. Likewise, the flow induced principal stresses can be related to the principal refraction indices. For example, in a simple shear flow this relation can be written as [5]

$$(N_1 - N_2) = \frac{2C}{\sin 2x}\tau_{12} = \frac{2C}{\sin 2x}\eta\dot{\gamma} \tag{12.8}$$

where x is the orientation of the principal axes in a simple shear flow.

Figure 12.9 [6] shows the birefringence pattern for the flow of linear low-density polyethylene in a rectangular die.

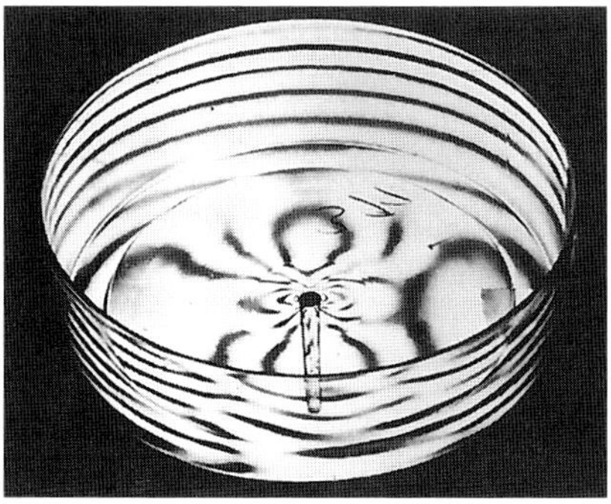

Figure 12.8 Transparent injection molded part viewed through a polariscope.

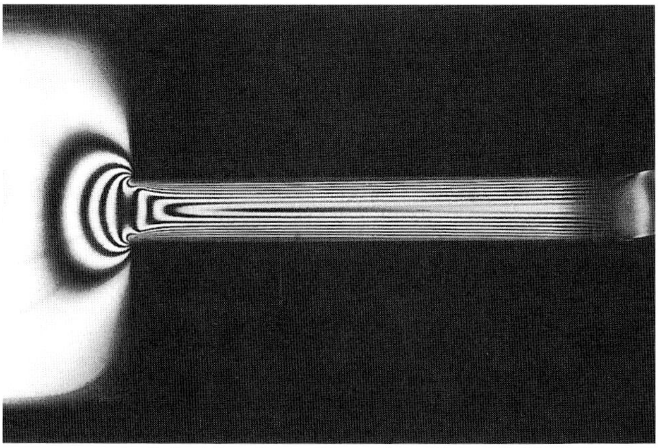

Figure 12.9 Birefringence pattern for flow of LLDPE in a rectangular die.

12.3 Transparency, Reflection, Absorption, and Transmittance

As rays of light pass through one media into another of a different refractive index, light will be scattered if the interface between the two materials shows discontinuities larger than the wavelength of visible light[1]. Hence, the transparency in semi-crystalline polymers is directly related to the crystallinity of the polymer. Since the characteristic size of the crystalline domains are larger than the wavelengths of visible light, and since the refractive index of the denser crystalline domains is higher compared to the amorphous regions, semi-crystalline polymers are not transparent; they are opaque or translucent. Similarly, high impact polystyrene— which is actually formed by two amorphous components, polybutadiene rubber particles[2] and polystyrene — appears white and translucent due to the different indices of refraction of the two materials. However, filled polymers can be made transparent if the filler size is smaller than the wavelength of visible light. Figure 12.10 shows various types of high impact polystyrene. The two types of HIPS shown in the lower part of the figure have polybutadiene particles that are smaller than the wavelength of visible light, making them transparent.

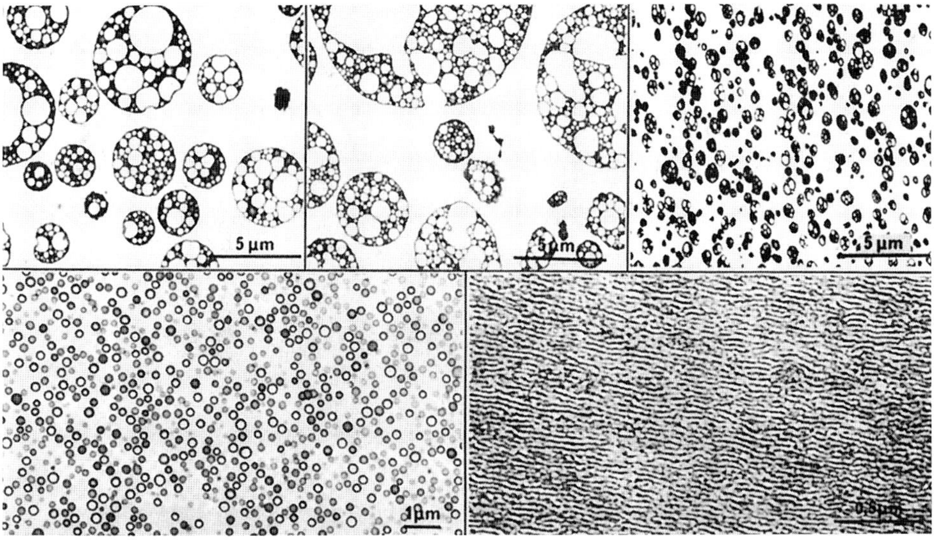

Figure 12.10 Morphology of polybutadiene particles in a polystyrene matrix for different types of high impact polystyrene.

1 The wavelength of visible light runs between 400 and 700 nm (0.4 and 0.7 µm).

2 The characteristic size of a rubber particle in high impact polystyrene is 1-10µm.

The concept of absorption and transmittance can be illustrated using the schematic and notation shown in Fig. 12.11. The figure plots the intensity of a light ray as it strikes and travels through an infinite plate of thickness d. For simplicity, the angle of incidence, θ_i, is 0°. The initial intensity of the incoming light beam, I, drops to I_0 as a fraction ρ_0 of the incident beam is reflected out. The reflected light beam can be computed using

$$I_r = \rho_0 I \tag{12.9}$$

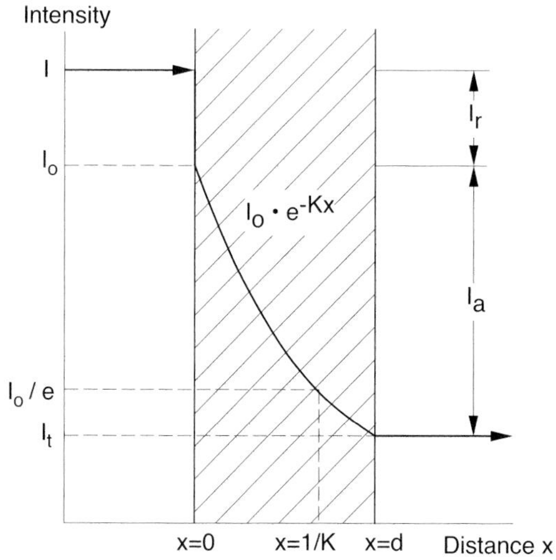

Figure 12.11 Schematic of light transmission through a plate.

The fraction of the incident beam which is reflected can be computed using Beer's law:

$$\rho_0 = \frac{(N-1)^2 + \chi^2}{(N+1)^2 + \chi^2} \tag{12.10}$$

Here, χ is the *absorption index* described by

$$\chi = \frac{K\lambda}{4\pi} \tag{12.11}$$

where λ is the wavelength of the incident light beam and K the *coefficient of absorption.*

The fraction of the beam that does penetrate into the material continues to drop due to absorption as it travels through the plate. The intensity fraction of the incident beam as it is transmitted through the material can be computed using *Bourger's law,*

$$T(x) = I_0 e^{-Kx} \tag{12.12}$$

where K is the *coefficient of absorption*. The intensity fraction of the incident beam transmitted to the rear surface of the plate can now be computed using

$$\tau = (1 - \rho_0)e^{-Kd} \tag{12.13}$$

However, as illustrated in Fig. 12.12, part of the beam is reflected back by the rear surface of the plate and is subsequently reflected and absorbed several times as it travels

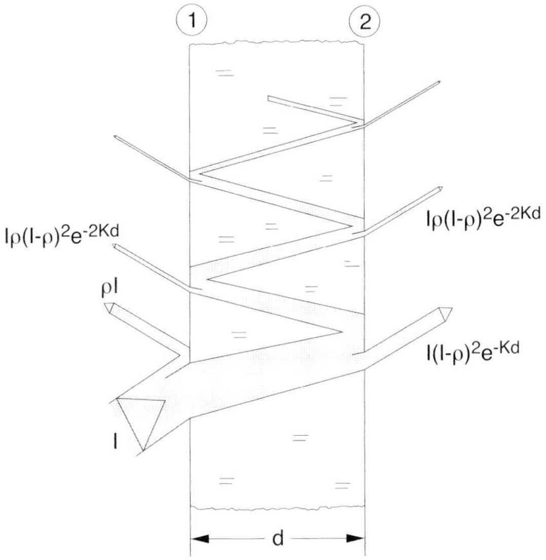

Figure 12.12 Schematic of light reflectance, absorption and transmission through a plate.

between the front and back surfaces of the plate. The infinite sum of transmitted rays can be approximated by

$$\tau = \frac{(1 - \rho_0)^2 e^{-Kd}}{1 - \rho_0^2 e^{-2Kd}} \tag{12.14}$$

and the total reflected rays can be approximated with

$$\rho = \rho_0\left(1 + \tau e^{-Kd}\right) \tag{12.15}$$

The fraction of incident beam absorbed by the material and transformed into heat inside the material is calculated using

$$\alpha = 1 - \tau - \rho \tag{12.16}$$

Figure 12.3 shows this relationship as a function of dimensionless thickness Kd.

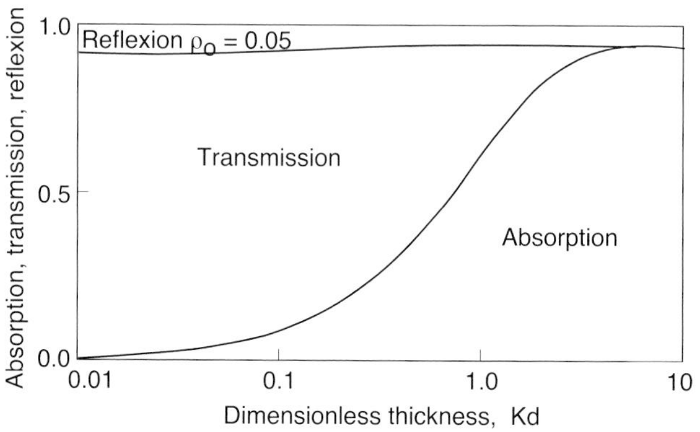

Figure 12.13 Reflection, transmission and absorption as a function of thickness.

The above analysis is complicated further for the case where the incident angle is no longer 0º. For such a case, and for materials with low coefficient of absorption, the amount of visible light reflected can be computed using Fresnel's equation [7],

$$I_r = \frac{1}{2}\left(\frac{\sin^2(\theta_i - \theta_r)}{\sin^2(\theta_i + \theta_r)} + \frac{\tan^2(\theta_i - \theta_r)}{\tan^2(\theta_i + \theta_r)}\right)I_0 \tag{12.17}$$

which can be written as

$$I_r = \rho I_0 \tag{12.18}$$

Plots of ρ as a function of incidence angle are shown in Fig. 12.14 for various refraction indices.

For the case with $\theta_i = 0^\circ$ the equation can be rewritten in terms of transmittance, T, as

$$T = \left(1 - \frac{(N-1)^2}{N^2 + 1}\right) \tag{12.19}$$

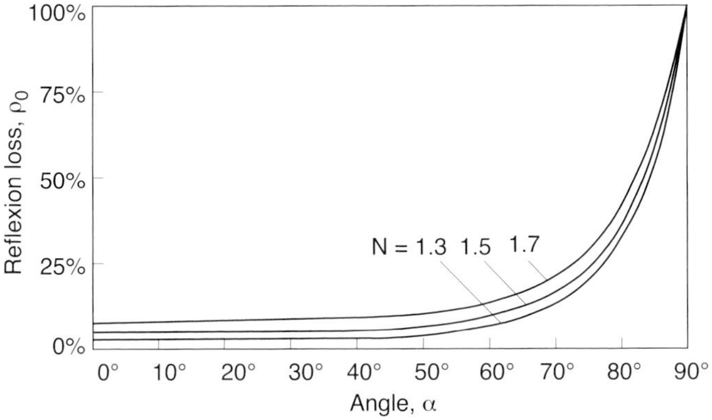

Figure 12.14 Influence of incidence angle on reflection losses.

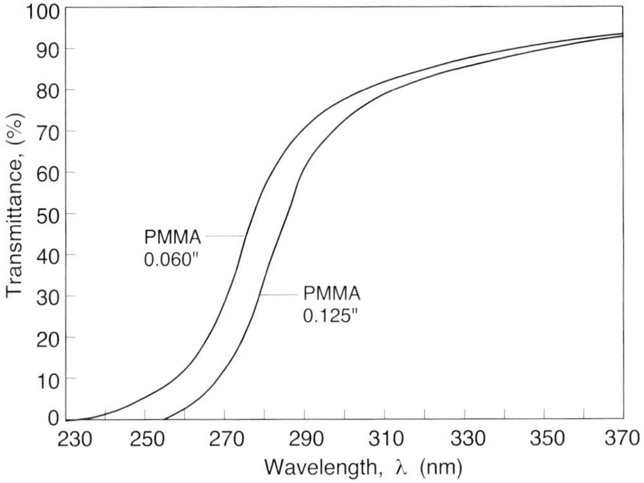

Figure 12.15 Ultraviolet light transmission through PMMA.

which is the fraction of the incident light that is transmitted through the material. For example, a PMMA with a refractive index of 1.49 would at best have a transmittance of 0.92 or 92%. The transmittance becomes less as the wavelength of the incident light decreases, as shown for PMMA in Fig. 12.15. The figure also demonstrates the higher absorption of the thicker sheet.

The transmissivity of polymers can be improved by altering their chemical composition. For example, the transmissivity of PMMA can be improved by substituting hydrogen atoms by fluorine atoms. The improvement is clearly demonstrated in Fig. 12.16[3]. Such modifications bring polymers a step closer to materials appropriate for usage in fiber optic applications[4]. Nucleating agents can also be used to improve the transmissivity of semi-crystalline polymers. A large number of nuclei will reduce the average spherulite size to values below the wavelength of visible light.

The haziness or luminous transmittance of a transparent polymer is measured using the standard ASTM D 1003 test, and the transparency of a thin polymer film is measured using the ASTM D 1746 test. The *haze measurement* (ASTM D 1003) is the most popular measurement for film and sheet quality control and specification purposes.

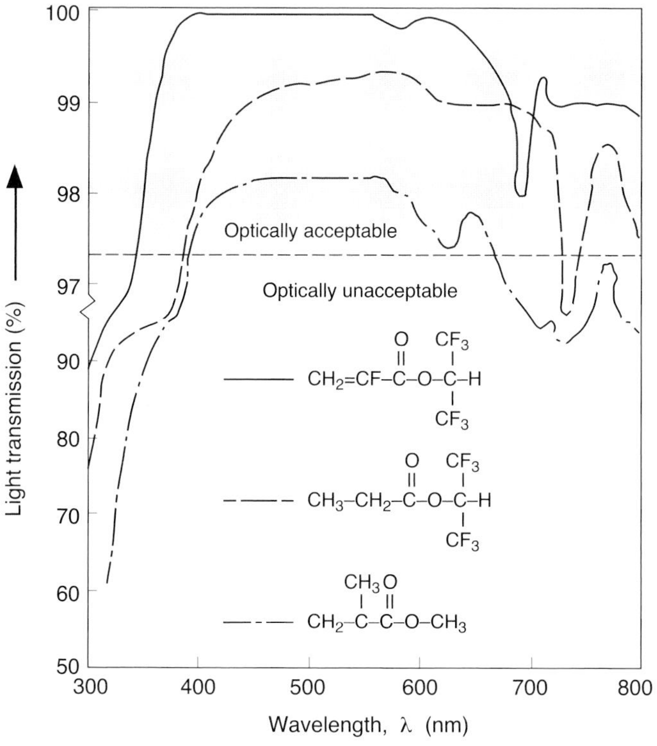

Figure 12.16 Effect of fluorine modification on the transmissivity of light through PMMA.

3 Courtesy of Hoechst, Germany.

4 Their ability to withstand shock and vibration and cost savings during manufacturing make some amorphous polymers important materials for fiber optics applications. However, in unmodified polymer fibers, the initial light intensity drops to 50% after only 100 m, whereas when using glass fibers the intensity drops to 50% after 3000 m.

12.4 Gloss

Strictly speaking, all of the above theory is valid only if the surface of the material is perfectly smooth. However, the reflectivity of a polymer component is greatly influenced by the quality of the surface of the mold or die used to make the part.

Specular gloss can be measured using the ASTM D 2457 standard technique which describes a part by the quality of its surface. A *glossmeter* or *lustremeter* is usually composed of a light source and a photometer as shown in schematic diagram in Fig. 12.17 [8]. These types of glossmeters are called *goniophotometers*. As shown in the figure, the specimen is illuminated with a light source from an angle α, and the photometer reads the light intensity from the specimen from a variable angle β. The angle α should be chosen according to the glossiness of the surface. For example, for transparent films values for α range from 20° for high gloss, 45° for intermediate and 60° low gloss. For opaque specimens ASTM test E 97 should be used. Figure 12.18 presents plots of reflective intensity as a function of photometer orientation for several surfaces with various degrees of gloss illuminated by a light source oriented at a 45° angle from the surface. The figure shows how the intensity distribution is narrow and sharp at 45° for a glossy surface, and the distribution becomes wider as the surface becomes matte. The color of the surface also plays a significant role on the intensity distribution read by the photometer as it sweeps through various angular positions. Figure 12.19 shows plots for a black and a white surface with the same degree of glossiness. The specular gloss is used as a measurement of the glossy appearance of films. However, gloss values of opaque and transparent films should not be compared with each other.

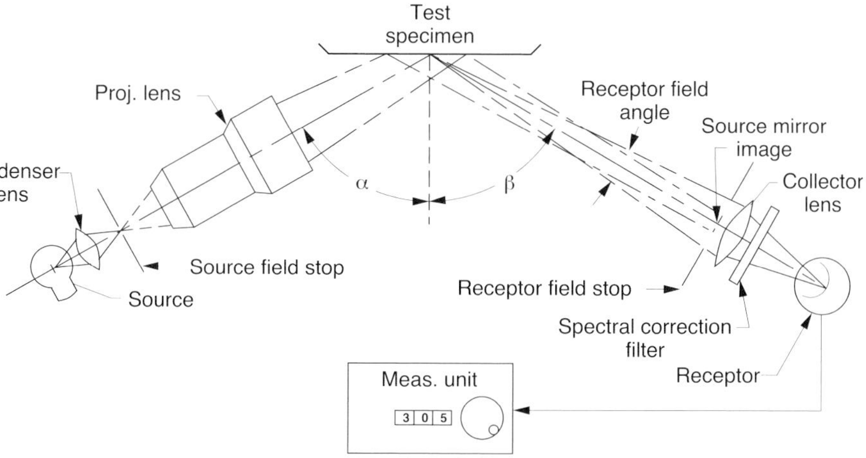

Figure 12.17 Schematic diagram of a glossmeter.

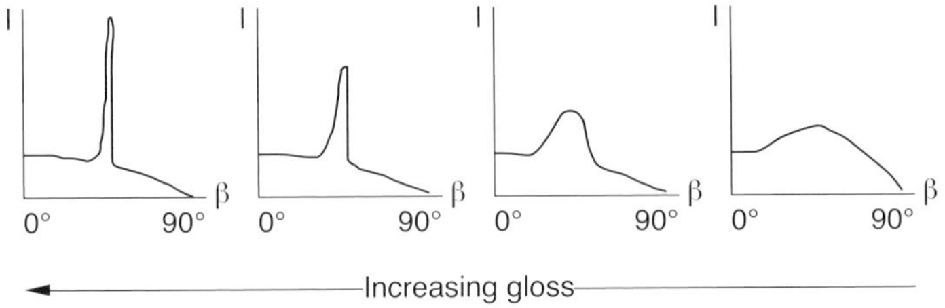

Figure 12.18 Reflective intensity as a function of photometer orientation for specimens with various degrees of surface gloss.

Figure 12.19 Reflective intensity as a function of photometer orientation for black and white specimens with equal surface gloss.

12.5 Color

The surface quality of a part is not only determined by how smooth or glossy it is but also by its color. Color is often one of the most important specifications for a part. In the following discussion it will be assumed that the color is homogeneous throughout the surface. This assumption is linked to processing, where efficient mixing must take place to disperse and distribute the pigments that will give the part color.

Color can always be described by combinations of basic red, green, and blue[5]. Hence, to quantitatively evaluate or measure a color, one must filter the intensity of the three basic colors. A schematic diagram of a color measurement device is shown in Fig. 12.20. Here, a specimen is lit in a diffuse manner using a photometric sphere, and the light reflected from the specimen is passed through red, green, and blue filters. The intensity coming from the three filters are allocated the variables X, Y, and Z for red, green, and blue, respectively. The variables X, Y, and Z are usually referred to as *tristimulus values*.

5 Most color measurement techniques are based on the CIE system. CIE is the *Commission International de l'Eclairage*, or International Commission on Illumination.

Another form

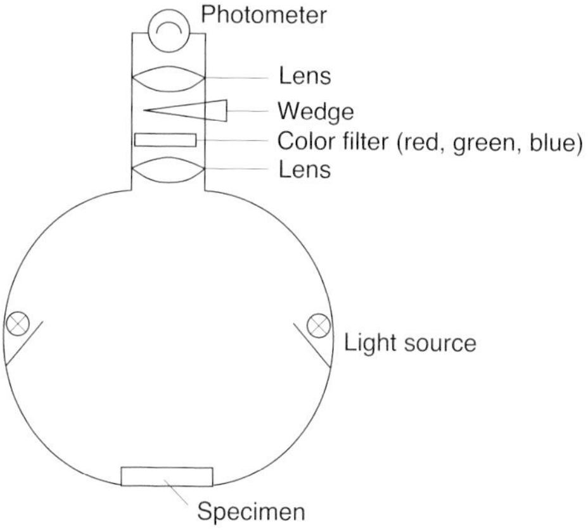

Figure 12.20 Schematic diagram of a colorimeter.

of measuring color is to have an observer compare two surfaces. One surface is the sample under consideration illuminated with a white light. The other surface is a white screen illuminated by light coming from three basic red, green, and blue sources. By varying the intensity of the three light sources, the color of the two surfaces are matched. This is shown schematically in Fig. 12.21 [9]. Here too, the intensities of red, green, and blue are represented with X, Y, and Z, respectively.

The resulting data is better analyzed by normalizing the individual intensities as

$$x = \frac{X}{X + Y + Z} \tag{12.20}$$

$$y = \frac{Y}{X + Y + Z} \tag{12.21}$$

$$z = \frac{Z}{X + Y + Z} \tag{12.22}$$

The parameters X, Y, and Z, usually termed *trichromatic coefficients*, are plotted on a three-dimensional graph that contains the whole spectrum of visible light, as shown in Fig. 12.22. This graph is usually referred to as a *chromaticity diagram*. The standard techniques that make use of the chromaticity diagram are ASTM E 308-90 and DIN 5033. Three points in the diagram have been standardized:

- Radiation from a black body at 2848 K corresponding to a tungsten filament light and denoted by A in the diagram;
- Sunlight, denoted by B; and
- North sky light, denoted by C.

It is important to note that colors plotted on the chromaticity diagram are only described by their *hue* and *saturation*. The *luminance* factor is plotted in the z direction of the diagram. Hence, all neutral colors such as black, gray, and white lie on point C of the diagram.

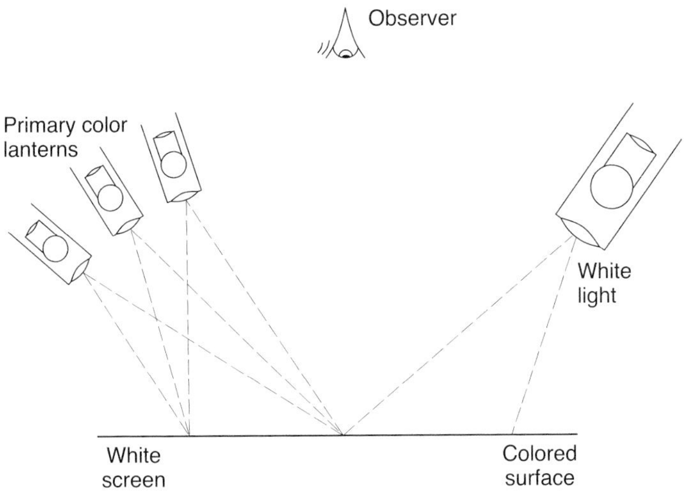

Figure 12.21 Schematic diagram of a visual colorimeter.

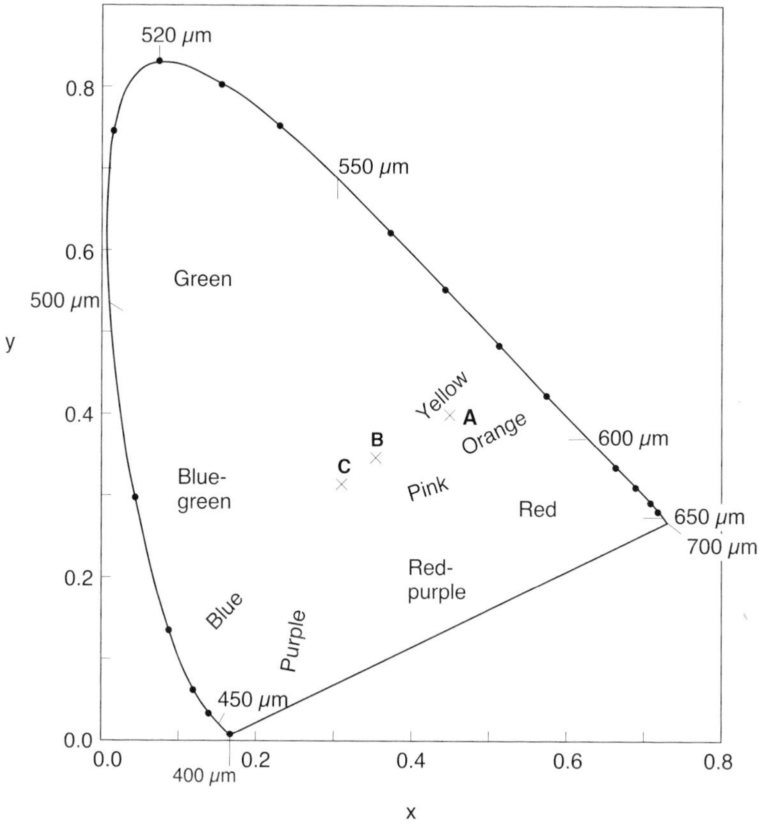

Figure 12.22 Chromaticity diagram with approximate color locations.

12.6 Infrared Spectroscopy

Infrared spectroscopy has developed into one of the most important techniques used to identify polymeric materials. It is based on the interaction between matter and electromagnetic radiation of wavelengths between 1 and 50 μm. The atoms in a molecule vibrate in a characteristic mode, which is usually called a fundamental frequency. Thus, each molecule has a set group of characteristic frequencies which can be used as a diagnostics tool to detect the presence of distinct groups. Table 12.2 [10] presents the absorption wavelength for several chemical groups. The range for most commercially available infrared spectroscopes is between 2 and 25 μm. Hence, the spectrum taken between 2 and 25 μm serves as a fingerprint for that specific polymer, such as shown in Fig. 12.23 for polycarbonate.

An *infrared spectrometer* to measure the absorption spectrum of a material is schematically represented in Fig.12.24. It consists of an infrared light source that can sweep through a certain wavelength range, and that is split in two beams: one that serves as a reference and the other that is passed through the test specimen. The comparison of the two gives the absorption spectrum, such as shown in Fig. 12.23.

Using infrared spectroscopy, also helps in quantitatively evaluating the effects of weathering (e.g., by measuring the increase of the absorption band of the COOH group, or by monitoring the water intake over time). One can also use the technique to follow reaction kinetics during polymerization.

Table 12.2 Absorption Wavelengths for Various Groups

Group	Wavelength region (μm)
O-H	2.74
N-H	3.00
C-H	3.36
C-O	9.67
C-C	11.49
C=O	5.80
C=N	5.94
C=C	6.07
C=S	6.57

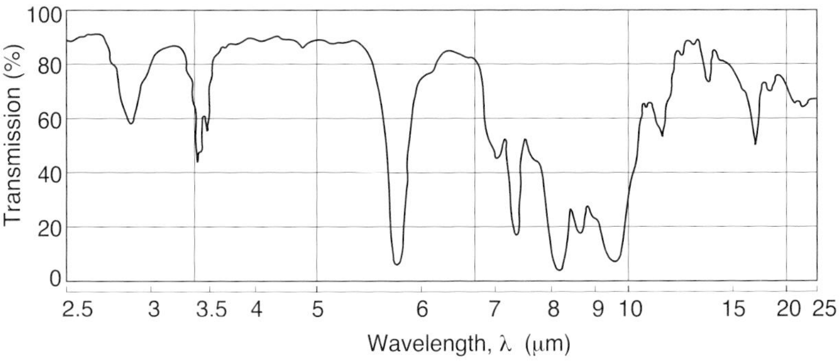

Figure 12.23 Infrared spectrum of a polycarbonate film.

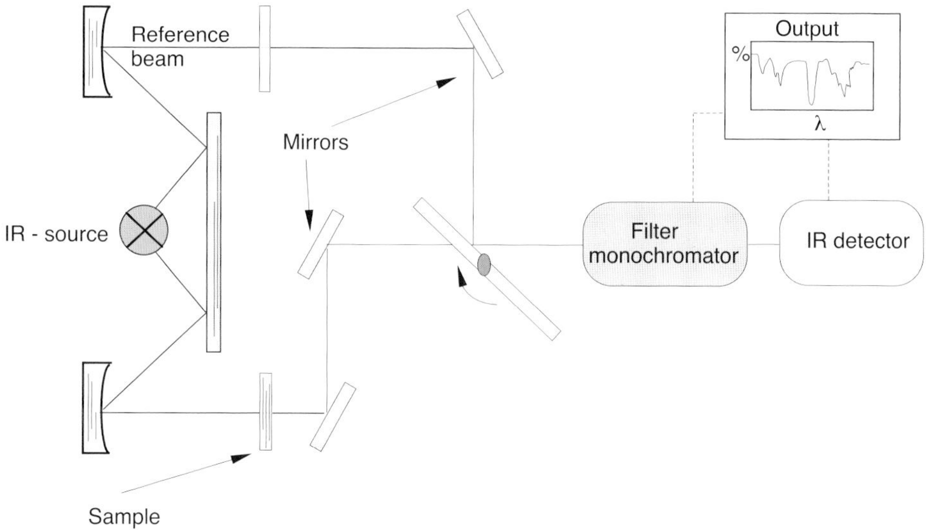

Figure 12.24 Schematic diagram of an infrared spectrometer.

12.7 Infrared Pyrometry

Today it is possible to measure the temperature at the surface of a polymer melt or component using an infrared probe. Infrared pyrometry is based on Planck's law, which describes the spectral distribution of blackbody radiation by

$$I_\lambda = \frac{c_1}{\pi\lambda^5\left[\exp\left(\frac{c_2}{\lambda T_\lambda}\right) - 1\right]} \tag{12.23}$$

where λ is the mean effective wavelength utilized by an IR pyrometer, I_λ is the amount of spectral radiance emitted by a blackbody with a spectral radiance temperature of T_λ (K), and $c_1(= 3.742 \times 10^8\ \mathrm{W}\mu\mathrm{m}^4/\mathrm{m}^2)$and $c_2(= 1.439 \times 10^4\ \mu\mathrm{m}^0\mathrm{K})$ are Planck's first and second radiation constants, respectively. Within the range of typical polymer processing temperatures, Planck's law can be further simplified as Wien's law, i.e., $\exp(c_2/\lambda T) >> 1$ which is given by

$$I_\lambda = \frac{c_1}{\pi\lambda^5 \exp\left(\frac{c_2}{\lambda T_\lambda}\right)} \tag{12.24}$$

Equations 12.23 or 12.24 can be used to convert the measured radiance from an IR pyrometer, I_λ, into a temperature, T_λ, for a blackbody with an emissivity of 1.0. However, when measuring the temperature of a non-blackbody whose emissivity is ε_l ($\neq 1.0$), the true surface temperature of the non-blackbody, T_s, and the spectral radiance temperature, T_λ, can be related to each other using Eq. 12.24 by the following equation

$$I_\lambda = \frac{c_1}{\pi\lambda^5 \exp\left(\dfrac{c_2}{\lambda T_\lambda}\right)} = \frac{\varepsilon_\lambda c_1}{\pi\lambda^5 \exp\left(\dfrac{c_2}{\lambda T_s}\right)} \qquad (12.25)$$

The true surface temperature of a polymer melt or component can then be calculated from rearranging Eq. 12.25 to,

$$\frac{1}{T_s} = \frac{1}{T_\lambda} + \frac{\lambda}{c_2}\ln(\varepsilon_\lambda) \qquad (12.26)$$

The spectral surface emissivity of a material can be obtained by either direct measurement [11] or by calculation using both Kirchhoff's law and the radiation energy balance equation (i.e., Eq. 12.16). Kirchhoff's law basically states that the emissivity and the absorptivity of the material are equal (i.e., $\varepsilon_\lambda = \alpha_\lambda$). Therefore, if the reflectivity and transmissivity of a semi-transparent polymer are known, then the emissivity of a polymer can be computed from Eq. 12.16 as

$$\varepsilon_\lambda = 1 - \tau_\lambda - \rho_\lambda \qquad (12.27)$$

However, if the polymer specimen is thick enough, we can eliminate transmissivity from Eq. 12.27, taking the measurements at a wavelength at which polymers absorb strongly; that is, at a wavelength which has a strong dip in its spectrum. To do this we select proper narrow-band filters to control the bandwidth and mean effective wavelength of the pyrometer. This concept is clearly demonstrated in Fig. 12.25 [12], which shows the transmissivity spectrum of polyethylene and polytetrafluoroethylene. It is easy to spot the bands of low transmissivity such as the one around 6.8 μm (CH_3 band) in the PE spectrum. For most polymer films one can assume a reflectivity value, ρ, of 0.05 [13].

As indicated from the preceding description, if the mean effective wavelength of an IR pyrometer is controlled at a bandwidth where the target polymer has a low transmissivity, then the IR pyrometer can measure the surface temperature of the target polymer. More recently, attempts have been made to measure the subsurface temperature in a polymer melt stream in a non-invasive manner using IR probes [14, 15]. For such measurements, the mean effective wavelength of an IR pyrometer is controlled at a bandwidth where the transmissivity of the polymer is relatively high (i.e., the polymer has a low radiation absorbing behavior), then the IR pyrometer is able to capture subsurface radiation and provide bulk temperature information of the polymer. Since the captured bulk temperatures contain detailed thermal information about the volume of the polymer, it is possible to retrieve the detailed temperature profile from the captured bulk temperatures using an inverse radiation technique [16]. The inverse radiation technique basically deciphers a set of bulk radiation measured at several different mean effective wavelengths utilized by an IR pyrometer, to retrieve the temperature distribution within the polymer. The theoretical background of the inverse radiation technique provides the foundation of a new generation IR pyrometer which will be very useful for process monitoring during polymer processing.

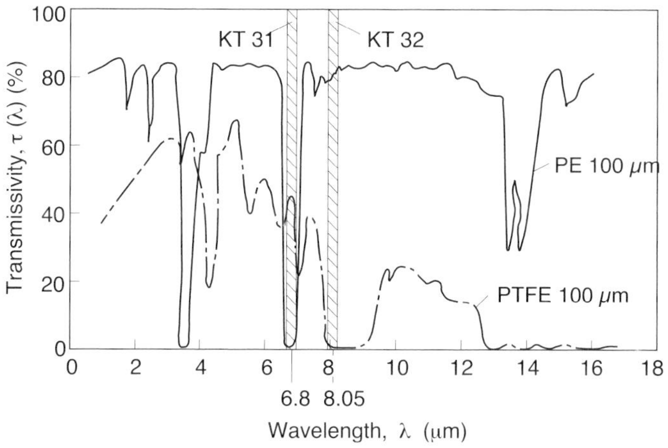

Figure 12.25 Transmissivity of a polyethylene and a polytetrafluoroethylene.

12.8 Heating with Infrared Radiation

Although the heating of polymer sheets for thermoforming or other membrane stretching processes belongs to the field of radiative heat transfer, it is largely an optical problem. In such processes, an infrared heater is used to radiate infrared rays onto the surface of the sheet under consideration. Much of the process that takes place can be analyzed using the theory of absorption and transmittance discussed in Section 12.3. For example, the absorption properties largely depend on pigments or spherulite size. Figure 12.26 [17] shows a plot of the inverse of the absorption coefficient (penetration depth) of polypropylene with two different spherulite sizes as a function of the wavelength. Figure 12.27 [15] presents a similar graph for polystyrene without pigmentation, translucent polystyrene with a blue pigment, and opaque polystyrene with a white pigment. The absorption of polymers also varies according to the temperature of the radiative heater. Figure 12.28 [15] shows this dependence for various polymer sheets of 1 mm thickness and variable pigment. It is interesting to note that all polymers observe a similar absorption at a heater temperature of 1200 K.

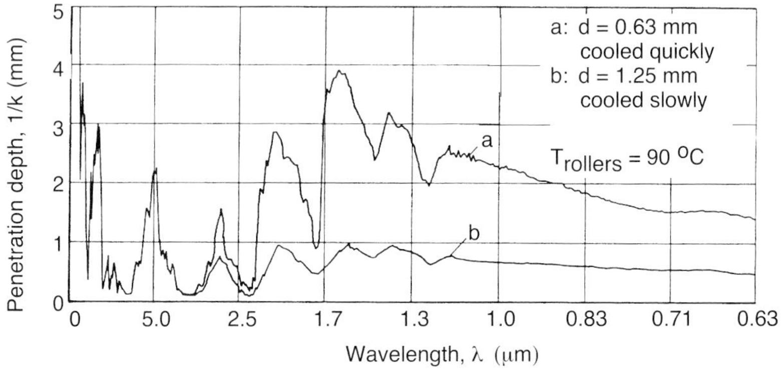

Figure 12.26 Penetration depth of infrared rays into polypropylenes with different morphology.

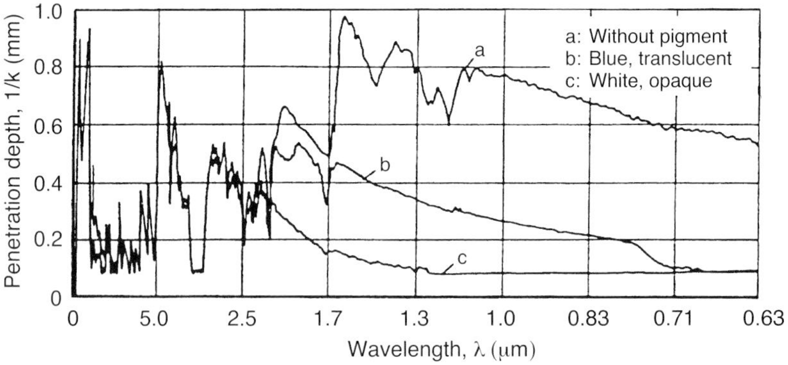

Figure 12.27 Penetration depth of infrared rays into polystyrenes with different pigmentation.

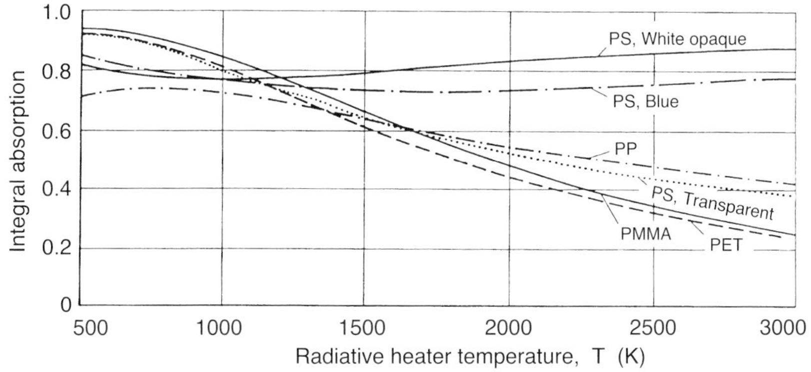

Figure 12.28 Integral absorption as a function of heater temperature for various polymers.

References

1. ASTM Vol 08.02, Plastics (II): ASTM-D4093, (1994).
2. Janeschitz-Kriegl, H., *Polymer Melt Rheology and Flow Birefringence,* Elsevier, Amsterdam, (1984).
3. Isayev, A.I., *Polym. Eng. Sci., 23,* 271, (1983).
4. Wimberger-Friedl, R., *Polym. Eng. Sci., 30,* 813, (1990).
5. Lodge, A.S., *Trans. Faraday Soc., 52,* 120, (1956).
6. Sornberger, G., Quantin, J.C., Fajolle, R., Vergnes, B., and Agassant, J.F., *J. Non-Newt. Fluid Mech., 23,* 123, (1987).
7. Bartoe, W.F., *Engineering Design for Plastics,* E. Baer, Ed., Chapter 8, R.E. Krieger Publishing Co., (1975).
8. ASTM Vol 08.02, Plastics (II): ASTM-D2457, (1994).
9. Adams, J.M., *The Science of Surface Coatings,* Chapter 12, Ed. H.W. Chatfield, D. van Nostrand Co., Inc. Princeton, (1962).
10. Chapman, D., *The Science of Surface Coatings,* Chapter 19, Ed. H.W. Chatfield, D. van Nostrand Co., Inc. Princeton, (1962).
11. Dewitt, D.P., and Richmond, J. C., "Thermal Radiative Properties of Materials," Chapter 2 in *Theory and Practice of Radiation Thermometry,* D.P. Dewitt and G.D. Nutter, Eds., Wiley, New York, (1988).
12. Heimann, W., Communication, Firma W. Heimann, Wiesbaden, Germany.
13. Haberstroh, E., Ph.D. Thesis, IKV, RWTH-Aachen, Germany, (1981).
14. Rietveld, J.X., and Lai, G.-Y., *SPE ANTEC, 38,* 2192, (1992).
15. Lai, G.-Y., Ph.D. Thesis, University of Wisconsin-Madison, Madison, WI, (1993).
16. Rietveld, J.X., and Lai, G.-Y., *SPE ANTEC, 40,* 836, (1994).
17. Weinand, D., Ph.D. Thesis, IKV, RWTH-Aachen, Germany, (1987).

13

Permeability Properties of Polymers

Because of their low density, polymers are relatively permeable by gases and liquids. A more in-depth knowledge of permeability is necessary when dealing with packaging applications and with corrosive protection coatings. The material transport of gases and liquids through polymers consists of various steps. They are:

- Absorption of the diffusing material at the interface of the polymer, a process also known as *adsorption*,
- *Diffusion* of the attacking medium through the polymer, and
- Delivery or secretion of the diffused material through the polymer interface, also known as *desorption*.

With polymeric materials these processes can occur only if the following rules are fulfilled:

- The molecules of the permeating materials are inert,
- The polymer represents a homogeneous continuum, and
- The polymer has no cracks or voids which channel the permeating material.

In practical cases, such conditions are often not present. Nevertheless, this chapter shall start with these "ideal cases," since they allow for useful estimates and serve as learning tools for these processes.

13.1 Sorption

We talk about adsorption when environmental materials are deposited on the surface of solids. Interface forces retain colliding molecules for a certain time. Possible causes include van der Waals' forces in the case of physical adsorption, chemical affinity (chemical sorption), or electrostatic forces. With polymers, we have to take into account all of these possibilities.

A gradient in concentration of the permeating substance inside the material results in a transport of that substance which we call *molecular diffusion*. The cause of molecular diffusion is the thermal motion of molecules that permit the foreign molecule to move along the concentration gradient using the intermolecular and intramolecular spaces. However, the possibility to migrate essentially depends on the size of the migrating molecule.

The rate of permeation for the case shown schematically in Fig. 13.1 is defined as the mass of penetrating gas or liquid that passes through a polymer membrane per unit time. The rate of permeation, $\dot{m}$, can be defined using Fick's first law of diffusion as

$$\dot{m} = -DA\rho\frac{dc}{dx} \tag{13.1}$$

where D is defined as the *diffusion coefficient,* A is the area and ρ the density. If the diffusion coefficient is constant Eq. 13.1 can be easily integrated to give

$$\dot{m} = -DA\rho(c_1 - c_2)/L \tag{13.2}$$

The equilibrium concentrations c_1 and c_2 can be calculated using the pressure, p, and the *sorption equilibrium parameter*, S:

$$c = Sp \tag{13.3}$$

which is often referred to as *Henry's law*.

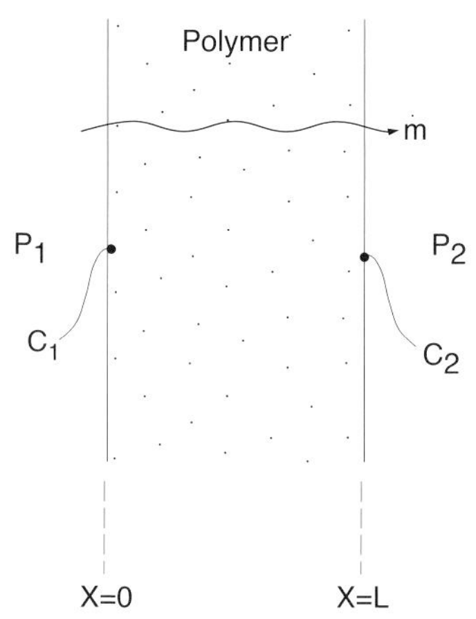

Figure 13.1 Schematic diagram of permeability through a film.

The sorption equilibrium constant, also referred to as *solubility constant*, is almost the same for all polymer materials. However, it does depend largely on the type of gas and on the boiling, T_b, or critical temperatures, T_{cr}, of the gas, such as shown in Fig. 13.2.

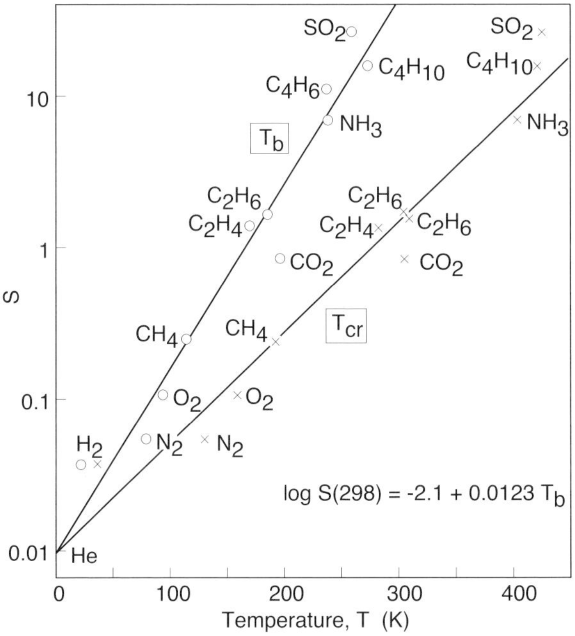

Figure 13.2 Solubility (cm^3/cm^3) of gas in natural rubber at 25 °C and 1 bar as a function of the critical and the boiling temperatures.

13.2 Diffusion and Permeation

Diffusion, however, is only one part of permeation. First, the permeating substance has to infiltrate the surface of the membrane; it has to be absorbed by the membrane. Similarly, the permeating substance has to be desorbed on the opposite side of the membrane. Combining Eq. 13.2 and 13.3 we can calculate the sorption equilibrium using

$$\dot{m} = -DS\rho A(p_1 - p_2)/L \tag{13.4}$$

where the product of the sorption equilibrium parameter and the diffusion coefficient is defined as the *permeability* of a material

$$P = DS = \frac{\dot{m}L}{A\Delta p \rho} \tag{13.5}$$

Equation 13.5 does not take into account the influence of pressure on the permeability of the material and is only valid for dilute solutions. The *Henry-Langmuir model* takes into account the influence of pressure and works very well for amorphous thermoplastics. It is written as

$$P = DS\left(1 + \frac{KR'}{1 + b\Delta p}\right) \tag{13.6}$$

where $K = c'_H b / S$, with c'_H being a saturation capacity constant and b an affinity coefficient. The constant R' represents the degree of mobility, $R'=0$ for complete immobility and $R'=1$ for total mobility. Table 13.1 [1] presents permeability of various gases at room temperature through several polymer films. In the case of multi-layered films commonly used as packaging material, we can calculate the permeation coefficient P_C for the composite membrane using

$$\frac{1}{P_C} = \frac{1}{L_C}\sum_{i=1}^{i=n}\frac{L_i}{P_i} \tag{13.7}$$

Table 13.1 Permeability of Various Gases Through Several Polymer Films

	Permeability (cm^3 mil/100 in^2/24 h/atm)		
Polymer	CO_2	O_2	H_2O
PET	12-20	5–10	2-4
OPET	6	3	1
PVC	4.75–40	8–15	2-3
HDPE	300	100	0.5
LDPE		425	1-1.5
PP	450	150	0.5
EVOH	0.05-0.4	0.05-0.2	1-5
PVDC	1	0.15	0.1

Sorption, diffusion, and permeation are processes activated by heat and as expected follow an Arrhenius type behavior. Thus, we can write

$$S = S_0 e^{-\Delta H_S / RT} \tag{13.8}$$

$$D = D_0 e^{-E_D / RT} \quad \text{and} \tag{13.9}$$

$$P = P_0 e^{-E_p/RT} \tag{13.10}$$

where ΔH_S is the enthalpy of sorption, E_D and E_p are diffusion and permeation activation energies, R is the ideal gas constant, and T is the absolute temperature. The Arrhenius behavior of sorption, diffusion and permeability coefficients, as a function of temperature for polyethylene and methyl bromine at 600 mm of Hg are shown in Fig. 13.3 [2]. Figure 13.4 [3] presents the permeability of water vapor through several polymers as a function of temperature. It should be noted that permeability properties drastically change once the temperature goes above the glass transition temperature. This is demonstrated in Table 13.2 [4], which presents Arrhenius constants for diffusion of selected polymers and CH_3OH.

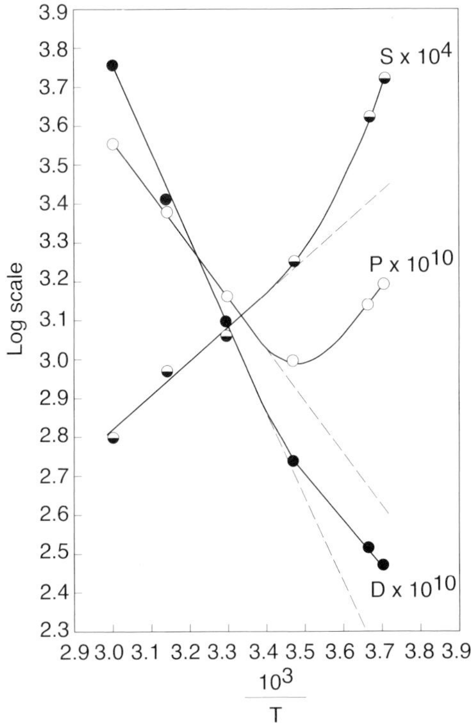

Figure 13.3 Sorption, diffusion, and permeability coefficients, as a function of temperature for polyethylene and methyl bromine at 600 mm of Hg.

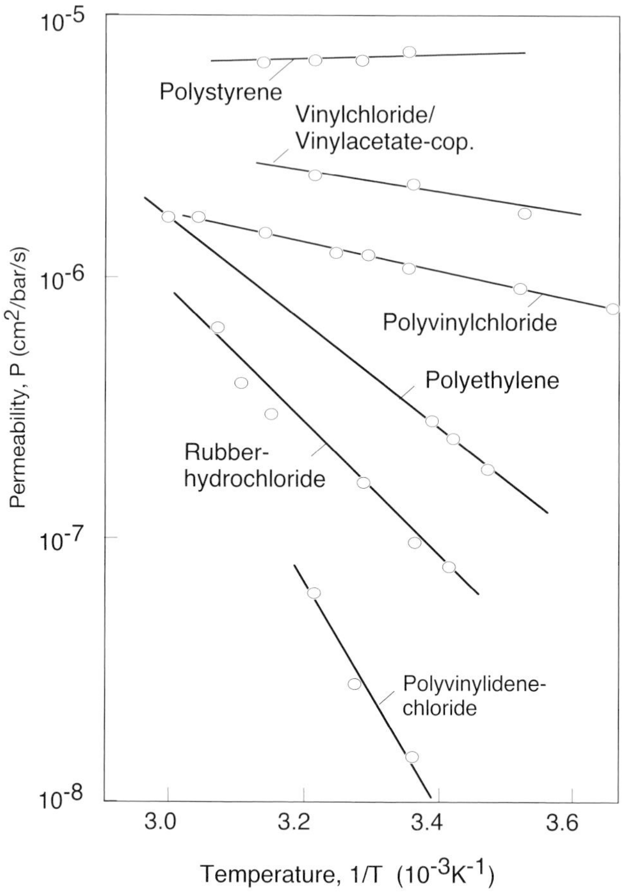

Figure 13.4 Permeability of water vapor as a function of temperature through various polymer films.

The diffusion activation energy E_D depends on the temperature, the size of the gas molecule d_x , and the glass transition temperature of the polymer. This relationship is well represented in Fig. 13.5 [1] with the size of nitrogen molecules, d_{N2} as a reference. Table 13.2 contains values of the effective cross section size of important gas molecules. Using Fig. 13.5 with the values from Table 13.1 and using the equations presented in Table 13.3 the *diffusion coefficient, D*, for several polymers and gases can be calculated.

Table 13.4 also demonstrates that permeability properties are dependent on the degree of crystallinity. Figure 13.6 presents the permeability of polyethylene films of different densities as a function of temperature. Again, the Arrhenius relation becomes evident.

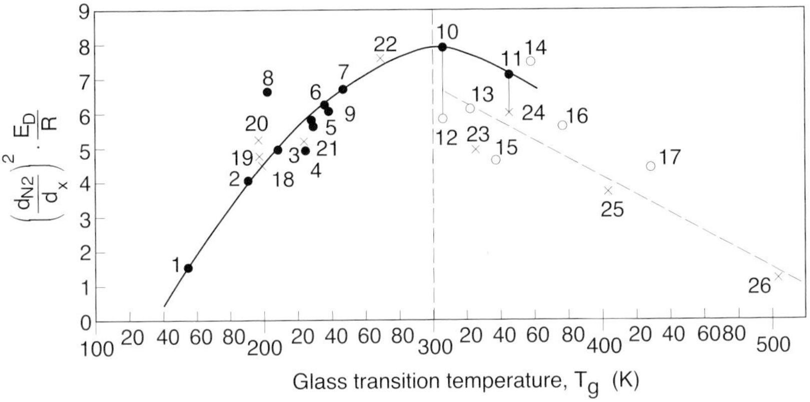

Figure 13.5 Graph to determine the diffusion activation energy E_D as a function of glass transition temperature and size of the gas molecule d_X, using the size of a nitrogen molecule, d_{N2}, as a reference. Rubbery polymers (•): 1= Silicone rubber, 2= Polybutadiene, 3= Natural rubber, 4= Butadiene/Acrylonitrile K 80/20, 5= Butadiene/Acrylonitrile K 73/27, 6= Butadiene/Acrylonitrile K 68/32, 7= Butadiene/Acrylonitrile K 61/39, 8= Butyl rubber, 9= Polyurethane rubber, 10= Polyvinyl acetate (r), 11= Polyethylene terephthalate (r). Glassy polymers (O): 12= Polyvinyl acetate (g), 13= Vinylchloride/vinyl acetate copolymer, 14= Polyvinyl chloride, 15= Polymethyl methacrylate, 16= Polystyrene, 17= Polycarbonate. Semi-crystalline polymers (X): 18= High-density polyethylene, 19= Low density polyethylene, 20= Polymethylene oxide, 21= Gutta percha, 22= Polypropylene, 23= Polychlorotrifluoroethylene, 24= Polyethyleneterephthalate, 25= Polytetraflourethylene, 26= Poly(2,6-diphenylphenyleneoxide).

Table 13.2 Diffusion Constants Below and Above the Glass Transition Temperature

Polymer	T_g (°C)	D_0 (cm²/s)		E_D (Kcal/mol)	
		$T < T_g$	$T > T_g$	$T < T_g$	$T > T_g$
Polymethylmethacrylate	90	0.37	110	12.4	21.6
Polystyrene	88	0.33	37	9.7	17.5
Polyvinyl acetate	30	0.02	300	7.6	20.5

Table 13.3 Important Properties of Gases

Gas	d (nm)	V_{cr} (cm^3)	T_b (K)	T_{cr} (K)	dN_2/dx
He	0.255	58	4.3	5.3	0.67
H_2O	0.370	56	373	647	0.97
H_2	0.282	65	20	33	0.74
Ne	0.282	42	27	44.5	0.74
NH_3	0.290	72.5	240	406	0.76
O_2	0.347	74	90	55	0.91
Ar	0.354	75	87.5	151	0.93
CH_3OH	0.363	118	338	513	0.96
Kr	0.366	92	121	209	0.96
CO	0.369	93	82	133	0.97
CH_4	0.376	99.5	112	191	0.99
N_2	0.380	90	77	126	1.00
CO_2	0.380	94	195	304	1.00
Xe	0.405	119	164	290	1.06
SO_2	0.411	122	263	431	1.08
C_2H_4	0.416	124	175	283	1.09
CH_3Cl	0.418	143	249	416	1.10
C_2H_6	0.444	148	185	305	1.17
CH_2Cl_2	0.490	193	313	510	1.28
C_3H_8	0.512	200	231	370	1.34
C_6H_6	0.535	260	353	562	1.41

Table 13.4 Equations to Compute D Using Data from Table 13.1 and Table 13.2[a]

Elastomers	$\log D = \frac{E_D}{2.3R}\left(\frac{1}{T} - \frac{1}{T_R}\right) - 4$
Amorphous thermoplastics	$\log D = \frac{E_D}{2.3R}\left(\frac{1}{T} - \frac{1}{T_R}\right) - 5$
Semi-crystalline thermoplastics	$\log D = \left(\frac{E_D}{2.3R}\left(\frac{1}{T} - \frac{1}{T_R}\right) - 5\right)(1 - x)$

[a] T_R = 435K and X is the degree of crystallinity.

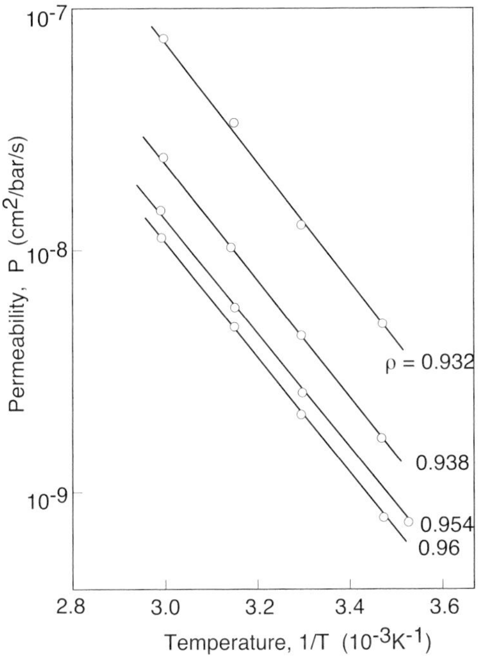

Figure 13.6 Permeation of nitrogen through polyethylene films of various densities.

13.3 Measuring S, D, and P

The *permeability* P of a gas through a polymer can be measured directly by determining the transport of mass through a membrane per unit time.

The *sorption constant* S can be measured by placing a saturated sample into an environment which allows the sample to desorb and measure the loss of weight. As shown in Fig. 13.7, it is common to plot the ratio of concentration of absorbed substance $c(t)$ to saturation coefficient c_∞ with respect to the root of time.

The *diffusion coefficient* D is determined using sorption curves as the one shown in Fig. 13.7. Using the slope of the curve, a, we can compute the diffusion coefficient as

$$D = \frac{\pi}{16} L^2 a^2 \tag{13.11}$$

where L is the thickness of the membrane.

Another method uses the lag time, t_0, from the beginning of the permeation process until the equilibrium permeation has occurred, as shown in Fig. 13.8. Here, the diffusion coefficient is calculated using

$$D = \frac{L^2}{6t_0} \tag{13.12}$$

The most important techniques used to determine gas permeability of polymers are the ISO 2556, DIN 53 380 and ASTM D 1434 standard tests.

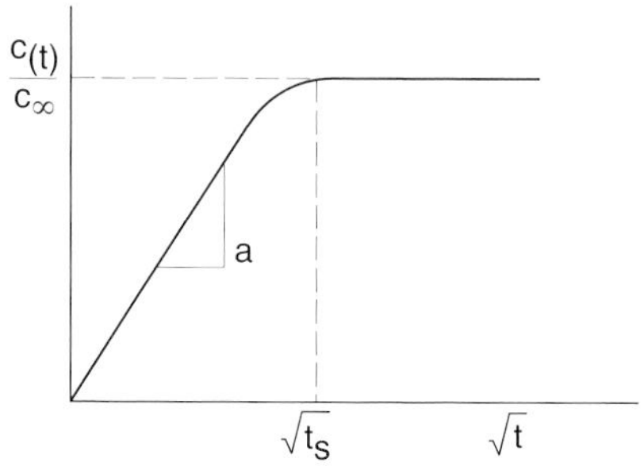

Figure 13.7 Schematic diagram of sorption as a function time.

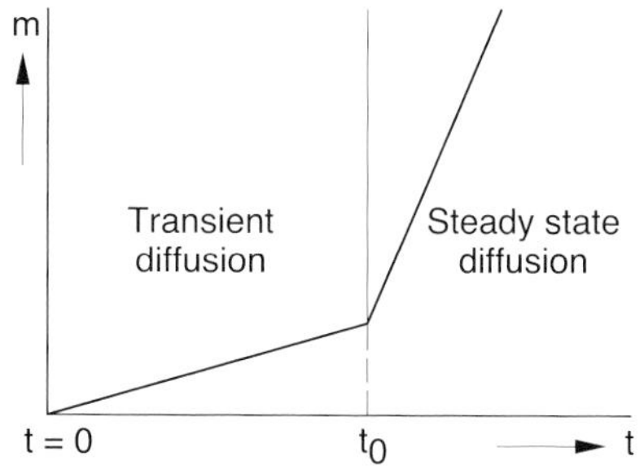

Figure 13.8 Schematic diagram of diffusion as a function of time.

13.4 Corrosion of Polymers and Cracking [5]

In contrast to metallic corrosion, where electrochemical corrosion mechanisms are dominant, several mechanisms play a role in the degradation of polymers. Attacks may occur by physical or chemical means or by combinations of both.

Even without a chemical reaction, the purely physical effect of a surrounding medium can adversely affect the properties of a polymer. Due to the low density of polymers, every surrounding medium that has moveable molecules will infiltrate or permeate the polymer. Experiments have shown that polymer samples under high hydrostatic pressures have even been permeated by silicon oils, which are completely inert at low pressures. The infiltration of silicone oil caused stress cracks and embrittlement in amorphous thermoplastics in the regions of low density, such as particle boundaries, filler material interfaces, and general surface imperfections. If we consider imperfections or particles of characteristic size L, we can perform an energy balance and conclude that the critical strain, ε_{crit}, at which a crack will occur is given by [6]

$$\varepsilon_{\text{crit}} \sim \sqrt{\frac{\gamma}{EL}} \tag{13.13}$$

where E represents Young's modulus and γ the adhesion tension between the individual particles. Crack formation and propagation is shown schematically in Fig. 13.9 [7]. Figure 13.10[1] shows an electron micrograph of permeating media through the inter-spherulitic boundaries of polypropylene.

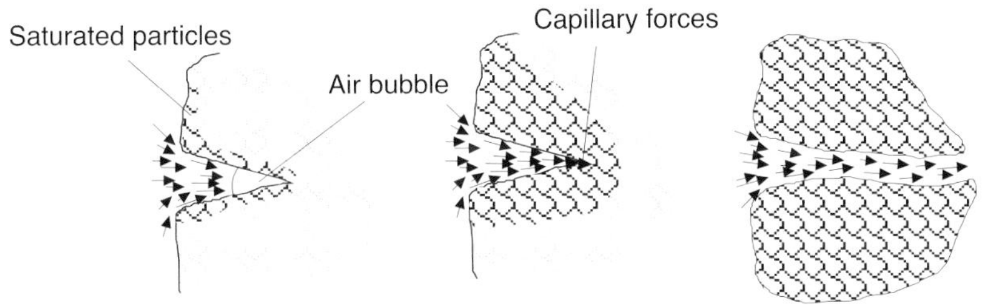

Figure 13.9 Schematic diagram of crack formation and propagation during diffusion.

Desorption, schematically shown in Fig. 13.11, is also undesirable for polymeric components. Similar to soil, which cracks as it dries out too quickly, the stresses that arise as the medium desorbs from the polymer give rise to cracks which may lead to failure of the component. As the absorbed media desorbs, the polymer component shrinks according to the loss of volume. However, inner layers which remain saturated do not shrink, leading to residual stress build-up similar to that which occurs with a cooling component with high temperature gradients. The schematic of the residual stress build-up

1 Courtesy IKV Aachen.

and concentration of the absorbed media is shown in Fig. 13.12 [8]. The stress history at the edge and center of a desorbing film is shown in Fig. 13.13. The stresses that arise during desorption are easily three times larger than during absorption. The maximum stress, which occurs at the outer edge of the part can be calculated using

$$\sigma_{\max} = \varepsilon_{\text{saturation}} \frac{E}{1-\nu} \tag{13.14}$$

The volume change in the immediate surface of the component is caused by the desorption process. The desorpting materials can be auxiliary agents for processing such as coloring agents, softeners, stabilizers, and lubricants, as well as low molecular components of the polymer.

Figure 13.10 Electron micrograph of permeating media through the interspherulitic boundaries of polypropylene.

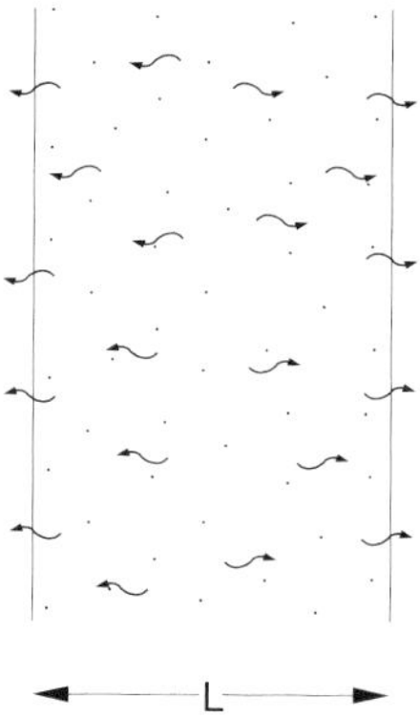

Figure 13.11 Schematic diagram of desorption from a plate.

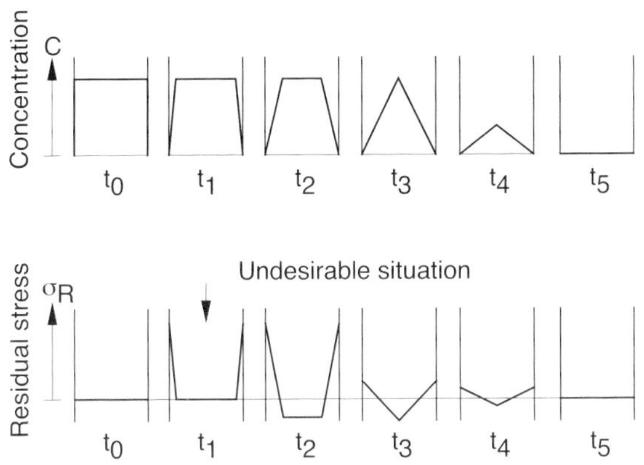

Figure 13.12 Schematic concentration (C) and residual stress (σ_R) as function of time inside a plate during desorption.

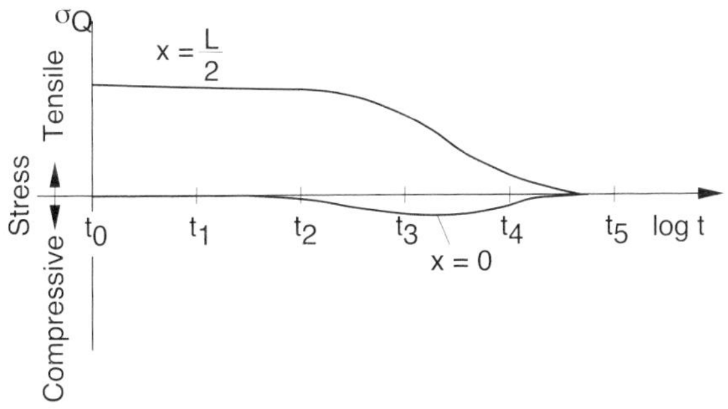

Figure 13.13 Residual stresses inside a plate during desorption.

13.5 Diffusion of Polymer Molecules and Self-diffusion

The ability to infiltrate the surface of a host material decreases with molecular size. Molecules of $M > 5 \times 10^3$ can hardly diffuse through a porous-free membrane. Self-diffusion is when a molecule moves, say in the melt, during crystallization. Also, when bonding rubber, the so-called tack is explained by the self-diffusion of the molecules. The diffusion coefficient for self-diffusion is of the order of

$$D \sim \frac{T}{\eta} \tag{13.15}$$

where T is the temperature and η the viscosity of the melt.

References

1. Rosato, D., and Rosato, D.V., *Blow Molding Handbook*, Hanser Publishers, Munich, (1989).
2. Rogers, C.E., *Engineering Design for Plastics,* Ed. E. Baer, Chapter 9, Robert E. Krieger Publishing Company, Huntington, (1975).
3. Knappe, W., *VDI-Berichte, 68*, 29, (1963).
4. van Krevelen, D.W., and Hoftyzer, P.J., *Properties of Polymers*, 2nd Ed., Elsevier, Amsterdam, (1976).
5. Menges, G., and Löwer, K., *Metallic Corrosion: Proceedings, 8th ICMC*, 2202, Mainz (1981).
6. Menges, G., *Kunststoffe, 63*, 95, (1973).
7. Menges, G. and Suchanek, H., *Kunststoffe, Fortschrittsberichte*, Vol. 3, Hanser Publishers, Munich, (1976).
8. Pütz, D., Ph.D. Thesis, IKV, RWTH-Aachen, Germany, (1977).

14

Acoustic Properties of Polymers

Sound waves, similar to light waves and electromagnetic waves, can be reflected, absorbed, and transmitted when they strike the surface of a body. The transmission of sound waves through polymeric parts is of particular interest to the design engineer. Of importance is the absorption of sound and the speed at which acoustic waves travel through a body, for example in a pipe, in the form of longitudinal, transversal, and bending modes of deformation.

14.1 Speed of Sound

The speed at which sound is transmitted through a solid barrier is proportional to Young's modulus of the material, E, but inversely proportional to its density, ρ. For sound waves transmitted through a rod, in the longitudinal direction, the speed of sound can be computed as

$$C_L^{\mathrm{rod}} = \sqrt{\frac{E}{\rho}} \tag{14.1}$$

Similarly, the transmission speed of sound waves through a plate along its surface direction can be computed as

$$C_L^{\mathrm{plate}} = \sqrt{\frac{E}{\rho\left(1-\nu^2\right)}} \tag{14.2}$$

where ν is Poisson's ratio. The transmission speed of sound waves through an infinite three-dimensional body can be computed using

$$C_{3D} = \sqrt{\frac{E(1-\nu)}{\rho(1+\nu)(1-2\nu)}} \tag{14.3}$$

The transmission of sound waves transversely though a plate (in shear) can be computed using

$$C_T^{\text{plate}} = \sqrt{\frac{G}{\rho}} \tag{14.4}$$

where G is the modulus of rigidity of the material. The transmission speed of sound waves with a frequency f which cause a bending excitation in plates can be computed using

$$C_B^{\text{plate}} = \sqrt{\frac{2\pi f h}{2\sqrt{3}} \frac{E}{\rho(1-\nu^2)}} \tag{14.5}$$

where h is the thickness of the plate.

The speed of sound through a material is dependent on its state. For example, sound waves travel much slower through a polymer melt than through a polymer in the glassy state. Table 14.1 presents orders of magnitude of the speed of sound through polymers in the glassy and rubbery states for various modes of transmission. One can see that the speed of sound through a polymer in the rubbery state is 100 times slower than that through a polymer in a glassy state.

In the melt state, the speed of sound drops with increasing temperature due to density increase. Figure 14.1 [1] presents plots of speed of sound through several polymer melts as a function of temperature. On the other hand, speed of sound increases with pressure as clearly demonstrated in Fig. 14.2 [1].

Table 14.1 Order of Magnitude of Properties Related to Sound Transmission

Modulus (MPa)	Speed of sound (m/s)
	Glassy (ν = 0.3)
E ∪ 10^3 to 10^4	C_L ∪ 2000
	C_T ∪ 1000
	C_{3D} ∪ 2000
	Rubbery (ν = 0.5)
E ∪ 1 to 10^2	C_L ∪ 10 to 400
	C_T ∪ 6 to 200
	C_{3D} ∪ 2000

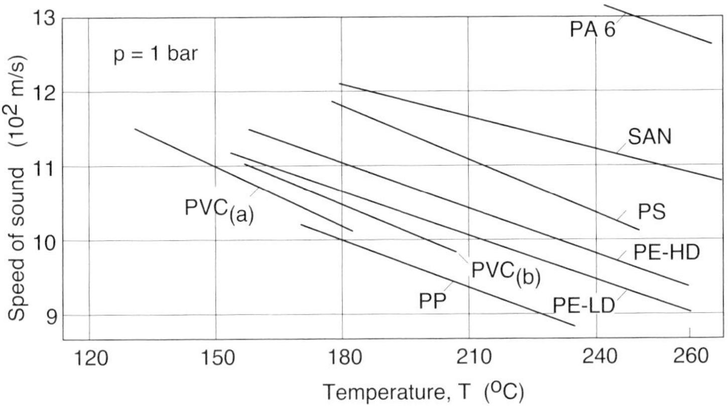

Figure 14.1 Speed of sound as a function of temperature through various polymers.

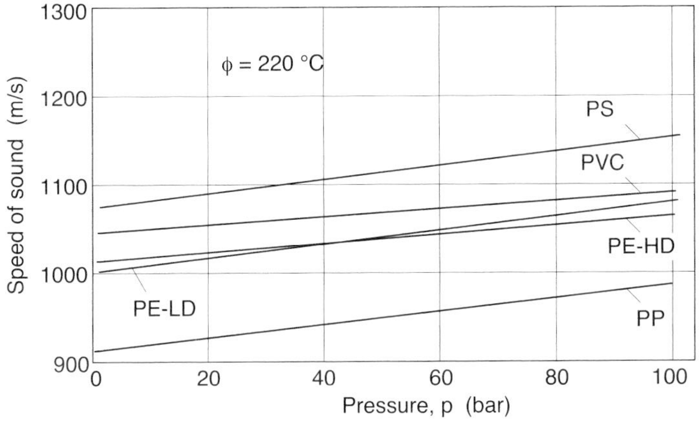

Figure 14.2 Speed of sound as a function of pressure through various polymers.

14.2 Sound Reflection

Sound reflection is an essential property for practical noise reduction. This can be illustrated using the schematic in Fig. 14.3. As sound waves traveling through media 1 strike the surface of media 2, the fraction of sound waves reflected back into media 1 is computed using

$$R = \frac{Z_2 - Z_1}{Z_2 + Z_1} \tag{14.6}$$

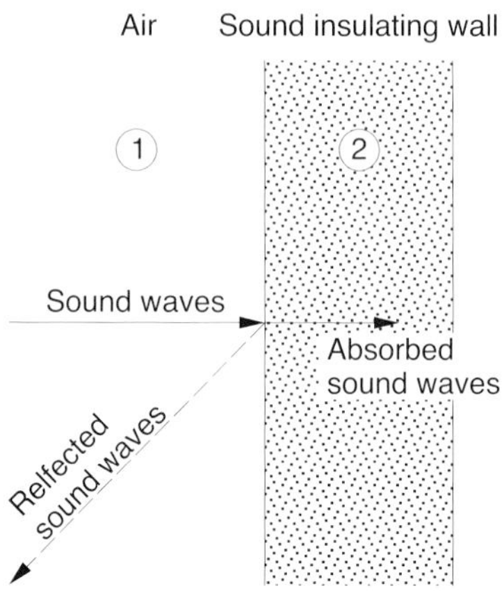

Figure 14.3 Schematic diagram of sound transmission through a plate.

where Z is the impedance or wave resistance, and is defined by

$$Z = \rho C_w \tag{14.7}$$

where C_w is the sound wave speed.

In order to obtain high sound reflection, the mass of the media 2 must be high compared to the mass of media 1 such that $Z_2 >> Z_1$. The mass of insulating sound walls can be increased with the use of fillers such as plasticized PVC with barium sulfate or by spraying similar anti noise compounds on the insulating walls. Using the mass of the insulating wall, another equation used to compute sound reflectance is

$$R = 20\log\left(\frac{\pi f M}{Z_0}\right) \tag{14.8}$$

where f is frequency of the sound wave, M the mass of the insulating wall and, Z_0 is the impedance of air. However, doubling the thickness of a wall (media 2) results in only 6 dB of additional sound reduction. It is common practice to use composite plates as insulating walls. This is only effective if walls whose flexural resonance frequencies do not coincide with the frequency of the sound waves are avoided.

14.3 Sound Absorption

Similar to sound reflection, sound absorption is an essential property for practical noise insulation. Materials which have the same characteristic impedance as air $(Z_1 \cup Z_2)$ are the best sound-absorbent materials. The sound waves that are not reflected back out into media 1, penetrate media 2 or the sound insulating wall. Sound waves that penetrate a polymer medium are damped out similar to that of mechanical vibrations. Hence, sound absorption also depends on the magnitude of the loss tangent $\tan\delta$, or logarithmic decrement Δ, described in Chapter 9. Table 14.2 presents orders of magnitude for the logarithmic decrement for several types of materials. As expected, elastomers and amorphous polymers have the highest sound absorption properties, whereas metals have the lowest.

Table 14.2 Damping Properties for Various Materials

Material	Temperature range	Logarithmic decrement Δ
Amorphous polymers	$T<T_g$	0.01 to 0.1
Amorphous polymers	$T>T_g$	0.1 to 1
Elastomers		0.1 to 1
Semi-crystalline polymers	$T_g<T<T_m$	≈ 0.1
Fiber reinforced polymers	$T_g<T<T_m$	< 0.01
Wood	$T<T_g$	0.01 to 0.02
Ceramic and glass	$T<T_g$	0.001 to 0.01
Metals	$T<T_m$	<0.0001

In a material, sound absorption takes place by transforming acoustic waves into heat. Since foamed polymers have an impedance of the same order as air, they are poor reflectors of acoustic waves. This makes them ideal to eliminate multiple reflections of sound waves in *acoustic* or *sound proof rooms*. Figure 14.4 [2] presents the sound absorption coefficient for several foamed polymers as a function of the sound wave frequency. It should be noted that the speed at which sound travels in foamed materials is similar to that of the solid polymers, since foaming affects the stiffness and the density in the same proportion.

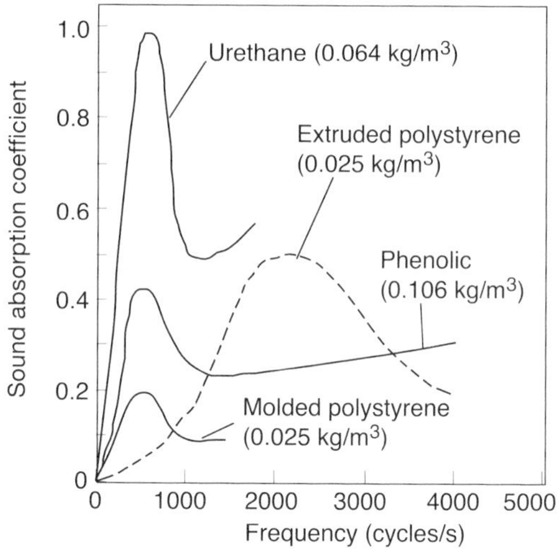

Figure 14.4 Sound absorption coefficients as a function of frequency for various foams.

When compared to wood, even semi-crystalline polymers are considered "sound-proof" materials. Materials with the glass transition temperature lower than room temperature are particularly suitable as damping materials. Commonly used for this purpose are thermoplastics and weakly cross-linked elastomers. Elastomer mats are often adhered on one or both sides of sheet metal, preventing resonance flexural vibrations of the sheet metal such as in automotive applications.

References

1. Offergeld, H., Ph.D Thesis, IKV, RWTH-Aachen, (1990).
2. Griffin, J.D., and Skochdopole, R. E., *Engineering Design for Plastics*, E. Baer, (Ed.), Chapter 15, Robert E. Krieger Publishing Company, Huntington, (1975).

Appendix

Table I Guide Values of the Physical Properties of Plastics

Polymer	Abbreviation	Density g/cm³	Density lb/in³	Mechanical properties: Tensile strength N/mm²	Tensile strength psi
Low density polyethylene	PE-LD	0.914/0.928	0.0329–0.0330	8/23	1140/3270
High density polyethylene	PE-HD	0.94/0.96	0.0338–0.0345	18/35	2560/4980
EVA	EVA	0.92/0.95	0.0331–0.0341	10/20	1420/2840
Polypropylene	PP	0.90/0.907	0.0324–0.0327	21/37	2990/5260
Polybutene-1	PB	0.905/0.920	0.0325–0.0331	30/38	4270/5400
Polyisobutylene	PIB	0.91/0.93	0.0327–0.0334	2/6	284/853
Poly-4-methylpent-1-ene	PMP	0.83	0.0298	25/28	3560/3980
Ionomers	–	0.94	0.0338	21/35	2990/4980
Rigid PVC	PVC-U	1.38/1.55	0.0496–0.0557	50/75	7110/10670
Plasticized PVC	PVC-P	1.16/1.35	0.0417–0.0486	10/25	1420/3560
Polystyrene	PS	1.05	0.0378	45/65	6400/9240
Styrene/acrylonitrile copolymer	SAN	1.08	0.0392	75	10670
Styrene/polybutadiene graft polymer	SB	1.05	0.0378	26/38	3700/5400
Acrylonitrile/polybut./styrene graft polymer	ABS	1.04/1.06	0.0374–0.0381	32/45	4550/6400
AN/AN elastomers/styrene graft polymer	ASA	1.04	0.0374	32	4550
Polymethylmethacrylate	PMMA	1.17/1.20	0.0421–0.0431	50/77	7110/10950
Polyvinylcarbazole	PVK	1.19	0.0428	20/30	2840/4270
Polyacetal	POM	1.41/1.42	0.0507–0.0511	62/70	8820/9960
Polytetrafluoroethylene	PTFE	2.15/2.20	0.0774–0.0791	25/36	3560/5120
Tetrafluoroethylene/hexafluoropropylene copolymer	FEP	2.12/2.17	0.0763–0.0781	22/28	3130/3980
Polytrifluorochlorethylene	PCTFE	2.10/2.12	0.0755–0.0762	32/40	4550/5690
Ethylene/tetrafluoroethylene	E/TFE	1.7	0.0611	35/54	4980/7680
Polyamide 6	PA 6	1.13	0.0406	70/85	9960/12090
Polyamide 66	PA 66	1.14	0.0410	77/84	10950/11950
Polyamide 11	PA 11	1.04	0.0374	56	7960
Polyamide 12	PA 12	1.02	0.0367	56/65	7960/9240
Polyamide 6-3-T	PA-6-3-T	1.12	0.0403	70/84	9960/11950
Polycarbonate	PC	1.2	0.0432	56/67	7960/9530
Polyethyleneterephthalate	PET	1.37	0.0492	47	6680
Polybutyleneterephthalate	PBT	1.31	0.0471	40	5690
Polyphenyleneether modified	PPE	1.06	0.0381	55/68	7820/9670
Polysulfone	PSU	1.24	0.0446	50/100	7110/14200
Polyphenylenesulfide	PPS	1.34	0.0483	75	10670
Polyarylsulfone	PAS	1.36	0.0490	90	12800
Polyethersulfone	PES	1.37	0.0492	85	12090
Polyarylether	PAE	1.14	0.0411	53	7540
Phenol/formaldehyde, grade 31	PF	1.4	0.0504	25	3560
Urea/formaldehyde, grade 131	UF	1.5	0.0540	30	4270
Melamine/formaldehyde, grade 152	MF	1.5	0.0540	30	4270
Unsaturated polyester resin, grade 802	UP	2.0	0.0720	30	4270
Polydiallylphthalate (GF) molding compound	PDAP	1.51/1.78	0.0543–0.0640	40/75	5690/10670
Silicone resin molding compound	SI	1,8/1.9	0.0648–0.0684	28/46	3980/6540
Polyimide molding	PI	1.43	0.0515	75/100	10570/14200
Epoxy resin, grade 891	EP	1.9	0.0683	30/40	4270/5690
Polyurethane casting resin	PU	1.05	0.0378	70/80	9960/11380
Thermoplastic PU-elastomers	PU	1.20	0.0432	30/40	4270/5690
Linear polyurethane (U_{50})	PU	1.21	0.0435	30 (σ_s)	4270 (σ_y)
Vulcanized fiber	VF	1.1/1.45	0.0396–0.0522	85/100	12090/14200
Celluloseacetate, grade 432	CA	1.30	0.0468	38/(σ_s)	5400 (σ_y)
Cellulosepropionate	CP	1.19/1.23	0.0429–0.0452	14/55	7990/7820
Celluloseacetobutyrate, grade 413	CAB	1.18	0.0425	26 (σ_s)	3600 (σ_y)

Elongation at break %	Tensile modulus of elasticity		Ball indentation hardness		Impact strength kJ/m²	Notched impact strength	
	N/mm²	kpsi	10-s-value	10-s-value psi		kJ/m²	ft lb/ in of notch
300/1000	200/500	28.4/71.1	13/20	1850/2840	no break	no break	–
100/1000	700/1400	99.6/199	40/65	5690/9240	no break	no break	–
600/900	7/120	0.99/17.1	–	–	no break	no break	no break
20/800	11000/1300	156/185	36/70	5120/9960	no break	3/17	0.5/20
250/280	250/350	35.6/49.8	30/38	4270/5400	no break	4/no break	no break
> 1000	–	–	–	–	no break	no break	no break
13/22	1100/1500	156/213	–	–	–	–	0.4/0.6
250/500	180/210	25.6/29.9	–	–	–	–	6/15
10/50	1000/3500	142/498	75/155	10670/22000	no break/> 20	2/50	0.4/20
170/400	–	–	–	–	no break	no break	–
3/4	3200/3250	455/462	120/130	17100/18500	5/20	2/2.5	0.25/0.6
5	3600	512	130/140	18500/19900	8/20	2/3	0.35/0.5
25/60	1800/2500	256/356	80/130	11380/18500	10/80	5/13	no break
15/30	1900/2700	270/384	80/120	11380/17100	70/no break	7/20	2.5/12
40	1800	256	75	10670	no break	18	6/8
2/10	2700/3200	384/455	180/200	25600/28400	18	2	0.3/0.5
–	3500	498	200	28400	5	2	–
25/70	2800/3200	398/455	150/170	21300/24200	100	8	1/2.3
350/550	410	58.3	27/35	3840/4980	no break	13/15	3.0
250/330	350	49.8	30/32	4270/4550	–		no break
120/175	1050/2100	149/299	65/70	9240/9960	no break	8/10	2.5/2.8
400/500	1100	156	65	9240	–		no break
200/300	1400	199	75	10670	no break	no break	3.0
150/300	2000	284	100	14200	no break	15/20	2.1
500	1000	142	75	10670	no break	30/40	1.8
300	1600	228	75	10670	no break	10/20	2/5.5
70/150	2000	284	160	22800	no break	13	–
100/130	2100/2400	299/341	110	15600	no break	20/30	12/18
50/300	3100	441	200	28400	no break	4	0.8/1.0
15	2000	284	180	25600	no break	4	0.8/1.0
50/60	2500	356	–	–	no break	–	4
25/30	2600/2750	370/391	–	–	–	–	1.3
3	3400	484	–	–	–	–	0.3
13	2600	370	–	–	–	–	1/2
30/80	2450	348	–	–	–	–	1.6
25/90	2250	320	–	–	–	–	8.0
0.4/0.8	5600/12000	796/1710	250/320	35600/45500	> 6	> 1.5	0.2/0.6
0.5/1.0	7000/10500	996/1490	260/350	39000/49800	> 6.5	> 2.5	0.5/0.4
0.6/0.9	4900/9100	697/1294	260/410	37000/58300	> 7.0	> 1.5	0.2/0.3
0.6/1.2	14000/20000	1990/2840	240	34100	> 4.5	> 3.0	0.5/16
–	9800/15500	1394/2200	–	–	–	–	0.4/15
–	6000/12000	853/1710	–	–	–	–	0.3/0.8
4/9	23000/28000	3270/3980	–	–	–	–	0.5/1.0
4	21500	3060	–	–	> 8	> 3	2/30
3/6	4000	569	–	–	–	–	0.4
400/450	700	99.6	–	–	no break	no break	no break
35 (ε_s)	1000	140	–	–	no break	3	–
–	–	–	80/140	11380/19900	20/120	–	–
3 (ε_s)	2200	313	50	7110	65	15	2.5
30/100	420/1500	59,7/213	47/79	6680/11240	no break	6/20	1.5
4 (ε_s)	1600	228	35/43	4980/6120	no break	30/35	4/5

(continued on next page)

Table I *(cont.)* Guide Values of the Physical Properties of Plastics

Polymer	Abbre-viation	Density g/cm³	Density lb/in³
Low density polyethylene	PE-LD	0.914/0.928	0.0329–0.0330
High density polyethylene	PE-HD	0.94/0.96	0.0338–0.0345
EVA	EVA	0.92/0.95	0.0331–0.0341
Polypropylene	PP	0.90/0.907	0.0324–0.0327
Polybutene-1	PB	0.905/0.920	0.0325–0.0331
Polyisobutylene	PIB	0.91/0.93	0.0327–0.0334
Poly-4-methylpent-1-ene	PMP	0.83	0.0298
Ionomers	–	0.94	0.0338
Rigid PVC	PVC-U	1.38/1.55	0.0496–0.0557
Plasticized PVC	PVC-P	1.16/1.35	0.0417–0.0486
Polystyrene	PS	1.05	0.0378
Styrene/acrylonitrile copolymer	SAN	1.08	0.0392
Styrene/polybutadiene graft polymer	SB	1.05	0.0378
Acrylonitrile/polybut./styrene graft polymer	ABS	1.04/1.06	0.0374–0.0381
AN/AN elastomers/styrene graft polymer	ASA	1.04	0.0374
Polymethylmethacrylate	PMMA	1.17/1.20	0.0421–0.0431
Polyvinylcarbazole	PVK	1.19	0.0428
Polyacetal	POM	1.41/1.42	0.0507–0.0511
Polytetrafluoroethylene	PTFE	2.15/2.20	0.0774–0.0791
Tetrafluoroethylene/hexafluoropropylene copolymer	FEP	2.12/2.17	0.0763–0.0781
Polytrifluorochlorethylene	PCTFE	2.10/2.12	0.0755–0.0762
Ethylene/tetrafluoroethylene	E/TFE	1.7	0.0611
Polyamide 6	PA 6	1.13	0.0406
Polyamide 66	PA 66	1.14	0.0410
Polyamide 11	PA 11	1.04	0.0374
Polyamide 12	PA 12	1.02	0.0367
Polyamide 6-3-T	PA-6-3-T	1.12	0.0403
Polycarbonate	PC	1.2	0.0432
Polyethyleneterephthalate	PET	1.37	0.0492
Polybutyleneterephthalate	PBT	1.31	0.0471
Polyphenyleneether modified	PPE	1.06	0.0381
Polysulfone	PSU	1.24	0.0446
Polyphenylenesulfide	PPS	1.34	0.0483
Polyarylsulfone	PAS	1.36	0.0490
Polyethersulfone	PES	1.37	0.0492
Polyarylether	PAE	1.14	0.0411
Phenol/formaldehyde, grade 31	PF	1.4	0.0504
Urea/formaldehyde, grade 131	UF	1.5	0.0540
Melamine/formaldehyde, grade 152	MF	1.5	0.0540
Unsaturated polyester resin, grade 802	UP	2.0	0.0720
Polydiallylphthalate (GF) molding compound	PDAP	1.51/1.78	0.0543–0.0640
Silicone resin molding compound	SI	1.8/1.9	0.0648–0.0684
Polyimide molding	PI	1.43	0.0515
Epoxy resin, grade 891	EP	1.9	0.0683
Polyurethane casting resin	PU	1.05	0.0378
Thermoplastic PU-elastomers	PU	1.20	0.0432
Linear polyurethane (U_{50})	PU	1.21	0.0435
Vulcanized fiber	VF	1.1/1.45	0.0396–0.0522
Celluloseacetate, grade 432	CA	1.30	0.0468
Cellulosepropionate	CP	1.19/1.23	0.0429–0.0452
Celluloseacetobutyrate, grade 413	CAB	1.18	0.0425

Optical properties		Water absorption	
Refractive index n_D^{20}	Transparency	mg (4 d)	% (24 h)
1.51	transparent	< 0.01	< 0.01
1.53	opaque	< 0.01	< 0.01
–	transparent/opaque	–	0.05/0.13
1.49	transparent/opaque	< 0.01	0.01/0.03
–	opaque	< 0.01	< 0.02
–	opaque	< 0.01	< 0.01
1.46	opaque	–	0.01
1.51	transparent	–	0.1/1.4
1.52/1.55	transparent/opaque	3/18	0.04/0.4
–	transparent/opaque	6/30	0.15/0.75
1.59	transparent	–	0.03/0.1
1.57	transparent	–	0.2/0.3
–	opaque	–	0.05/0.6
–	opaque	–	0.2/0.45
–	translucent/opaque	–	–
1.49	transparent	35/45	0.1/0.4
–	opaque	0.5	0.1/0.2
1.48	opaque	20/30	0.22/0.25
1.35	opaque	–	0
1.34	transparent/translucent	–	< 0.1
1.43	translucent/opaque	–	0
1.40	transparent/opaque		0.03
1.53	translucent/opaque	–	1.3/1.9
1.53	translucent/opaque	–	1.5
1.52	translucent/opaque	–	0.3
–	translucent/opaque	–	0.25
1.53	transparent	–	0.4
1.58	transparent	10	0.16
–	transparent/opaque	18/20	0.30
–	opaque	–	0.08
–	opaque	–	0.06
1.63	transparent/opaque	–	0.02
–	opaque	–	0.02
1.67	opaque	–	1.8
1.65	transparent	–	0.43
–	translucent/opaque	–	0.25
–	opaque	< 150	0.3/1.2
–	opaque	< 300	0.4/0.8
–	opaque	< 250	0.1/0.6
–	opaque	< 45	0.03/0.5
–	opaque	–	0.12/0.35
–	opaque	–	0.2
–	opaque	–	0.32
–	opaque	< 30	0.05/0.2
–	transparent	–	0.1/0.2
–	translucent/opaque	–	0.7/0.9
–	translucent/opaque	130	–
–	opaque	–	7/9
1.50	transparent	130	6
1.47	transparent	40/60	1.2/2.8
1.47	transparent	40/60	0.9/3.2

(continued on next page)

Table I *(cont.)* Guide Values of the Physical Properties of Plastics

Polymer	Abbreviation	Density g/cm³	Density lb/in³	Thermal properties: Service temperature max./short time °C	°F	max./continuous °C	°F
Low density polyethylene	PE-LD	0.914/0.928	0.0329–0.0330	80/90	176/194	60/75	140/167
High density polyethylene	PE-HD	0.94/0.96	0.0338–0.0345	90/120	194/248	70/80	158/176
EVA	EVA	0.92/0.95	0.0331–0.0341	65	149	55	131
Polypropylene	PP	0.90/0.907	0.0324–0.0327	140	284	100	212
Polybutene-1	PB	0.905/0.920	0.0325–0.0331	130	266	90	194
Polyisobutylene	PIB	0.91/0.93	0.0327–0.0334	80	176	65	149
Poly-4-methylpent-1-ene	PMP	0.83	0.0298	180	356	120	248
Ionomers	–	0.94	0.0338	120	248	100	212
Rigid PVC	PVC-U	1.38/1.55	0.0496–0.0557	75/100	167/212	65/85	149/185
Plasticized PVC	PVC-P	1.16/1.35	0.0417–0.0486	55/65	131/149	50/55	122/131
Polystyrene	PS	1.05	0.0378	60/80	140/176	50/70	122/158
Styrene/acrylonitrile copolymer	SAN	1.08	0.0392	95	203	85	185
Styrene/polybutadiene graft	SB	1.05	0.0378	60/80	140/176	50/70	122/158
Acrylonitrile/polybut./styrene graft polymer	ABS	1.04/1.06	0.0374–0.0381	85/100	188/212	75/85	167/185
AN/AN elastomers/styrene graft polymer	ASA	1.04	0.0374	85/90	188/194	70/75	158/167
Polymethylmethacrylate	PMMA	1.17/1.20	0.0421–0.0431	85/100	188/212	65/90	149/194
Polyvinylcarbazole	PVK	1.19	0.0428	170	338	160	320
Polyacetal	POM	1.41/1.42	0.0507–0.0511	110/140	230	90/110	194/230
Polytetrafluoroethylene	PTFE	2.15/2.20	0.0774–0.0791	300	572	250	482
Tetrafluoroethylene hexafluoropropylene copolymer	FEP	2.12/2.17	0.0763–0.0781	250	482	205	401
Polytrifluorochlorethylene	PCTFE	2.10/2.12	0.0755–0.0762	180	356	150	302
Ethylene tetrafluoroethylene	E/TFE	1.7	0.0611	220	428	150	302
Polyamide 6	PA 6	1.13	0.0406	140/180	284/356	80/100	176/212
Polyamide 66	PA 66	1.14	0.0410	170/200	338/392	80/120	176/248
Polyamide 11	PA 11	1.04	0.0374	140/150	284/302	70/80	158/176
Polyamide 12	PA 12	1.02	0.0367	140/150	284/302	70/80	158/176
Polyamide 6-3-T	PA-6-3-T	1.12	0.0403	130/140	266/284	80/100	176/212
Polycarbonate	PC	1.2	0.0432	160	320	135	275
Polyethyleneterephthalate	PET	1.37	0.0492	200	392	100	212
Polybutyleneterephthalate	PBT	1.31	0.0471	165	329	100	212
Polyphenyleneether modified	PPE	1.06	0.0381	150	302	80	176
Polysulfone	PSU	1.24	0.0446	200	392	150	302
Polyphenylensulfide	PPS	1.34	0.0483	300	572	200	392
Polyarylsulfone	PAS	1.36	0.0490	300	572	260	500
Polyethersulfone	PES	1.37	0.0492	260	500	200	392
Polyarylether	PAE	1.14	0.0411	160	320	120	248
Phenol/formaldehyde, grade 31	PF	1.4	0.0504	140	284	110	230
Urea/formaldehyde, grade 131	UF	1.5	0.0540	100	212	70	158
Melamine/formaldehyde, grade 152	MF	1.5	0.0540	120	248	80	176
Unsaturated polyester resin, grade 802	UP	2.0	0.0720	200	392	150	302
Polydiallylphthalate (GF) molding compound	PDAP	1.51/1.78	0.0543–0.0640	190/250	374/482	150/180	302/356
Silicone resin molding compound	SI	1.8/1.9	0.0648–0.0684	250	482	170/180	338/356
Polyimide molding	PI	1.43	0.0515	400	752	260	500
Epoxy resin, grade 891	EP	1.9	0.0683	180	356	130	266
Polyurethane casting resin	PU	1.05	0.0378	100	212	80	176
Thermoplastic PU-elastomers	PU	1.20	0.0432	110	230	80	176
Linear polyurethane (U_{50})	PU	1.21	0.0435	80	176	60	140
Vulcanized fiber	VF	1.1/1.45	0.0396–0.0522	180	356	105	221
Celluloseacetate, grade 432	CA	1.30	0.0468	80	176	70	158
Cellulosepropionate	CP	1.19/1.23	0.0429–0.0452	80/120	176/248	60/115	140/239
Celluloseacetobutyrate, grade 413	CAB	1.18	0.0425	80/120	176/248	60/115	140/239

nin./continuous		Heat deflection temperature				Coefficient of linear expansion		Thermal conductivity		Specific heat	
		°C		°F							
		VSP (Vicat 5 kg)	1.86/ 0.45 N/ mm²	VSP (Vicat lb)	264/ 66 psi	$K^{-1} \cdot 10^6$	in/in/°F $\cdot 10^{-6}$	W/mK	BTU in/ ft² h °F	kJ/ kgK	BTU/ lb °F
'C	°F										
–50	–58	–	35	–	95	250	140	0.32/0.40	2.2/2.8	2.1/2.5	8.8/10.5
–50	–58	60/70	50	140/158	122	200	110	0.38/0.51	2.6/3.5	2.1/2.7	8.8/11.5
–60	–76	–	34/62	–	93/144	160/200	90/110	0.35	2.4	2.3	9.5
)/–30	32/–22	85/100	45/120	185/212	113/248	150	83	0.17/0.22	1.2/1.5	2.0	8.3
)	32	70	60/110	158	140/230	150	83	0.20	1.4	1.8	7.5
–50	–58	–	–	–	–	120	67	0.12/0.20	0.8/1.4	–	–
)	32	–	–	–	–	117	65	0.17	1.2	2.18	9.1
–50	–58	–	38/45	–	100/113	120	67	0.24	1.7	2.20	9.2
–5	21	75/110	60/82	167/230	140/180	70/80	39/45	0.14/0.17	1.0/1.2	0.85/0.9	3.55/3.75
)/–20	32/–4	40	–	104	–	150/210	83/110	0.15	1.05	0.9/1.8	3.75/7.5
–10	14	78/99	110/80	172/210	230/176	70	39	0.18	1.25	1.3	5.4
–20	–4	–	104/90	–	219/194	80	45	0.18	1.25	1.3	5.4
–20	–4	77/95	104/82	171/203	219/180	70	39	0.18	1.25	1.3	5.4
–40	–40	95/110	80/120	203/230	176/248	60/110	33/61	0.18	1.25	1.3	5.4
–40	–40	92	100/110	198	212/230	80/110	44/61	0.18	1.25	1.3	5.4
–40	–40	70/100	60/100	158/212	140/212	70	39	0.18	1.25	1.47	6.15
–100	–148	180	–	356	–	–	–	0.29	2.0	–	–
–60	–76	160/173	110/170	320/344	230/338	90/110	50/61	0.25/0.30	1.7/2.1	1.46	6.1
–200	–328	–	–/121	–	–/250	100	56	0.25	1.7	1.0	4.20
–100	–148	–	–/70	–	–/158	80	45	0.25	1.7	1.12	4.65
–40	–40	–	–/126	–	–/259	60	33	0.22	1.5	0.9	3.75
–190	–310	–	71/104	–	160/219	40	22	0.23	1.6	0.9	3.75
–30	–22	180	80/190	356	176/374	80	44	0.29	2.0	1.7	7.1
–30	–22	200	105/200	392	221/392	80	44	0.23	1.6	1.7	7.1
–70	–94	175	150/130	347	302/266	130	72	0.23	1.6	1.26	5.25
–70	–94	165	140/150	329	284/302	150	83	0.23	1.6	1.26	5.25
–70	–94	145	140/80	293	284/176	80	45	0.23	1.6	1.6	6.70
–100	–148	138	130/145	280	266/293	60/70	33/39	0.21	1.45	1.17	4.90
–20	–4	188	–	280	–	70	39	0.24	1.65	1.05	4.40
–30	–22	178	50/190	352	122/374	60	33	0.21	1.45	1.30	5.40
–30	–22	148	100/140	298	212/284	60	33	0.23	1.60	1.40	5.85
–100	–148	–	175/180	–	347/356	54	30	0.28	1.95	1.30	5.40
–	–	–	137/–	–	277/–	55	31	0.25	1.70	–	–
–	–	–	–	–	–	47	26	0.16	1.10	–	–
–	–	–	–	–	–	55	31	0.18	1.25	1.10	4.6
–	–	–	150/160	–	302/320	65	36	0.26	1.80	1.46	6.1
–	–	–	150/190	–	302/374	30/50	17/28	0.35	2.40	1.30	5.40
–	–	–	130/–	–	266/–	50/60	28/33	0.40	2.75	1.20	5.0
–	–	–	180/–	–	356/–	50/60	28/33	0.50	3.45	1.20	5.0
–	–	–	230/–	–	446/–	20/40	11/22	0.70	4.85	1.20	5.0
–50	–58	–	220/–	–	428/–	10/35	55/19	0.60	4.15	–	5.0
–50	–58	–	480/–	–	896/–	20/50	11/28	0.3/0.4	2.05/2.75	0.8/0.9	3.35/3.75
–200	–239	–	240/–	–	464/–	50/63	28/35	0.6/0.65	4.15/4.50	–	–
–	–	–	200/–	–	392/–	11/35	6.1/19	0.88	6.1	0.8	3.35
–	–	–	90/–	–	194/–	10/20	5.5/11	0.58	4.0	1.76	7.30
–40	–40	–	–	–	–	150	83	1.7	1.15	0.5	2.10
–15	3	100	–	212	–	210	12	1.8	1.25	0.4	1.65
–30	–22	–	–	–	–	–	–	–	–	–	–
–40	–40	50/63	90/–	122/144	194/–	120	67	0.22	1.50	1.6	6.7
–40	–40	100	73/98	212	163/208	110/130	61/72	0.21	1.45	1.7	7.1
–40	–40	60/75	62/71	140/167	144/160	120	67	0.21	1.45	1.6	6.7

(continued on next page)

Table I *(cont.)* Guide Values of the Physical Properties of Plastics

Polymer	DIN 7728 Bl. 1	Density g/cm³	Density lb/in³	Electrical properties: Volume resistivity Ω cm	Electrical properties: Surface resistance Ω
Low density polyethylene	PE-LD	0.914/0.928	0.0329–0.0330	$>10^{17}$	10^{14}
High density polyethylene	PE-HD	0.94/0.96	0.0338–0.0345	$>10^{17}$	10^{14}
EVA	EVA	0.92/0.95	0.0331–0.0341	$<10^{15}$	10^{13}
Polypropylene	PP	0.90/0.907	0.0324–0.0327	$>10^{17}$	10^{13}
Polybutene-1	PB	0.905/0.920	0.0325–0.0331	$>10^{17}$	10^{13}
Polyisobutylene	PIB	0.91/0.93	0.0327–0.0334	$>10^{15}$	10^{13}
Poly-4-methylpent-1-ene	PMP	0.83	0.0298	$>10^{16}$	10^{13}
Ionomers	–	0.94	0.0338	$>10^{16}$	10^{13}
Rigid PVC	PVC-U	1.38/1.55	0.0496–0.0557	$>10^{15}$	10^{13}
Plasticized PVC	PVC-P	1.16/1.35	0.0417–0.0486	$>10^{11}$	10^{11}
Polystyrene	PS	1.05	0.0378	$>10^{16}$	$>10^{13}$
Styrene/acrylonitrile copolymer	SAN	1.08	0.0392	$>10^{16}$	$>10^{13}$
Styrene/polybutadiene graft polymer	SB	1.05	0.0378	$>10^{16}$	$>10^{13}$
Acrylonitrile/polybut./styrene graft polymer	ABS	1.04/1.06	0.0374–0.0381	$>10^{15}$	$>10^{13}$
AN/AN elastomers/styrene graft polymer	ASA	1.04	0.0374	$>10^{15}$	$>10^{13}$
Polymethylmethacrylate	PMMA	1.17/1.20	0.0421–0.0431	$>10^{15}$	10^{15}
Polyvinylcarbazole	PVK	1.19	0.0428	$>10^{16}$	10^{14}
Polyacetal	POM	1.41/1.42	0.0507–0.0511	$>10^{15}$	10^{13}
Polytetrafluoroethylene	PTFE	2.15/2.20	0.0774–0.0791	$>10^{18}$	10^{17}
Tetrafluoroethylene/hexafluoropropylene copolymer	FEP	2.12/2.17	0.0763–0.0781	$>10^{18}$	10^{16}
Polytrifluorochloroethylene	PCTFE	2.10/2.12	0.0755–0.0762	$>10^{18}$	10^{16}
Ethylene/tetrafluoroethylene	E/TFE	1.7	0.0611	$>10^{16}$	10^{13}
Polyamide 6	PA 6	1.13	0.0406	10^{12}	10^{10}
Polyamide 66	PA 66	1.14	0.0410	10^{12}	10^{10}
Polyamide 11	PA 11	1.04	0.0374	10^{13}	10^{11}
Polyamide 12	PA 12	1.02	0.0367	10^{13}	10^{11}
Polyamide 6-3-T	PA-6-3-T	1.12	0.0403	10^{11}	10^{10}
Polycarbonate	PC	1.2	0.0432	$>10^{17}$	$>10^{15}$
Polyethyleneterephthalate	PET	1.37	0.0492	10^{16}	10^{16}
Polybutyleneterephthalate	PBT	1.31	0.0471	10^{16}	10^{13}
Polyphenyleneether modified	PPE	1.06	0.0381	10^{16}	10^{14}
Polysulfone	PSU	1.24	0.0446	$>10^{16}$	–
Polyphenylenesulfide	PPS	1.34	0.0483	$>10^{16}$	–
Polyarylsulfone	PAS	1.36	0.0490	$>10^{16}$	–
Polyethersulfone	PES	1.37	0.0492	10^{17}	–
Polyarylether	PAE	1.14	0.0411	$>10^{10}$	–
Phenol/formaldehyde, grade 31	PF	1.4	0.0504	10^{11}	$>10^{8}$
Urea/formaldehyde, grade 131	UF	1.5	0.0540	10^{11}	$>10^{10}$
Melamine/formaldehyde, grade 152	MF	1.5	0.0540	11^{11}	$>10^{8}$
Unsaturated polyester resin, grade 802	UP	2.0	0.0720	$>10^{12}$	$>10^{10}$
Polydiallylphthalate (GF) molding compound	PDAP	1.51/1.78	0.0543–0.0640	$10^{13}/10^{16}$	10^{13}
Silicone resin molding compound	SI	1.8/1.9	0.0648–0.0684	10^{14}	10^{12}
Polyimide molding	PI	1.43	0.0515	$>10^{16}$	$>10^{15}$
Epoxy resin, grade 891	EP	1.9	0.0683	$>10^{14}$	$>10^{12}$
Polyurethane casting resin	PU	1.05	0.0378	10^{16}	10^{14}
Thermoplastic PU-elastomers	PU	1.20	0.0432	10^{12}	10^{11}
Linear polyurethane (U_{50})	PU	1.21	0.0435	10^{13}	10^{12}
Vulcanized fiber	VF	1.1/1.45	0.0396–0.0522	10^{10}	10^{8}
Celluloseacetate, grade 432	CA	1.30	0.0468	10^{13}	10^{12}
Cellulosepropionate	CP	1.19/1.23	0.0429–0.0452	10^{16}	10^{14}
Celluloseacetobutyrate, grade 413	CAB	1.18	0.0425	10^{16}	10^{14}

Dielectric constant		Dissipation (power) factor tan δ		Dielectric strength		Tracking resistance		
50 Hz	10^6 Hz	50 Hz	10^6 Hz	kV/25 µm	kV/cm	KA	KB	KC
2.29	2.28	$1.5 \cdot 10^4$	$0.8 \cdot 10^{-4}$	> 700	–	3b	> 600	> 600
2.35	2.34	$2.4 \cdot 10^{-4}$	$2.0 \cdot 10^{-4}$	> 700	–	3c	> 600	> 600
2.5/3.2	2.6/3.2	0.003/0.02	0.03/0.05	–	620/780	–	–	–
2.27	2.25	$< 4 \cdot 10^{-4}$	$< 5 \cdot 10^{-4}$	800	500/650	3c	> 600	> 600
2.5	2.2	$7 \cdot 10^{-4}$	$6 \cdot 10^{-4}$	700	–	3c	> 600	> 600
2.3	–	0.0004	–	230	–	3c	> 600	> 600
2.12	2.12	$7 \cdot 10^{-5}$	$3 \cdot 10^{-5}$	280	700	3c	> 600	> 600
.5	3.0	0.011	0.015	200/400	350/500	2/3b	600	600
/8	4/4.5	0.08	0.12	150/300	300/400	–	–	–
2.5	2.5	$1/4 \cdot 10^{-4}$	$0.5/4 \cdot 10^{-4}$	500	300/700	1/2	140	150/250
2.6/3.4	2.6/3.1	$6/8 \cdot 10^{-3}$	$7/10 \cdot 10^{-3}$	500	400/500	1/2	160	150/260
2.4/4.7	2.4/3.8	$4/20 \cdot 10^{-4}$	$4/20 \cdot 10^{-4}$	500	300/600	2	> 600	> 600
2.4/5	2.4/3.8	$3/8 \cdot 10^{-3}$	$2/15 \cdot 10^{-3}$	400	350/500	3a	> 600	> 600
/4	3/3.5	0.02/0.05	0.02/0.03	350	360/400	3a	> 600	> 600
.3/3.9	2.2/3.2	0.04/0.06	0.004/0.04	300	400/500	3c	> 600	> 600
–	3	$6/10 \cdot 10^{-4}$	$6/10 \cdot 10^{-4}$	500	–	3b	> 600	> 600
.7	3.7	0.005	0.005	700	380/500	3b	> 600	> 600
< 2.1	< 2.1	$< 2 \cdot 10^{-4}$	$< 2 \cdot 10^{-4}$	500	480	3c	> 600	> 600
2.1	2.1	$< 2 \cdot 10^{-4}$	$< 7 \cdot 10^{-4}$	500	550	3c	> 600	> 600
2.3/2.8	2.3/2.5	$1 \cdot 10^{-3}$	$2 \cdot 10^{-2}$	500	550	3c	> 600	> 600
.6	2.6	$8 \cdot 10^{-4}$	$5 \cdot 10^{-3}$	380	400	3c	> 600	> 600
.8	3.4	0.01	0.03	350	400	3b	> 600	> 600
.0	4.0	0.14	0.08	400	600	3b	> 600	> 600
.7	3.5	0.06	0.04	300	425	3b	> 600	> 600
.2	3.1	0.04	0.03	300	450	3b	> 600	> 600
.0	3.0	0.03	0.04	250	350	3b	> 600	> 600
.0	2.9	$7 \cdot 10^{-4}$	$1 \cdot 10^{-2}$	350	380	1	120/160	260/300
.0	4.0	$2 \cdot 10^{-3}$	$2 \cdot 10^{-2}$	500	420	2	–	–
.0	3.0	$2 \cdot 10^{-3}$	$2 \cdot 10^{-2}$	500	420	3b	420	380
.6	2.6	$4 \cdot 10^{-4}$	$9 \cdot 10^{-4}$	500	450	1	300	300
.1	3.0	$8 \cdot 10^{-4}$	$3 \cdot 10^{-3}$	–	425	1	175	175
.1	3.2	$4 \cdot 10^{-4}$	$7 \cdot 10^{-4}$	–	595	–	–	–
.9	3.7	$3 \cdot 10^{-3}$	$13 \cdot 10^{-3}$	–	350	–	–	–
.5	3.5	$1 \cdot 10^{-3}$	$6 \cdot 10^{-3}$	–	400	–	–	–
.14	3.10	$6 \cdot 10^{-3}$	$7 \cdot 10^{-3}$	–	430	–	–	–
	4.5	0.1	0.03	50/100	300/400	1	140/180	125/175
	7	0.04	0.3	80/150	300/400	3a	> 400	> 600
	8	0.06	0.03	80/150	290/300	3b	> 500	> 600
	5	0.04	0.02	120	250/530	3c	> 600	> 600
.2	4	0.04	0.03	–	400	3c	> 600	> 600
	3.5	0.03	0.02	–	200/400	3c	> 600	> 600
.5	3.4	$2 \cdot 10^{-3}$	$5 \cdot 10^{-3}$	–	560	1	> 300	> 380
.5/5	3.5/5	0.001	0.01	–	300/400	3c	> 300	200/600
6	3.4	0.05	0.05	–	240	3c	–	–
.5	5.6	0.03	0.06	–	300/600	3a	> 600	> 600
.8	4.0	0.12	0.07	330	–	–	–	–
	–	0.08	–	70/180	–	–	–	–
.8	4.6	0.02	0.03	320	400	3a	> 600	> 600
2	3.7	0.01	0.03	350	400	3a	> 600	> 600
.7	3.5	0.006	0.021	380	400	3a	> 600	> 600

Table II Permeability of Films Made from Various Polymers

Polymer	Temper-ature	Film thickness	Water vapor	N_2	Air	O_2	CO_2	H_2	Ar	He
	SI									
	°C	µm [1]	g/cm² day	cm³/m² day bar						
PE-LD	23	100	1	700	1100	2000	10000	8000	–	–
PE-HD ($\rho = 0.95$ g/cm³, unstretched)	25	40	0.9	525	754	1890	7150	6000	–	–
	30	40	1.7	720	960	2270	8600	7600	–	–
	40	40	3.5	1220	1660	3560	13100	11400	–	–
	50	40	8.1	2140	2650	5650	19500	16800	–	–
PE-HD ($\rho = 0.95$ g/cm³, stretched)	25	40	1.0	430	680	1210	5900	5000	–	–
	30	40	1.6	560	830	1530	7200	6000	–	–
	40	40	4.3	1050	1490	2650	11200	9400	–	–
	50	40	10.5	1870	2670	4650	18100	14800	–	–
E/VA copolymer, VAC 20%	23	100	455	1400	–	4000	17000	–	–	–
Polypropylene (unstretched)	25	40	2.1	430	700	1900	6100	17700	1480	19200
	30	40	3.2	600	960	2500	8400	18200	2100	21700
	40	40	7.4	1280	1820	5100	14800	28100	4100	29800
	50	40	19.0	2800	3600	9200	27300	46600	8000	43500
Polypropylene (stretched)	25	40	0.81	200	350	1000	3300	6700	–	7300
	30	40	1.2	260	480	1200	3900	8200	–	8500
	40	40	3.3	560	940	2300	7050	12300	–	12100
	50	40	8.4	1200	1850	4150	13200	19800	–	17800
PVC-U (unstretched	20	40	7.6	12	28	87	200	–	–	–
PVC-U (stretched)	20	40	4.4	13	13	43	110	–	–	–
PVC-P	20	40	20	350	550	1500	8500	–	–	–
Polyvinylidenechloride	25	25	0.1/0.2	1.8/2.3	5/10	1.7/11	60/700	630/1400	–	–
Polystyrene (stretched)	25	50	14.0	27	80	235	800	1260	–	–
Polyacetal	20	40	2.5	10	16	50	96	420	–	–
PFEP copolymers	40	25	2.0	375	–	3000	6500	2000	–	–
CTFE	40	25	0.38/0.85	39	–	110/230	250/620	3400/5200	–	–
E/TFE copolymers	23	25	0.6	470	–	1560	3800	–	–	–
E/CTFE copolymers	23	25	9.0	150	–	39.0	1700	–	–	–
PVF	23	25	50.0	3.8	–	4.7	170	900	–	–
Polyamide 6	25	25	80/110	14	–	40	200	1500	–	–
Polyamide 66	25	25	15/30	11	–	80	140	–	–	–
Polyamide 11	25	25	1.5/4.0	50	–	540	2400	5000	–	–
Polyamide 12	25	25	0.35	200/280	–	800/1400	2600/5300	–	–	–
Polycarbonate	23	25	4	680	–	4000	14500	22000	–	–
Polyethyleneterephthalate (stretched)	23	25	0.6	9/15	–	80/110	200/340	1500	–	–
Polysulfone	23	25	6	630	–	3600	15000	28000	–	–
PU elastomers	23	25	13/25	550/1600	–	1000/4500	6000/22000	–	–	–
Polyimide	23	25	25	94	–	390	700	3800	–	–
Celluloseacetate	25	25	150/600	470/630	–	1800/2300	13000/15000	14000	–	–
Celluloseacetobutyrate	25	25	460/600	3800	–	15000	94000	–	–	–

[1]) 1 µm = 0.0394 mil.

Table III General Properties of Selected Polymeric Materials

ABS - Acrylonitrile-Butadiene-Styrene Copolymer

Characteristics: Glossy, high stiffness, impact resistance, scratch resistance, high dimensional stability, ability to be galvanized, poor environmental resistance.

Identified by: Opaque, smells sweet during burn test, yellow flame with soot.

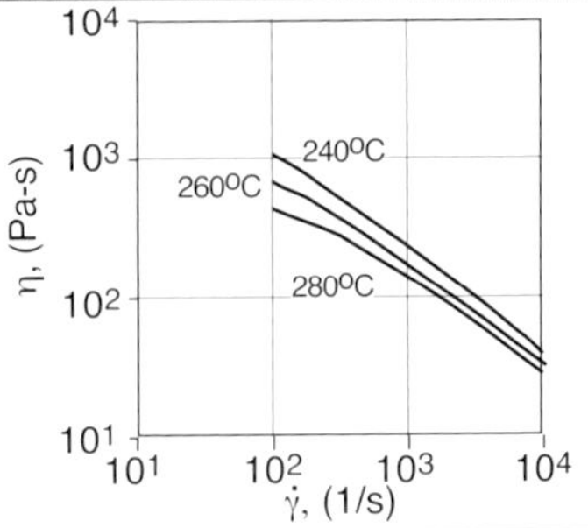

$\rho = 1.03 - 1.07\ g/cm^3$

$C_p = 1{,}470\ J/kg/K$

$k = 0.3\ W/m/K$

$T_{max}^{short\ term} = 80 - 95^oC$

$T_{max}^{long\ term} = 75 - 85^oC$

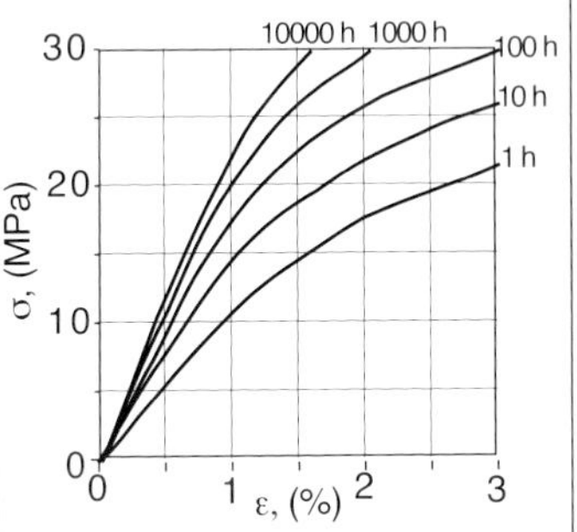

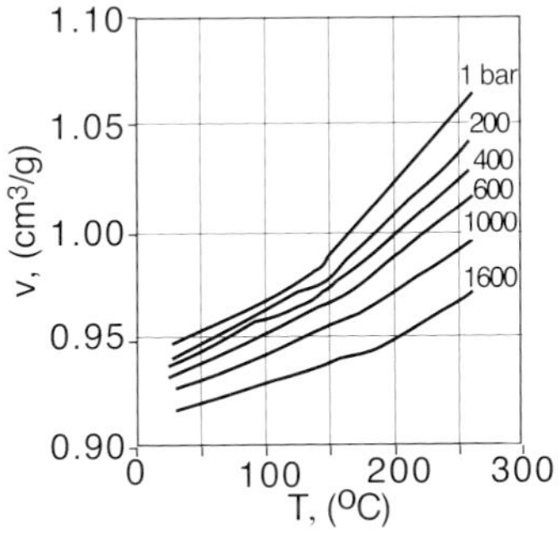

PA6 - Polyamide 6

Characteristics: Abrasion resistance, impact resistance, aborbs water up to 9.5%, often found as a fiber or filament, often found reinforced.

Identified by: Opaque, cream color, smells of burnt hair during burn test, yellow flame with blue halo.

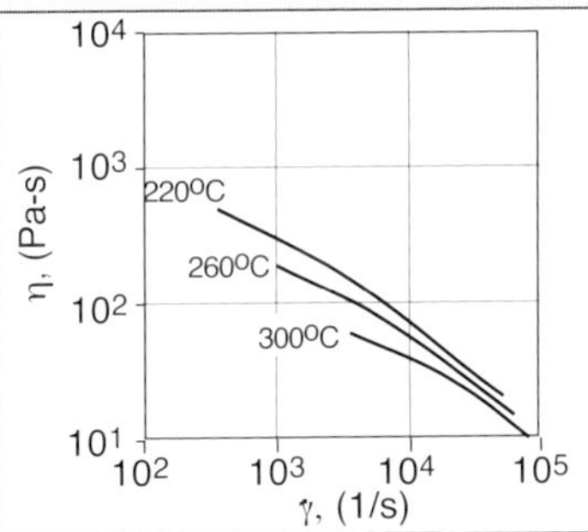

$\rho = 1.12 - 1.15\ g/cm^3$

$C_p = 1670 J/kg/K$

$k = 0.24\ W/m/K$

$T_{max}^{short\ term} = 140 - 160^oC$

$T_{max}^{long\ term} = 80 - 100^oC$

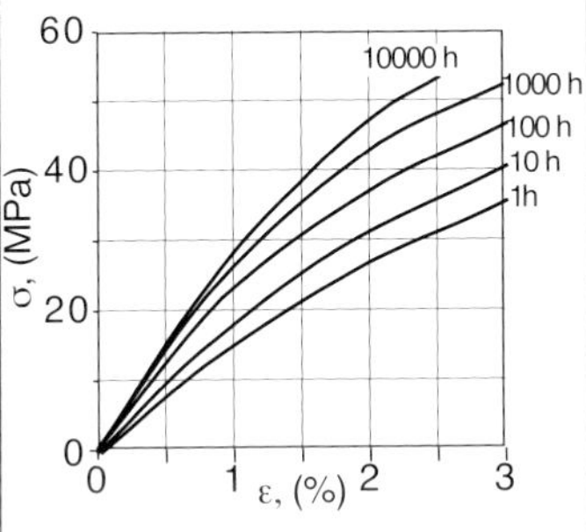

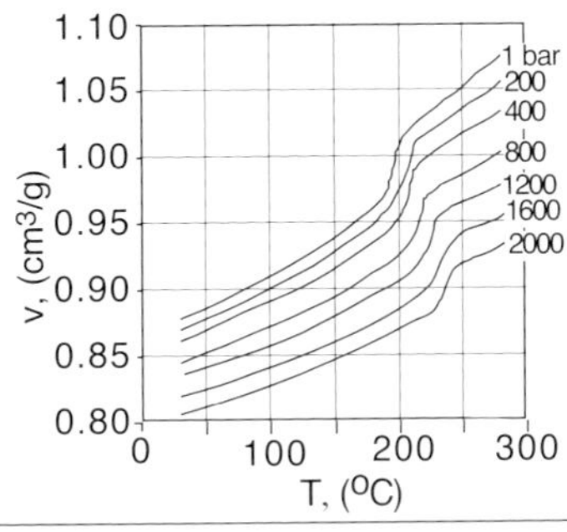

Table III General Properties of Selected Polymeric Materials (*continued*)

PA66 - Polyamide 66

Characteristics: Abrasion resistance, impact resistance, aging and heat resistance, aborbs water up to 8.5%, often found as a fiber or filament, often found reinforced.

Identified by: Opaque, cream color, smells of burnt hair during burn test, yellow flame with blue halo.

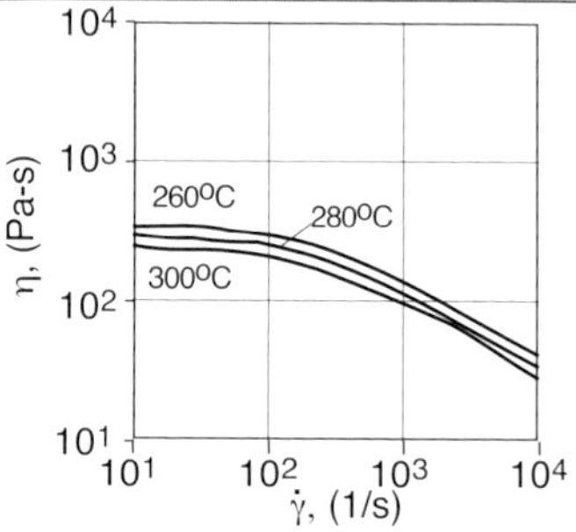

$\rho = 1.13 - 1.16\ g/cm^3$

$C_p = 1670 J/kg/K$

$k = 0.24\ W/m/K$

$T_{max}^{short\ term} = 140 - 170^oC$

$T_{max}^{long\ term} = 80 - 100^oC$

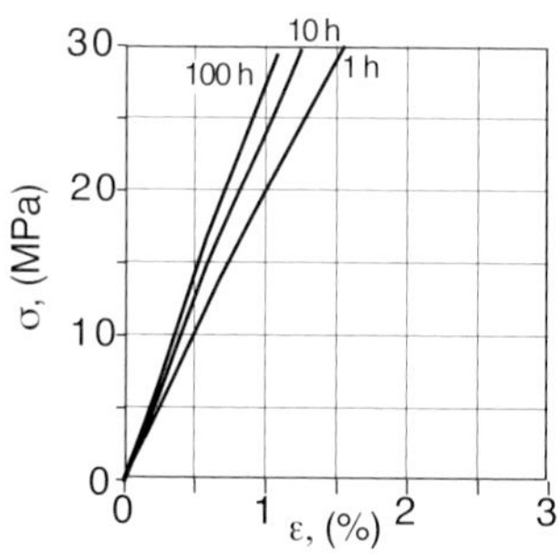

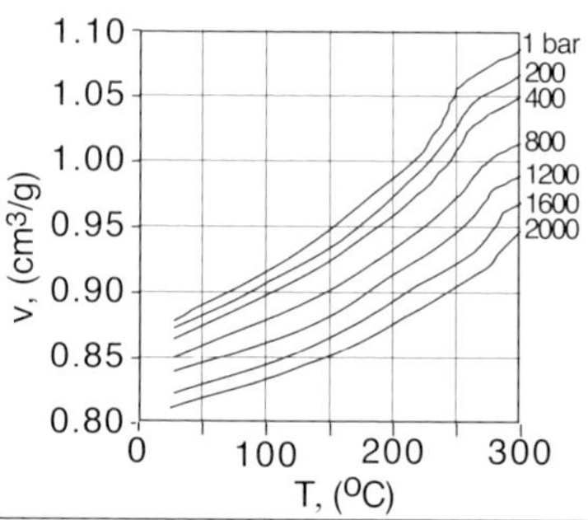

PA66 GF30 - Polyamide 66 -30% Glass Reinforced

Characteristics: Abrasion resistance, high stiffness, impact resistance, aging and heat resistance.

Identified by: Opaque, cream color, smells of burnt hair during burn test, yellow flame with blue halo.

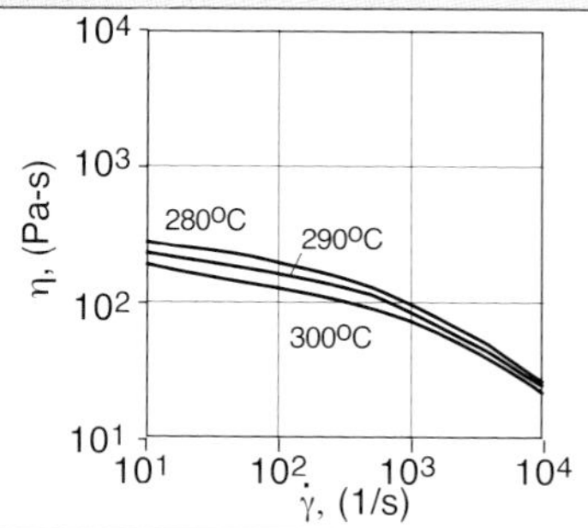

$\rho = 1.38\ g/cm^3$

$C_p = 1{,}260 J/kg/K$

$k = 0.52\ W/m/K$

$T_{max}^{short\ term} = 140 - 170^oC$

$T_{max}^{long\ term} = 80 - 100^oC$

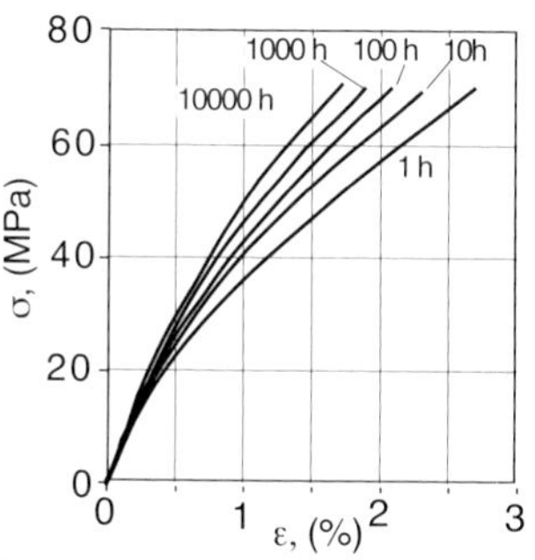

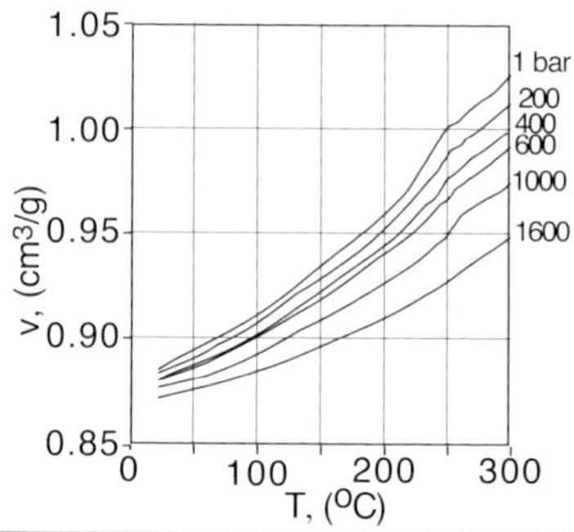

Table III General Properties of Selected Polymeric Materials (*continued*)

PC - Polycarbonate

Characteristics: Glossy, high optical properties, high dimesional stability, impact resistance, high thermal stability.

Identified by: Transparent, dark yellow flame during burn test, soot forming, self extinguishing.

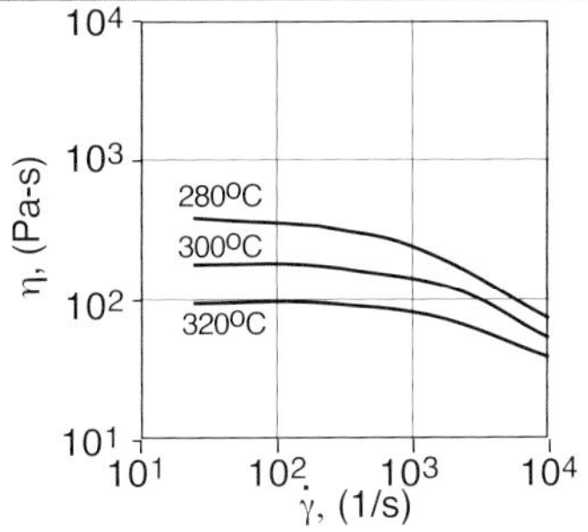

$\rho = 1.20 - 1.24\ g/cm^3$

$C_p = 1{,}260 J/kg/K$

$k = 0.2\ W/m/K$

$T_{max}^{short\ term} = 135^o C$

$T_{max}^{long\ term} = 100^o C$

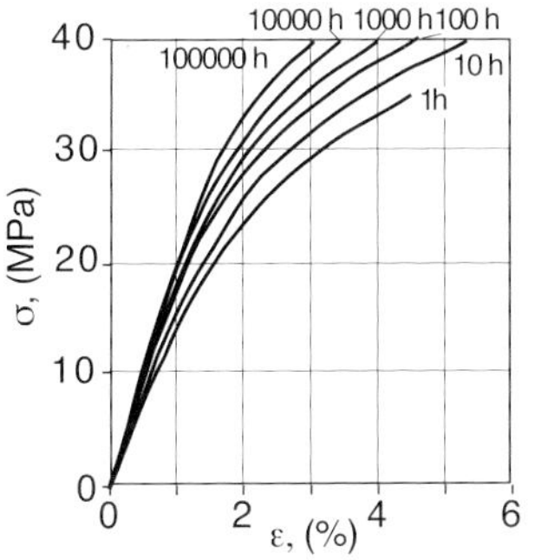

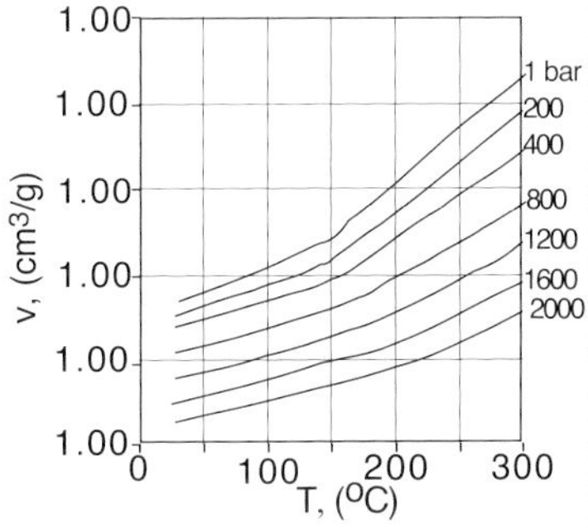

PMMA - Polymethylmethacrylate

Characteristics: Scratch resistance, stiffness, brittleness, good chemical and environmental stability, succeptible to stress cracking.

Identified by: Transparent, flamability, blue flame during burn test, fruity and sweet smell during burn test.

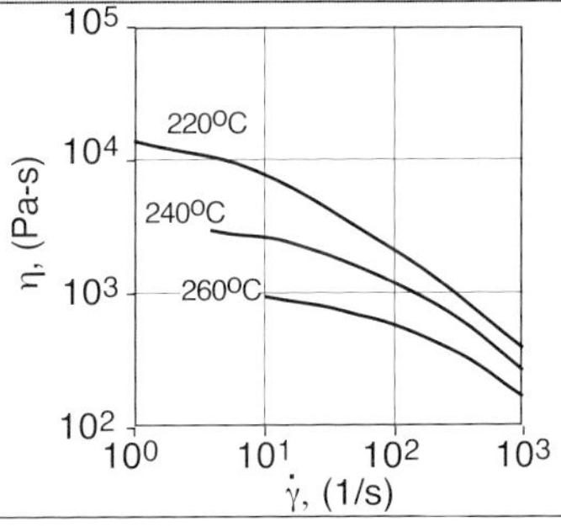

$\rho = 1.15 - 1.19\ g/cm^3$

$C_p = 1{,}470 J/kg/K$

$k = 0.2\ W/m/K$

$T_{max}^{short\ term} = 85 - 95^o C$

$T_{max}^{long\ term} = 65 - 80^o C$

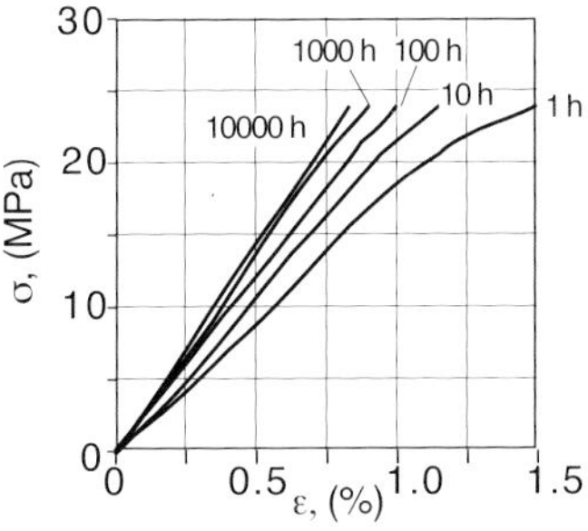

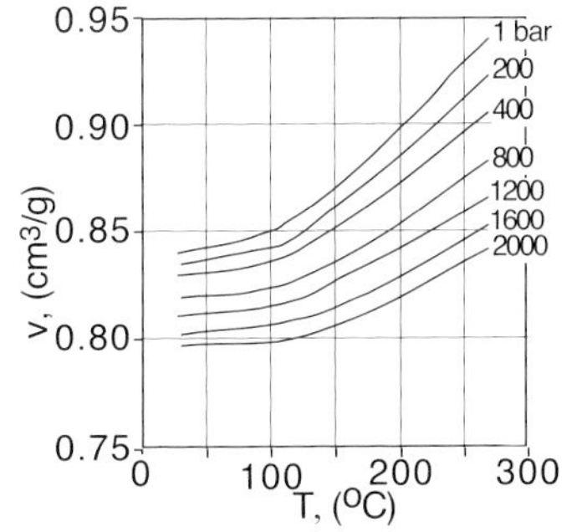

Table III General Properties of Selected Polymeric Materials (*continued*)

POM - Polyoxymethylene (Polyacetal)

Characteristics: High strength and stiffness, highly flexible, low friction coefficient, high dimensional stability, low succeptibility to stress cracking, sensitive to UV light.

Identified by: Opaque, white color, smell of formaldehyde during burn test, blue flame during burn test.

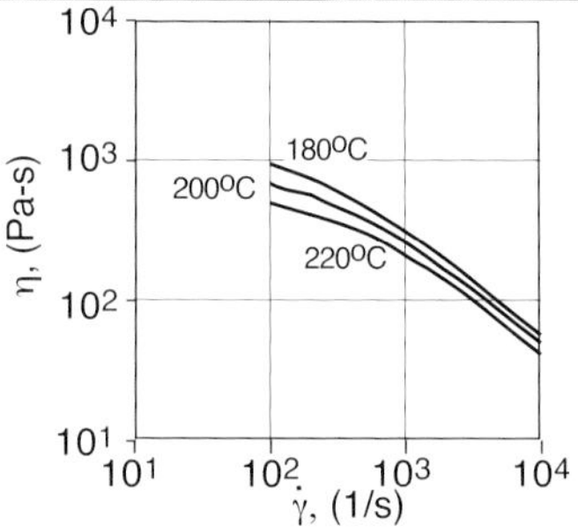

$\rho = 1.41 - 1.43\ g/cm^3$

$C_p = 1,470 J/kg/K$

$k = 0.2\ W/m/K$

$T_{max}^{short\ term} = 110 - 140^oC$

$T_{max}^{long\ term} = 90 - 100^oC$

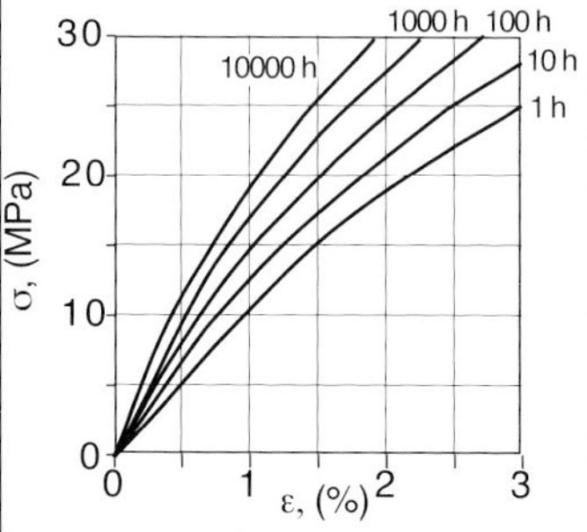

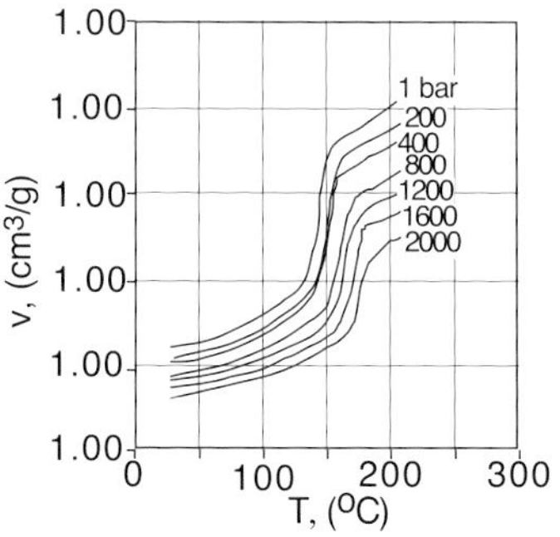

PP - Polypropylene

Characteristics: Low strength and stiffness, low dimensional stability, low stress cracking, homopolymer is brittle at low temperatures, copolymerized with ethylene reduces glass transition temperature, often filled and reinforced, chemical resistance, inexpensive.

Identified by: Translucent, smell of parafin during burn test, bright blue flame during burn test.

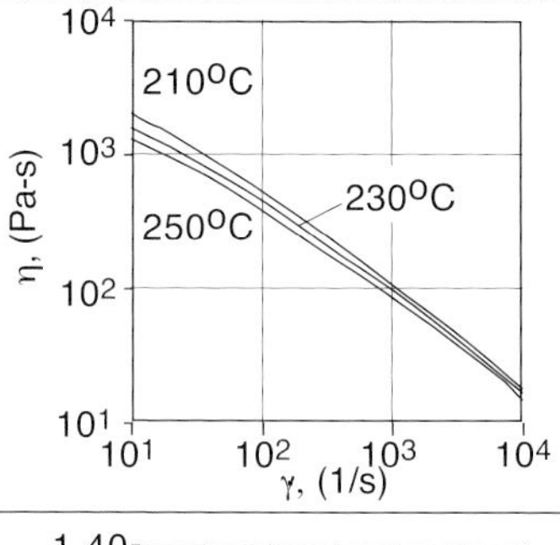

$\rho = 0.90 - 0.91\ g/cm^3$

$C_p = 1,930 J/kg/K$

$k = 0.24\ W/m/K$

$T_{max}^{short\ term} = 130^oC$

$T_{max}^{long\ term} = 90^oC$

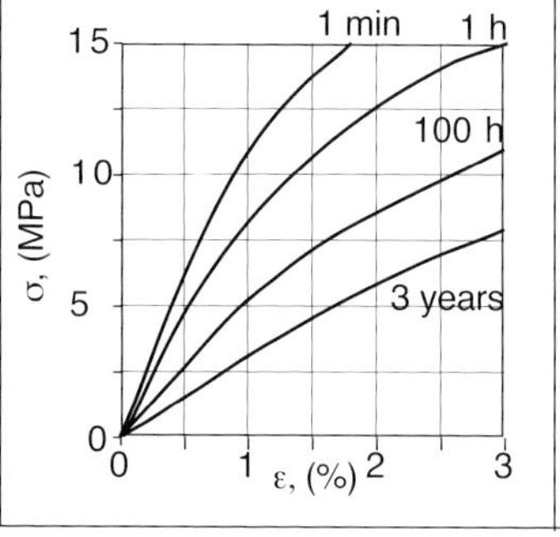

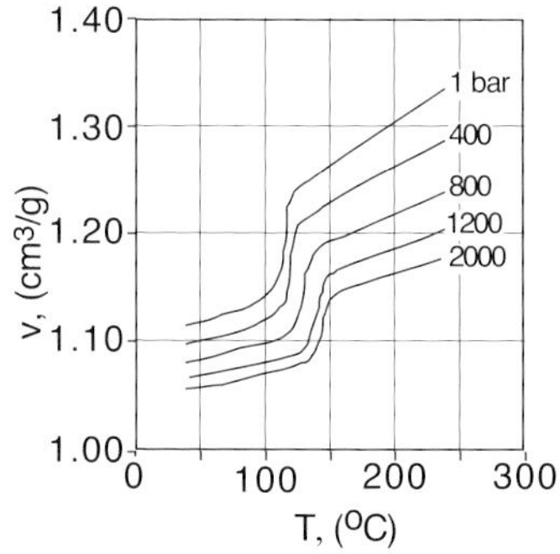

Table III General Properties of Selected Polymeric Materials (*continued*)

PS - Polystyrene

Characteristics: High stiffness, brittle, glassy smooth surface, good dimensional stability, succeptibility to stress cracking, limited chemical stability.

Identified by: Transparent, sweet smell (styrene) during burn test, yellow flame and soot during burn test.

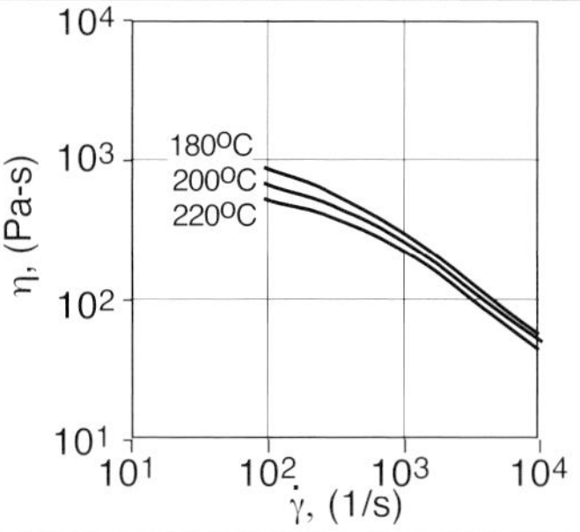

$\rho = 1.05\ g/cm^3$

$C_p = 1{,}340 J/kg/K$

$k = 0.15\ W/m/K$

$T_{\max}^{short\ term} = 90^o C$

$T_{\max}^{long\ term} = 80^o C$

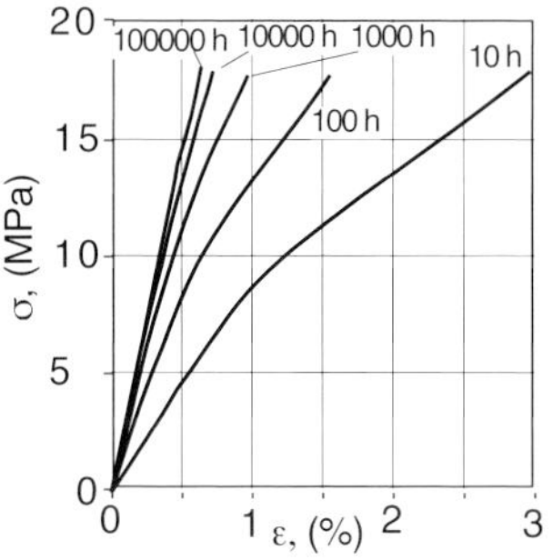

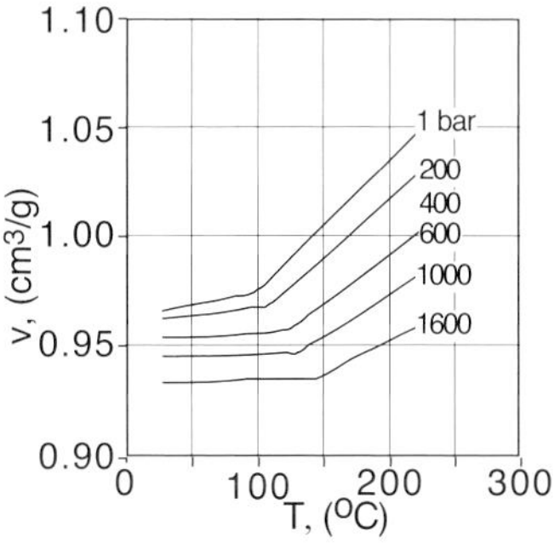

PVC-R - Rigid Polyvinylchloride

Characteristics: Low toughness at low temperature, good dimensional stability, good environmental resistance, inexpensive, large variability in properties when plasticized.

Identified by: Transparent to opaque, poisonous fumes during burn test, hydrochloric acid during burn test, self extinguishing.

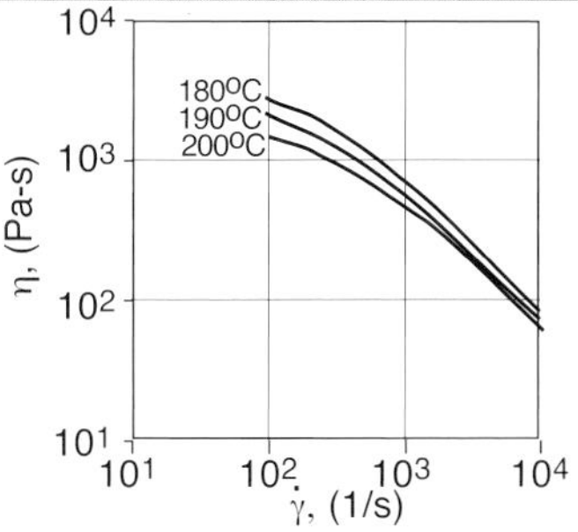

$\rho = 1.38 - 1.55\ g/cm^3$

$C_p = 1{,}000 J/kg/K$

$k = 0.16\ W/m/K$

$T_{\max}^{short\ term} = 70^o C$

$T_{\max}^{long\ term} = 60^o C$

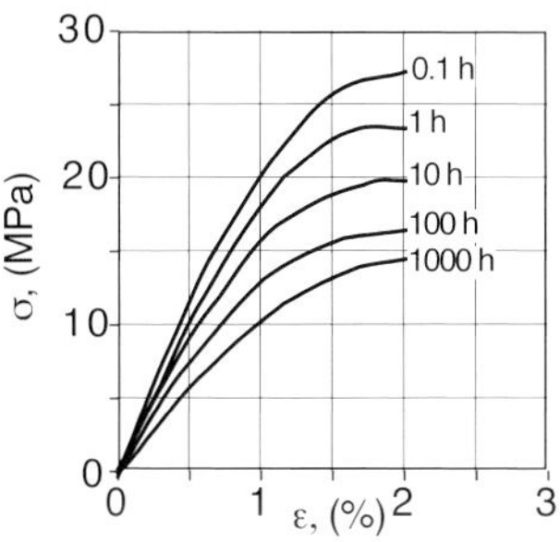

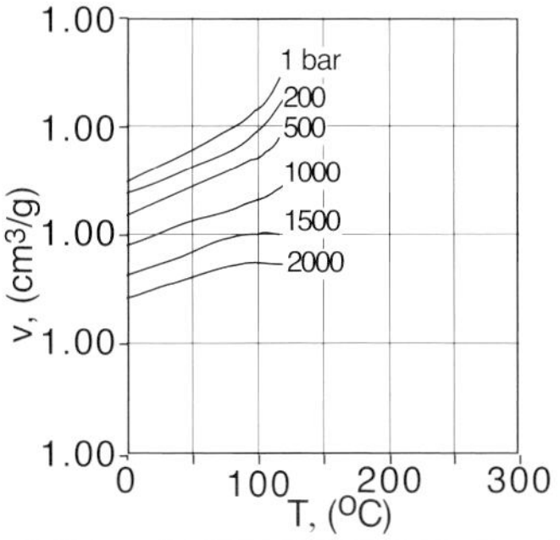

Table III General Properties of Selected Polymeric Materials (*continued*)

PE-HD - High Density Polyethylene

Characteristics: low density, low strength and stiffness, low dimensional stability, high chemical resistance, inexpensive, high degree of crystallinity (60-80%).

Identified by: Milky color, smell of parafin during burn test, yellow flame, drips.

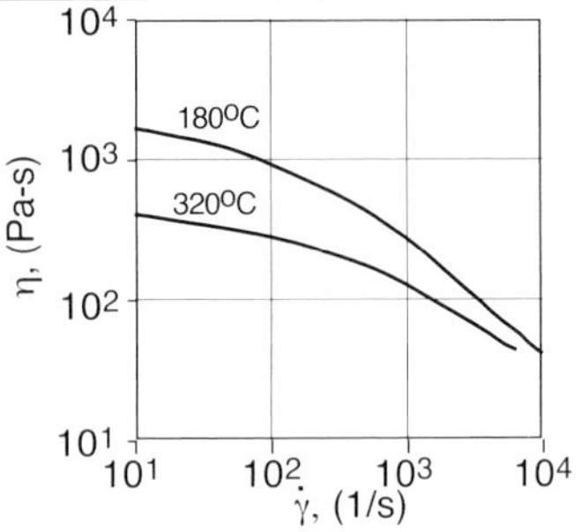

$\rho = 0.940 - 0.960\ g/cm^3$

$C_p = 2,300\ J/kg/K$

$k = 0.63\ W/m/K$

$T_{max}^{short\ term} = 80 - 110^oC$

$T_{max}^{long\ term} = 60 - 80^oC$

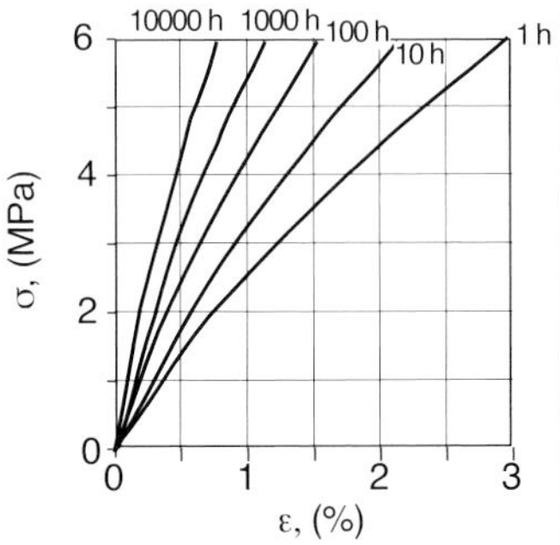

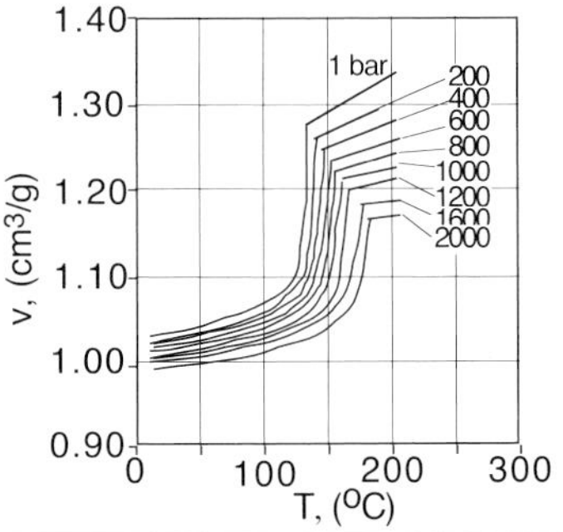

PE-LD - Low Density Polyethylene

Characteristics: low density, easily processed, high flow stability, low dimensional stability, inexpensive, low degree of crystallinity (40-55%).

Identified by: Translucent, smell of parafin during burn test, yellow flame, drips.

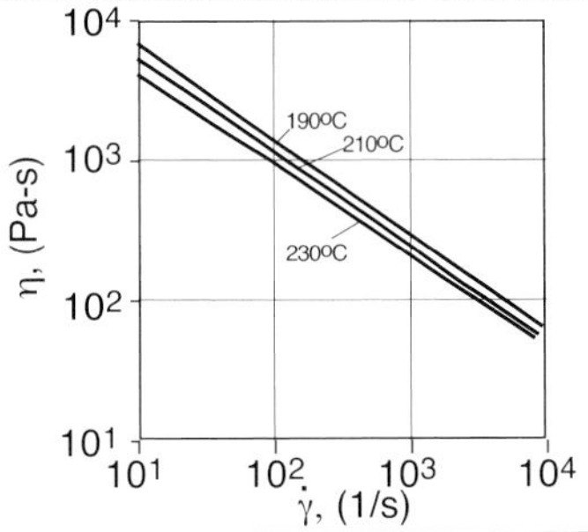

$\rho = 0.914 - 0.928\ g/cm^3$

$C_p = 2,300\ J/kg/K$

$k = 0.33\ W/m/K$

$T_{max}^{short\ term} = 80 - 90^oC$

$T_{max}^{long\ term} = 60 - 70^oC$

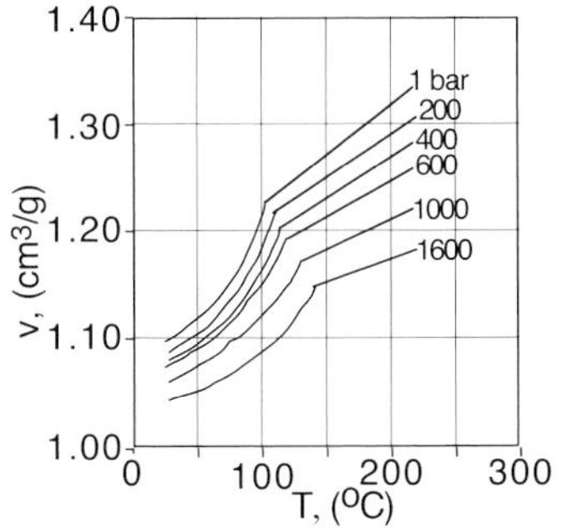

Table IV Energy Equation

Tensor notation $\rho C_v \frac{DT}{Dt} = \left(\nabla \cdot k \nabla T\right) + \frac{1}{2}\mu\left(\underline{\underline{\dot{\gamma}}} : \underline{\underline{\dot{\gamma}}}\right) + \dot{Q}$
Expanded equation in rectangular coordinates (x,y,z): $\rho C_v \left(\frac{\partial T}{\partial t} + v_x \frac{\partial T}{\partial x} + v_y \frac{\partial T}{\partial y} + v_z \frac{\partial T}{\partial z}\right) = K\left(\frac{\partial^2 T}{\partial x^2} + \frac{\partial^2 T}{\partial y^2} + \frac{\partial^2 T}{\partial z^2}\right)$ $+2\mu\left(\left(\frac{\partial v_x}{\partial x}\right)^2 + \left(\frac{\partial v_y}{\partial y}\right)^2 + \left(\frac{\partial v_z}{\partial z}\right)^2\right) + \mu\left(\left(\frac{\partial v_x}{\partial y} + \frac{\partial v_y}{\partial x}\right)^2 + \left(\frac{\partial v_x}{\partial z} + \frac{\partial v_z}{\partial x}\right)^2 + \left(\frac{\partial v_y}{\partial z} + \frac{\partial v_z}{\partial y}\right)^2\right) + \dot{Q}$
Expanded equation in cylindrical coordinates (r,θ,z): $\rho C_v \left(\frac{\partial T}{\partial t} + v_r \frac{\partial T}{\partial r} + \frac{v_\theta}{r}\frac{\partial T}{\partial \theta} + v_z \frac{\partial T}{\partial z}\right) = K\left(\frac{1}{r}\frac{\partial}{\partial r}\left(r\frac{\partial T}{\partial r}\right) + \frac{1}{r^2}\frac{\partial^2 T}{\partial \theta^2} + \frac{\partial^2 T}{\partial z^2}\right)$ $+2\mu\left(\left(\frac{\partial v_r}{\partial r}\right)^2 + \left(\frac{1}{r}\left(\frac{\partial v_\theta}{\partial \theta} + v_r\right)\right)^2 + \left(\frac{\partial v_z}{\partial z}\right)^2\right) + \mu\left(\left(\frac{\partial v_\theta}{\partial z} + \frac{1}{r}\frac{\partial v_z}{\partial \theta}\right)^2 + \left(\frac{\partial v_z}{\partial r} + \frac{\partial v_r}{\partial z}\right)^2 + \left(\frac{1}{r}\frac{\partial v_r}{\partial \theta} + r\frac{\partial}{\partial r}\left(\frac{v_\theta}{r}\right)\right)^2\right) + \dot{Q}$

Table V Continuity Equation in Compressible fluids.

Tenson notation $\nabla \cdot \underline{v} = 0$
Expanded equation in rectangular coordinates $\frac{\partial v_x}{\partial x} + \frac{\partial v_y}{\partial y} + \frac{\partial v_z}{\partial z} = 0$
Expanded equation in cylindrical coordinates (r,θ,z) $\frac{1}{r}\frac{\partial}{\partial r}(r v_r) + \frac{1}{r}\frac{\partial v_\theta}{\partial \theta} + \frac{\partial v_z}{\partial z} = 0$

Table VI Equation of Motion in Terms of $\underline{\underline{\tau}}$

Tensor notation

$$\rho \frac{D\mathbf{v}}{Dt} = -\nabla p + [\nabla \cdot \underline{\underline{t}}] + \rho \mathbf{g}$$

Expanded equation in rectangular coordinates (x,y,z)

$$\rho\left(\frac{\partial v_x}{\partial t} + v_x \frac{\partial v_x}{\partial x} + v_y \frac{\partial v_x}{\partial y} + v_z \frac{\partial v_x}{\partial z}\right) = -\frac{\partial p}{\partial x}\left(\frac{\partial \tau_{xx}}{\partial x} + \frac{\partial \tau_{yx}}{\partial y} + \frac{\partial \tau_{zx}}{\partial z}\right) + \rho g_x$$

$$\rho\left(\frac{\partial v_y}{\partial t} + v_x \frac{\partial v_y}{\partial x} + v_y \frac{\partial v_y}{\partial y} + v_z \frac{\partial v_y}{\partial z}\right) = -\frac{\partial p}{\partial y} + \left(\frac{\partial \tau_{xy}}{\partial x} + \frac{\partial \tau_{yy}}{\partial y} + \frac{\partial \tau_{zy}}{\partial z}\right) + \rho g_y$$

$$\rho\left(\frac{\partial v_z}{\partial t} + v_x \frac{\partial v_z}{\partial x} + v_y \frac{\partial v_z}{\partial y} + v_z \frac{\partial v_z}{\partial z}\right) = -\frac{\partial p}{\partial z} + \left(\frac{\partial \tau_{xz}}{\partial x} + \frac{\partial \tau_{yz}}{\partial y} + \frac{\partial \tau_{zz}}{\partial z}\right) + \rho g_z$$

Expanded equation in cylindrical coordinates (r,θ,z)

$$\rho\left(\frac{\partial v_r}{\partial t} + v_r \frac{\partial v_r}{\partial r} + \frac{v_\theta}{r}\frac{\partial v_r}{\partial \theta} - \frac{v_\theta^2}{r} + v_z \frac{\partial v_r}{\partial z}\right) = -\frac{\partial p}{\partial r} + \left(\frac{1}{r}\frac{\partial}{\partial x}(r\tau_{rr}) + \frac{1}{r}\frac{\partial \tau_{r\theta}}{\partial \theta} - \frac{\tau_{\theta\theta}}{r} + \frac{\partial \tau_{rz}}{\partial z}\right) + \rho g_r$$

$$\rho\left(\frac{\partial v_\theta}{\partial t} + v_r \frac{\partial v_\theta}{\partial r} + \frac{v_\theta}{r}\frac{\partial v_\theta}{\partial \theta} + \frac{v_r v_\theta}{r} + v_z \frac{\partial v_\theta}{\partial z}\right) = -\frac{1}{r}\frac{\partial p}{\partial \theta} + \left(\frac{1}{r^2}\frac{\partial}{\partial r}(r^2\tau_{r\theta}) + \frac{1}{r}\frac{\partial \tau_{\theta\theta}}{\partial \theta} + \frac{\partial \tau_{\theta z}}{\partial z}\right) + \rho g_\theta$$

$$\rho\left(\frac{\partial v_z}{\partial t} + v_r \frac{\partial v_z}{\partial r} + \frac{v_\theta}{r}\frac{\partial v_z}{\partial \theta} + v_z \frac{\partial v_z}{\partial z}\right) = -\frac{\partial p}{\partial z} + \left(\frac{1}{r}\frac{\partial}{\partial r}(r\tau_{rz}) + \frac{1}{r}\frac{\partial \tau_{\theta z}}{\partial \theta} + \frac{\partial \tau_{zz}}{\partial z}\right) + \rho g_z$$

Table VII Equation of Motion in terms of **v** **(Navier-Stokes Equations),**

Tensor notation

$$\rho \frac{D\mathbf{v}}{Dt} = -\nabla p + \mu \nabla^2 \underline{\mathrm{v}} + \rho \underline{g}$$

Expanded equation in rectangular coordinates (x,y,z)

$$\rho\left(\frac{\partial v_x}{\partial t} + v_x \frac{\partial v_x}{\partial x} + v_y \frac{\partial v_x}{\partial y} + v_z \frac{\partial v_x}{\partial z}\right) = -\frac{\partial p}{\partial x} + \mu\left(\frac{\partial^2 v_x}{\partial x^2} + \frac{\partial^2 v_x}{\partial y^2} + \frac{\partial^2 v_x}{\partial z^2}\right) + \rho g_x$$

$$\rho\left(\frac{\partial v_y}{\partial t} + v_y \frac{\partial v_y}{\partial x} + v_y \frac{\partial v_y}{\partial y} + v_z \frac{\partial v_y}{\partial z}\right) = -\frac{\partial p}{\partial y} + \mu\left(\frac{\partial^2 v_y}{\partial x^2} + \frac{\partial^2 v_y}{\partial y^2} + \frac{\partial^2 v_y}{\partial z^2}\right) + \rho g_y$$

$$\rho\left(\frac{\partial v_z}{\partial t} + v_x \frac{\partial v_z}{\partial x} + v_y \frac{\partial v_z}{\partial y} + v_z \frac{\partial v_z}{\partial z}\right) = -\frac{\partial p}{\partial z} + \mu\left(\frac{\partial^2 v_z}{\partial x^2} + \frac{\partial^2 v_z}{\partial y^2} + \frac{\partial^2 v_z}{\partial z^2}\right) + \rho g_z$$

Expanded equation in cylindrical coordinates (r,θ,z)

$$\rho\left(\frac{\partial v_r}{\partial t} + v_r \frac{\partial v_r}{\partial r} + \frac{v_\theta}{r}\frac{\partial v_r}{\partial \theta} - \frac{v_\theta^2}{r} + v_z \frac{\partial v_r}{\partial z}\right) = -\frac{\partial p}{\partial r} + \mu\left(\frac{\partial}{\partial r}\left(\frac{1}{r}\frac{\partial}{\partial r}(r v_r)\right) + \frac{1}{r^2}\frac{\partial^2 v_r}{\partial \theta^2} - \frac{2}{r^2}\frac{\partial v_\theta}{\partial \theta} + \frac{\partial^2 v_r}{\partial z^2}\right) + \rho g_r$$

$$\rho\left(\frac{\partial v_\theta}{\partial t} + v_r \frac{\partial v_\theta}{\partial r} + \frac{v_\theta}{r}\frac{\partial v_\theta}{\partial \theta} + \frac{v_r v_\theta}{r} + v_z \frac{\partial v_\theta}{\partial z}\right) = -\frac{1}{r}\frac{\partial p}{\partial \theta} + \mu\left(\frac{\partial}{\partial r}\left(\frac{1}{r}\frac{\partial}{\partial r}(r v_\theta)\right) + \frac{1}{r^2}\frac{\partial^2 v_\theta}{\partial \theta^2} - \frac{2}{r^2}\frac{\partial v_r}{\partial \theta} + \frac{\partial^2 v_\theta}{\partial z^2}\right) + \rho g_\theta$$

$$\rho\left(\frac{\partial v_z}{\partial t} + v_r \frac{\partial v_z}{\partial r} + \frac{v_\theta}{r}\frac{\partial v_z}{\partial \theta} + v_z \frac{\partial v_z}{\partial z}\right) = -\frac{\partial p}{\partial z} + \mu\left(\frac{1}{r}\frac{\partial}{\partial r}\left(r\frac{\partial v_z}{\partial r}\right) + \frac{1}{r^2}\frac{\partial^2 v_z}{\partial \theta^2} + \frac{\partial^2 v_z}{\partial z^2}\right) + \rho g_z$$

Subject Index

D

E

F

G

H

I

O

P

Q

R

U

V

W

X

Y

Z

Author Index

L

M

N

O

P

Q

R

Y

Z